Valentina Zharkova

Electron and Proton Kinetics and Dynamics in Flaring Atmospheres

Related Titles

Spatschek, K.-H.

High Temperature Plasmas
Theory and Mathematical Tools for Laser and Fusion Plasmas

2012
ISBN: 978-3-527-41041-5

Smirnov, B. M.

Fundamentals of Ionized Gases
Basic Topics in Plasma Physics

2011
ISBN: 978-3-527-41085-9

Rehder, D.

Chemistry in Space
From Interstellar Matter to the Origin of Life

2010
ISBN: 978-3-527-32689-1

Smirnov, B. M.

Gluster Processes in Gases and Plasmas

2010
ISBN: 978-3-527-40943-3

Stock, R. (ed.)

Encyclopedia of Applied High Energy and Particle Physics

2009
ISBN: 978-3-527-40691-3

Smirnov, B. M.

Plasma Processes and Plasma Kinetics
Worked-Out Problems for Science and Technology

2007
ISBN: 978-3-527-40681-4

Stahler, S. W., Palla, F.

The Formation of Stars

2004
ISBN: 978-3-527-40559-6

Foukal, P. V,

Solar Astrophysics

2004
ISBN: 978-3-527-40374-5

Woods, L. C,

Physics of Plasmas

2004
ISBN: 978-3-527-40461-2

Valentina Zharkova

Electron and Proton Kinetics and Dynamics in Flaring Atmospheres

WILEY-VCH Verlag GmbH & Co. KGaA

The Author

Prof. Valentina Zharkova
Bradford University
Computing and Mathematics
Horton Building D1.10
UK – Bradford BD7 1DP
v.v.zharkova@Bradford.ac.uk

Library of Congress Card No.:
applied for

British Library Cataloguing-in-Publication Data:
A catalogue record for this book is available from the British Library.

Bibliographic information published by the Deutsche Nationalbibliothek
The Deutsche Nationalbibliothek lists this publication in the Deutsche Nationalbibliografie; detailed bibliographic data are available on the Internet at http://dnb.d-nb.de.

Cover Design Grafik-Design Schulz, Fußgönheim
Typesetting le-tex publishing services GmbH, Leipzig
Printing and Binding Markono Print Media Pte Ltd, Singapore

Printed in Singapore
Printed on acid-free paper

Print ISBN 978-3-527-40847-4
ePDF ISBN 978-3-527-63968-7
ePub ISBN 978-3-527-63967-0
Mobi ISBN 978-3-527-63969-4
oBook ISBN 978-3-527-63966-3

Contents

Preface

Charged particles, electrons, protons, ions and neutral atoms are invisible but very powerful participants in all processes in plasmas of the Sun, stars, magnetospheres, interplanetary space and laboratory experiments. Their presence in theoretical research is very often masked behind macro descriptions of the plasma status by means of temperature, density, electric and magnetic fields and so on.

All of these are defined by some sort of ensembles of particles whose various properties (e.g. velocities, charges, masses, numbers or excitation-state status) define the macro parameters which are good descriptors of a plasma's status in equilibrium. However, in many events on the Sun or stars or in interplanetary space, the atmospheres are well beyond equilibrium. The subject of this book is the investigation of processes of non-equilibrium in flaring atmospheres with a consideration of particle kinetics, dynamics and radiative processes.

The author's PhD thesis, titled "Radiative transfer in solar quiescent prominences with filamentary structure", investigated non-equilibrium radiative processes in cool, steady atmospheres and their effects on hydrogen lines and continuous emission. The research was done under the supervision of the late Prof. Nina Morozhenko (Solar Division, Main Astronomical Observatory, Ukraine), a researcher of the highest caliber, who taught well how to properly conduct research and test hypotheses with theoretical predictions. After completing her PhD, the author applied the approach used and knowledge gained during the writing of her thesis to the development of physical concepts in such dynamic events as solar flares.

The research in particle kinetics was initiated at the Astronomy Unit, State Kiev University, by the prominent plasma physicist Prof. Nikolaj Kotsarenko (1941–1993), former head of the Space Physics and Astronomy division in the Physics Department, National University of Kiev, Ukraine. This work was also kindly supported by the theoretical group Theory and Diagnostics of Physical Processes in Solar Flares, led by Prof. Boris Somov of Moscow Sternberg Astronomical Institution, Moscow State University, Russia, and the researchers comprising the group who now carry out their research at various institutions around the world. During annual gatherings of this group, the researchers had many fruitful talks and discussions, which helped the author to make significant progress in her knowledge and understanding of the complex physical processes of particle acceleration and precipitation in solar flares.

My acquaintance with Prof. John Brown, Astronomer Royal for Scotland, University of Glasgow, Scotland, and his famous group, which consisted of Prof. Gordon Emslie, Drs. D. Alexander, A. MacKinnon and other researchers, gave the present author a better understanding of particle kinetics and dynamics developed by various groups in Russia, the United Kingdom and the United States. Very frequently our research seminars and talks sparked extensive debates which motivated further research to clarify the argued points. Such discussions helped the author to build, brick by brick, her knowledge and understanding of such complex phenomena as the physical processes in solar flares, for which the author is enormously grateful.

The idea of this book was conceived at one of the RHESSI workshops frequently devoted to particle acceleration and precipitation in flaring atmospheres on the Sun and their diagnostics from multi-wavelength observations. Particle kinetics is a rather complex topic which needs to be taught to younger scientists so that they may continue the research begun four decades ago with the pioneering works of Prof. Sergey Syrovatsky (Moscow Physical-Technical Institution, Russia), Prof. John Brown (Glasgow University, UK) and Dr. Olga Shmeleva (IZMIRAN, Russia).

The author is also very grateful to her PhD students, who were engaged in the study of various aspects of particle kinetics: Dr. Victor Kobylinskij V.A. (funded by Kiev University, 1989–1993), Dr. Dmitry Syniavskij (Kiev University, Ukraine) (funded by Kiev University, Ukraine, 1990–1994) and Dr. Mykola Gordovskyy (Bradford University, UK), whose study was funded by the Engineering and Physical Sciences Research Council (2002–2005). The students' dedication to and thorough knowledge of their topics significantly advanced the subject to new levels of understanding, and their knowledge of the topic is reflected in the current book.

The research carried out with her students helped the author to produce a strong synergy between research in kinetics and the dynamics of solar flares and the helioseismology of the solar interior behind these events. The author wishes to acknowledge a very fruitful collaboration with Dr. Alexander Kosovichev (Stanford University, USA), which led to the discovery of sunquakes. These are seismic responses of the solar interior to processes occurring in solar flares. They were reported in a paper in Nature on 27 May 1998 and gained worldwide media coverage on 28 May 1998 by all major TV and radio stations and newspapers.

The author is very grateful to her younger collaborators: Dr. Taras Siversky, my former post-doctoral research assistant employed on a research grant funded by the Science, Technology and Facilities Council (2007–2009); Dr. Sergey Zharkov (son), employed on the European Framework 5 Grant EGSO (2002–2005), currently a Research Fellow at Mullard Space Science Laboratory (MSSL), University College London (UCL), UK; and Dr. Sarah Matthews, a Reader at MSSL, UCL, who helped the author to significantly advance the topics of particle acceleration and precipitation in flaring atmospheres, the generation of seismic responses (sunquakes) associated with solar flares, and the determination of the connection of these processes with the phenomenon of solar flares covering atmospheric heights from the corona to the solar interior.

The author also wishes to acknowledge the Russian collaborators from the Institute of Solar-Terrestrial Physics, Irkutsk, Russia (Prof. A. Altyntsev, Drs. L. Kashapova and N. Meshalkina), whose contribution to our joint research within the Royal Society Joint International Grant (2009–2011) made a reality of recent papers comparing our kinetic and dynamic simulations with multi-wavelength observations, which ultimately became an important part of this book.

And last but not least, the author appreciates the support of her family and partner which allowed her to stay focused on this project and complete the book.

The author hopes that this book will help researchers who are just beginning their study of the physical phenomena of flaring atmospheres on the Sun.

Bradford, January 2012 *Valentina Zharkova*

Color Plates

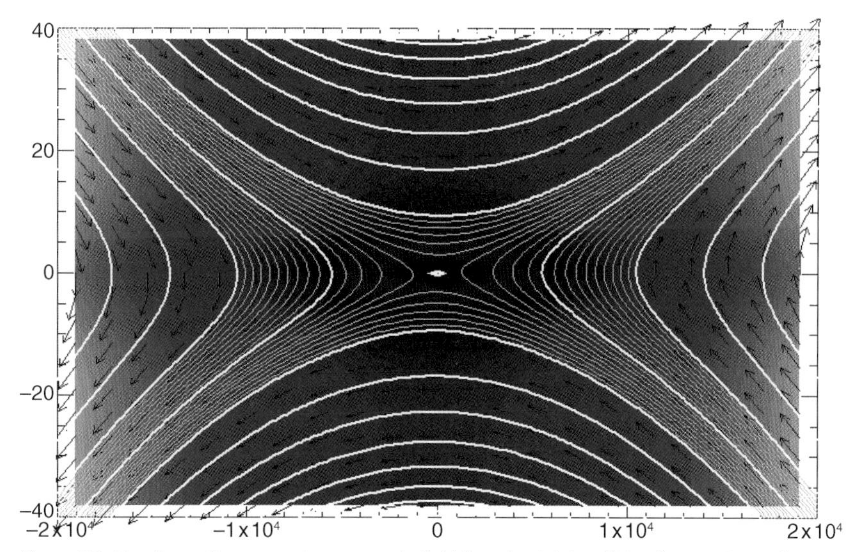

Figure 2.1 Topology of reconnecting magnetic field lines in vicinity of X null point $B_y = \pm\alpha B_x$.

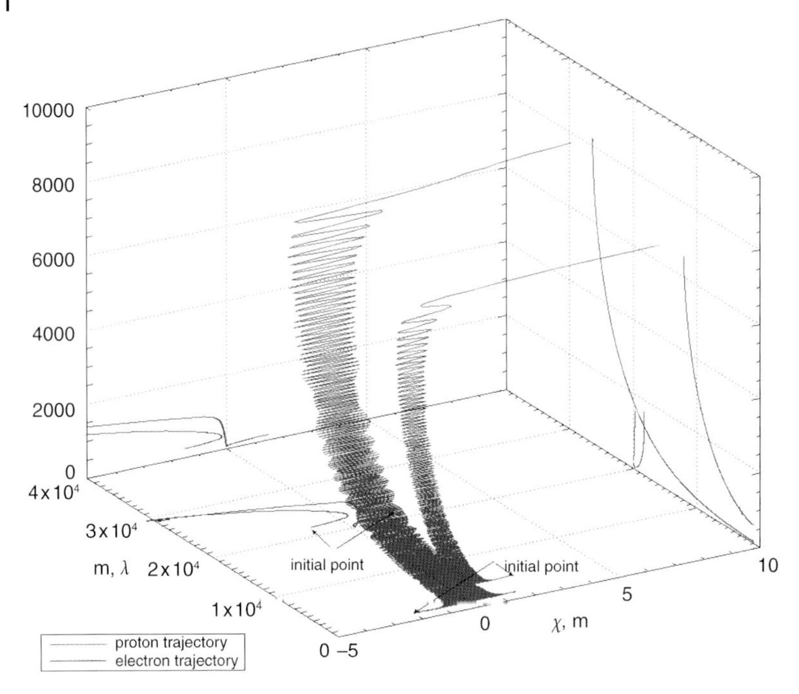

Figure 2.9 Proton (blue lines) and electron (purple lines) trajectories in a 3-D RCS with a lateral magnetic field of $B_x = 10\,\mathrm{G}$ and a drift electric field of $E_y = 100\,\mathrm{V\,m^{-1}}$. From Zharkova and Agapitov (2009).

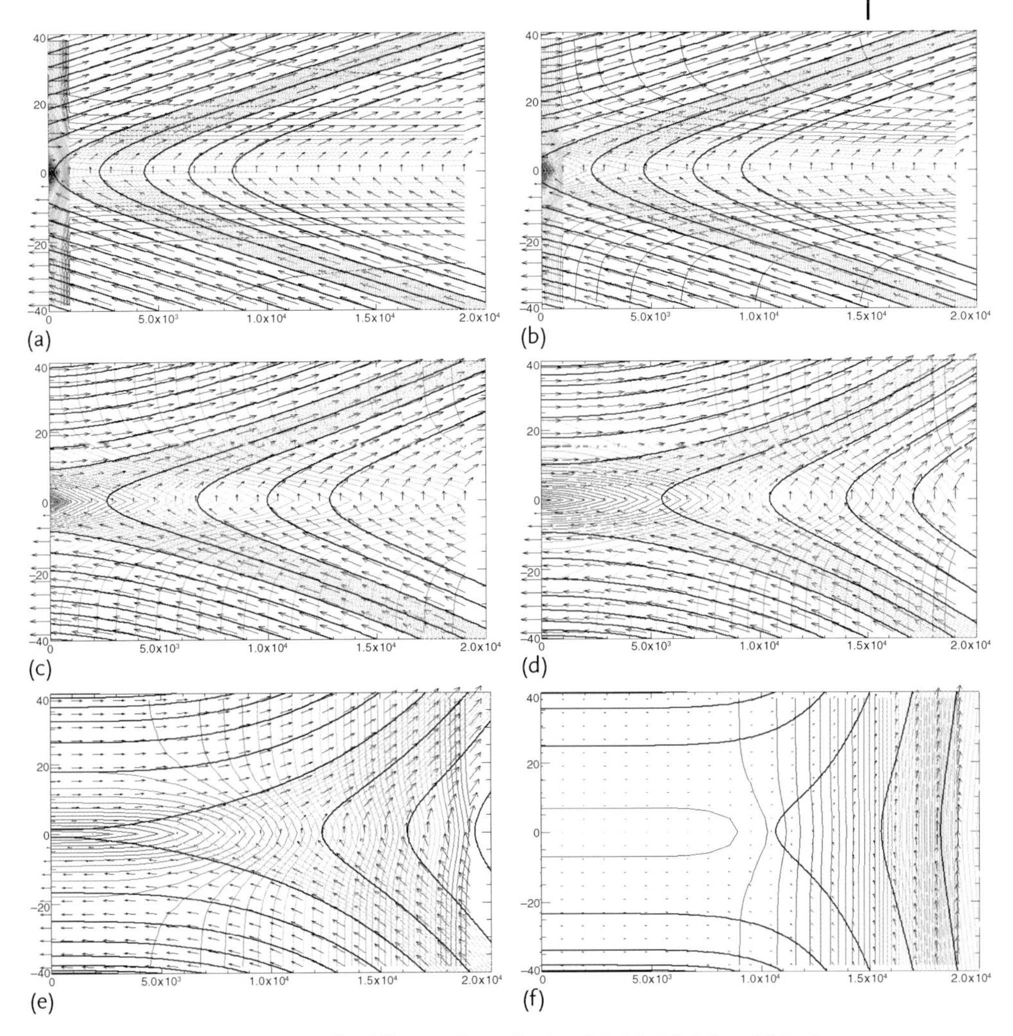

Figure 2.11 Views of current sheet for different values of α: 0.5, 0.7, 1.0, 1.5, 2.0, and 5.0, plotted by rows from (a) to (f). From Zharkova and Agapitov (2009).

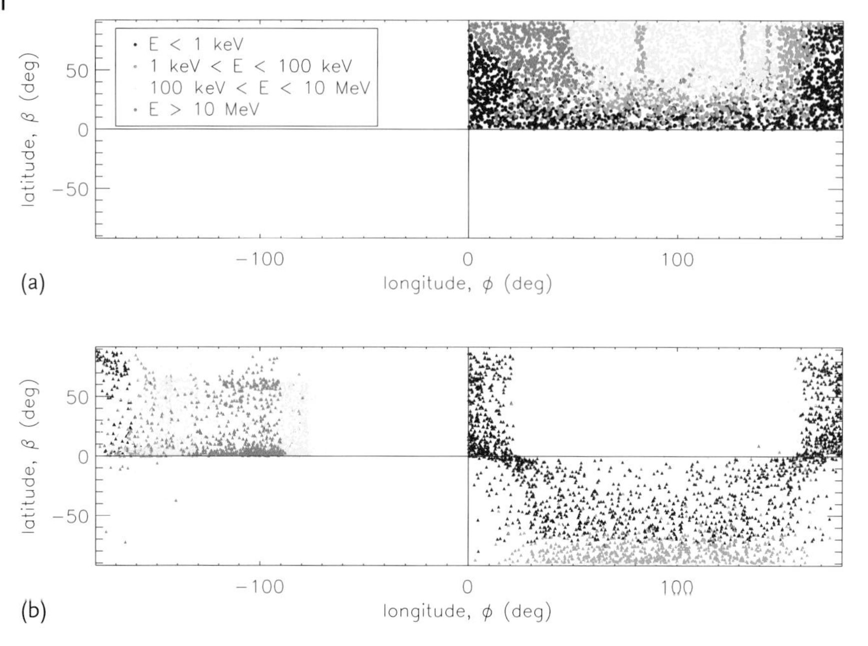

Figure 2.16 Positions of particles at initial (a) and final (b) times for *spine* reconnection, color-coded according to their final energies. From Dalla and Browning (2006).

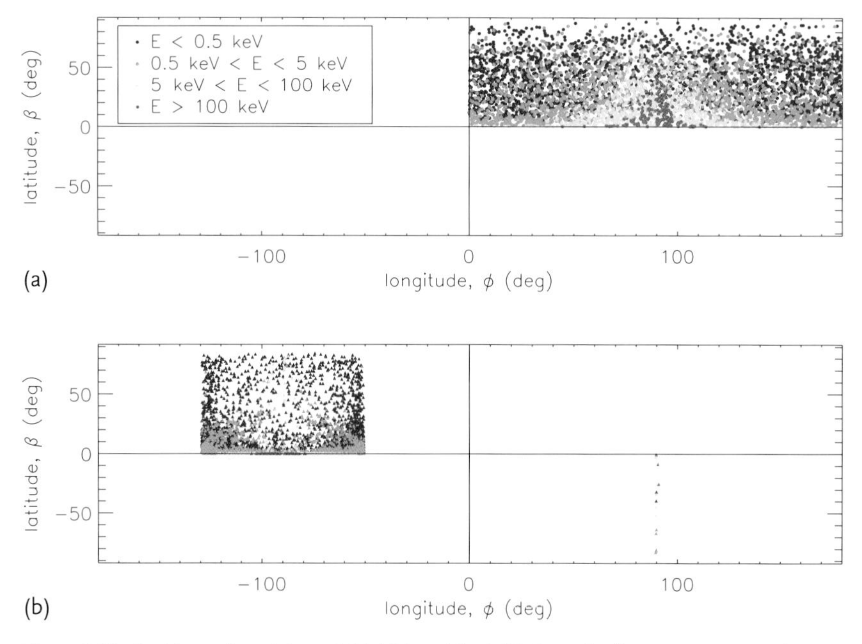

Figure 2.17 Positions of particles at initial (a) and final (b) times for *fan* reconnection, color-coded according to their final energies. From Dalla and Browning (2008).

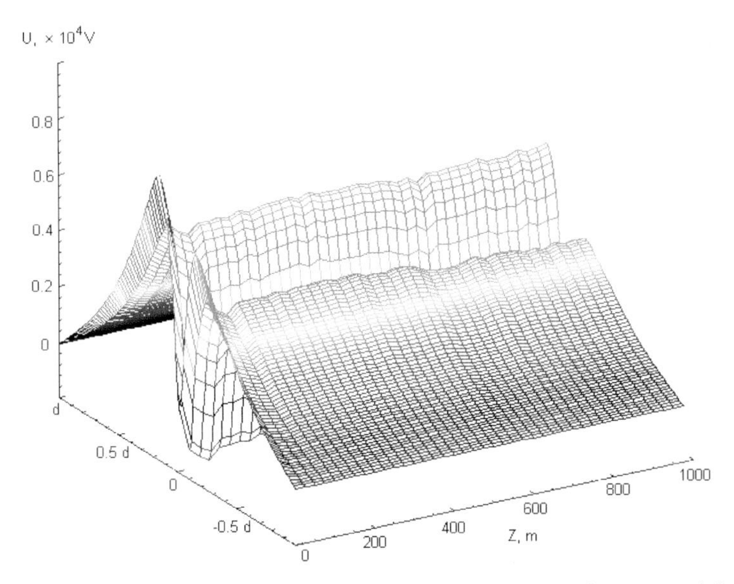

Figure 2.18 Polarization electric field induced by the separation of protons and electrons accelerated in a 3-D current sheet with $B_0 = 10\,\mathrm{G}$, at various distances z from the null point. From Zharkova and Agapitov (2009).

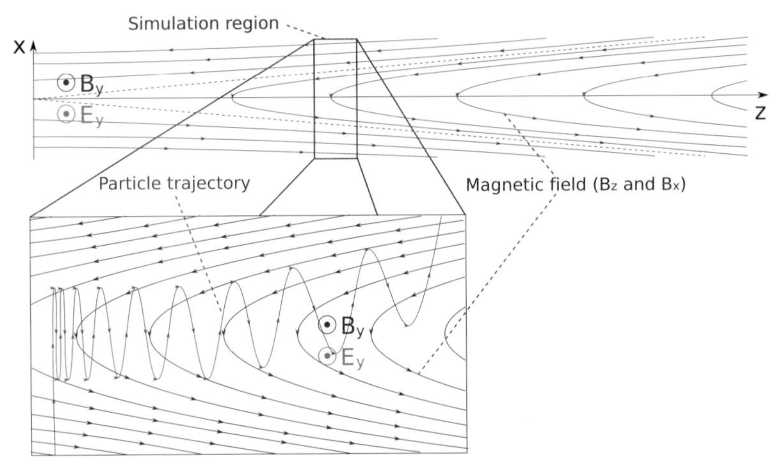

Figure 2.19 Magnetic field topology and electric field used in PIC simulation model. The vertical rectangle shows the size of the simulation box, which is shifted along the z-axis to account for the different magnitudes of tangential magnetic field component B_x.

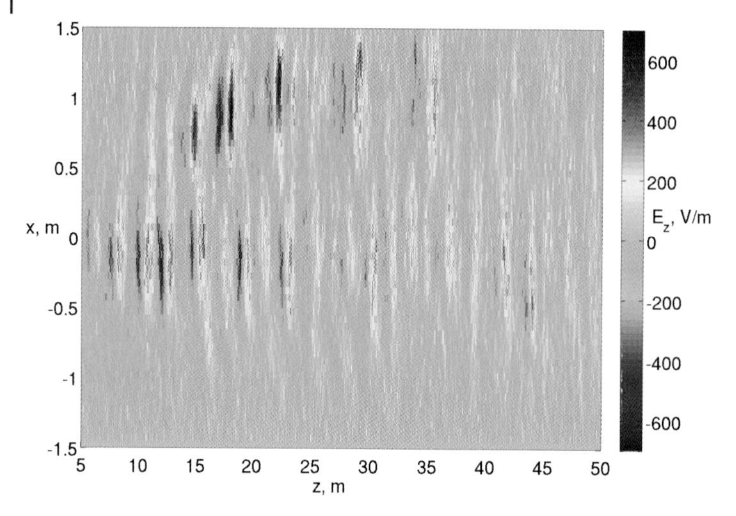

Figure 2.29 Electric field $\tilde{E}_z$ induced by particles in PIC simulation ($B_{z0} = 10^{-3}$ T, $B_{y0} = 10^{-4}$ T, $B_{x0} = 4 \times 10^{-5}$ T, $E_{y0} = 250\,\mathrm{V\,m^{-1}}$, $m_p/m_e = 10$, $n = 10^6\,\mathrm{cm^{-3}}$).

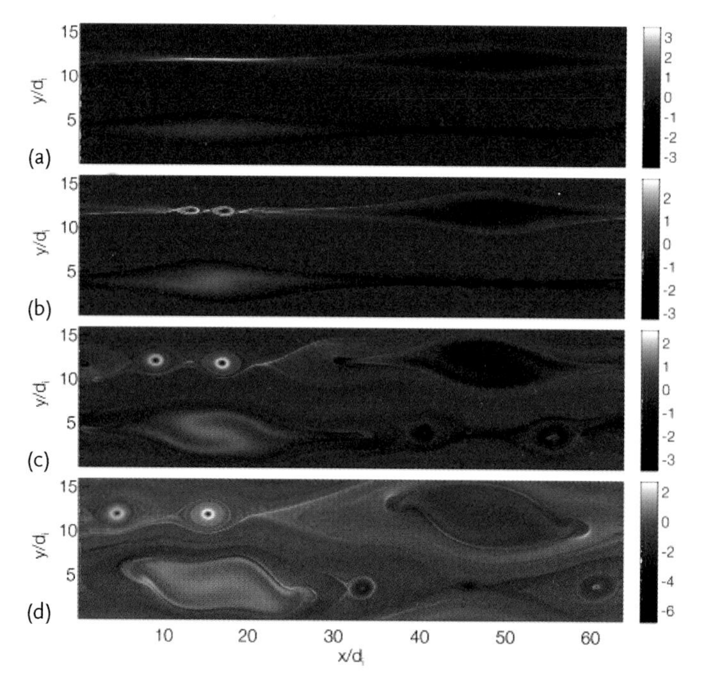

Figure 2.31 Out-of-plane current density in *2.5-D PIC* simulations of reconnection with a guiding field, for four consecutive time intervals from (a) to (b). The environment contains two current sheets and is characterized by the appearance of islands formed by the tearing instability. At the start there are only (a) large islands, followed at later times (b–d) by the formation of smaller magnetic islands. Note that such islands do not appear in 2-D PIC simulations of the same environment but without a guiding field. From Drake *et al.* (2006b).

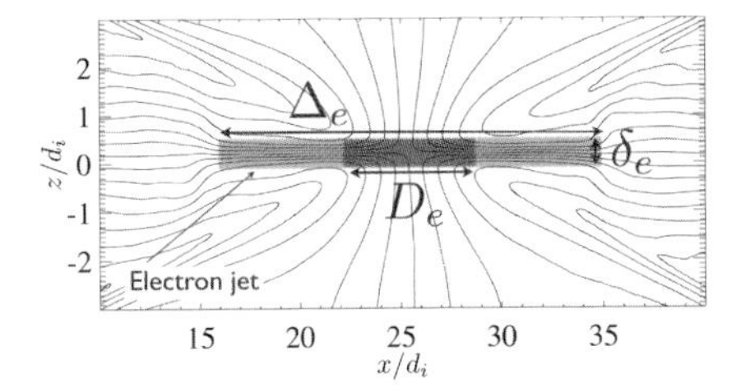

Figure 2.33 Multiscale structure of electron diffusion region around an X-type null point of total length Δ_e and thickness δ_e, as derived from 3-D PIC simulations. The solid lines show the electron trajectories for the time $t\omega_{ci} = 80$ (ω_{ci} is the ion cyclotron frequency). In the inner region of length D_e there is a steady inflow of electrons and a strong out-of-plane current; the outer region is characterized by electron outflow jets. From Karimabadi *et al.* (2007).

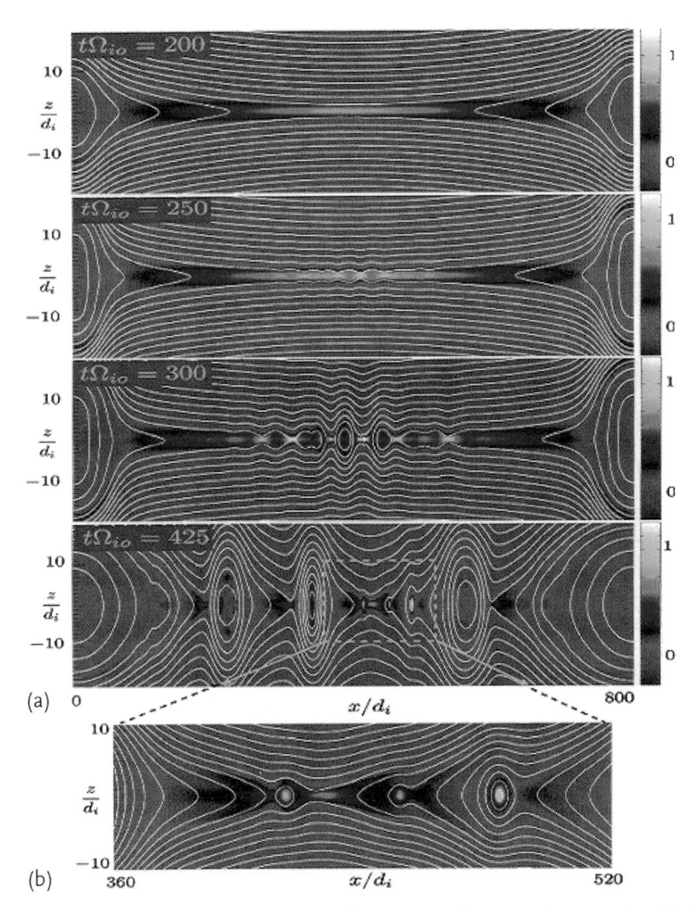

Figure 2.34 Evolution of the current density over a region of extent $800d_i$ (where d_i is the ion inertial length), for different times t. White lines are the magnetic flux surfaces. A close-up of a region in which the formation of new islands is occurring is shown in (e). From Daughton *et al.* (2009).

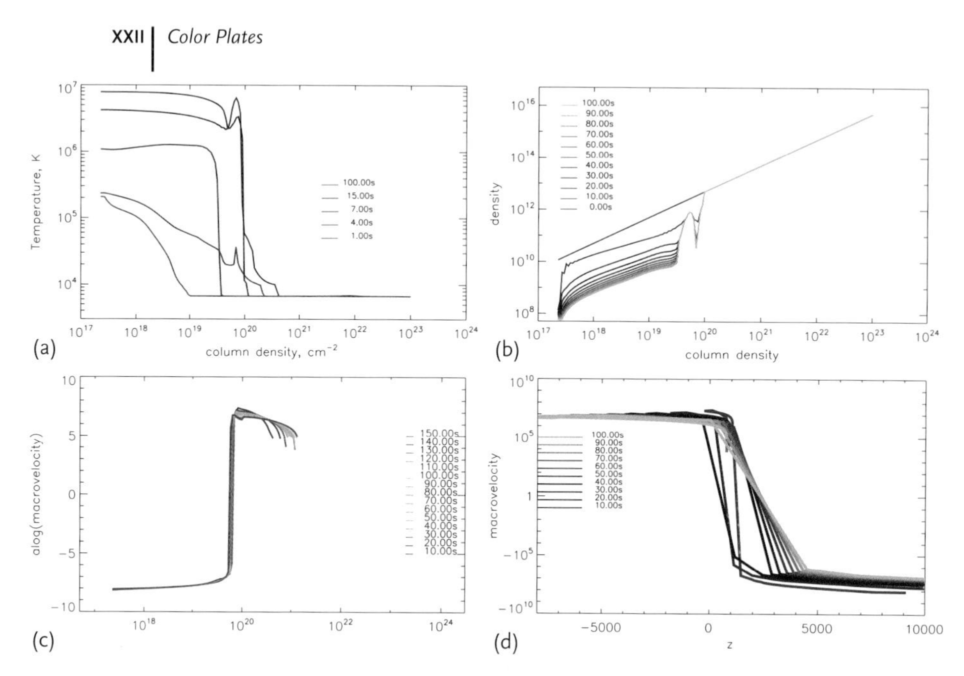

Figure 6.5 Snapshots of hydrodynamic models to electron-beam injection with parameters $F_0 = 2 \times 10^{12}$ erg cm^{-2} s^{-1}: electron temperature (a), ambient density (b), and macrovelocity profiles (c,d) vs. linear height above photosphere.

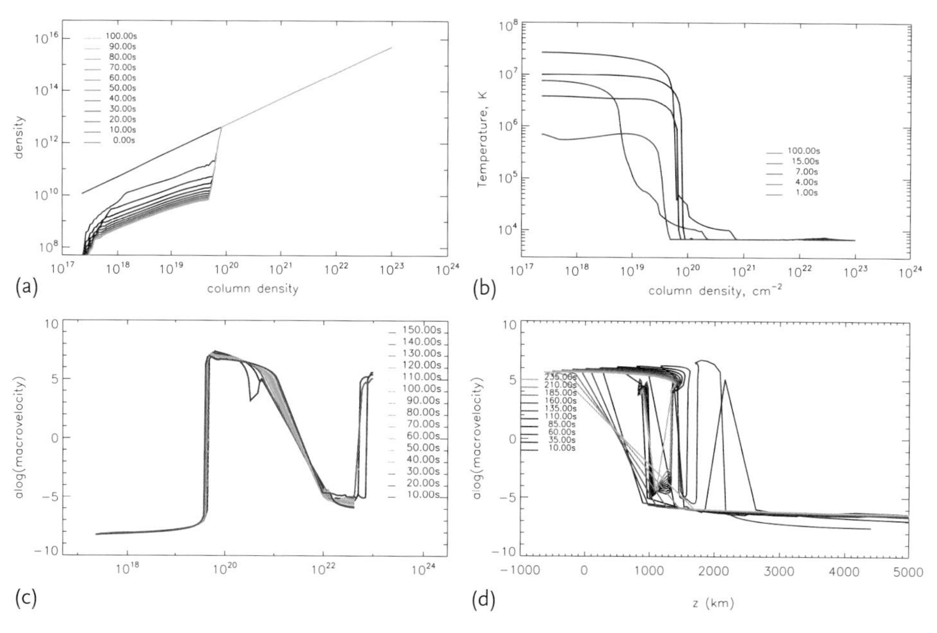

Figure 6.6 Snapshots of hydrodynamic models to the mixed beam injection (70% protons and 30% electrons in energy flux) with parameters $F_0 = 2 \times 10^{13}$ erg cm^{-2} s^{-1}: electron temperature (a), ambient density (c), and macrovelocity profiles (c,d) vs. linear height above photosphere.

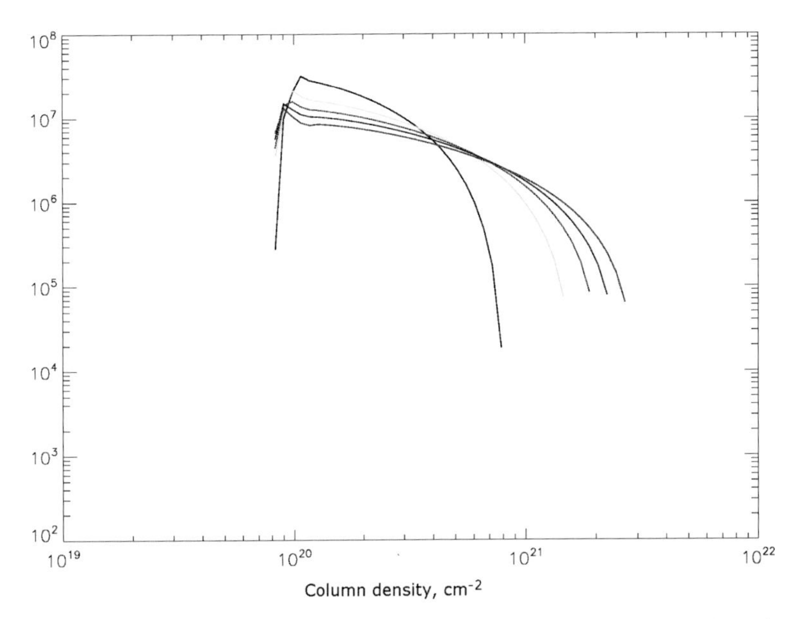

Figure 6.8 Close-up of Figure 6.7 of the temporal variations of macrovelocity in centimeters per second in a lower-temperature condensation, or a hydrodynamic shock (y-axis) vs. column depth in cm^{-2} (x-axis), appearing in response to injection of mixed proton/electron beams (Figure 6.7d–f) after 10 s (black line), 30 s (red line), 50 s (burgundy line), 70 s (navy line), and 100 s (blue line).

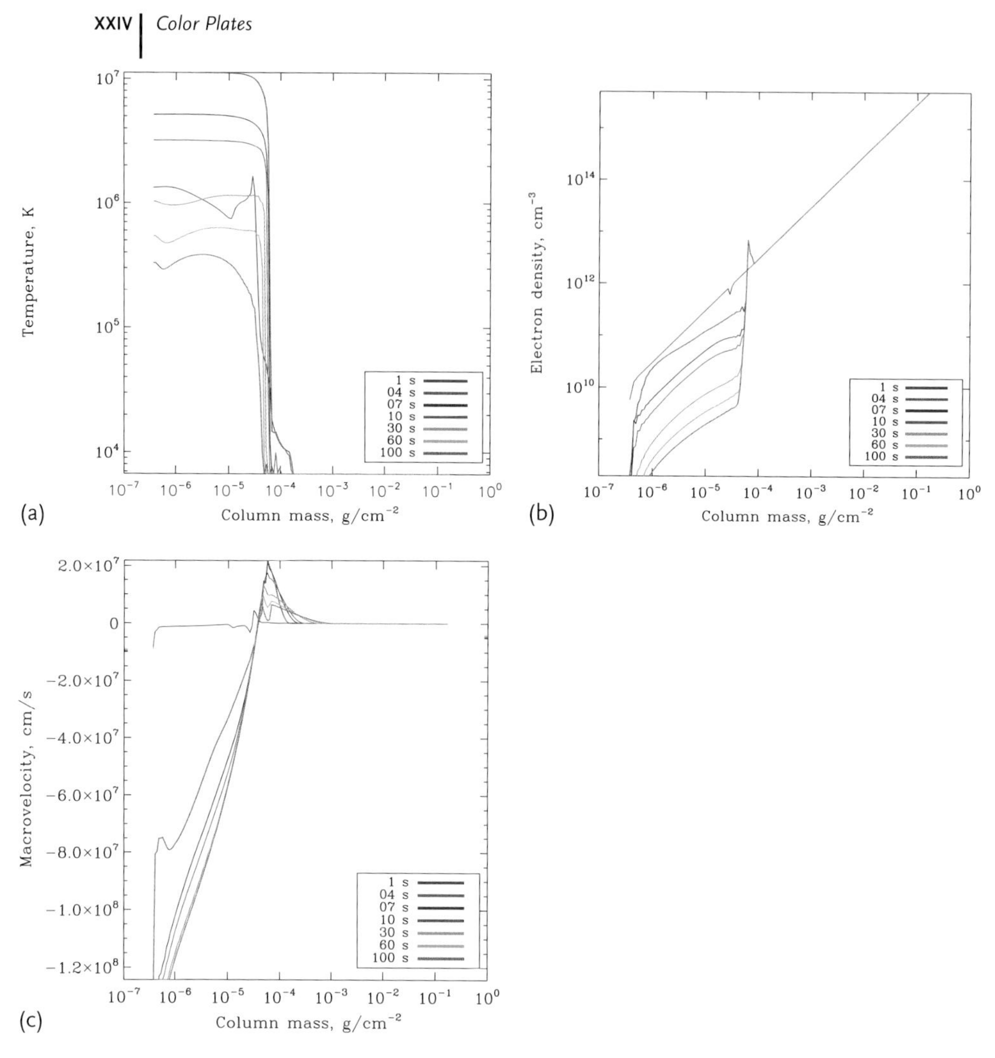

Figure 6.14 Temperature (a), density (b), and macrovelocity (c) variations in the main flare event calculated from a hydrodynamic response to the injection of an electron beam with parameters derived from HXR emission.

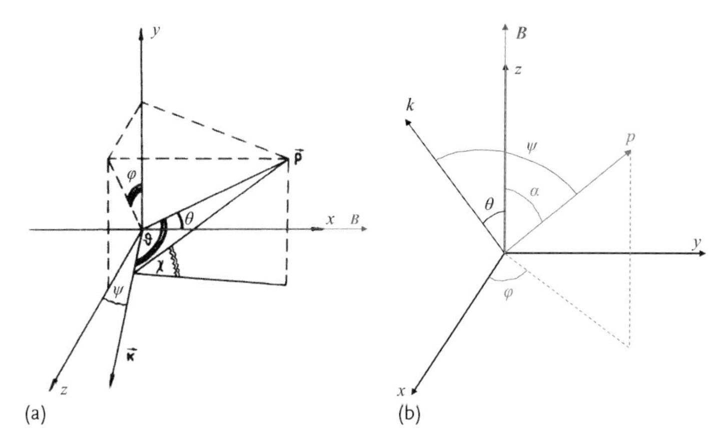

Figure 7.1 Geometry of observation of HXR emission from a flare under viewing angle ψ with respect to electron and photon momenta and magnetic field direction: (a) geometry considered by Nocera *et al.* (1985) and Zharkova *et al.* (1995); (b) geometry accepted by Syniavskii and Zharkova (1994) and Zharkova *et al.* (2010). See text for more details.

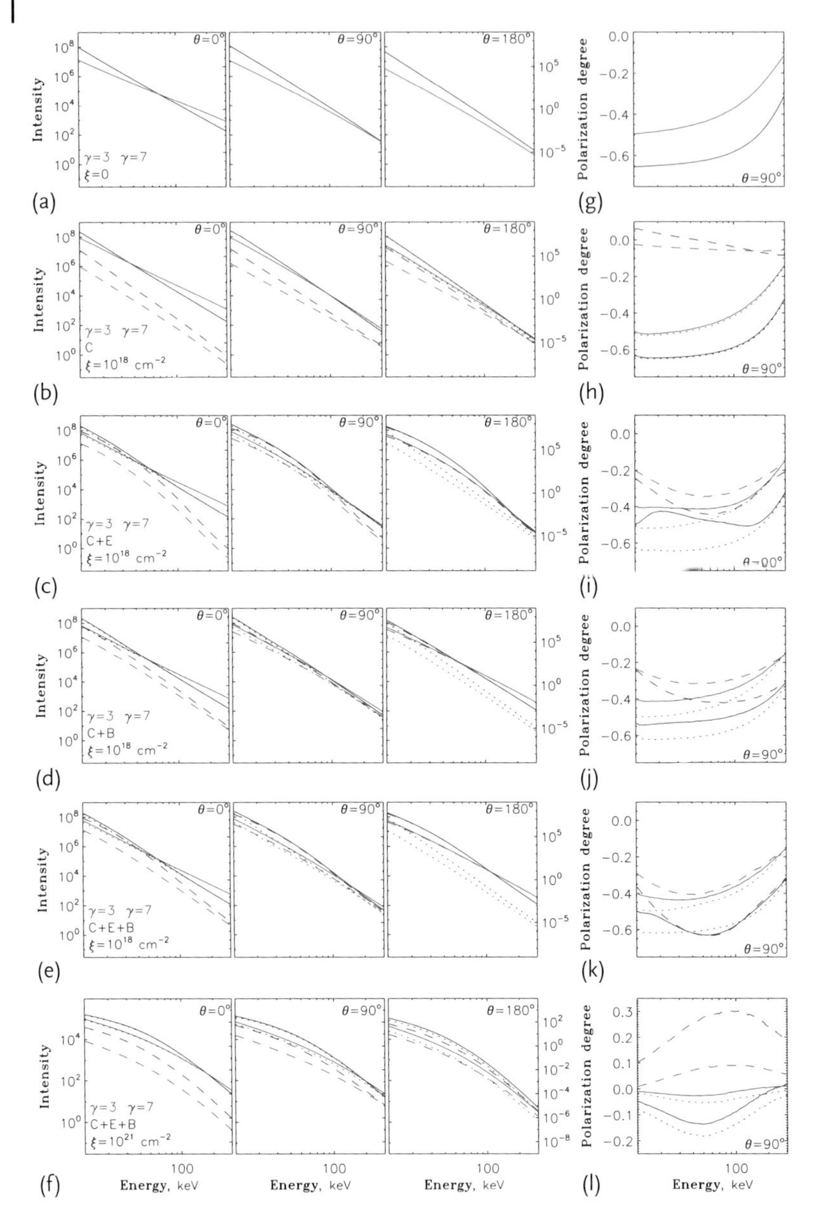

Figure 7.8 Intensity (in relative units) and polarization of the HXR emission for different energies. The initial power-law indices of the accelerated particles are $\gamma = 3$ (red lines) and $\gamma = 7$ (blue lines). The intensity plots for $\gamma = 3$ and $\gamma = 7$ are drawn using a different scaling; the intensity values are shown in the left ($\gamma = 3$) and right ($\gamma = 7$) margins. Dotted line: emission of downward propagating particles (with $\mu > 0$); dashed line: emission of upward propagating particles (with $\mu < 0$); solid line: total emission (downward + upward). The different panels (a–f) correspond to the different depths (ξ) indicated in the legends. The factors taken into account are collisions (C), self-induced electric field (E), and convergence of magnetic field (B).

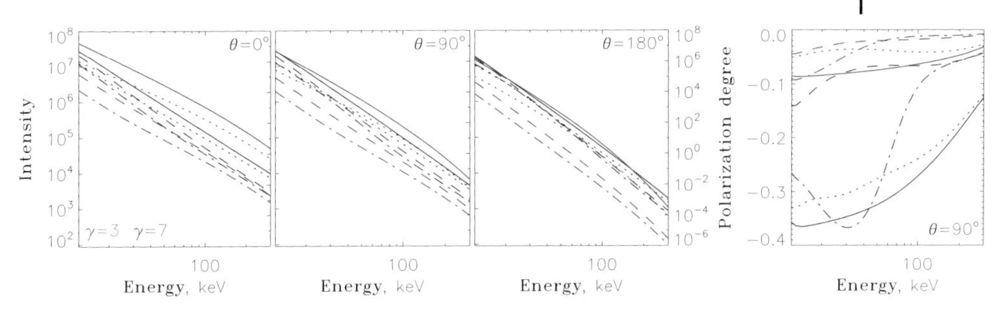

Figure 7.9 Integrated by column depth intensity (in relative units) and polarization of HXR emission for different energies. The initial power-law indices of the beam electrons are $\gamma = 3$ (red lines) and $\gamma = 7$ (blue lines). The intensity plots for $\gamma = 3$ and $\gamma = 7$ are drawn by using different scaling: the intensity magnitudes are shown in the left ($\gamma = 3$) and right ($\gamma = 7$) margins. Different lines correspond to the different simulation models: solid line – pure collisions (C), dotted line – collisions and return current (C+E), dashed line – collisions and converging magnetic field (C+B), dash-dotted line – all factors taken into account (C+E+B).

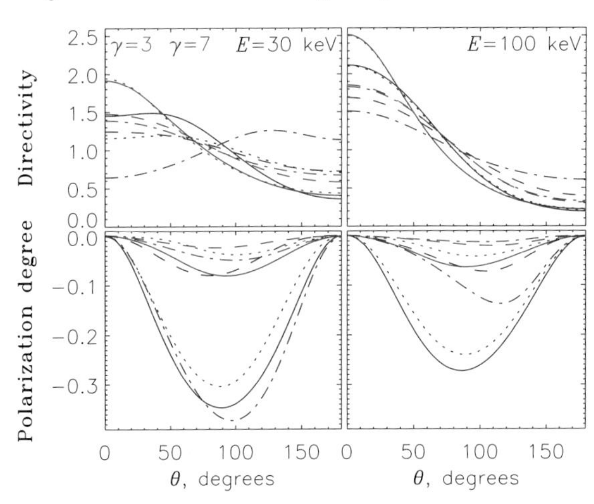

Figure 7.10 Directivity and polarization of hard X-ray emission for different directions of observations. The emission parameters are integrated over all layers of coronal magnetic tube. The initial power-law indices of the accelerated particles are $\gamma = 3$ (red lines) and $\gamma = 7$ (blue lines). The different lines correspond to the different simulation models (see Figure 7.9).

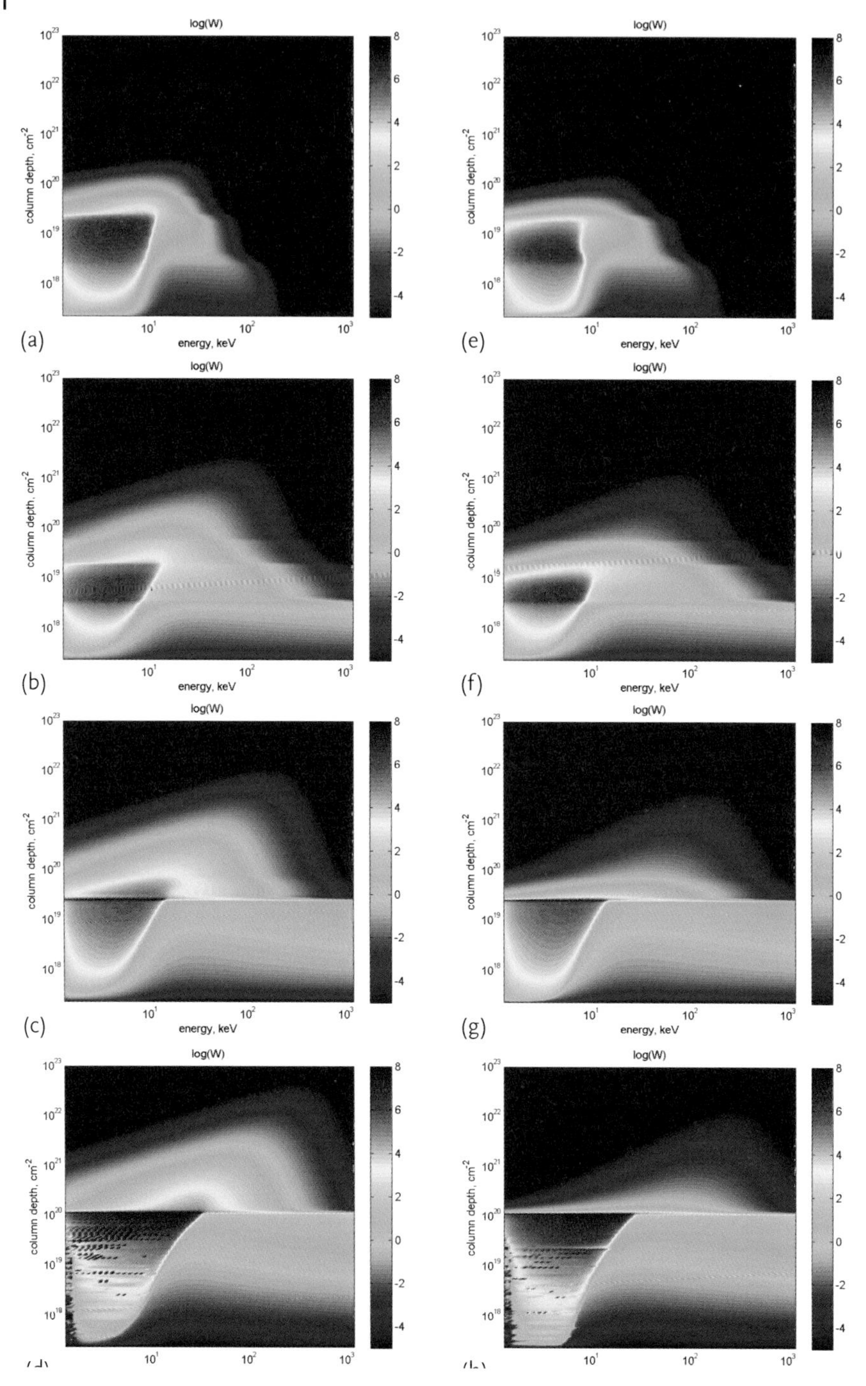

Figure 9.5 3-D density of Langmuir wave energy (erg cm^{-4} s) vs. column depth simulated without (a) and with (b) a self-induced electric field for beams with the following parameters: (a,b) $\delta = 7$ and $F_0 = 10^{10}$ erg cm^{-2} s^{-1}; (b,f) $\delta = 3$ and $F_0 = 10^{10}$ erg cm^{-2} s^{-1}; (c,g) $\delta = 3$ and $F_0 = 10^{11}$ erg cm^{-2} s^{-1}; (d,h) $\delta = 3$ and $F_0 = 10^{12}$ erg cm^{-2} s^{-1}.

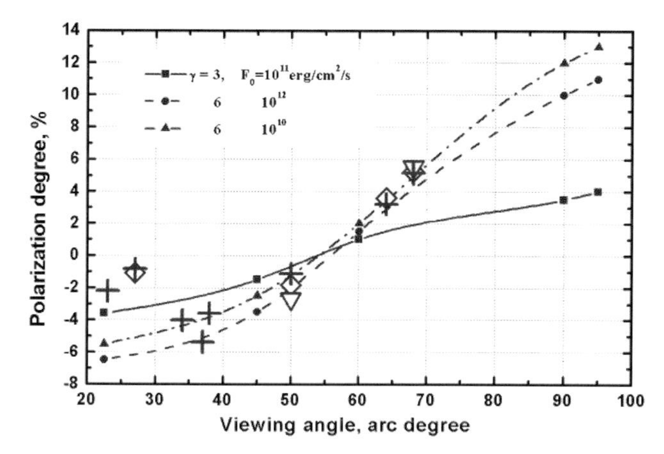

Figure 11.7 The H_α linear polarization as a function of position angle ψ caused by electron beams with a spectral index of $\gamma = 7$, $F_0 = 10^{10}$ erg cm^{-2} s^{-1} (dashed line with triangles), $\gamma = 7$, $F_0 = 10^{12}$ (dashed line with circles), $\gamma = 4$, $F_0 = 10^{11}$ erg cm^{-2} s^{-1} (solid line with squares). The crosses, diamonds, and triangles are the observations for different moustaches, discussed in text.

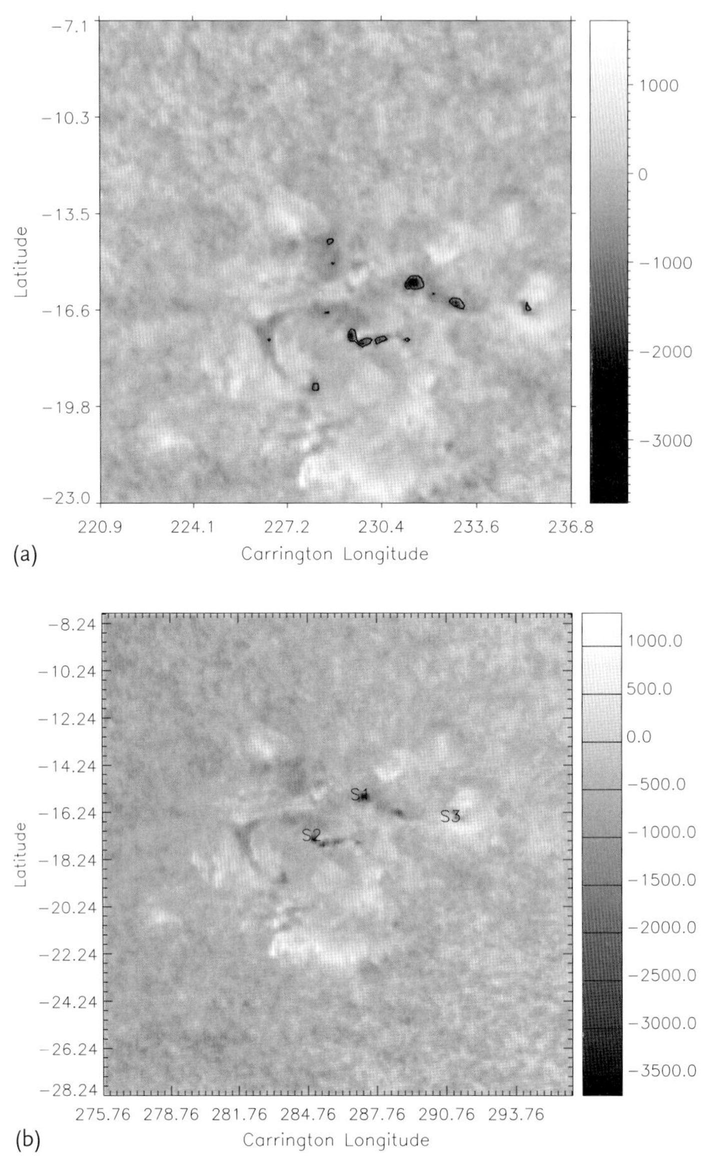

Figure 12.7 Locations of 11 Doppler sources (red contours) with downward motion higher than 1 km s^{-1} (a) but only three of them with detectable seismic responses in locations specified in columns 1–3 of Table 12.1 (b).

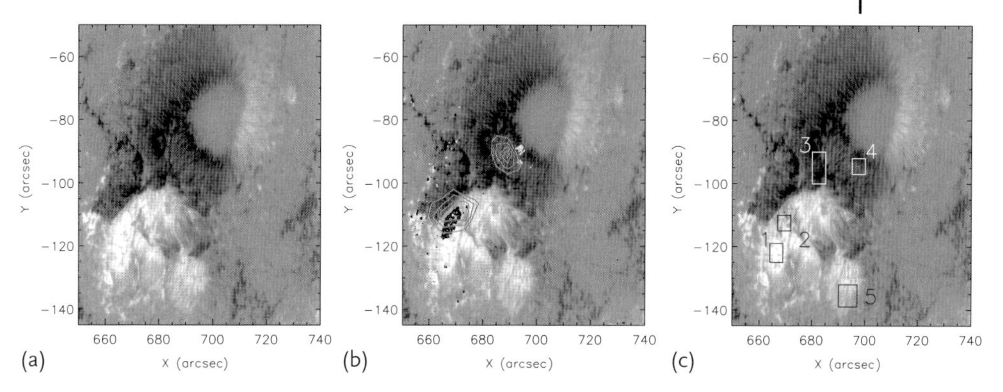

Figure 12.10 Longitudinal magnetic field image showing the location of 50–100 keV HXR emission seen by RHESSI (red contours), the magnetic reversals (black and white contours), and the regions where magnetic variations and photospheric velocities were measured. Magnetic variations were measured in regions 1–4 and are shown in Figure 12.11. Horizontal photospheric velocities were measured in all five regions, with region 5 providing a reference location outside of the main flare disturbance. (a) 22:01:43; (b) 22:09:42; (c) 22:09:42.

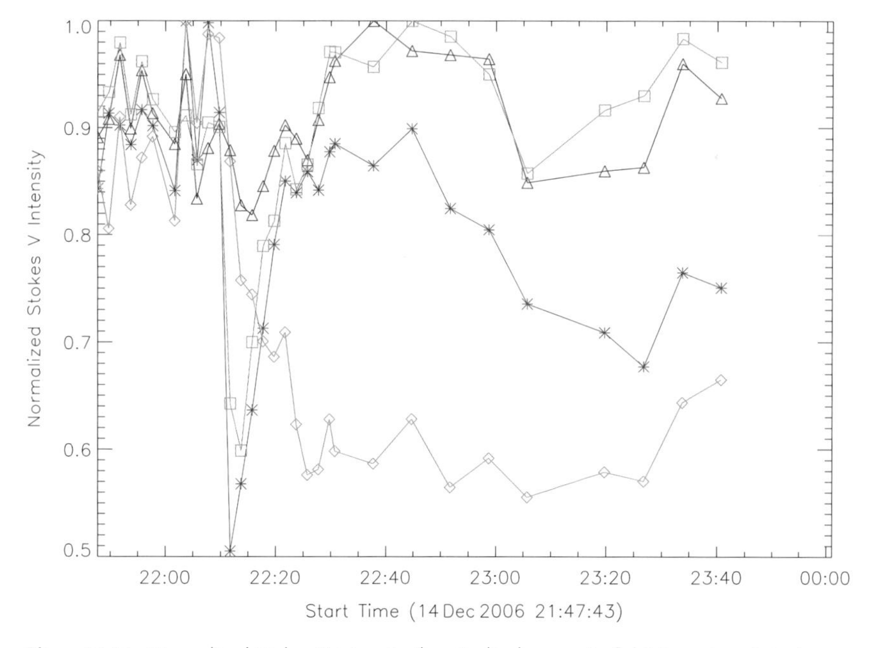

Figure 12.11 Normalized Stokes V intensity (longitudinal magnetic field) in regions 1 (red asterisks), 2 (green diamonds), 3 (blue triangles) and 4 (orange squares) as indicated in Figure 12.10.

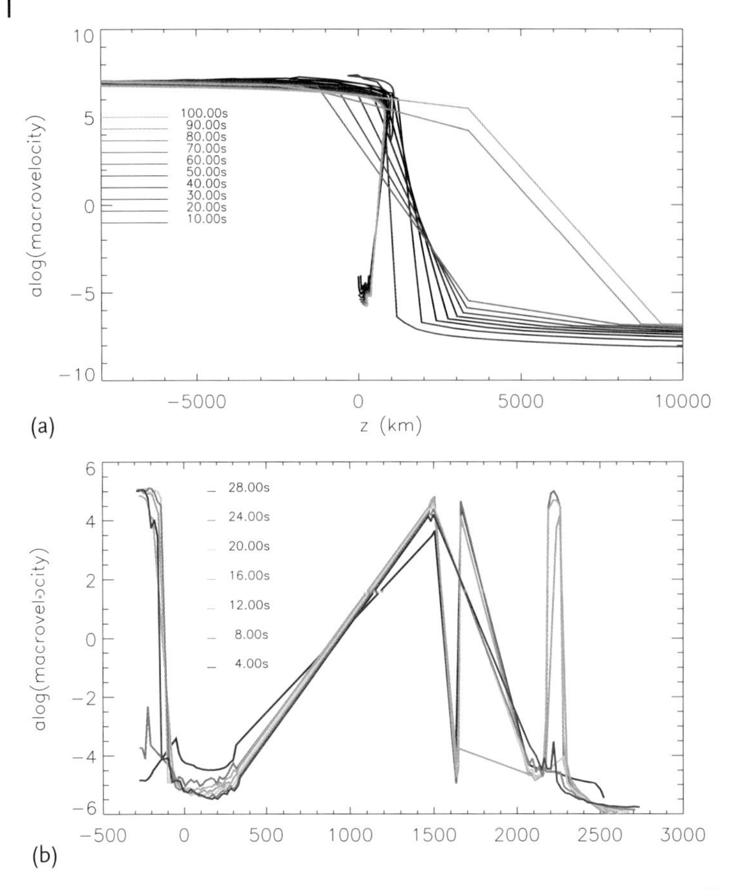

Figure 12.15 Macrovelocity profiles with shocks vs. linear depth above the photosphere formed at different times after the injection of the following agents: (a) a pure electron beam with parameters 2×10^{13} erg cm^{-2} s^{-1}, $\gamma = 4.9$; (b) 70% of proton beam and 30% of electron beam with total energy flux of 3×10^{13} erg cm^{-2} s^{-1}.

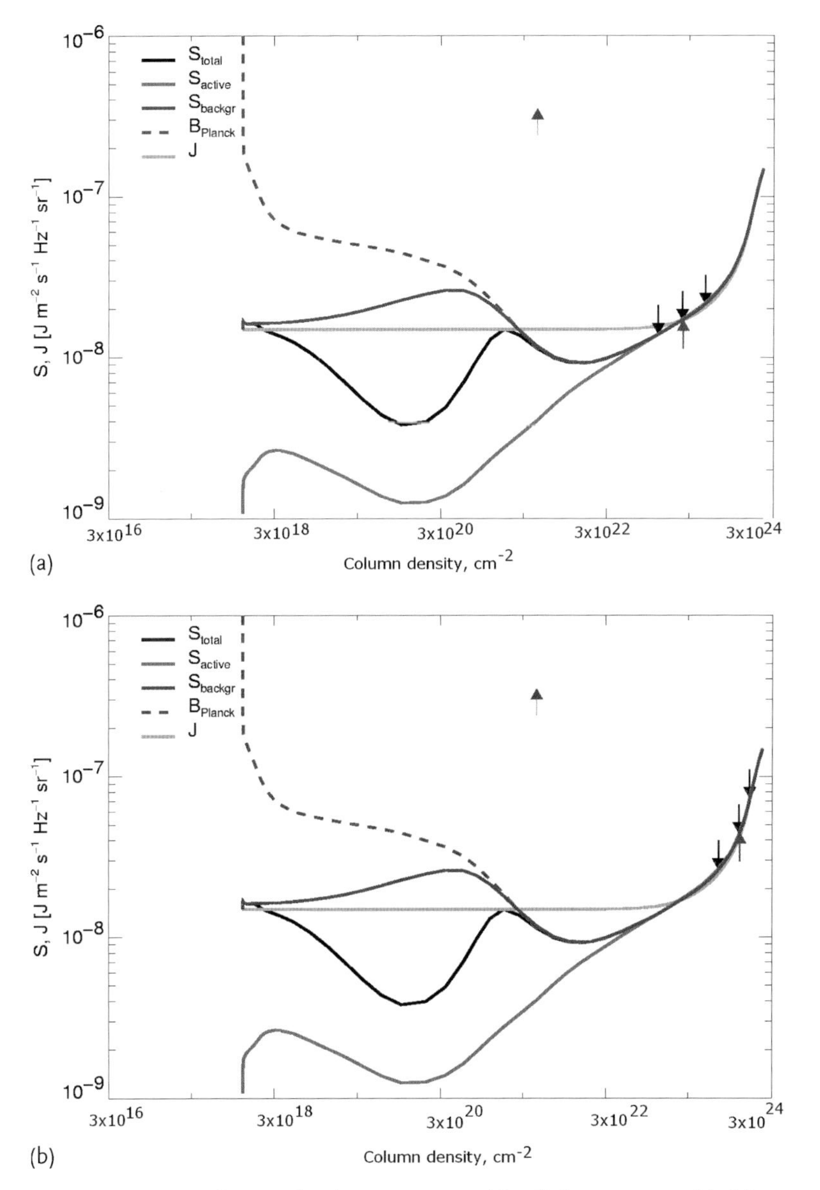

Figure 12.16 Source function distributions calculated for quiet atmosphere (a) and flaring atmosphere (b) heated by a hard intense electron beam ($\gamma = 3$, $F_0 = 10^{12}$ erg cm^{-2} s^{-1}) for Ni-line transition 6768 Å (S_{active}, gray line), background elements (S_{backgr}, blue line), Planck function (S_{planck}, red dashed line) and total for all elements (S_{total}, black line) as well as the mean intensity J (yellow line) simulated using the full NLTE MULTI-based) approach for a full coronal abundance of elements and some molecules (CO, C_2, CH, CN, 23 in total) (Uitenbroek, 2001; Zharkova and Kosovichev, 2002).

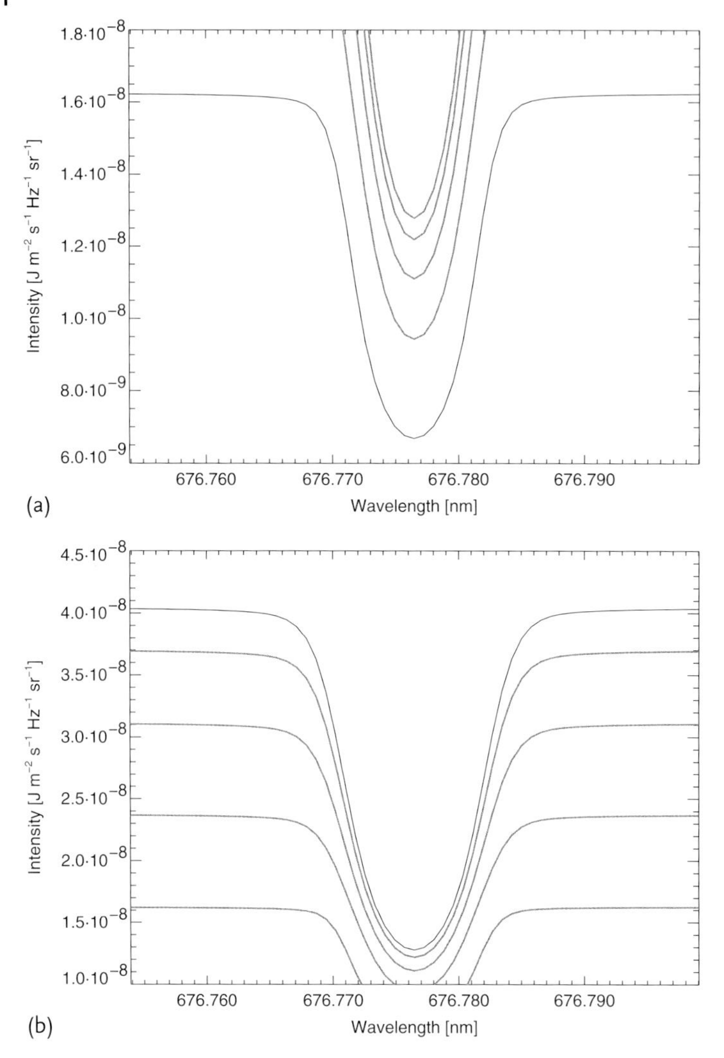

Figure 12.17 Ni-line profiles calculated for hydrodynamic models with initial energy fluxes of 10^{11} erg cm^{-2} s^{-1} and a spectral index of 4 using full non-LTE code (Uitenbroek, 2001) and nonthermal hydrogen ionization degree caused by beam electrons at 0 s (a) and 10 s (b) after their injection. For comparison, the line profiles are plotted for 2, 4, 6, and 8 s in each graph from bottom to top, respectively.

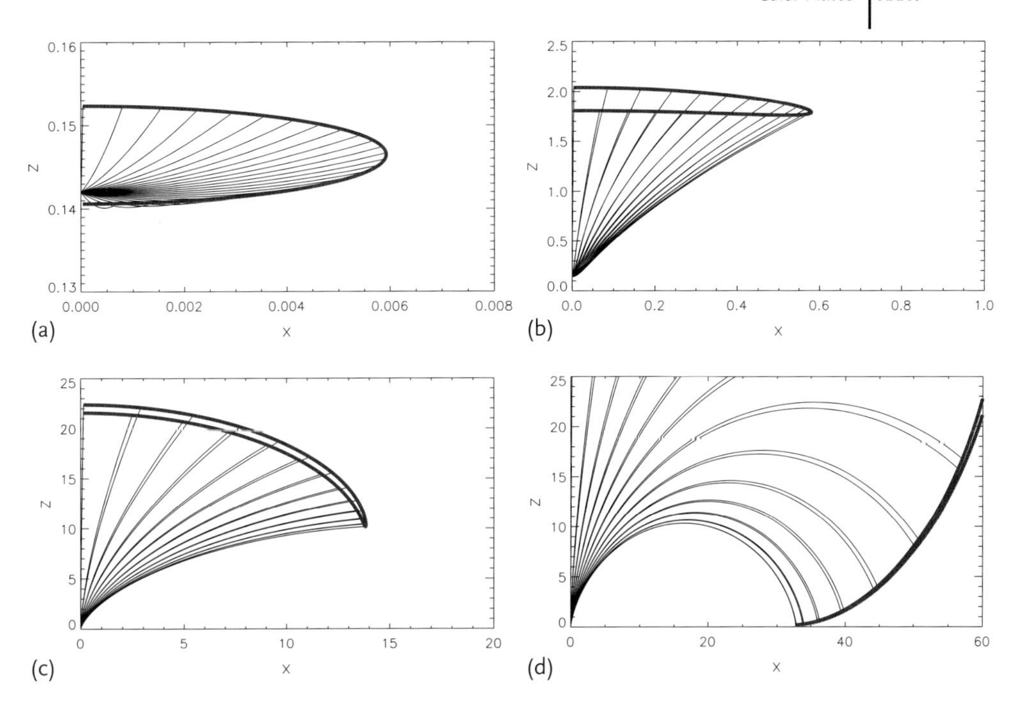

Figure 12.18 Wavefield generated by a high-frequency spherical point source, at 6.5 mHz, located near the surface as calculated for a simple polytrope model taken at different times: (a) 0.17 min; (b) 0.33 min; (c) 11.67 min; (d) 25.00 min. Depth is plotted along the y-axis, with 0.14 Mm approximately corresponding to the surface. The rays emanating from the source in the positive y-direction are plotted in black. The wavefront indicating the arrival of the perturbation is plotted in red.

1
Observational Phenomena of Solar Flares

1.1
Observational Constraints

Recent progress in hard X-ray (HXR) observations with *RHESSI* of light curves in numerous energy bands (Lin *et al.*, 2002), in combination with advanced imaging (Hurford *et al.*, 2002) and inversion techniques (Kontar *et al.*, 2011), allows us to reconsider the current models of particle acceleration and the effect of their transport on observational characteristics.

1.2
Hard X-Ray Light Curves and Spectra

1.2.1
Light Curves

Spatially integrated HXR light curves obtained by RHESSI confirm many of the temporal features established by previous observations: sharp increases (bursts) of HXR intensity over a relatively short ($\sim 0.5-5$ s) time scale accompanied by a more slowly varying component with a time scale of up to a few tens of minutes (Holman *et al.*, 2011).

The appearance of both sharp HXR bursts and steady increases in HXR intensity suggests that electrons are accelerated on two fundamentally distinct time scales: a rapid acceleration to high, bremsstrahlung-emitting energies and a more stationary process that maintains the high-energy electron flux to produce steady HXR emission for a substantial fraction of an hour or even longer.

At lower photon energies ϵ, where *thermal* bremsstrahlung dominates the total emission, emission at higher photon energies is weighted more heavily by plasma at high temperatures T. As a result, the decrease in conductive cooling time with temperature ($\tau \sim T^{-5/2}$) leads to emissions at higher energies peaking sooner (Aschwanden, 2007), and hence the light curve's peaking progressively earlier with an increase in photon energy. At higher energies, where *nonthermal* bremsstrahlung dominates, the relative timing of the emission at different en-

Electron and Proton Kinetics and Dynamics in Flaring Atmospheres, First Edition. Valentina Zharkova.

ergies depends on both the reduced "time of flight" for higher-energy electrons before they impact upon the thick target of the lower atmosphere (which tends to *advance* high-energy emissions relative to low-energy ones; Aschwanden and Schwartz, 1996; Brown *et al.*, 1998) and upon the decrease in collision frequency with energy (which tends to *delay* high-energy emissions relative to low-energy ones – Aschwanden *et al.*, 1997).

1.2.2
Photon and Electron Energy Spectra

HXR emission during flares typically shows a very steep spectrum at lower energies $\epsilon \sim 10\,\text{keV}$, indicative of a thermal process. It must be noted, however, that the frequent assumption (Hurford *et al.*, 2003a) of an *iso*thermal plasma is neither expected on the basis of simple physics nor required by observations (e.g., Brown, 1974). Indeed, Aschwanden (2007) has shown that the assumption of an isothermal source is *inconsistent* with the observations of temporal variations of the HXR spectrum in flares observed by RHESSI.

At higher energies, the spectrum tends to flatten to a roughly power-law form, with a spectral index γ typically in the range ~ 3–5. There is evidence for spectral flattening at higher energies ~ 500–$1000\,\text{keV}$, likely due to the increasing contribution of electron–electron bremsstrahlung (Kontar *et al.*, 2007). Sometimes, for the most energetic flares, there is noticeable emission up to a few hundred megaelectronvolts (Hurford *et al.*, 2006; Kuznetsov *et al.*, 2006). In most powerful flares the photon spectra reveal double power laws with the spectral indices below ~ 30–$60\,\text{keV}$ being smaller (flatter) than those above this energy by 2–4 units (Grigis and Benz, 2006; Krucker *et al.*, 2008a).

RHESSI observations reveal a relationship between spectral hardness and time. Values of the "local" spectral index $\gamma(\epsilon) = -d \log I(\epsilon)/d \log \epsilon$ are generally lowest around the peak of the event – a so-called "soft-hard-soft" evolution (Fletcher and Hudson, 2002; Grigis and Benz, 2008), but the decrease before and increase after this time do not occur at the same rate for all ϵ (see Figure 1.1 from Kontar and MacKinnon, 2005). A flare typically starts with a spectral index γ that is strongly dependent on energy, with the minimum value of γ showing a tendency to grow with time.

Nonthermal emission in the corona is identified in the impulsive phase by its softer spectrum (Mariska and McTiernan, 1999; Petrosian *et al.*, 2002), consistent with the small column depth of the coronal part of the source. The absence of a significant amount of (energy-dependent) collisional losses in this relatively thin target should result in a spectrum two powers steeper than the target-averaged spectrum (Brown, 1971; Datlowe and Lin, 1973; Hudson, 1972).

It should be noted that even if the accelerated electrons have a power-law energy spectrum $F_0(E_0) \sim E_0^{-\delta}$, characteristic energies associated with either the electron transport or the radiation physics may produce deviations from the power-law behavior in the observed spatially integrated photon spectrum. Further, as elaborated upon by Kontar *et al.* (2011), anisotropy in the mean electron distribution, com-

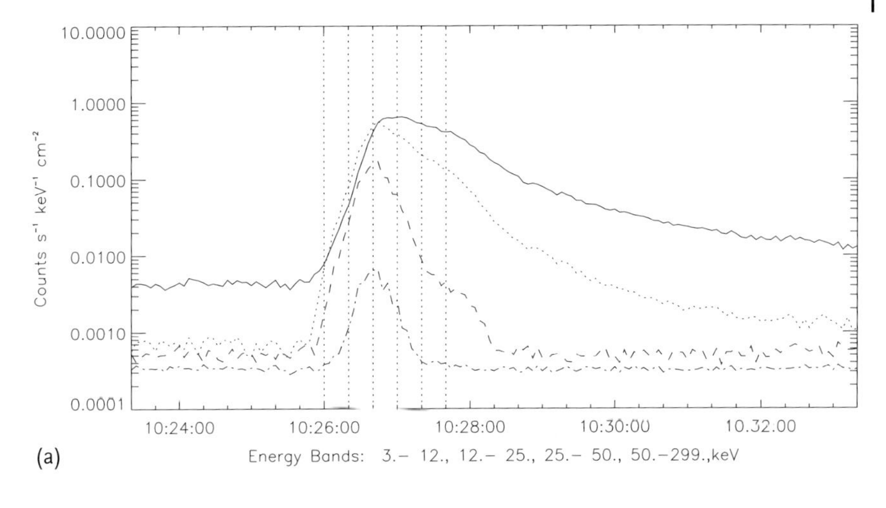

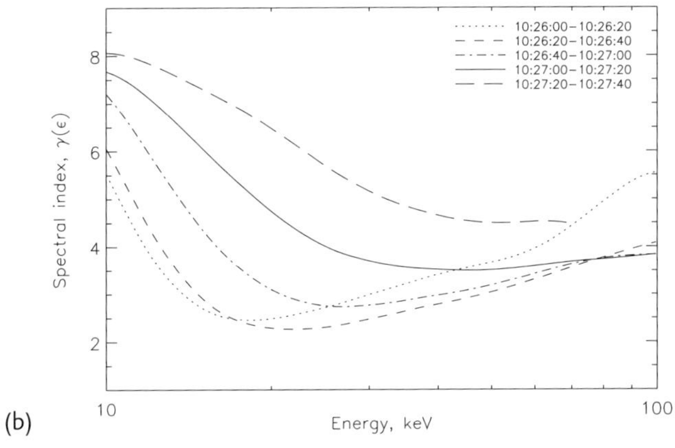

Figure 1.1 (a) Temporal variation (4 s cadence) of count rates in seven front RHESSI segments for the 26 February 2002 solar flare (10:26 UT). The vertical lines show five 20 s accumulation intervals for spectral analysis. (b) Temporal variation of energy-dependent photon spectral index $\gamma(\epsilon) = -d \log I(\epsilon)/d \log \epsilon$. Each line corresponds to one time interval. From Kontar and MacKinnon (2005).

bined with the intrinsic directivity of the bremsstrahlung emission process, produces an anisotropic distribution of primary photons. Compton backscattering of photons from the photosphere ("X-ray albedo", Bai and Ramaty, 1978; Langer and Petrosian, 1977) not only influences the observed photon spectrum but also has a diagnostic potential for determining the electron angular distribution (Kontar and Brown, 2006).

1.2.3 Electron Numbers

As pointed out by, for example, Tandberg-Hanssen and Emslie (1988), the total injected electron flux depends critically on the value of the low-energy cutoff E_c. A brief discussion of this issue is in order. Historically, a low-energy cutoff E_c was assumed simply in order to keep the injected power $\int_{E_c}^{\infty} E_0 F_0(E_0) d E_0$ finite (e.g., Brown, 1971; Holman *et al.*, 2003). To determine whether or not such a cutoff is actually *required* by observations, it is essential to adopt a nonparametric approach to interpreting the photon spectrum $I(\epsilon)$ – that is, to infer from $I(\epsilon)$ what range of mean electron source functions $\overline{F}(E)$ (Brown *et al.*, 2003) allow a statistically acceptable fit to $I(\epsilon)$. Kašparová *et al.* (2005a), in their analysis of the 20 August 2002 flare, have shown that a low-energy cutoff, or even a gap, in the mean source electron spectrum exists if the observed spectrum is considered as primary bremsstrahlung only; however, such cutoffs or gaps disappear if an albedo correction (for an isotropic primary source) is applied to the observed photon spectrum. Finally, Emslie *et al.* (2003) has shown that, when allowance is made for warm target effects in the electron energy loss rate, the injected electron energy $\int_{E_c}^{\infty} E_0 F_0(E_0) d E_0$ corresponding to a pure power-law photon spectrum can be finite even if no low-energy cutoff exists.

It appears, therefore, that the actual value of E_c, if one exists at all, should be below 10 keV. In view of the steep spectra commonly observed during flares, extending the spectrum to lower and lower energies requires an ever-increasing number of accelerated electrons, thus imposing even more significant constraints on the electron-acceleration mechanism in flares.

This having been said, electron fluxes deduced from observed photon spectra vary between 10^9 and 10^{12} erg cm^{-2} s^{-1}. For example, Figure 1.2 shows the photon spectrum for the 28 October 2003 flare as recorded by RHESSI. Assuming that the nonthermal HXR emission is thick-target bremsstrahlung (Brown, 1971), one can straightforwardly derive the differential injected nonthermal electron flux (electrons s^{-1} keV^{-1}) using a forward-fitting method (see Holman *et al.*, 2003). For this event, one obtains a total nonthermal electron production rate of $F_e = 1.7 \times 10^{36}$ s^{-1}, corresponding to an injected power of $P_e = 1.5 \times 10^{29}$ erg s^{-1}. For a nominal flare area $\sim 10^{17}$–10^{18} cm^2, such a particle rate corresponds to a flux of $F_e \sim 10^{19}$ cm^2 s^{-1} and, hence, since the average electron velocity $v \sim (c/3) \sim 10^{10}$ cm s^{-1}, to a beam density of $n_b \sim 10^9$ cm^{-3}. This is a substantial fraction of the ambient density $n \sim 10^{10}$ cm^{-3}.

The total electric current corresponding to such a particle acceleration rate, *if the acceleration is unidirectional*, is $I \sim 3 \times 10^{17}$ A, with a current density of $j \sim 1$ A cm^{-2}. Such high current densities, especially since they are introduced over a time scale of seconds, produce unacceptably large inductive electric and magnetic fields, unless the acceleration is near-isotropic, as in stochastic acceleration models (Miller *et al.*, 1996; Petrosian and Liu, 2004), the source has a very fine structure (Holman, 1985), or *the beam current is effectively neutralized by a cospatial return*

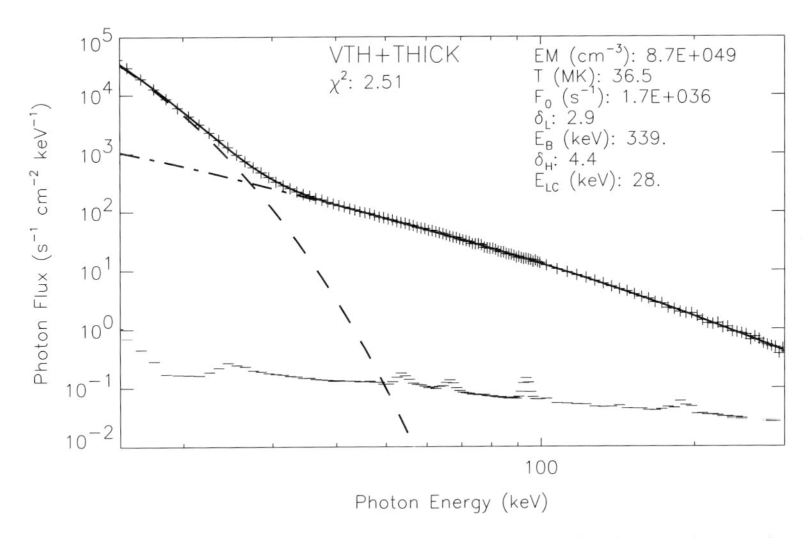

Figure 1.2 RHESSI HXR photon spectrum for solar event on 28 October 2003 (11:06:20–11:06:40 UT). The photon flux (plus signs) is fitted with a thermal bremsstrahlung contribution from an isothermal plasma (dashed curve) plus a double power-law nonthermal electron flux component with a low-energy cutoff (dashed–dotted curve). The best-fit parameters are shown in the plot.

current (see Brown and Bingham, 1984; Emslie, 1980; Knight and Sturrock, 1977; Larosa and Emslie, 1989; van den Oord, 1990; Zharkova and Gordovskyy, 2006).

Finally, it must be noted that replenishment within a short time scale of flares of the particles in an acceleration region (Emslie and Hénoux, 1995) has to be accounted for by any acceleration or transport mechanisms.

1.3 Light Curves and Energy Spectra of Gamma-Rays

1.3.1 γ-Ray Light Curves

Due to the limited sensitivities of GR detectors, information regarding short times scales for ion acceleration is less stringent than for the time scales of energetic electrons described in Section 2.1. Before the RHESSI launch, no detailed analysis had been performed systematically of the temporal profiles of prompt γ-ray line (GRL) emissions, although simultaneous peaking within ± 1 s of the emission in the GRL domain and of HXR emission has been reported for a few events (e.g., Kane *et al.*, 1986). This indicates that ion acceleration to a few megaelectronvolts must happen within a very short time scale of less than 1 s. It also indicates that in most cases the ions must interact with the ambient plasma in a dense region in order to produce nuclear lines (Vilmer *et al.*, 2011). These indications can be a reflection of either pure acceleration or acceleration and transport processes that need to be further investigated.

Recent analysis of RHESSI observations have provided the first information about the temporal evolution of prompt γ-ray lines (derived from spectral analysis) (Lin *et al.*, 2003; Share *et al.*, 2003; and Figure 4.7 in Chapter 4). When compared with the temporal evolution of the X-ray flux at 150 keV, it shows that the HXRs and γ-ray evolutions are roughly similar, indicating a common origin of the accelerated electrons and ions, but that there is for this event a small delay of around 10 s in the maximum of the X-ray and γ-ray flux. This has been interpreted by Dauphin and Vilmer (2007) and discussed by Vilmer *et al.* (2011) in terms of transport of energetic electrons and ions.

1.3.2
Energy Spectra and Abundances of Ions in Flares

As reviewed by Vilmer *et al.* (2011), the interactions of energetic ions in the 1 to 100 MeV/nuc range produce a complete γ-ray line spectrum. While narrow γ-ray lines result from the bombardment of ambient nuclei by accelerated protons and α particles, broad lines occur from the inverse reactions, in which accelerated carbon (C) or even heavier nuclei collide with ambient hydrogen (H) or helium (He) nuclei. Deexcitation GRLs provide information on flare ions of energies above 2 MeV/nuc. The shape of the ion distribution below this energy is unknown, but it is of particular interest for the total ion content.

Quantitative results on γ-ray line observations were obtained for more than 20 GRL events. The accelerated ion spectra were found to extend as unbroken power laws down to at least 2 MeV/nuc, if a reasonable ambient Ne/O abundance ratio was used (for details see Chapter 4 in this volume). While for the 19 solar maximum missions (SMM) flares measured prior to the launch of RHESSI the average spectral index was around -4.3, events observed with RHESSI tend to have much harder slopes (Lin *et al.*, 2003; Share and Murphy, 2006).

Even though most of the energy contained in ions resides in protons and α-particles, the crucial constraints for acceleration processes also arise from the estimations of the abundances of heavier accelerated ions. Information on the abundances of flare-accelerated ions was deduced for the SMM/GRS (γ-ray spectrometer) flares as well as for a few flares observed by RHESSI. Noticeable enhancements are generally deduced for the numbers of alpha particles as well as of accelerated ^{3}He isotopes compared to the standard coronal ratios of element abundances including ^{4}He. Furthermore, accelerated heavy ions, such as Ne, Mg, and Fe, are also generally found to be overabundant with respect to their coronal composition.

1.3.3
Ion Numbers

Since the first detection of γ-ray lines in 1972, information on energy content in ions has been obtained for more than 20 events. The analysis carried out from 19 events observed by SMM/GRS and a few events observed with RHESSI show that the energy contained in > 1 MeV ions ranges from 10^{29} to 10^{32} ergs for GRL flares. This showed that the energy contained in > 1 MeV ions may be comparable to the energy contained in subrelativistic electrons. There are still large uncertainties in the determination of these quantities and a large dispersion of the relative electron (20 keV) and ion (1 MeV/nuc) energy contents from one flare to another.

The question of relative ion and electron acceleration in solar flares can also be addressed by comparing the ratios of ion production to relativistic electron production with energies of > 300 keV. This has been done using both SMM/GRS observations and recent RHESSI observations (see, e.g., Chupp and Benz, 1994; Shih *et al.*, 2009). A good correlation has been found between the total energy content in protons above 30 MeV and the total energy content in electrons above 300 keV, suggesting that high-energy electrons and ions are directly linked by acceleration processes. This link can be even more direct than for the production ratio of subrelativistic electrons and ions due to a larger difference between electron energy contents above 20 keV and ion energy contents above 1 MeV/nuc. These differences need to be accounted for in the proton- and ion-acceleration models.

1.4 Geometry of Hard X-Ray and Gamma-Ray Sources

The imaging capabilities of RHESSI, combined with its high spectral resolution, allow us to resolve in detail coronal sources (e.g., Masuda *et al.*, 1994) and footpoints occurring in the same flare (Figure 1.3) and hence to study the acceleration processes that give rise to such sources.

The coronal source often appears before the main flare HXR increase and the appearance of footpoints. In the impulsive phase, the coronal HXR emission is, generally, well correlated in both time and spectrum with the footpoints (Battaglia and Benz, 2006; Emslie *et al.*, 2003). These observations suggest a strong coupling between the corona and chromosphere during flares, a coupling that is presumably related to transport of accelerated particles from one region to another.

1.4.1 Differences in Footpoint Spectral Indices

An interesting RHESSI observation is the approximate equality of the spectral indices in different footpoints of the same loop. Emslie *et al.* (2003) reported differences $\Delta\gamma \sim 0.3$–0.4 between the spectral indices of the two dominant footpoints in the 23 July 2002 event. In a few smaller events analyzed by Battaglia and Benz (2006), $\Delta\gamma$ is even smaller and indeed is significant in only one out of five cases. However, other observations of X-class flares reveal a much stronger difference (up to 5) between the spectral indices of footpoints of the same loop (Battaglia and Benz, 2006; Krucker *et al.*, 2007; Takakura *et al.*, 1995).

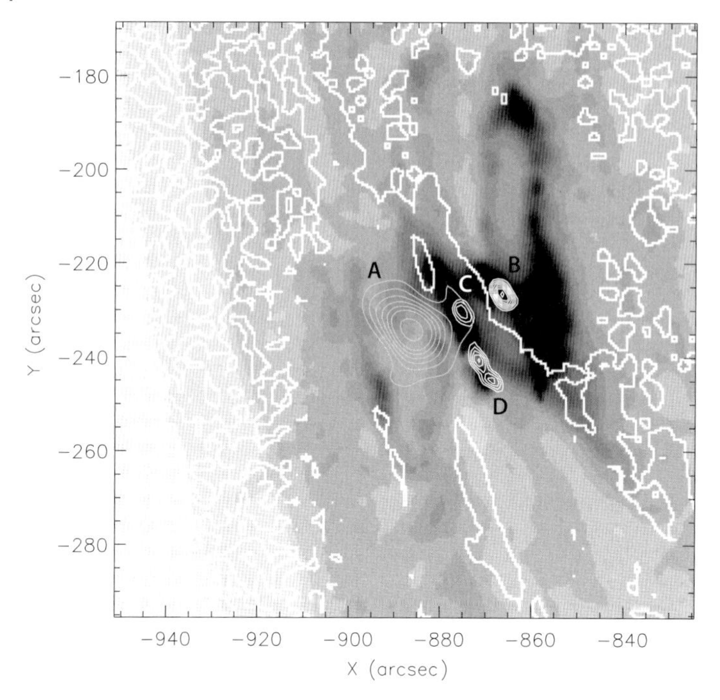

Figure 1.3 RHESSI HXR images of flare of 23 July 2003 (thin white contours) taken at 00:28 UT, overlaid on MDI (Michelson Doppler Imager) neutral-line magnetograms (thick white contours) and H_α images (negatives) taken at 00:28:45 UT. One extended (coronal A) and three compact (footpoints B, C, and D) HXR sources are evident. From Zharkova *et al.* (2005a).

More surprising than the occasional difference between the spectral indices of footpoints are the differences, or lack thereof, between the spectra of the coronal source and the footpoints. The spectral index difference $\Delta\gamma = 2.0$ predicted (Brown, 1971) between thin- and (collision-dominated) thick-target sources is *not* confirmed in a number of observations (Fletcher *et al.*, 2011). Differences $\Delta\gamma$ smaller than 2 could in principle be explained by invoking a coronal target that is intermediate between thin- and thick-target conditions; such spectral differences could also be partially accounted for by the transport effects of precipitating electrons, e.g., by the effect of a self-induced electric field of precipitating electrons (Zharkova and Gordovskyy, 2006). However, $\Delta\gamma$ ranges from 0.59 to 3.68 (Grigis and Benz, 2006; Krucker *et al.*, 2007), which includes values beyond the capability of simple models to explain. For example, Figure 1.4 (after Battaglia and Benz, 2006) displays the photon energy spectra of two footpoints and a coronal source, obtained for the same time interval during the flare of 13 July 2005. The footpoint nonthermal power-law indices γ are near-identical (2.7 ± 0.1), while the spectral index of the nonthermal emission for the coronal source is substantially steeper: $\gamma = 5.6 \pm 0.1$.

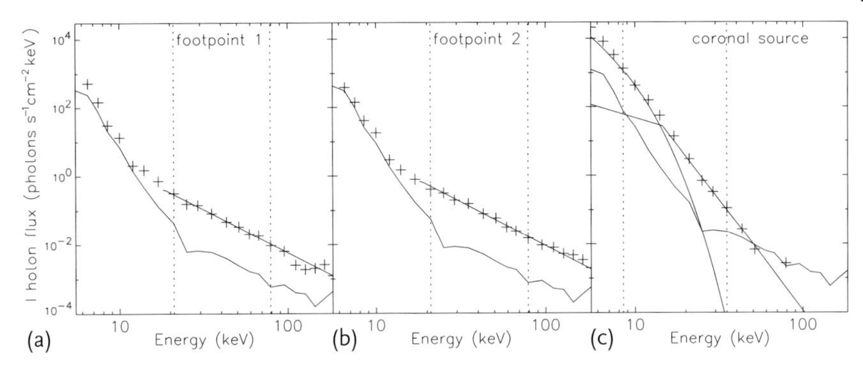

Figure 1.4 Comparison of spectra of footpoint and coronal sources in flare of 13 July 2005. (a,b) footpoint spectra. A power law has been fitted between the dotted vertical lines; (c) spectrum of coronal source. A power law plus an isothermal spectrum has been fitted between the dotted lines. From Battaglia and Benz (2006).

1.4.2 Hard X-Ray and Gamma-Ray Source Locations

Gamma-ray imaging is a unique capability of RHESSI. Gamma-ray images were first observed by RHESSI for the flare of 23 July 2002. These observations (Hurford *et al.*, 2003a) established that the HXR sources in this flare were spatially separated by several arcseconds from the 2.223 MeV neutron-capture γ-ray source. A similar, but smaller, separation was also detected for the "Halloween" flare on 28 October 2003 (Hurford *et al.*, 2006). This separation of HXR and γ-ray sources must be accounted for by acceleration models.

1.5 Pre- and Postflare Hard X-Ray and Radio Emission

Providing additional information to complement that provided by HXRs, radio emission is another channel of information on nonthermal electrons in flares; coherent radio emission and HXRs are often observed simultaneously. Radio emission is caused by gyrosynchrotron radiation of mildly relativistic electrons; it correlates well in *time* with HXRs, but, in general, not in *space* (Krucker *et al.*, 2008b; White *et al.*, 2011). Moreover, the most intense radio emission in flares at meter and decimeter wavelengths originates not from single particles, but from various plasma waves, that is, from coherent radiation processes.

An association of RHESSI HXRs with coherent radio emission in the meter and decimeter band, for 201 flares with flare classes larger than C5, has been surveyed by Benz *et al.* (2005). They found that coherent radio emission occurred before the onset of HXR emission in 9% of the flares studied. Most of the radio emission was type III radiation at the plasma frequency and its first harmonic, emitted by a beam of electrons moving upward in the solar atmosphere into regions of progressively

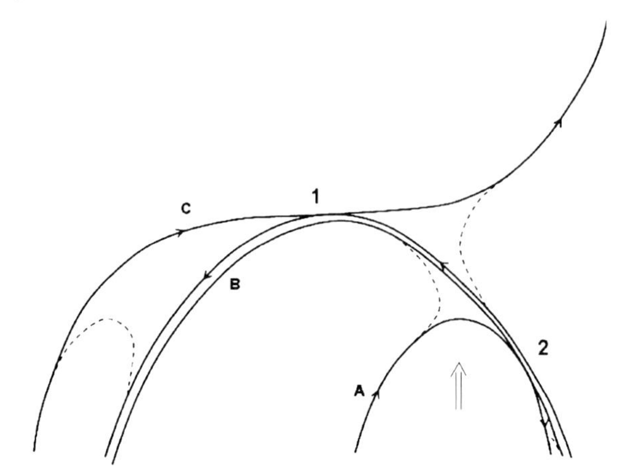

Figure 1.5 Schematic drawings illustrating the presence of two simultaneous reconnection sites (labeled 1 and 2) in a flare. Loop system A is the main driver, and reconnection 1 is the main energy release site. Reconnected field lines are indicated by dashed curves. From Benz *et al.* (2005).

lower density. In a few cases, the researchers also found a pulsating continuum, possibly caused by cyclotron maser emission. Both types of emission indicate the presence of electron acceleration in the preflare phase, which is often accompanied by X-ray emission from a thermal source in the corona.

Coronal sources can be observed to emit HXRs even before HXR footpoints appear. Nonthermal emission at centimeter wavelengths, suggesting the presence of relativistic electrons, has also been reported in such coronal sources (Asai *et al.*, 2006). One can conclude from these observations that although the energization process that operates in the preflare phase involves mostly heating of coronal plasma, its contribution to particle acceleration cannot be excluded. Comparison of radio and HXR emissions by Benz *et al.* (2006) also revealed, in some events, coherent radio emission *after the HXR emission had stopped*. This phenomenon was reported in only 5% of the flares studied.

The kinds of emission that occur at higher frequencies, such as decimetric narrowband spikes, pulsations, and stationary type IV events, correlate more frequently with the HXR flux and thus appear to be more directly related to the acceleration process (Arzner and Benz, 2005). While there is a good *association* between coherent radio emission and HXRs, a strong *correlation* in the details of the time profile is less frequent, so that coherent emission is not a reliable proxy for the main flare energy release. This can be accounted for by invoking multiple reconnection sites connected by common field lines along which accelerated particles propagate and serve as a trigger for distant accelerations (see Figure 1.5 in Benz *et al.*, 2005).

1.6
Magnetic Field Changes Associated with Flares

Sharp temporal increases of HXR emissions are often closely correlated in time with variations of the magnetic field measured in the photosphere (Kosovichev and Zharkova, 2001; Sudol and Harvey, 2005; Zharkova *et al.*, 2005a). For the flare of 23 July 2002, the magnetic flux change over the flare duration was about 1.2×10^{21} Mx; note that the magnetic flux in the areas not spanned by the magnetic inversion line do not show significant variations above the noise level (Figure 1.6). Magnetic field changes occurring around an apparent magnetic neutral line (AMNL), in general (Figure 1.6), and in the locations of three HXR footpoint sources of the flare of 23 July 2002 (see Figure 9 in Zharkova *et al.*, 2005a) are irreversible, or steplike, for example the magnetic field reaches a new level of the steady state and does not return to a preflare value (Kosovichev and Zharkova, 2001; Sudol and Harvey, 2005; Zharkova *et al.*, 2005a).

The magnetic flux in the other areas of either positive or negative polarities, not including the magnetic inversion lines, does not show noticeable variations above the noise level. A cross-correlation analysis with a time lag between the temporal magnetic variations covering AMNL and the HXR light curves for the flare of 23 July 2002 reveals a noticeable positive correlation of 05–06 with a time lag no larger than 1–2 min for for all energy bands (Figure 1.7). This observation strengthens the belief, on theoretical grounds, that irreversible changes in the magnetic field are responsible for the initiation and development of flare phenomena.

1.6.1
Local Magnetic Field Variations

In order to detect magnetic field variations in the locations of HXR emission, the precise RHESSI HXR images in the 40–80 keV band with the four HXR sources appearing during the course of the flare similar to the locations in the paper by Krucker *et al.* (2003) were overlaid onto MDI magnetograms (Zharkova *et al.*, 2005a). The magnetic flux variations were extracted from the maximum areas covered by this HXR emission for each minute before and after the flare onset for four HXR sources (A, B, C, D) detected in this flare.

Source A is found not to be associated with any magnetic field changes, despite its appearance in the 12–25 keV band 1 min earlier than the other footpoint sources B, C, and D, but still 1 min later after a start of the magnetic changes. Source A is likely a projection of the top of the loop (see the TRACE image overlaid onto the H_α image, Figure 1.3, bottom); the loop is embedded into the photosphere at the locations close to the footpoints (sources B, C, and D). Source A could be a good candidate for the site of primary electron acceleration in this flare because of the HXR emission timing and energy and its relation to the radio emission discussed by Krucker *et al.* (2003). However, it does not show any direct connection with the magnetic field locations, possibly because of its occurrence on the loop's top and the tilt of the loop toward the limb.

(a)

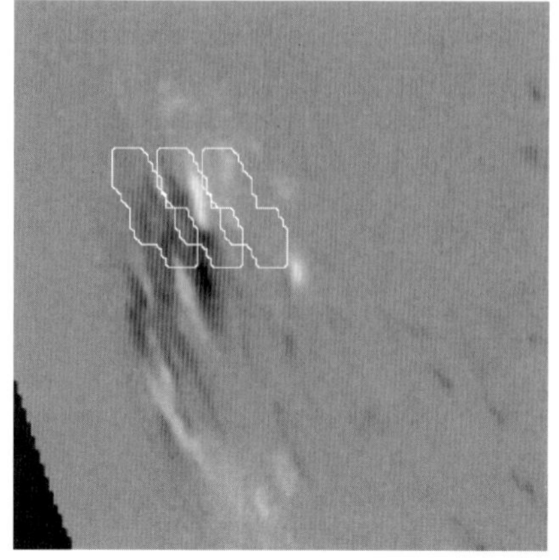

(b)

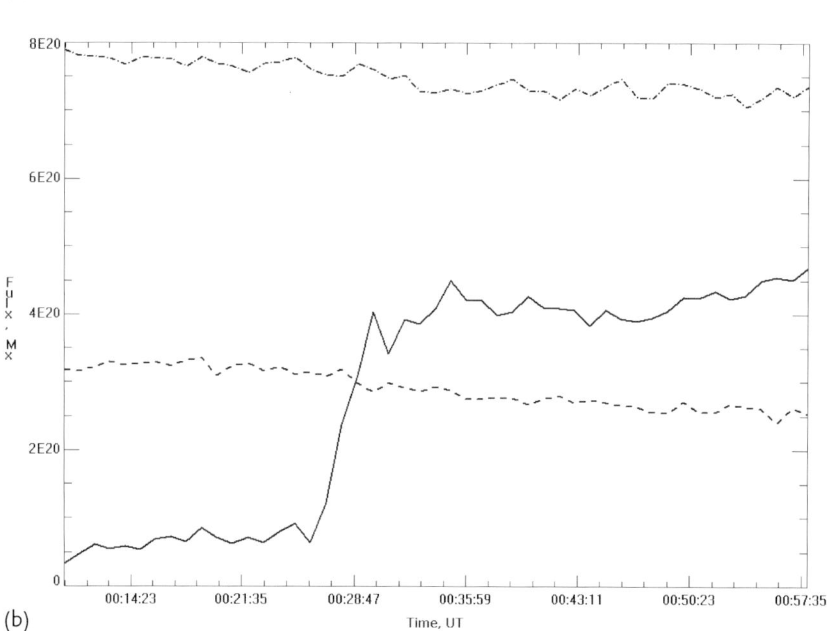

Figure 1.6 Total magnetic flux variations (in Mx) from 23 July 2002 flare: (a) mask locations on magnetograms: mask 1 (left) – negative polarity only, mask 2 (middle) – including magnetic inversion (neutral line); mask 3 (right) – positive magnetic polarity only; (b) corresponding total magnetic flux variations (dashed: left mask, solid: middle mask, dashed–dotted: right mask).

The temporal magnetic field variations in footpoint sources B, C, and D are presented in Figure 1.8 (bottom to top, respectively).

Sources B, C, and D are clearly the footpoint locations, and hence their connection with the magnetic field changes are much more pronounced. The changes in footpoints B and D are fully irreversible; the magnetic field in them decreases from

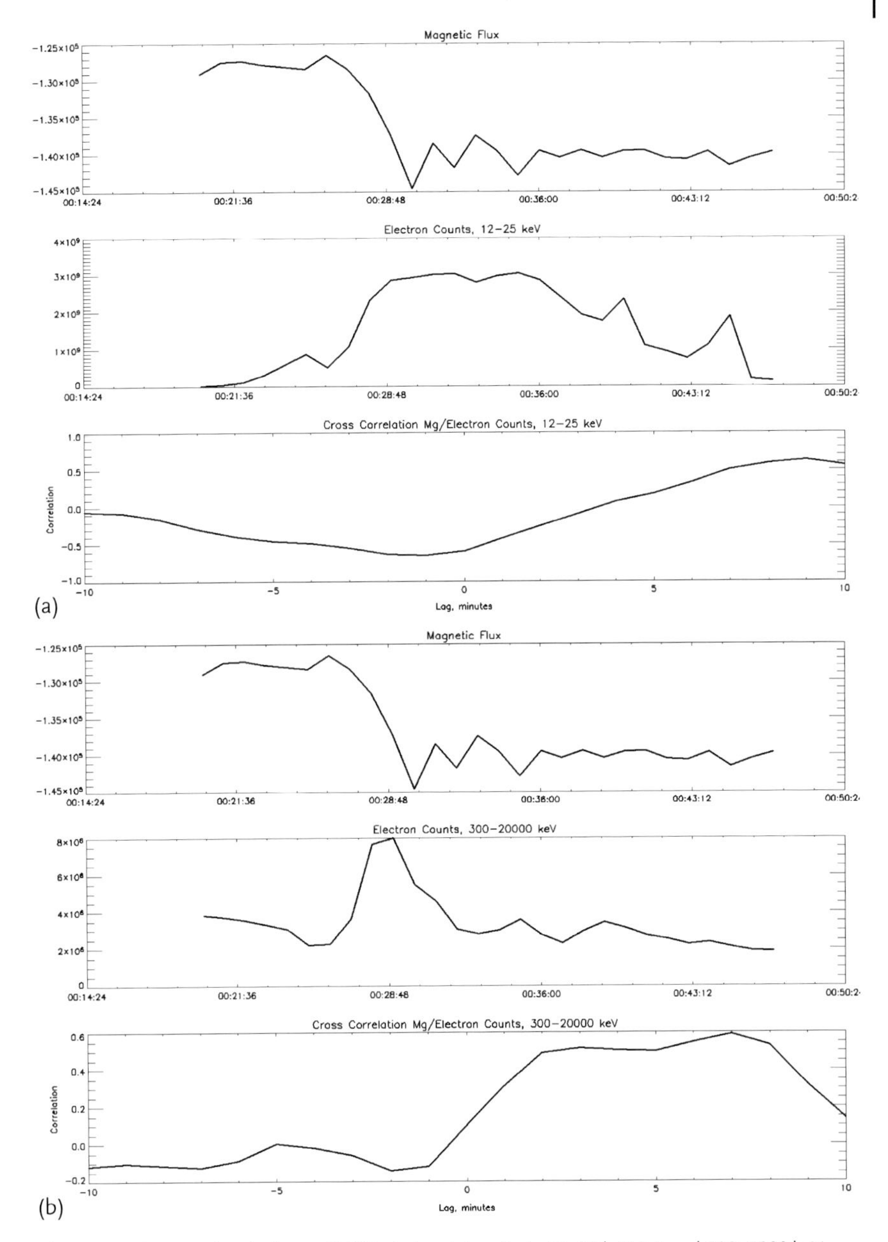

Figure 1.7 Temporal variations of HXR photon intensity in 12–50 keV (a) and 300–2000 keV (b) bands of magnetic field flux in a rectangle including the magnetic inversion line and their relevant correlation coefficients. Adapted from Zharkova *et al.* (2005a).

−300 to −420 G. Source D has these changes from 00:26:00 UT for 2 min; they are then followed at 00:29:00 UT by the magnetic changes in sources B and C lasting for 2 and 4 min, respectively.

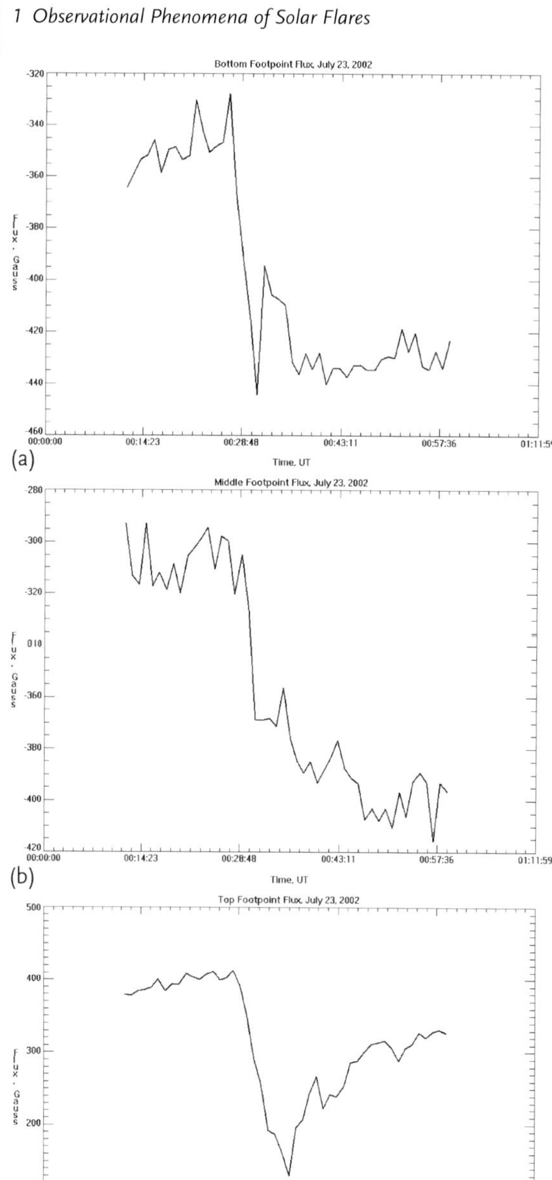

Figure 1.8 Magnetic flux variations in the locations of HXR sources D (a), B (b), and C (c). Sources A, B, C, and D are defined in Figure 1.3.

In source C the magnetic changes are mostly reversible, and the magnetic field first desreases from 400 to 100 G and returns to about 300 G after the flare (Figure 1.8c). Since in source C the difference between the new level of magnetic field, achieved after all changes were stopped, and the preflare level is close to the magnitude of irreversible changes measured in footpoints B and D, we can conclude that these changes are partially irreversible. The major part of these magnetic variations in source C are reversible and of the transient type, similar to those found by Paterson and Zirin (1981) that are not real magnetic field changes but caused by a short-term increase of the line intensity resulting from the nonthermal excitation of the upper level of the Ni atom in the transition 6768 A by the precipitation of high-energy electrons (see for details Zharkova and Kosovichev, 2002).

1.7
UV and Optical Emission

Often there is a close (up to 1 s) temporal correlation in spectral observations of HXR, UV, and optical emissions appearing from a flare (Asai *et al.*, 2002; Bianda *et al.*, 2005; Cheng *et al.*, 1987; Hiei, 1987; Kurokawa *et al.*, 1988). These fast temporal fluctuations in HXR and optical emissions are usually attributed to the propagation of beams of accelerated particles (electrons or protons) and to the dissipation of their energy in lower layers of the solar atmosphere via nonthermal hydrogen excitation and ionization processes and the heating of ambient plasma (Allred *et al.*, 2005; Hénoux, 1991b; Hudson, 1972; Metcalf *et al.*, 2003; Ricchiazzi and Canfield, 1983; Zharkova and Kobylinskii, 1993; Zharkova *et al.*, 2007). This investigation of nonthermal hydrogen emission uncovers some details of particle precipitation into deeper atmospheric levels. However, it raises many other questions related to the exact scenario of the interaction of particles with neutral ambient atoms and the physical condition of the ambient plasma, that is, how these spectral changes are related to the magnetic structures leading to flares.

Special attention has been paid by many authors to the investigation of rapid variations of the H_α line intensity and its correlation with HXR flux during the impulsive phase of chromospheric flares, for example, Trottet *et al.* (2000) and references therein. This often includes a comparison of the spatial distribution of HXR sources and H_α flare kernels (Asai *et al.*, 2002; Kashapova *et al.*, 2006, 2007) and white light emission (Kosovichev, 2007c) showing that many H_α kernels can brighten in succession during the evolution of flare ribbons. These brightenings, frequently accompanied by white light flare (Kosovichev, 2007c), allow one to follow the process of consequent energy release in a very complicated magnetic structure of a flare and to establish that some H_α kernels coincide with HXR sources while many others do not.

In a recent multiwavelength investigation of a few solar flares (Kashapova *et al.*, 2007; Kašparová *et al.*, 2005a,b; Zharkova *et al.*, 2007) considerable attention was paid to the locations of fast changes of H_α intensity. In agreement with the predictions of some solar flare models, the HXR sources were located on the external edges of the H_α emission and found to be connected with chromospheric plasma heated by nonthermal electrons. But the fast changes of H_α intensity are normally placed not only inside HXR sources, as expected, if they are the signatures of the chromospheric response to the electron bombardment, but also outside of them. This fact may indicate that the response of the lower atmosphere to flare energy release is not restricted to sites of propagation of accelerated electrons only but can be associated with some other agents (Kašparová *et al.*, 2005a; Zharkova *et al.*, 2007).

1.8 Seismic Responses

The detection of significant seismic emissions from solar flares, or sunquakes (Donea *et al.*, 1999; Kosovichev and Zharkova, 1998), is one of the main discoveries in solar physics in the past decade. Helioseismology of sunquakes, circular or dipole/quadruple waves propagating along the solar surface outward from an impulsive flare for $\sim$ 30–90 min after the impulsive phase, offers us the opportunity to explore various physical processes in the lower solar atmosphere and its interior associated with flare phenomena.

Kosovichev and Zharkova (1998) made the first identification of a sunquake in the form of circular ripples emanating from the X2.6 flare of 9 July 1996 and spreading horizontally for up to 120 Mm from the flare location, being visible for up to 55 min after the flare start. The momentum required to provide the observed ripples was assumed to be delivered by a hydrodynamic shock appearing at the injection of a very hard ($\gamma = 3$) and intense ($F_0 = 10^{12}$ erg cm^{-2} s^{-1}) electron beam (Kosovichev and Zharkova, 1998).

However, a comparison of the observed ripples with a theoretical model of seismic ripples (Kosovichev and Zharkova, 1995, 1998) revealed that the momentum required to produce the observed seismic response ($\sim 2 \times 10^{22}$ g cm s^{-1}) was one order of magnitude higher than those of $\sim 10^{21}$ g cm s^{-1} observed from the plasma downflows in MDI dopplergrams. Also, the travel time of this shock to the photosphere was more than 2 min, while the start time, at which the helioseismic response first seen in the time-distance diagrams (where the ridge crosses the *Y*-axis), coincides closely (within a minute) with the time of the HXR impulse. This raises a question about some additional sources that can deliver within a very short time scale, coinciding with the start of HXR impulse, a required momentum to the solar photosphere.

Later, another method, computational seismic holography, was applied to the MDI observations of this flare to image its seismic source (Donea *et al.*, 1999) centered on the composite umbra of the $\beta\gamma\delta$-configuration sunspot at the heart of the active region AR7978. The source was clearly visible in the 2.5–4.5 mHz spectrum and even more pronounced in the 5–7 mHz spectrum. These results indicated that the sunquakes, or ripples, could also be accompanied by seismic emission in the mHz range whose intensity varies significantly with depth in a flaring atmosphere, suggesting that different physical processes take place at different atmospheric levels.

The next four quakes were observed nearly 8 years later during the descending phase of solar cycle 23 for the X17 flare of 28 October 2003 and the X10-class flare of 29 October 2003 using a holographic analysis (Donea and Lindsey, 2005). The acoustic signatures of the 28 October 2003 flare were found to be less energetic than that of the X2.6 flare of 9 July 1996, but still detectable. Similar to the first sunquake, in all four seismic sources the emission from lower frequencies (2 mHz) occupies smaller areas.

These holographic observations were also confirmed by the time-distance approach (Kosovichev, 2006; Zharkova and Zharkov, 2007; Zharkova *et al.*, 2005b). For the 28 October flare only 3 out of 11 seismic sources had detectable ripples with wave propagation times within the datacube of a 120 Mm flare and, thus, different horizontal velocities varying from source to source (Zharkova and Zharkov, 2007). Two seismic sources (S3 and S3) were close to the HXR and γ-ray sources observed by RHESSI, while the third largest one was cospatial with the location of the strongest HXR emission up to energies of 100 MeV. The seismic ripples in these sources are found to be anisotropic, as reported by Kosovichev (2006), manifested by the noncircular shaped (elliptical) ripples possibly caused by nonvertical initial impacts. The seismic waves, or ripples, observed in flares can be produced by the hydrodynamic shocks caused by mixed high-energy proton beams, quasithermal proton flows, or electron beams (Zharkova and Zharkov, 2007).

Donea and Lindsey (2005) anticipated that a significant seismic activity could also be excited by relatively small flares (either C- or a M-class flares) and provide enough energy to be measured as seismic sources in relatively localized footpoints. This was recently confirmed by numerous observations using the holographic method (Beşliu-Ionescu *et al.*, 2006a,b; Moradi *et al.*, 2007). These sources of seismic emission associated with medium-sized flares were visible in the 2.5–3.5 mHz range and even more pronounced in the 5–7 mHz range and often found to be cospatial with white-light flares (Beşliu-Ionescu *et al.*, 2006b; Donea *et al.*, 2006). In all cases the seismic sources were close to the locations of HXR emission, although they revealed a delay of up to 3 min from the start of the HXR emission for the flares of 6 and 10 April 2001 (Beşliu-Ionescu *et al.*, 2006b; Martínez-Oliveros *et al.*, 2008b).

The time-distance diagram technique was further applied to some X-class flares, allowing us to add another three flares (16 July 2004, 15 January 2005, and 23 July 2002) to the list of observed seismic ripples associated with flares (Kosovichev, 2006, 2007a). Kosovichev reported multiple sunquakes produced by each of these flares, which originated simultaneously in different spatial positions coinciding with the locations of strong downward motions detected by MDI dopplergrams and occurring within 1 min of the initiation of the HXR emission. This supports the suggestion that these sunquakes are somehow related to the precipitation of high-energy particles and their energy deposition into a flaring atmosphere and the solar interior beneath.

1.9 Critical Issues

Particle acceleration during solar flares has two stages: an impulsive one lasting from a few to tens of seconds and a gradual one lasting for tens of minutes. Accelerated particles (electrons and ions) are observed *in situ* in the interplanetary space at the same time as their radiative signatures in the solar atmosphere. The latter are comprised of multiwavelength observations in HXR bremsstrahlung continuous emission produced by high-energy electron scattering by ambient plasma particles, γ-ray continuous emission produced by proton and ion scattering, and gyrosynchrotron emission in a microwave range produced by the motion of high-energy electrons in the magnetic field (Benz *et al.*, 2005; Miller *et al.*, 1996, 1997).

Continuous γ-ray emission is sometimes accompanied by γ-ray line emission caused by interactions of accelerated ions with energies $\geq$ 1 MeV nuc^{-1} and ambient nuclei producing nuclei, neutrons, and positrons for nonrelativistic energies and pions and high-energy nuclei for relativistic energies of accelerated particles (Vilmer *et al.*, 2011). These kinds of high-energy emission are accompanied by soft X-rays and lower-frequency radio emission reflecting thermal radiation, EUV, UV, and optical emission appearing in close temporal correlation to high-energy emission (Fletcher *et al.*, 2011).

While it is well accepted that the gradual phase is a result of ambient plasma heating by processes associated with solar flares (particles, waves, or shocks), the impulsive phase is considered to be a reflection of particle acceleration processes. Some flares are considered to be *electron-dominated*, others *proton-dominated*, depending on the emission they produce. Moreover, emission observations allow us to establish a *time scale within which this emission is to be produced* and *at what rate*, for example, how many particles per second must be accelerated to account for observed intensities. This emission also allows us to deduce *minimal and maximal energies* of accelerated particles, *their energy spectra evolution* for the duration of a flare, and *their associations with appearances in lower-energy emission in lines and continua* occurring in deeper flaring atmospheres. These points are summarized below (Zharkova *et al.*, 2011a).

1. *Some flares are accompanied by coronal mass ejections and filament eruptions and some are not.* This seems to reflect the fact that the main energy release in helmet-type loops occurs through magnetic field dissipation during a magnetic line reconnection of the upper part of loops leading the filament eruptions upward and collapsing current sheets downward and with the lower set of loops forming a two-ribbon flare.
2. *The temporal evolution of flare emission reveals an impulsive phase with a number of short impulses on top of a steady increase of HXR emission followed by a gradual phase with a strong increase of emission in EUV, UV, and optical range sometimes accompanied by sunquakes* (Fletcher *et al.*, 2011). This item clearly points to the impulsive nature of the particle acceleration process in flares, to a very short time scale of this acceleration, and to the fact that particle transport processes

significantly change the observed photon distribution in HXRs without changing this time scale very much.

3. *A loop top source and footpoint sources appear within different time scales* (Fletcher *et al.*, 2011). This item is a clear indication that the primary energy release process is often related to magnetic reconnection in a helmetlike loop structure with a diffusion region, or reconnecting current sheet, that is the primary energy release occurring on top, where the primary particle acceleration occurs. The emission in footpoints must be a mixture of all three processes: acceleration on top followed by particle transport and emission transport into the loop legs, or footpoints.
4. *Within a very short timescale superthermal plasma appears in the coronal sources increasing their temperatures to tens or hundreds of million degrees* (Fletcher *et al.*, 2011). This is related to the previous item and is likely to be the property of, mainly, acceleration and, partially, transport.
5. *Spectral indices of electrons can differ or be the same as those of protons* (Holman *et al.*, 2011; Vilmer *et al.*, 2011). This item is a mixed property of particle acceleration and transport because some acceleration models report the difference in spectral indices of electrons and protons, while radiation and particle transport can change these indices even further (Kontar *et al.*, 2011; Vilmer *et al.*, 2011).
6. *Spectral indices of electrons and protons in the interplanetary space are often the same as those detected in the solar atmosphere* (Holman *et al.*, 2011; Vilmer *et al.*, 2011). This item must be a pure property of particle acceleration and its relation to the primary energy release scenario.
7. *There is a close temporal correlation between HXR and MW emission of within 1 s (or even closer)* (Kontar *et al.*, 2011; White *et al.*, 2011). This item has the properties of both particle acceleration and transport.
8. *There is a temporal delay of 2–20 s in the appearance of γ-ray emission* (Fletcher *et al.*, 2011). This point can be a result of the joint effect of acceleration and transport.
9. *Sometimes the emission in two footpoints is produced by power-law electrons, while in other cases in one footpoint the emission is produced by power-law electrons and in the other by quasithermal ones* (Battaglia and Benz, 2006; Krucker *et al.*, 2007). This must be caused by the topological properties of particle acceleration, which produce different populations of accelerated particles with different energy distributions.
10. *HXR and γ-ray sources are spatially separated by a few arcseconds (into the opposite legs of the same loop)* (Vilmer *et al.*, 2011). This again must be a property of either the magnetic field topology or a reconnecting current sheet.
11. *Brighter HXR footpoints correspond to less bright MW emission in the same footpoint, and fainter HXR emission corresponds to much brighter MW emission* (White *et al.*, 2011). This can be related more to particle transport than to particle acceleration.
12. *For stronger flares there are often double power laws observed in HXR photon spectra with a spectrum flattening toward lower energies* (Holman *et al.*, 2011). This is assumed to be a property of particle transport combined with acceleration.

13. *The temporal evolution of HXR spectra in energies below 35 keV reveals a soft-hard-soft, or sometimes soft-hard-harder, behavior* (Fletcher *et al.*, 2011; Holman *et al.*, 2011). This is assumed to be the property of particle transport combined with acceleration.
14. *Typical electron numbers accelerated in flares are about* 10^{34}–10^{37} *electrons* s^{-1} (Holman *et al.*, 2011). This must be related to particle acceleration as well as to particle and emission transport.
15. *UV and optical emissions reveal close (under 1 s) temporal correlation with HXR emission followed by the appearance of blue shifts in EUV and UV emission showing plasma upflows combined with red shifts in* H_α *and optical emission accompanied on some occasions with white-light emission lasting for tens of minutes.* The spatial correlation between the different kinds of emission is not always one-to-one, possibly reflecting a combination of geometric effects combined with different physical mechanisms applicable to various atmospheric depths. These effects are clear manifestations of energy and radiation transports.
16. *Seismic emission and ripples (sunquakes) often accompany flares of M and X classes, revealing the strongest ripples and emission in the flares of very moderate classes in soft X-rays.* Some sunquakes are cospatial with HXR sources and have close temporal correlation or temporal delay of a few minutes while some sunquakes are cospatial with γ-ray sources (Zharkova, 2008b), or with none of the above (Matthews *et al.*, 2011; Zharkov *et al.*, 2011b). This imposes very strong constraints on the energy transport mechanisms to lower atmospheric levels during flaring events.

2
Particle Acceleration in Flares

2.1
Models of Particle Acceleration

In this section we first review some basic concepts, such as acceleration by a direct electric field in a plasma environment (Section 2.1.1) and magnetic reconnection (Section 2.1.2), in which magnetic energy is released into various forms, including the acceleration of charged particles, both electrons and ions. Recognizing the strong observational evidence for the formation of current sheets (localized regions in which a strong magnetic shear is present) in solar flares, our initial discussion relates to traditional two-dimensional (2-D) current sheets, with subsequent elaboration to the three-dimensional (3-D) case (Section 2.1.3). Two likely products of the fundamental reconnection process are magnetohydrodynamic (MHD) shocks and stochastic MHD turbulence, and particle acceleration by these agents is discussed in Section 2.1.4.

2.1.1
Basic Physics

The physics of particle acceleration is in principle rather simple – a charged particle gains energy when moving in an electric field in its rest frame. Thus the electric field $\boldsymbol{E}$ may be a large-scale externally imposed field, a $\boldsymbol{V} \times \boldsymbol{B}$ field associated with particles crossing magnetic field lines, or a collective field associated with the environment in which the particle finds itself (e.g., a collisional Coulomb field or a field associated with a level of plasma wave energy). The richness of these various means of creating local electric fields is evident in the richness of particle acceleration models in magnetized plasmas.

One of the most basic concepts is that of the *Dreicer field* (Chen, 1974). Consider an electron subject to an externally imposed large-scale electric field E, plus the frictional force due to collisions with stationary ambient particles. The equation of motion for such a particle is (in one dimension)

$$m \frac{dv}{dt} = eE - f_c v \,, \tag{2.1}$$

Electron and Proton Kinetics and Dynamics in Flaring Atmospheres, First Edition. Valentina Zharkova.

where m and e are the electron mass and (absolute value of) charge and f_c is the collision frequency. Since for Coulomb collisions $f_c \sim v^{-3}$ (e.g., Chen, 1974), one can write

$$m\frac{dv}{dt} = e\left[E - E_D\left(\frac{v_{th}}{v}\right)^2\right] , \tag{2.2}$$

where v_{th} is the electron thermal speed and we have defined the Dreicer field by

$$E_D = \frac{f_c(v_{th})v_{th}}{e} ; \tag{2.3}$$

numerically, $E_D \simeq 10^{-8} n\,(\mathrm{cm}^{-3})/T\,(\mathrm{K})\,\mathrm{V\,cm}^{-1}$.

If $E < E_D$, then only high-velocity particles $[v > v_{crit} = v_{th}(E_D/E)^{1/2}]$ see a net positive force and gain velocity. This increased velocity reduces the drag force, leading to even greater net acceleration and, eventually, runaway acceleration. In contrast, particles with $v < v_{crit}$ suffer a net *retardation* force and are not accelerated. If $E > E_D$, then all particles with velocity in excess of the thermal speed v_{th} are accelerated, and the efficient collisional repopulation of the accelerated electrons leads effectively to acceleration of *all* the electrons in the distribution.

2.1.2
Magnetic Reconnection Models Associated with Flares

It is generally accepted that the energy release in solar flares occurs through reconstruction of a magnetic field, caused by the change of connectivity of magnetic field lines during a magnetic *reconnection*. The electric field associated with this changing magnetic field or with the associated driven currents leads to particle acceleration.

Reconnection is a fundamental process defined by the magnetohydrodynamics of a magnetized plasma (Priest and Forbes, 2000; Somov, 2000). The magnetic diffusion equation (see, e.g., Eq. (3.91) of Tandberg-Hanssen and Emslie, 1988) is

$$\frac{\partial \boldsymbol{B}}{\partial t} = \nabla \times (\boldsymbol{v} \times \boldsymbol{B}) + \frac{\eta c^2}{4\pi}\nabla^2 \boldsymbol{B} , \tag{2.4}$$

where c is the speed of light, $\boldsymbol{B}$ the magnetic field, $\boldsymbol{v}$ the fluid velocity, and η the resistivity. The order-of-magnitude ratio of the terms on the right side of this equation defines the *magnetic Reynolds number*

$$R_m = \frac{4\pi \ell V}{\eta c^2} ; \tag{2.5}$$

it measures the ratio of the advective to diffusive contributions to the change in the magnetic field. For typical solar coronal values of η and with $\ell \sim 10^9$ cm, $R_m \sim 10^{14}$; for such high magnetic Reynolds numbers, the plasma is effectively "frozen in" to the field and negligible change in the field topology, with its concomitant release of magnetic energy, can occur. A topological change in the magnetic field requires a breakdown in this ideal "frozen-in" flux condition; by Eq. (2.5) this can occur either in small-scale regions (high values of $\nabla^2 \boldsymbol{B}$) or in regions where the resistivity η is anomalously enhanced. As a result, magnetic reconnection fundamentally occurs in narrow boundary layers called *diffusion regions*.

2.1.2.1 **Basic 2-D MHD Theory of Magnetic Reconnection**

In the simple neutral sheet geometry originally proposed by Sweet and Parker (Sweet, 1969), oppositely directed magnetic field lines $\boldsymbol{B}$ in close proximity follow a *plasma inflow* $\boldsymbol{V}_\mathrm{i}$ oriented in the x-direction perpendicular to the field lines – see for example Figure 2.1 where the x-direction is vertical and the y-direction is horizontal. (Note that there does exist evidence for such reconnection geometries in flares – Aulanier *et al.*, 2000; Des Jardins *et al.*, 2009; Fletcher *et al.*, 2001; Sui and Holman, 2003). The increasing magnetic pressure in the localized region of field reversal is alleviated by reconnection in a small region of high $\nabla^2 \boldsymbol{B}$ near the origin (dimensions $2a$ in the y-direction $\times 2d$ in the d-direction), which allows *plasma outflows with velocity* V_o toward the sides of the diffusion region.

By mass continuity, the equation can be written as follows:

$$V_\mathrm{i} a = V_\mathrm{O} d \ . \tag{2.6}$$

Integrating the steady state momentum equation $\rho v_x \partial v_x / \partial x = -\partial p / \partial x$ from $x = 0$ ($v_x = 0, p = p_\mathrm{i}$) to $x = a$ ($v_x = V_\mathrm{o}, p = 0$) gives $V_\mathrm{o} = \sqrt{2 p_\mathrm{i} / \rho} = \sqrt{B^2 / 4\pi\rho} = V_\mathrm{A}$, the Alfvén speed. Finally, conservation of energy demands that the rate of influx

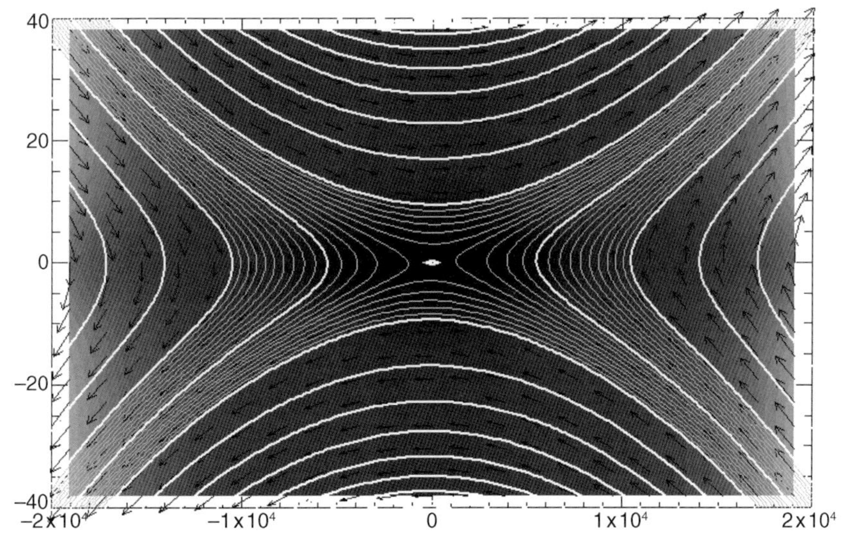

Figure 2.1 Topology of reconnecting magnetic field lines in vicinity of an X null point $B_y = \pm \alpha B_x$. For a color version of this figure, please see the color plates at the beginning of the book.

of magnetic energy be balanced by ohmic dissipation in the diffusion region:

$$2\frac{B^2}{8\pi}V_i(2a)^2 = \eta j^2(2a)^2 2d\ , \tag{2.7}$$

that is, $B^2/(8\pi)V_i = \eta j^2 d$. Substituting $j = cB/(4\pi d)$ (from Ampère's law) gives

$$V_i = \frac{\eta c^2}{2\pi d}\ . \tag{2.8}$$

Finally, solving Eqs. (2.6) and (2.8) for V_i and d gives

$$V_i = \sqrt{\frac{\eta c^2 V_A}{2\pi a}}\ ; \quad d = \sqrt{\frac{\eta c^2 a}{2\pi V_A}}\ . \tag{2.9}$$

In terms of the magnetic Reynolds number Eq. (2.5), with $\ell = 2a$,

$$\frac{V_i}{V_A} = \frac{d}{a} = \frac{2}{\sqrt{R_m}}\ . \tag{2.10}$$

Because of the $R_m^{-1/2}$ dependence on the inflow velocity and, hence, on a rate of magnetic energy dissipation, such a neutral current sheet scenario describes a "slow" magnetic reconnection regime. In an effort to increase the reconnection rate, Petschek (1964) proposed a model involving the formation of shocks propagating outward from the diffusion region. In this scenario the reconnection process occurs at a significantly faster rate, with $V_i \sim V_A$. Both these models were originally proposed with a classical value for the resistivity η, and both resulted in reconnection rates being too slow to explain a rapid energy release observed in flares. Later it turned out that resistivity in a reconnection region could be driven to anomalously high values through the presence in the current sheet of various plasma instabilities (see, for example, Bret, 2009; Dahlburg *et al.*, 1992, and references therein). These included the lower hybrid drift instability leading to the Kelvin–Helmholtz instability (Chen *et al.*, 1997; Landau and Lifshitz, 1960; Lapenta *et al.*, 2003), kink (Daughton, 1999; Pritchett *et al.*, 1996; Zhu and Winglee, 1996) or sausage (Büchner, 1996) instabilities, or magnetoacoustic waves (Schekochihin and Cowley, 2007). This revived hopes that the reconnection mechanisms could be a viable candidate for the energy release in flares (Priest, 1982; Somov, 2000; Syrovatskii, 1981).

Another simple 2-D diffusion region geometry involves magnetic field lines of the form $B_y^2 = \alpha^2 B_x^2$. Such a form is associated with two types of neutral point: an O type, elliptic, topology ($\alpha^2 < 0$; see Section 2.5) and an X type (see Figure 2.1), hyperbolic, topology ($\alpha^2 > 0$; discussed in Section 2.1.3).

2.1.2.2 Effect of the Hall Current on Reconnection Rates in 2-D Models

In conductive plasma within a diffusive region where reconnection occurs, the current sheet tends to diffuse outward at a slow rate with a time scale of $\tau_{\text{dif}} = a^2\varpi$, where $2a$ is the current sheet width and $\varpi = (\mu\sigma)^{-1}$ is the magnetic diffusivity, μ being the magnetic permeability and σ the plasma conductivity (Priest and Forbes,

2000). This diffusion process ohmically converts magnetic energy into heat. As discussed in Section 2.1.3, this time scale is too slow to account for the fast reconnection rates observed in the impulsive phase of flares unless some other processes, such as resistive instabilities or *Hall current* effects, are invoked to enhance a reconnection process.

The basic Sweet–Parker 2-D reconnection model discussed above can produce more realistic reconnection rates if, instead of a resistive MHD model, one uses a two-fluid MHD model (electrons and protons/ions), which includes in the equation of motion a Hall current (Sonnerup, 1979). Use of a two-fluid model means that the velocity vector $\boldsymbol{v}$ in Eq. (2.4) now represents the velocity $\boldsymbol{V}$ of the plasma as a whole, with an additional vector $\boldsymbol{V}_{\text{H}}$ representing the relative velocity between the electrons and the ions. The Hall current results from the fact that inside the current sheet the electrons and ions are magnetized to significantly different degrees with the spatial scales of c/ω_{pe} for electrons and c/ω_{pi} for protons (ω_{pe} and ω_{pi} are the respective plasma frequencies, which are proportional to the square root of the proton-to-electron mass ratio $\sqrt{m_{\text{p}}/m_{\text{e}}} \simeq 43$).

At length scales between c/ω_{pe} and c/ω_{pi} the motion of electrons and ions are decoupled, and the current generated from the relative motion of electrons and ions can lead to the formation of a polarization electric field (Zharkova and Agapitov, 2009), which in turn leads to the formation of various instabilities: whistler waves (Biskamp, 1997; Drake *et al.*, 1997; Ma and Bhattacharjee, 1996; Mandt *et al.*, 1994) or Langmuir waves (Siversky and Zharkova, 2009a). Inside such a region the reconnection rate becomes insensitive to electron inertia, allowing the reconnection rates to be strongly increased to levels large enough to account for those measured in solar flares (Huba and Rudakov, 2004). These results indicate a compelling need to consider a full kinetic approach to modeling magnetic reconnection processes in the diffusion region (for details, see Birn *et al.*, 2001, and references therein); this will be discussed in Sections 2.4 and 2.5.

2.1.2.3 Further Improvement of Reconnection Models

It must be recognized that magnetic reconnection is inherently a 3-D process and that the addition of the third dimension inherently introduces fundamentally new physics. (As an elementary example, 2-D models may predict unrealistic energy gains since the induced electric field $\boldsymbol{E} = (V/c) \times \boldsymbol{B}$ is infinite in extent in the invariant direction.) In three dimensions, the magnetic field has a considerably more complicated structure (Figure 2.2), involving *separatrices* and *separators* (surfaces separating different domains of magnetic connectivity, and lines of intersection of separatrix surfaces, respectively; see Démoulin *et al.*, 1996a,b; Priest and Démoulin, 1995). As shown by Priest and Démoulin (1995), magnetic null points are no longer required for reconnection to take place, and even when they do exist, magnetic nulls have a more complicated structure, typically involving a "fan" structure oriented around an axial "spine" (Priest *et al.*, 1997).

In summary, reconnection can proceed in a variety of different magnetic field geometries and topologies (Priest and Forbes, 2002). Furthermore, depending on

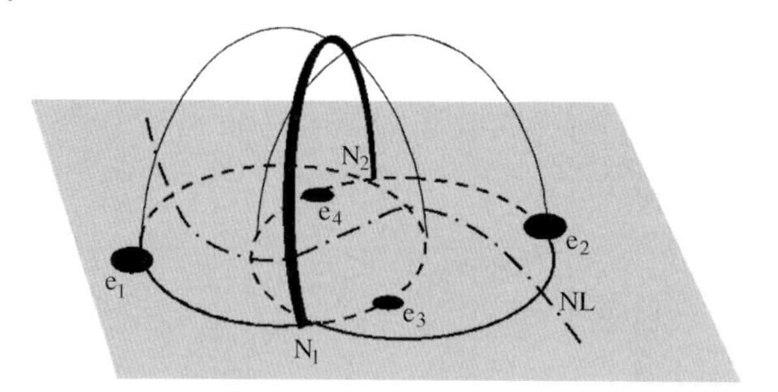

Figure 2.2 The magnetic field geometry formed by four pointlike sources of different magnetic polarities. The dotted–dashed line labeled "NL" is the neutral line. The *separatrix* surfaces are shown as thin domelike lines connecting the photospheric sources. The intersection of these surfaces is the *separator*, shown as a solid line. From Titov and Hornig (2002).

the magnetic field configuration, reconnection can be accompanied by a variety of other physical processes such as the generation of plasma waves, turbulence, and shocks, each occurring on different temporal and spatial scales. These additional processes not only directly accelerate particles (e.g., in regions of enhanced magnetic field, which act as moving "mirrors"), but can also feed back on the magnetic field topology and hence the ongoing reconnection rate.

2.1.3
Particle Acceleration in a Reconnecting Current Sheet

Two of the most popular magnetic field topologies associated with solar flares involve the formation of a current sheet (Gorbachev and Somov, 1989; Sui and Holman, 2003), located either at the top of a single helmetlike loop (Figure 2.3) or at the intersection of two interacting loops (Figure 2.4). Other more complicated topologies are essentially a combination of these two basic ones, and they are discussed in Section 2.2. This section provides an overview of the basic physics of particle acceleration in reconnecting current sheets (RCSs).

Let us consider a magnetic field topology as presented in the RCS geometries illustrated in Figures 2.3b and 2.4b. The tangential magnetic field component $B_z(x)$ is approximated by the linear function

$$B_z(x) = B_0\left(-\frac{x}{d}\right) , \tag{2.11}$$

where d is the RCS thickness (Litvinenko and Somov, 1993; Martens and Young, 1990). The magnitudes of the other two magnetic field components B_x and B_y are assumed to be constant:

$$|B_x| = B_{x0} , \quad |B_y| = B_{y0} . \tag{2.12}$$

Particles, effectively frozen in to the field lines, are advected toward the RCS plane (the x-axis) by the flow field $\boldsymbol{V}_\mathrm{i}$, where they are accelerated by the $(\boldsymbol{V}_\mathrm{i}/c) \times \boldsymbol{B}$

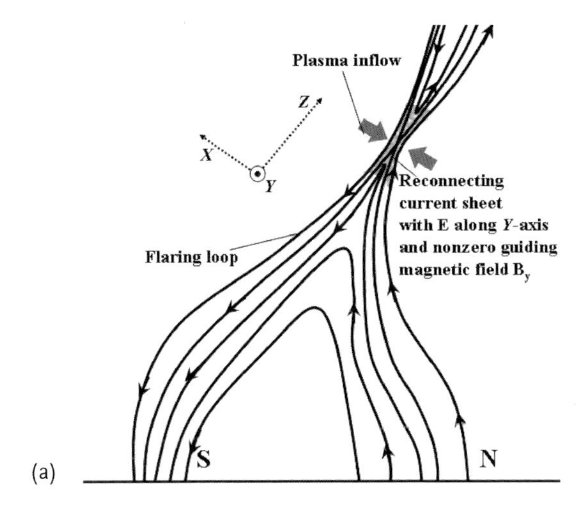

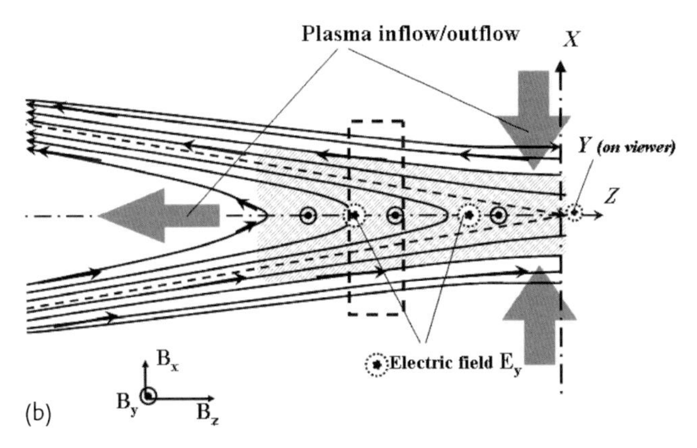

Figure 2.3 Current sheet location (a) and magnetic field topology (b) in a single helmet-type reconnecting nonneutral current sheet. The large arrows in (b) depict plasma inflows from the top and the bottom and outflows to the side. The *x* and *z* components of the magnetic field lie on the plane of the figure; a guiding magnetic field component B_y and a drift electric field E_y are perpendicular to the plane.

electric field (Litvinenko, 1996b; Litvinenko and Somov, 1993). As the reconnection process continues, thermal ambient particles drift into the RCS and get accelerated by a reconnection (drifted) electric field E_y directed along the guiding magnetic field B_y. The rate of particle acceleration is limited by the size of the thin diffusion region.

The drifted electric field E_y can be derived from Ampère's law as follows (Litvinenko and Somov, 1993; Martens and Young, 1990; Zharkova and Gordovskyy, 2004):

$$E_y = \frac{V_i B_{z0}}{c} - \frac{\eta c}{4\pi} \frac{\partial B_z}{\partial x} ; \qquad (2.13)$$

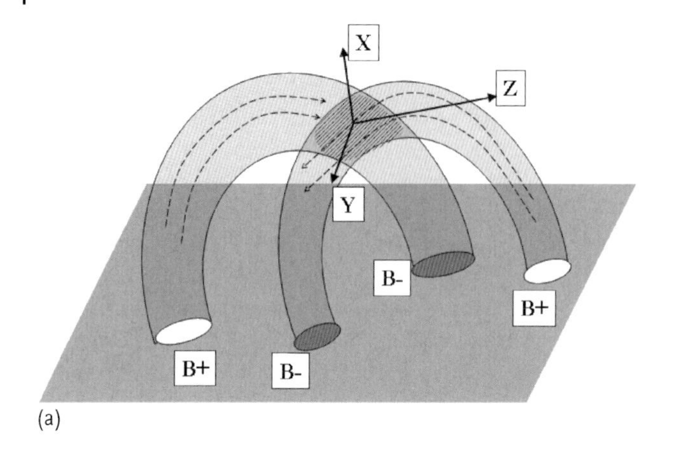

(a)

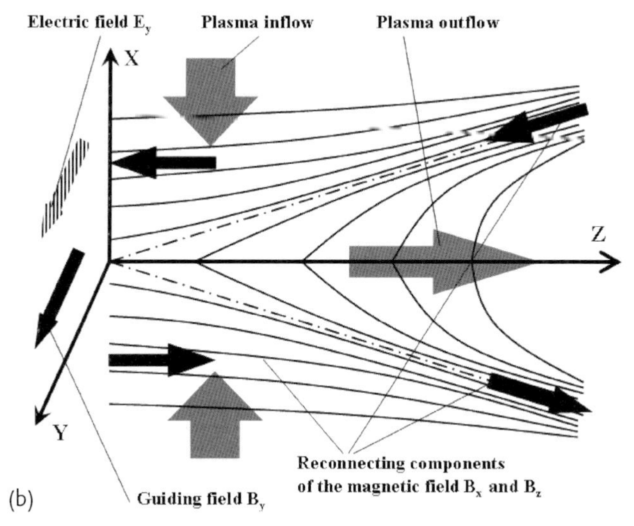

(b)

Figure 2.4 Current sheet location (a) and magnetic field topology (b) in a reconnecting nonneutral current sheet formed by two interacting loops. The large arrows depict the plasma inflows. As for Figure 2.3, the x and z components of the magnetic field lie on the plane of the figure; a guiding magnetic field component B_y and a drift electric field E_y are perpendicular to the plane.

where the first and second terms on the right-hand side of Eq. (2.13) dominate outside and inside the diffusion region, respectively (Somov, 2000). Priest (1982) has shown that the plasma inflow velocity V_i is smaller than the local Alfvén speed by about two orders of magnitude, that is, $V_i \sim 0.01 V_A$. Hence the electric field is

$$E_y \sim 0.01 \frac{B_{z0}^2}{c\sqrt{4\pi n m_p}} , \tag{2.14}$$

where m_p is the proton mass. Let us assume the following physical conditions in the current sheet: temperature $T = 10^6$ K, density $n = 10^{10}\,\text{cm}^{-3}$, and $B_{z0} \simeq$

100 G. Then $V_A \sim 2 \times 10^6\,\mathrm{m\,s^{-1}}$ and the thermal velocities are about $10^5\,\mathrm{m\,s^{-1}}$ for protons and $10^7\,\mathrm{m\,s^{-1}}$ for electrons. Taking the initial inflow velocity V_i to be the proton thermal speed, we obtain $E_y \simeq 2\,\mathrm{V\,cm^{-1}}$, which is roughly four orders of magnitude higher than the corresponding Dreicer field $\simeq 10^{-4}\,\mathrm{V\,cm^{-1}}$ (Litvinenko, 1996b; Martens and Young, 1990 see Section 2.1.1). This electric field is directed along the longitudinal magnetic field component B_y, allowing rather efficient particle acceleration in the y-direction.

The magnitudes of the magnetic field components B_x and B_y are chosen such that electrons remain magnetized in the vicinity of the midplane while protons are unmagnetized during the acceleration phase. In this case, the energy of the unmagnetized protons upon ejection can be estimated (Litvinenko, 1996b) as

$$\varepsilon = 2 m_\alpha c^2 \left(\frac{E_{y0}}{B_{x0}}\right)^2 , \tag{2.15}$$

where m_α is the particle mass. However, since the electrons are magnetized, they follow magnetic field lines and gain an energy (Litvinenko, 1996b):

$$\varepsilon = d \left| e E_{y0} \frac{B_{y0}}{B_{x0}} \right| , \tag{2.16}$$

where d is the current sheet thickness along the x-direction (see Figure 2.3b). Such energies are compatible with those of bremsstrahlung-emitting electrons.

Although such a model does meet the challenge of (i) a sufficiently high reconnection rate and (ii) sufficient energy gain by accelerated electrons, the number of accelerated particles is fundamentally limited by the small width, $2a$, of the current sheet. These deficiencies are addressed in the further development of this basic model to include 3-D considerations (Section 2.2.3) and a full kinetic particle-in-cell (PIC) approach to model portions of a 3-D current sheet (Sections 2.4 and 2.5).

2.1.4 Particle Acceleration by Shocks and Turbulence

In this section we review the basic physics involved in acceleration models that do not rely on a large-scale coherent electric field.

2.1.4.1 Shock Acceleration

Ever since the demonstration by several authors (Bell, 1978; Blandford and Ostriker, 1978; Cargill, 1991; Giacalone and Jokipii, 1996) that a very simple model of shock wave acceleration leads to a power law spectrum of the accelerated particles (in good agreement with observations of galactic cosmic rays), shock acceleration has been frequently invoked in both space and astrophysical plasma contexts. However, this simple model, though very elegant, has a significant number of shortcomings. Inclusion of losses (Coulomb at low energies and synchrotron at high energies) or the influence of accelerated particles on the shock structure (see, e.g.,

Amato and Blasi, 2005; Ellison and *et al.*, 2005) can cause significant deviations from a simple power law.

Moreover, shocks are unable to accelerate low-energy background particles efficiently; shock acceleration requires the injection of (fairly high energy) "seed" particles. Also, shock acceleration requires some scattering agents (most likely plasma waves or turbulence) to cause repeated passage of the particles through the shock front. In the case of quasiparallel shocks, in which the magnetic field is nearly parallel to the normal of the shock front, the rate of energy gain is governed by the pitch angle diffusion rate.

As pointed out by Wu (1983) and Leroy and Mangeney (1984), in the Earth's magnetosphere many shocks can be quasiperpendicular (i.e., the vector defining the normal to the shock is almost perpendicular to the shock velocity vector). In such a case the particles drift primarily along the surface of the shock and get accelerated more efficiently. However, acceleration in quasiperpendicular shocks suffers from the fundamental limitation that the injection of seed particles with very large velocities $v > u_{\rm sh}\xi$ (where $u_{\rm sh}$ is the shock speed and $\xi \gg 1$ is the ratio of parallel to perpendicular diffusion coefficients) is required (Amano and Hoshino, 2010). Moreover, although there may be some *indirect* morphological evidence for the existence of such shocks in flares (see, e.g., Sui and Holman, 2003), there is little *direct* evidence for the occurrence of these kinds of shocks near the top of flaring loops, where the acceleration seems to take place during the impulsive phase. Furthermore, such shocks typically do not appear in 3-D MHD simulations of the reconnection process.

In summary, many of the features that make acceleration of cosmic rays by shocks attractive are not directly applicable to the solar flare environment and as a result have received only limited attention in the literature.

2.1.4.2 **Stochastic Acceleration**

A main contender for particle acceleration mechanism in flares is stochastic acceleration by the magnetic field component of low-frequency magnetoacoustic waves (Miller *et al.*, 1997; Petrosian *et al.*, 2006; Schlickeiser and Miller, 1998).

In the originally proposed Fermi acceleration process (Davis, 1956; Fermi, 1949), particles of velocity v scatter coherently off agents moving toward each other with a velocity of $\pm u$. Since the velocity gain in a single collision is $2u$, the rate of velocity increase is $\Delta v/\Delta t \sim 2u/(L/v)$, where L is the separation of the scattering centers. Solving this for the case where a scatter occurs on a time scale significantly less than that required for L to decrease shows that the energy grows exponentially with time: $E \sim e^{4ut/L}$. For particles scattering off *randomly* moving agents, both energy gains and losses are possible. However, the increased likelihood of a head-on (energy-increasing) collision leads to a general (slower than exponential) increase in energy, at a rate proportional to $(u/v)^2 D_{\mu\mu}$, where $D_{\mu\mu}$ is the angular diffusion coefficient. This process is known as *second-order* Fermi acceleration.

In general, in a magnetized plasma characterized by a gyrofrequency Ω, a particle of velocity $\boldsymbol{v}$ will resonate with a wave of frequency ω and wavenumber $\boldsymbol{k}$ when

$$\omega = \boldsymbol{k} \cdot \boldsymbol{v} - \ell\Omega \; , \tag{2.17}$$

where ℓ is an integer. Particles with velocities near the Cherenkov resonance ($v \approx \omega / k_{\parallel}$) resonate with $\ell = 0$ waves. This process is known as *transit-time damping* (Hollweg, 1974, and references therein). It acts to diffuse the particle distribution $f(\boldsymbol{p})$ in momentum space, described by the Fokker–Planck equation

$$\frac{\partial f(\boldsymbol{p})}{\partial t} = \left(\frac{1}{2} \sum_{i,j} \frac{\partial}{\partial p_i \partial p_j} D_{ij} - \sum_{i} \frac{\partial}{\partial p_i} F_i \right) f(\boldsymbol{p}) + Q(\boldsymbol{p}) - S(\boldsymbol{p}) f(\boldsymbol{p}) \; . \tag{2.18}$$

The diffusion coefficient D_{ij} and the secular coefficient F_i ($i = 1, 2$ correspond to energy and angular terms, respectively) contain the essential physics: the action of both the accelerating waves and decelerating effects (e.g., Coulomb collisions); the quantity $Q(\boldsymbol{p})$ corresponds to the injection of *particle energy* and the term $S(\boldsymbol{p}) f(\boldsymbol{p})$ to escape.

For solar flares the most likely sources for the energization term $Q(\boldsymbol{p})$ are plasma wave turbulence (Hamilton and Petrosian, 1992; Pryadko and Petrosian, 1997) and cascading MHD turbulence (Miller *et al.*, 1996). In the first scenario, a presupposed level of plasma wave turbulence in the ambient plasma accelerates ambient particles of energy E stochastically with a rate of D_{EE}/E^2. It has been shown (e.g., Hamilton and Petrosian, 1992; Miller *et al.*, 1996) that for flare conditions plasma wave turbulence can accelerate the background particles to the required high energies within the very short time scale derived from HXR observations. More importantly, at low energies and, especially, in strongly magnetized plasmas, the acceleration rate D_{EE}/E^2 of the ambient electrons may significantly exceed the scattering rate $D_{\mu\mu}$ for particular pitch angles and directions of propagation of this turbulence (Pryadko and Petrosian, 1997, 1998), leading to an efficient acceleration. However, high-frequency waves near the plasma frequency are problematic as drivers for stochastic acceleration in flares because they would couple into decimeter radio waves and would be present at high levels in every flare, contrary to observations (Benz *et al.*, 2005).

In the second scenario, involving cascading MHD turbulence, moving magnetic compressions associated with fast-mode MHD waves propagating at an angle to the magnetic field constitute the scattering centers. The wavelength of the induced MHD turbulence is assumed to cascade through an inertial range from large to very small, stopping at the point where wave dissipation becomes significant due to transit-time damping. Resonance with one wave will result in an energy change, leading to resonance with a neighboring wave, and so on (Miller *et al.*, 1996). Electrons can thus be accelerated all the way from thermal to relativistic energies through a series of overlapping resonances with low-amplitude fast-mode waves in a continuum broadband spectrum.

In order to be efficiently accelerated, the test particle (TP) must have an initial speed that is greater than the speed of the magnetic compression, that is, on the order of the Alfvén speed. For the preflare solar corona, *electrons* have speeds comparable to, or greater than, the Alfvén speed, and so are efficiently accelerated immediately. On the other hand, in a low-β plasma the protons necessarily have an initial speed that is less than the Alfvén speed and so are not efficiently accelerated: the scattering centers converge before the TP has suffered a sufficient number of energy-enhancing collisions. Efficient acceleration of ions therefore requires preacceleration to Alfvén speeds. In the model of Miller and Roberts (1995), this preacceleration is accomplished through (parallel propagating) Alfvén waves, and it is essential to realize that *this process takes some time to accomplish*. During this preacceleration period, most of the energy is deposited in accelerated electrons; only after the protons are preaccelerated to the Alfvén speed are they effectively accelerated – they then gain the bulk of the released energy by virtue of their greater mass (Miller and Roberts, 1995).

In practice, since there is little difference *mathematically* between plasma wave turbulence and shock mechanisms (Jones, 1994), acceleration by both turbulence and shocks can be effectively combined. It should also be noted that stochastic acceleration generally predicts an accelerated spectrum that is not a strict power law but rather steepens with increasing energy (e.g., Miller *et al.*, 1996). However, a recent development in transit-time damping theory involves adding both the loss term $S(\boldsymbol{p})$ in Eq. (2.18) that allows electron escape and a source term $Q(\boldsymbol{p})$ due to the return current replenishing the electrons in the acceleration region. These additions cause the solution to become stationary within less than 1 s under coronal conditions (Grigis and Benz, 2006); furthermore, in the HXR spectral region of observed nonthermal energies the photon distribution is close to a power law, in agreement with observation. The predicted spectral index γ depends on the wave energy density and on the interaction time between the electrons and the waves. Additional trapping (e.g., by an electric potential) tends to harden the electron spectrum.

The main limitation of stochastic acceleration models at this juncture is the need to presuppose an "ad hoc" injection of the necessary waves (plasma or MHD). Furthermore, in the absence of pitch angle scattering, the stochastic acceleration process generally leads to a decrease in particle pitch angles. This in turn reduces the efficiency of the acceleration process since now only waves with very high parallel phase speeds, or those with pitch angles near 90°, can resonate with the particles being accelerated. Therefore, in order for significant stochastic acceleration to occur, the effective particle scattering rate (which is proportional to the collision frequency between particles and waves) must be greater than the frequency of Coulomb collisions between the charged particles. The latter condition imposes rather strict limitations on the growth rates of plasma waves or MHD turbulence, as appropriate.

2.2
Recent Theoretical Developments

In this section some recent promising developments in understanding the physics of particle acceleration in the magnetized plasma environment associated with solar flares are highlighted for all models using the TP approach.

In particular, our attention is turned to a critical aspect of all acceleration models – the need to recognize that, since a substantial fraction of the magnetic energy released in flares appears in the form of accelerated particles, these particles simply cannot be considered as an ensemble of TPs interacting with a *prescribed* electrodynamic environment. Rather, the electric and magnetic fields generated by the accelerated particles themselves are an essential element of this environment; the self-consistent inclusion of these fields represents a paramount consideration in acceleration models, without which unphysical and even paradoxical conclusions can result.

Hence, the limitations of existing acceleration models in this context are reviewed and recent attempts are discussed to include this electrodynamic feedback in the models by using full kinetic PIC models, in which the self-consistent electric and magnetic fields associated with the accelerated particles *do* form an essential element of the modeling.

2.2.1
Stochastic Acceleration

As discussed in Chapter 1, a low value of the low-energy rollover of the electron energy distribution suggested by RHESSI observations (Kontar *et al.*, 2005; Saint-Hilaire and Benz, 2005) has significant implications for the number of flare electrons. Indeed, for sufficiently low rollover energies, the number of electrons required in a collisional thick-target model (Brown, 1971) becomes so great that models in which the accelerated particles form a subset of the ambient distribution become untenable. The only viable mechanism of particle acceleration is then acceleration of the entire plasma within the acceleration region through a stochastic process that results in little or no net electrical current, to avoid the issues for HXR photons presented in Chapter 1.

Also high-frequency waves near the plasma frequency are problematic as drivers for stochastic acceleration (Section 2.1.4) because they would couple into decimeter radio waves and would be present at high levels in every flare, contrary to observations (Benz *et al.*, 2005). On the other hand, acceleration by transit-time damping does not suffer from this drawback – the driving waves have a frequency that is far below the plasma frequency.

2.2.1.1 Improvements to the Stochastic Acceleration Model

The basic transit-time damping stochastic acceleration mechanism has been improved by adding low-frequency fluctuating electric fields parallel to the magnetic field (Grigis and Benz, 2004, 2005). Such fields may originate from low-frequency and high-amplitude turbulence, such as kinetic Alfvén waves as discussed in Section 2.7.1 (see also Miller *et al.*, 1996; Pryadko and Petrosian, 1997). They can accelerate and decelerate electrons (and protons), leading to a net diffusion in the energy space (Arzner and Vlahos, 2004). As for other models for stochastic acceleration, the authors assume an *ad hoc* distribution of turbulence. In this model the nonthermal electron distributions in coronal sources "grow" out of the thermal population: the coronal source is initially purely thermal, then a soft nonthermal population develops, getting harder at the HXR peak and softening toward the end of the emission.

As discussed in Chapter 1, such a "soft–hard–soft" behavior has been observed in the coronal source (Battaglia and Benz, 2007); indeed, Grigis and Benz (2004) report RHESSI observations suggesting a point in the electron energy distribution where there is an energy at which the HXR flux does not change with time during an event; rather the spectrum rises and falls on either side of this *pivot point*, consistent with the predicted "soft–hard–soft" behavior of the photon spectrum. Note, however, that Zharkova and Gordovskyy (2006) have pointed out that such a "pivot-point" behavior in the variation of the HXR spectrum with time could also be associated with the self-induced return current electric field (e.g., Emslie, 1981; Knight and Sturrock, 1977) associated with an electron beam accelerated by a drifted electric field (Section 2.2.3).

2.2.1.2 **Stochastic Acceleration with Particle Feedback**

A numerical approach to stochastic acceleration was recently reported by Bykov and Fleishman (2009), who simulated electron acceleration by MHD waves but included the effect on the MHD turbulence (spectrum and intensity) caused by the injection of high-energy particle beams into the acceleration region. The initial MHD turbulence is prescribed and is assumed to be produced in the form of transverse motions with a Gaussian spectrum on the scale $2\pi/k_0$. Through mode coupling, this turbulence produces a corresponding level of longitudinal turbulence in a system of finite size R (where $Rk_0 > 1$). The model includes *only large-scale energy-conserving motions*, with the kinetics of particles on a smaller scale being determined by turbulent advection, valid only for electron energies up to ~ 1 MeV (Bykov and Fleishman, 2009). The phase-space diffusion coefficients D_{ij} in Eq. (2.18) are expressed in terms of the spectral functions that describe correlations between large-scale turbulent motions (for details, see Bykov and Toptygin, 1993).

The model assumes a continuous injection of monoenergetic particles (both electrons and protons) into the acceleration region. The initial phase of the acceleration is characterized by a linear growth regime that results in effective particle acceleration by the longitudinal large-scale turbulent motions and thus leads to a spectral hardening. However, because the accelerated particles eventually accumulate a considerable fraction of the turbulent energy, the efficiency of the acceleration decreases; this predominantly affects the higher-energy particles, leading to spectral softening.

2.2.2
Electron Acceleration in Collapsing Current Sheets

A typical helmet-type magnetic structure often observed in solar flares is depicted in Figure 2.5 (see, e.g., Sui and Holman, 2003) in which a prominence is destabilized due to photospheric footpoint motions and rises upward as an *eruptive prominence*. As the prominence rises, the underlying field lines are stretched, leading to the formation of a current sheet. We suppose that when the current exceeds a critical value, the resistivity is suddenly enhanced (say, by plasma waves excited by various instabilities) and rapid magnetic reconnection begins in the diffusion region (DR), as shown earlier in Figure 2.1.

MHD solutions for the large-scale magnetic configuration presented in Figure 2.5 allow us to investigate (Section 2.1.2) the reconnection process inside the diffusion region, which is characterized by plasma inflows with velocity V_0 in the x-direction (from the sides in Figure 2.5; from the top and bottom in Figure 2.1) and outflowing with velocities V_{out} in the z-direction (to the top and bottom of Figure 2.5; to the left or right sides of Figure 2.1). The outflows are generally very fast and create a rather complicated structure in the outer region at the RCS edge. The plasma confined on the newly reconnected field lines will start moving with the velocity V_1, both upward (not shown in the cartoon) and downward, where it will encounter a magnetic obstacle in the form of the underlying loop. If the distance to this obstacle is large, a collapsing magnetic trap (Figure 2.6b) is formed. Moreover, the downward outflow moves into a region of generally higher density and hence lower Alfvén speed. If a point is reached at which the outflow velocity V_{out} exceeds the local magnetoacoustic wave velocity, then a fast MHD shock, termed a "termination shock", can appear ahead of this magnetic obstacle (Somov and Kosugi, 1997), as shown by the dashed line in Figure 2.6a. In this sense the collapsing

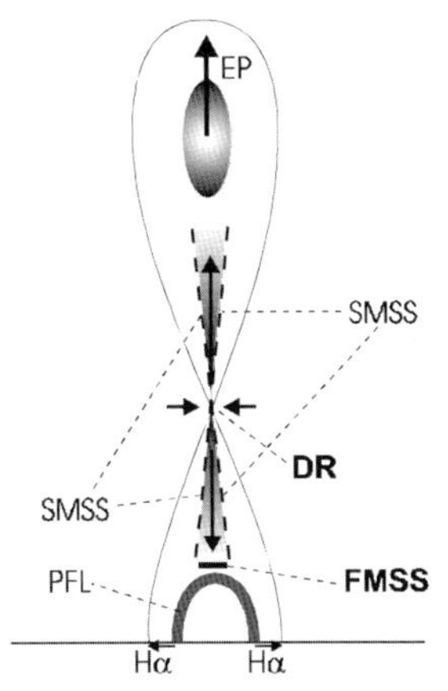

Figure 2.5 Sketch of flare scenario. EP: erupting prominence; DR: diffusion region; SMSS: slow-mode standing shock; FMSS: fast-mode standing (also called termination) shock; PFL: post-flare loop. The gray shaded areas with the arrows show the outflow jets. From Mann *et al.* (2006).

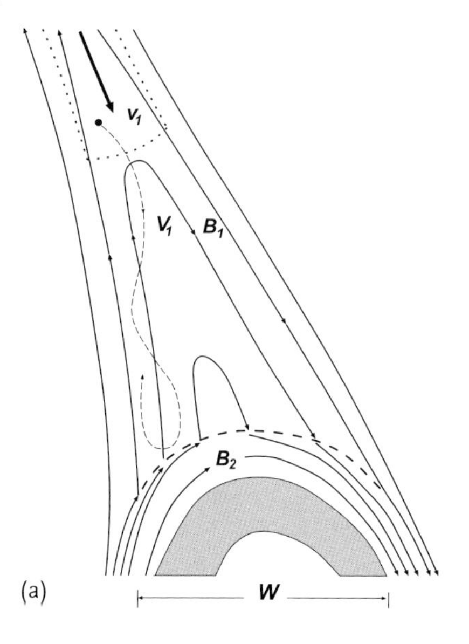

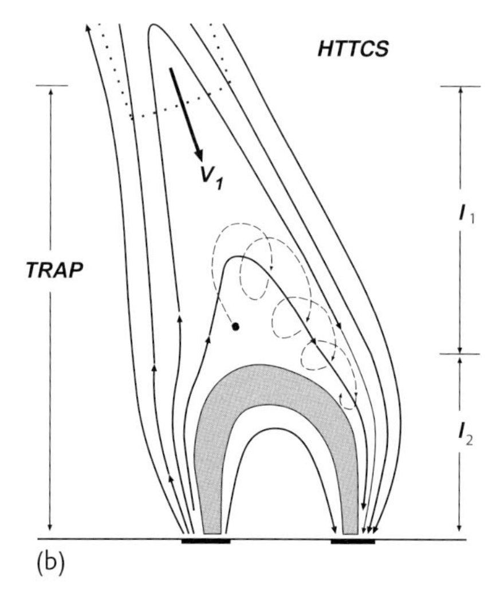

Figure 2.6 Schematic models of collapsing magnetic field lines occurring in MHD models of magnetic reconnection with (a) (dashed line) and without (b) a termination shock. From Somov and Kosugi (1997).

magnetic trap acceleration is complementary to the acceleration during a magnetic reconnection itself (Section 2.2.3) providing the secondary acceleration of electrons ejected from a current sheet.

2.2.2.1 **Acceleration in Collapsing Magnetic Traps**

Somov and Kosugi (1997), Karlický and Kosugi (2004), and Bogachev and Somov (2005, 2007) have carried out simulations of the acceleration of electrons in collapsing magnetic traps for particles with an initial Maxwellian distribution that are injected at random points of a fragmented current sheet geometry obtained using a 3-D MHD simulation.

As a result of its motion from the reconnection region toward the chromosphere, the length L of the magnetic trap decreases, leading to an increase in energy of the trapped particles via a first-order Fermi process associated with the conservation of the longitudinal invariant $\int p_{\|} ds$ (Benz, 2002). Also, as a result of the *transverse* contraction of the magnetic trap, particles are accelerated by the betatron mechanism. This transverse contraction can be described by the ratio $b(t) = B(t)/B_0$, which changes from 1 when $B(0) = B_0$ to $b_m = B_m/B_0$, where B_m is the magnetic field in the magnetic mirrors. Thus the particle momenta $p_{\|}$ and $p_{\perp}$ change as follows:

$$p_{\|} = p_{0\|}\ell^{-1} , \tag{2.19}$$

where $\ell = (L/L_0)$, and

$$p_{\perp} = p_{0\perp} b^{1/2} . \tag{2.20}$$

Both the Fermi and betatron accelerations are accompanied by a change in particle pitch angle – a decrease (Fermi) or increase (betatron). When these two mechanisms act simultaneously, the pitch angles of accelerated particle vary according to

$$\tan\alpha = \frac{p_{\perp}}{p_{\|}} = \ell b^{1/2} \tan\alpha_0 , \tag{2.21}$$

which can result in an increase or decrease of pitch angle, according to the relative importance of the two effects. The particle kinetic energy increases in the magnetic trap until it falls into the loss cone.

Simulations show that particles with an initially Maxwellian distribution gain energies via both acceleration mechanisms; they retain a quasithermal distribution but with much higher energies. The change of a total kinetic energy has the form of a burst of particles with energies up to a few megaelectronvolt. However, the magnetic energy is absorbed very rapidly by the electrons, so that a TP approach very rapidly ceases to be valid. A fuller treatment of this scenario therefore requires full kinetic simulations similar to those discussed in Section 2.4.

2.2.2.2 **Acceleration by a Termination Shock**

Due to the strong curvature of the magnetic field lines in the vicinity of the diffusion region in the helmet-type reconnection (Figure 2.5), slowly inflowing plasma runs away from the reconnection site as a hot jet (Zharkova *et al.*, 2011a). These oppositely directed jets are embedded between a pair of slow magnetosonic shocks. If the speed of this jet is super-Alfvénic, a fast magnetosonic shock, also called a *termination shock*, is established, as shown previously in Figure 2.6. The appearance of such shocks was predicted in the numerical simulations of Forbes (1986), Shibata *et al.* (1995), and Somov and Kosugi (1997).

Radio measurements (Aurass *et al.*, 2002; White *et al.*, 2011) of the 28 October 2003 event revealed the termination shock as a standing radio source with a (large) half-power source area of $A_s = 2.9 \times 10^{20}\,\mathrm{cm}^2$. A strong enhancement of the electromagnetic emission in the HXR and γ-ray range up to 10 MeV is simultaneously observed with the appearance of the radio signatures of the shock (Aurass and Mann, 2004). These observations indicate that the termination shock could well be the source of the highly energetic electrons needed for the generation of the HXR and γ-ray radiation.

A fast magnetosonic shock (FMSS) formed during the MHD simulations of a reconnection process is accompanied by a compression of both the magnetic field and the density. Thus, it represents a moving magnetic mirror at which charged particles can be reflected and accelerated through a process termed *shock drift acceleration* (SDA). This process can further accelerate energetic electrons up to 10 MeV (Mann *et al.*, 2006).

Generally, SDA represents reflections at the termination shocks. Analysis of such an acceleration is most straightforwardly performed in the *de Hoffmann–Teller frame,* the frame in which the electric field vanishes, so that the reflection takes place under conservation of both kinetic energy and magnetic moment. From such an analysis (Mann *et al.*, 2006, 2009) one obtains the relationships between the electron velocities $\beta = v/c$ parallel and perpendicular to the upstream magnetic field before (index i) and after (index r) the reflection:

$$\beta_{\mathrm{r},\parallel} = \frac{2\beta_\mathrm{s} - \beta_{\mathrm{i},\parallel}\left(1 + \beta_\mathrm{s}^2\right)}{1 - 2\beta_{\mathrm{i},\parallel}\beta_\mathrm{s} + \beta_\mathrm{s}^2} \tag{2.22}$$

and

$$\beta_{\mathrm{r},\perp} = \frac{(1 - \beta_\mathrm{s}^2)}{1 - 2\beta_{\mathrm{i},\parallel}\beta_\mathrm{s} + \beta_\mathrm{s}^2} \cdot \beta_{\mathrm{i},\perp}\,, \tag{2.23}$$

where $\beta_\mathrm{s} = v_\mathrm{s} \sec\theta/c$. Here v_s is the shock speed, c the velocity of light, and θ the angle between the shock normal and the upstream magnetic field. In addition, the reflection conditions

$$\beta_{\mathrm{i},\parallel} \le \beta_\mathrm{s}\,, \tag{2.24}$$

and for particle acceleration the following condition must be satisfied:

$$\beta_{\mathrm{i},\perp} \ge \frac{\tan\alpha_{\mathrm{lc}}}{\sqrt{1 - \beta_\mathrm{s}^2}} \cdot (\beta_\mathrm{s} - \beta_{\mathrm{i},\parallel})\,. \tag{2.25}$$

The loss-cone angle is defined by $\alpha_{\mathrm{lc}} = \sin^{-1}[(B_{\mathrm{up}}/B_{\mathrm{down}})^{1/2}]$, where B_{up} and B_{down} denote the magnitude of the magnetic field in the up- and downstream regions, respectively.

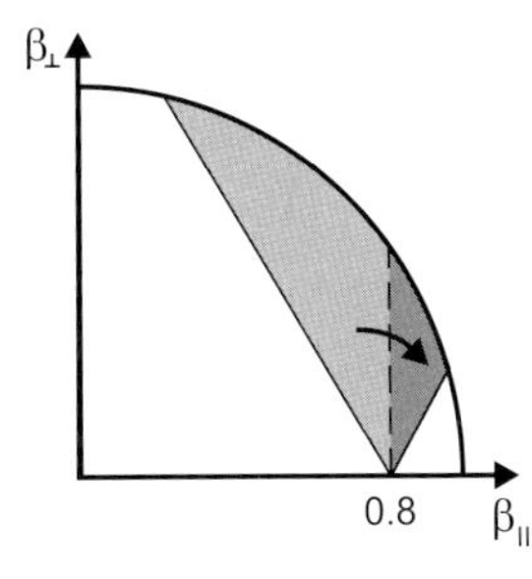

Figure 2.7 Illustration of SDA in the $\beta_\perp$–$\beta_\parallel$ plane for a shock with $\beta_s = 0.8$ and $\alpha_{lc} = 45°$. All particles initially located in the light gray area are transformed into the dark one.

SDA represents a transformation in the $\beta_\perp$–$\beta_\parallel$ plane, as illustrated in Figure 2.7, which shows that the accelerated distribution function is simply a *shifted loss-cone distribution*. The flux of the accelerated electrons parallel to the upstream magnetic field is defined by

$$F_{\mathrm{acc},\parallel} = 2\pi N_0 c^4 \int_0^\infty d\beta_\parallel \beta_\parallel \int_0^\infty d\beta_\perp \beta_\perp \cdot f_{\mathrm{acc}}(\beta_\parallel, \beta_\perp) \,, \tag{2.26}$$

and the differential flux is given by

$$j_{\mathrm{acc},\parallel} = \frac{dF_{\mathrm{acc},\parallel}}{dE} \,. \tag{2.27}$$

Figure 2.8 shows the differential fluxes obtained from the SDA model for various angles θ, assuming an initial distribution in the form of a Maxwellian with a temperature $T = 10\,\mathrm{MK}$. For comparison, the dotted line in Figure 2.8 repre-

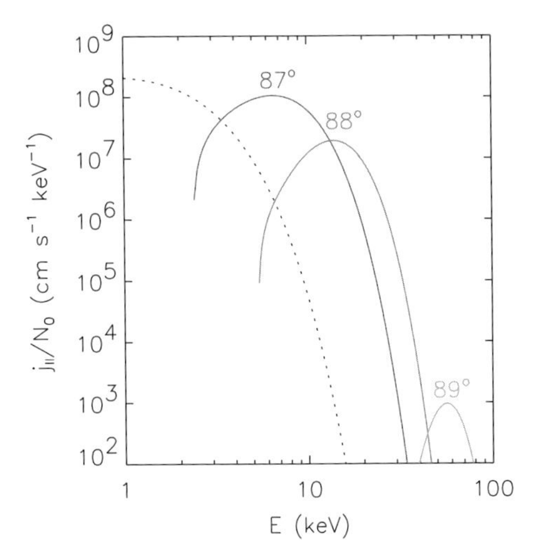

Figure 2.8 Differential flux of electrons accelerated by SDA for various shock angles. The dotted line represents the flux of a pure Maxwellian distribution.

sents a pure Maxwellian distribution with the same temperature. One can see that SDA indeed provides a significantly enhanced component of energetic electrons, particularly for large values of the angle θ. These analytically obtained results are qualitatively confirmed by numerical simulations performed for plasma conditions usually found at the Earth's bow shock (Krauss-Varban and Burgess, 1991; Krauss-Varban and Wu, 1989; Krauss-Varban *et al.*, 1989).

In order to compare these results with observations, the termination shock appears at 300 MHz, for the harmonic emission was considered for the solar event on 28 October 2003 (Mann *et al.*, 2009). Matching the harmonic emission to the observed 300 MHz frequency requires an electron number density of $2.8 \times 10^8\,\mathrm{cm}^{-3}$. From Figure 2.8 it can be seen that the flux is very sensitive to θ, so the authors (Mann *et al.*, 2009) model this curvature by a circle, and the resulting flux of accelerated electrons can be integrated over θ, giving an accelerated electron production rate of $F_e = 2 \times 10^{36}\,\mathrm{s}^{-1}$ and a related power of $P_e = 1 \times 10^{29}\,\mathrm{erg\,s}^{-1}$ above 20 keV. This agrees rather well with the results inferred from RHESSI observations (Warmuth *et al.*, 2007, 2009).

It must be stressed that a nonnegligible part of the energy of the outflow jet is transferred to accelerated electrons. Thus such a TP approach has its limitations: the termination shock acceleration process must be treated in a fully kinetic manner, in which the accelerated particles themselves are included in the conservation laws across the shock.

2.2.3 Particle Acceleration in a Single 3-D RCS with Complicated Magnetic Topology

2.2.3.1 Basic Equations

Following Section 2.1.3, we consider a highly stressed current sheet oriented in the x-direction, with a lateral field

$$B_z(x) = B_0 \tanh\left(-\frac{x}{d}\right) \tag{2.28}$$

and a transverse magnetic field component B_x that is dependent on z according to (Zharkova and Gordovskyy, 2005b,c):

$$B_x(z) = B_0 \left(\frac{z}{a}\right)^{\alpha}, \tag{2.29}$$

where $\alpha > 0$ (we consider an illustrative value of $\alpha = 1$). A weak, constant longitudinal component of the magnetic field B_y is also assumed to be present:

$$B_y = \kappa B_0, \tag{2.30}$$

where κ is an arbitrary coefficient that varies from 0 to 1. Particle trajectories are calculated from the relativistic equations of motion

$$\frac{d\boldsymbol{r}}{dt} = \frac{\boldsymbol{p}}{m_0\gamma} \tag{2.31}$$

$$\frac{d\boldsymbol{p}}{dt} = q\left(\boldsymbol{E} + \frac{1}{c}\frac{\boldsymbol{p}}{m_0\gamma} \times \boldsymbol{B}\right) , \tag{2.32}$$

where t is time, $\boldsymbol{r}$ and $\boldsymbol{p}$ are the particle position and momentum vectors, $\boldsymbol{E}$ and $\boldsymbol{B}$ are the electric and magnetic field vectors, q and m_0 are the charge and rest mass of the particle, c is the speed of light, and $\gamma = \sqrt{(p/m_0c)^2 + 1}$ is the Lorentz factor. The integration time steps are chosen to be much smaller than the corresponding gyro period, that is, $dt \le 0.1(m/q)B_z^{-1}$. For protons, this typically requires a time step of $\lesssim 10^{-7}$ s, while for electrons the required time step is substantially shorter, $\lesssim 4 \times 10^{-11}$ s. Again, similar to Section 2.1.3, in this simulation the drifted, or reconnection, electric field is assumed to be located in the region covering the whole diffusion region, which in the MHD approach varies from 10^2 to 10^6 m (Bulanov and Syrovatskii, 1976; Oreshina and Somov, 2000a; Somov, 1992). This, of course, contradicts the full PIC kinetic simulations (Drake and Lee, 1977; Furth *et al.*, 1963) carried out for a small diffusion region with a modified speed of light of 20 000 km s^{-1}, where the region with a reconnection field is rather small. Which approach is more correct remains to be determined in future simulations, but for the TP approach we choose to use the MHD scale.

It should be noted that in the above description, the electric and magnetic field vectors $\boldsymbol{E}$ and $\boldsymbol{B}$ have the *prescribed* quantities because the kinetic times of particle acceleration is a thousand times shorter than MHD characteristic times. Also the self-consistent electric and magnetic fields produced by the accelerated particles themselves are neglected. Such a TP approach is valid for up to only $\sim 10^5$ TPs (Birdsall and Langdon, 1985), and we shall return to this point in Section 2.3.

2.2.3.2 Particle Motion Inside a 3-D RCS

In the TP description used here, the electron trajectory and the direction of ejection are determined uniquely by the particle's initial coordinate, velocity, and the magnetic field geometry. Particles inflowing into a 3-D RCS are accelerated mostly near the lateral field reversal at the x-axis. In the presence of a nonzero B_y, electrons and protons are ejected in opposite directions from the midplane ($x = 0$); whether the particle is ejected to $x > 0$ or $x < 0$ depends on the sign of the particle charge and on the signs and magnitudes of B_x and B_y. If B_y is strong enough, then all protons (regardless of the side of the RCS they entered from) are ejected to the $x > 0$ semispace, while all electrons are ejected to the opposite semispace, as illustrated in Figure 2.9 (Zharkova and Agapitov, 2009). For particles of the same charge, there are two fundamental types of trajectories inside an RCS: particles that enter from and then are ejected into the *same* semispace are hereafter referred to as "bounced" particles; particles that are ejected into the *opposite* semispace are referred to as "transit" particles.

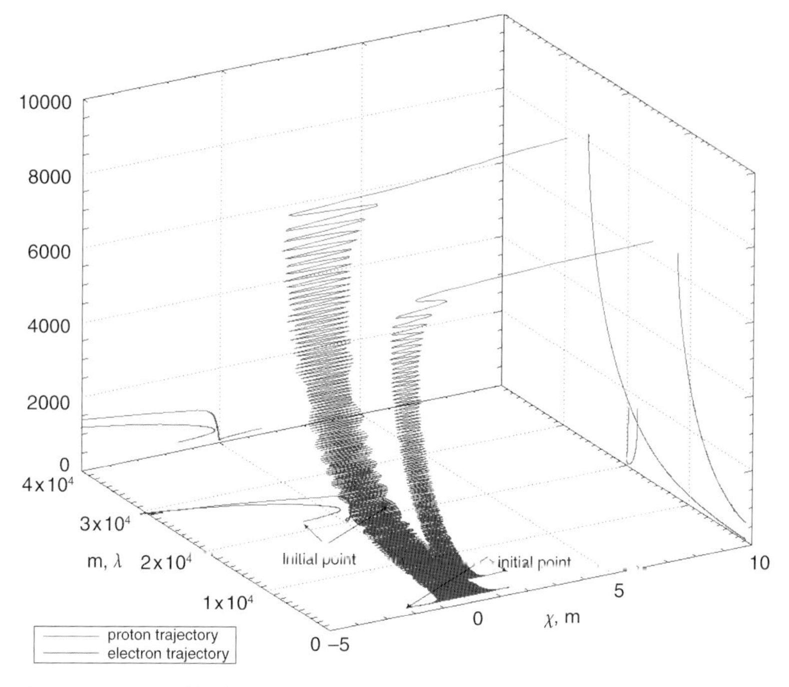

Figure 2.9 Proton (blue lines) and electron (purple lines) trajectories in a 3-D RCS with a lateral magnetic field $B_x = 10$ G and a drift electric field $E_y = 100\,\mathrm{V\,m^{-1}}$. From Zharkova and Agapitov (2009). For a color version of this figure, please see the color plates at the beginning of the book.

Electron/proton asymmetry in the particle trajectories is dependent on the magnitude of the guiding field (Zharkova and Gordovskyy, 2004). The asymmetry is *full* (equal to unity) if B_y is strong enough ($B_y/B_z \gtrsim 1.5 \times 10^{-2}$) that electrons are ejected into one semiplane and protons into the opposite one. If the guiding field is weak ($B_y/B_z \lesssim 10^{-6}$), then electrons and protons can be ejected into the *same* semiplane (Zharkova and Gordovskyy, 2004, 2005c). For a guiding field of intermediate magnitude, electrons and protons are ejected in a partially neutralized manner, with protons dominating in one semiplane and electrons in the other (Zharkova and Gordovskyy, 2005c). The result of this separation of particle species is the appearance of a polarization electric field between the protons at the edge and electrons at the midplane of the RCS. Such a field was anticipated by Litvinenko and Somov (1993); Oreshina and Somov (2000a) and has been investigated in some detail by Zharkova and Agapitov (2009).

The model produces power-law energy distributions with spectral indices $\delta = 1.5$ for protons and $\delta = 2$ for electrons. Such a result is confirmed by both analytic (Litvinenko, 1996a) and numerical approaches (Wood and Neukirch, 2005; Zharkova and Gordovskyy, 2005c). For comparison, the particle energy spectra calculated for the model RCS of Oreshina and Somov (2000a) and presented in Figure 2.10 show features beyond such simple power-law forms: there is a rapid rise in flux

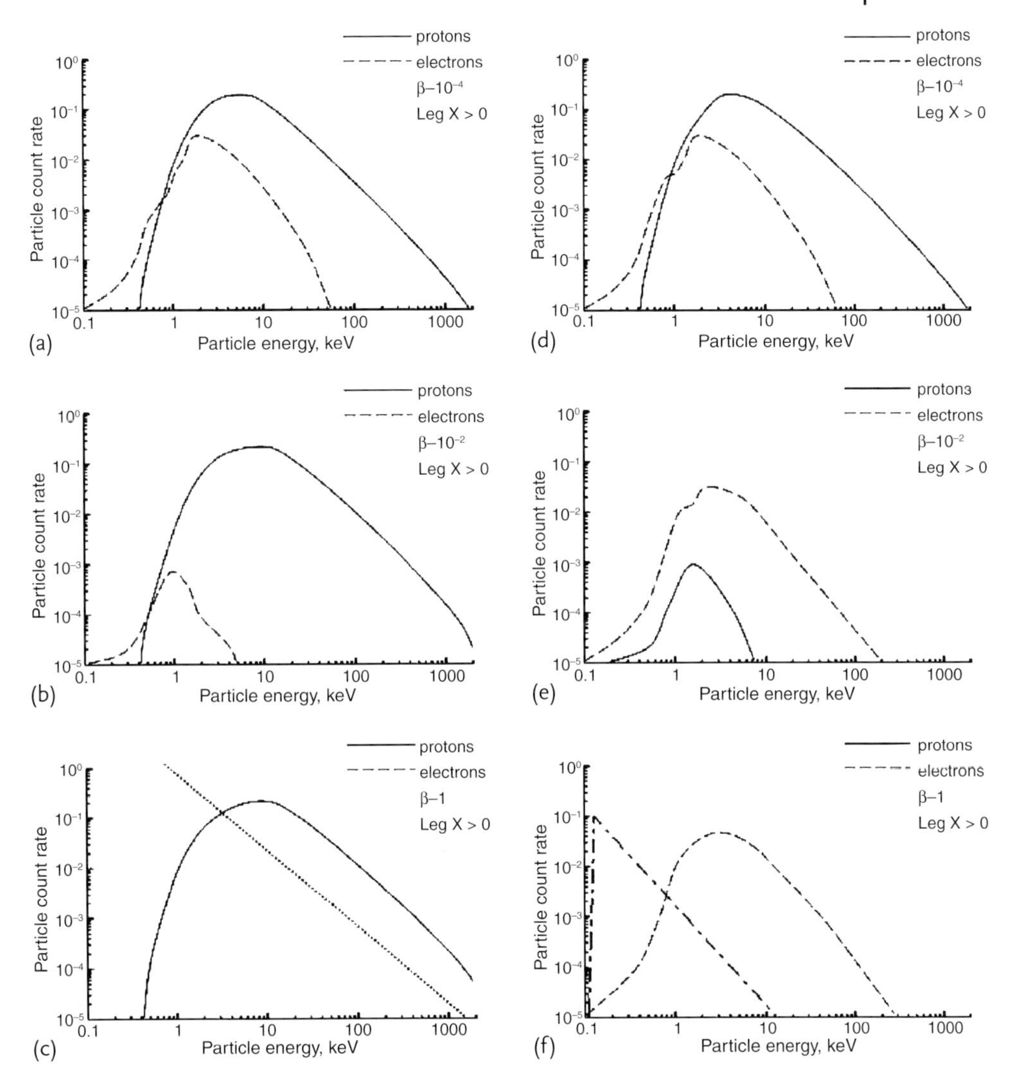

Figure 2.10 Energy spectra of electrons (dashed lines) and protons (solid lines) ejected from an RCS. Spectra (a–c) are calculated for $\beta(= B_y/B_0)$ equal to 10^{-4}, 10^{-2} and 1, respectively, and correspond to the semi-space $x > 0$. Spectra (d–f) are calculated for the same values of β, but for the semi-space $x < 0$. The estimated proton spectrum is presented as a dotted line in (c), and the estimated electron spectrum is shown as a dashed–dotted line in (f). From Zharkova and Gordovskyy (2005c).

to a maximum at $E = E_{\rm m}$; above this the spectrum is approximately power law. $E_{\rm m} \simeq 10$–12 keV for protons ($\simeq 2$–5 keV for electrons); these values increase slightly with the value of the parameter $\beta = B_y/B_z$. The energy $E_{\rm m}$ can be considered as the lower energy cutoff to the electron spectrum (Section 1.2.3), and the values obtained here agree quite well with the values deduced from RHESSI observations (Holman *et al.*, 2011).

The spectral index δ for the power-law part of the spectrum ($E > E_m$) has been found to vary slightly with the magnitude of the guiding field B_y. For a weak guiding field ($\beta = 10^{-4}$ – Figure 2.10a,d), the particle trajectories are nearly symmetrical (Zharkova and Gordovskyy, 2005c); the particles are ejected equally into both semispaces as approximately neutral beams. The particle energy spectra in both semispaces ($x < 0$ and $x > 0$) are very similar, with a sharp increase from zero to E_m, followed by a power law with spectral indices $\delta \simeq 1.8$ for protons and $\delta \simeq 2.2$ for electrons. The power-law shapes extend up to about 100 keV (electrons) or 1 MeV (protons).

For a moderate guiding field ($\beta = 10^{-2}$ – Figure 2.10b,e), the protons and electrons are also ejected into both semispaces; however, the symmetry of their trajectories is partially destroyed. Hence, most protons are ejected into the semispace $x > 0$ with $\delta \simeq 1.7$, while a few low-energy protons are ejected into the semispace $x < 0$ with very soft, thermal-like spectra ($\delta \simeq 4.8$). The opposite picture holds for electrons: the bulk are ejected into the semispace $x < 0$ with an energy spectrum of $\delta \simeq 2.0$, while a smaller number of low-energy electrons are ejected into the semispace $x > 0$ with a thermal-like spectrum having a maximum $E_m \approx 1$ keV.

Finally, if the guiding field is very strong ($\beta \simeq 1$ – Figure 2.10c,f), then the particle trajectories are completely asymmetric. Particles are ejected completely separately from the RCS midplane into opposite semispaces: all protons are ejected into the semispace $x > 0$, while all electrons are ejected into the semispace $x < 0$. Their energy distributions show a sharp increase of particle number up to around a few kiloelectronvolts followed by a power-law energy spectrum of index $\delta \simeq 1.5$ for protons and $\delta \simeq 1.8$ for electrons. (These spectral indices vary with different reconnection scenarios and with different phases of magnetic reconnection, as discussed in Section 2.2.6.)

As shown in Figure 2.11, during the reconnection process the shape of the RCS changes from a noncompressed one with a symmetric X-type null point to a strongly compressed shape with a well-defined elongation along the z-direction and a compression in the x-direction. The extent of this deformation depends on the value of the parameter α (Priest and Forbes, 2000; Somov, 2000) and possibly corresponds to different reconnection rates (Arber and Haynes, 2006). At the reconnection onset ($\alpha = 0$) the diffusion region is located very near the null point. As the reconnection progresses (α increases from 0 to 1), the diffusion region becomes much wider in both the x- and z-directions. As α increases further, the reconnection begins to slow down (Zharkova and Agapitov, 2009).

The spectral index of the accelerated particle energy spectra is associated with the parameter α. For an RCS with a density that increases exponentially with distance z and for a variation of density $n(z) \sim (z/a)^\lambda$, the following results were obtained (Zharkova and Agapitov, 2009). For electrons in a weak (strong) guiding field B_y:

$$\delta_{\text{e,weak}} = 1 + \frac{2+\lambda}{2\alpha}\,; \quad \delta_{\text{e,strong}} = \frac{1}{2} + \frac{1+\lambda}{2\alpha}\,, \tag{2.33}$$

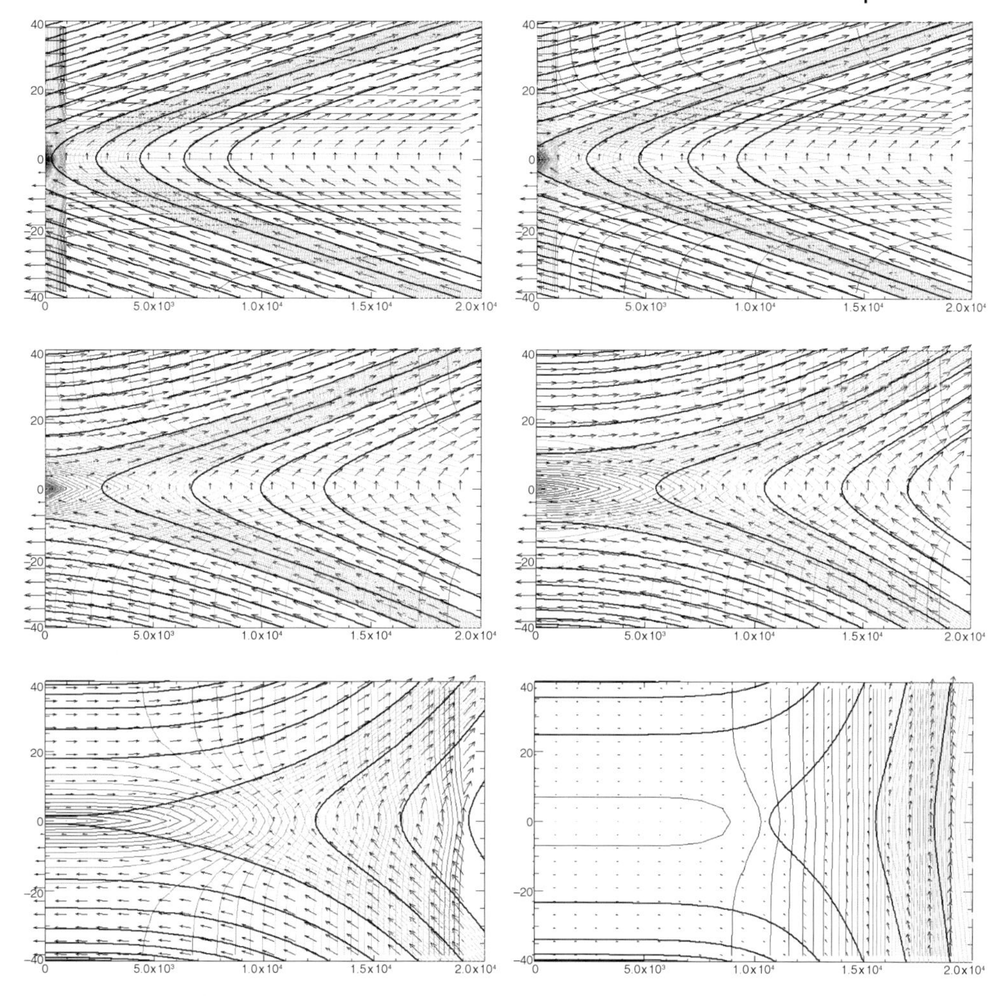

Figure 2.11 Views of current sheet for values of α of 0.5, 0.7, 1.0, 1.5, 2.0, and 5.0 (a–f). From Zharkova and Agapitov (2009). For a color version of this figure, please see the color plates at the beginning of the book.

while for protons in a guiding field of *any* strength:

$$\delta_{\mathrm{p}} = 1 + \frac{1+\lambda}{2\alpha} \,. \tag{2.34}$$

In principle, by simultaneously measuring the spectral indices of electrons and protons in the same flare one can deduce the magnetic field geometry under which these particles were accelerated (Zharkova and Agapitov, 2009; Zharkova and Gordovskyy, 2005c).

We again stress that all of these results are obtained using a TP approach in which the feedback of the electromagnetic fields associated with the accelerated particles is not taken into account. Such a limitation is addressed in the PIC approach discussed in Section 2.4.

2.2.4
Estimations of Accelerated Particle Parameters

2.2.4.1 Estimation of RCS Parameters

In order to evaluate the accelerated particle parameters in an adopted reconnection model, let us consider the flux conservation equations for outer (Eq. (2.35)) and inner (Eq. (2.35)) reconnection regions, the energy conservation Eq. (2.35c), and the Ampère law equations for the inner (Eq. (2.35d)) and outer (Eq. (2.35e)) regions (Gordovskyy *et al.*, 2005). For consistency this set must be completed by the expressions for velocities of protons V_p and electrons V_e accelerated by a super-Dreicer field. Although these particles have a wide range of velocity spectra, for the current estimations the average magnitudes (at lower-energy cutoffs) can be taken as per Eqs. (2.35f) and (2.35g) (see Zharkova and Gordovskyy, 2004). Hence, the set of equations can be written as follows:

$$(A - L)\,V_{in} = (a - l)\,V_{sj}\,, \tag{2.35a}$$

$$LV_{in} = l\,V_p\,, \tag{2.35b}$$

$$\frac{B_0^2}{2\mu}AV_{in} = \frac{1}{2}n_0 m_p V_p^3 l + \frac{1}{2}n_0 m_p V_{sj}^3 (a - l)\,, \tag{2.35c}$$

$$\frac{B_0}{l} = en_0(V_p + V_e)\,, \tag{2.35d}$$

$$\mathcal{E}_0 = \frac{V_{in}}{B_0}\,, \tag{2.35e}$$

$$V_p \simeq \pi\frac{E_0}{\bar{B}_x}\,, \tag{2.35f}$$

$$V_e \simeq \frac{e}{m_e}l\bar{B}_y\,. \tag{2.35g}$$

Here B_0, B_{tr}, and B_{lon} are the average magnitudes of tangential (changing the sign), transversal, and londitudinal (guiding) magnetic field components, respectively, $\mathcal{E}_0$ is the electric field in the inner region, A is the length of the outer region, n_0 is the plasma density, and V_{in} is the inflow velocity. These values are assumed to be taken from the reconnection models. Hence, the solution of this set provides us with the electric field $\mathcal{E}_0$, thickness l, and width L of the inner region, the thickness a of the outer region, and the velocity V_{sj} (separatrix jet velocity) of the neutral plasma outflow, respectively.

The parameters of the reconnection region as well as the average velocities of accelerated particles are estimated from a solution of the set of Eq. (2.35).

Assuming that the magnetic field components are $B_0 = 10^{-2}$ T, $B_{tr} = 10^{-4}$ T, and $B_{lon} = 10^{-4}$ T, the width of the outer region is $A = 10^5$ m, the density and velocity of the inflowing plasma are $n_0 = 10^{15}\,\mathrm{m^{-3}}$ and $V_{\mathrm{in}} = 10^5\,\mathrm{m\,s^{-1}}$, respectively, and the desired values are as follows: the electric field is $\mathcal{E}_0 = 1000\,\mathrm{V\,m^{-1}}$, the average velocities of protons and electrons accelerated by the super-Dreicer field are $V_{\mathrm{p}} = 3 \times 10^7\,\mathrm{m\,s^{-1}}$ ($\approx 5\,V_{\mathrm{A}}$) and $V_{\mathrm{e}} \approx 10^8\,\mathrm{m\,s^{-1}}$ ($\approx 17\,V_{\mathrm{A}}$), respectively, and the width and thickness of the diffusion region are $L = 6 \times 10^2$ m and $l \approx 3$ m, respectively. The thickness of the outer region is $a = 1.5 \times 10^3$ m, and the separatrix jet velocity is $V_{\mathrm{sj}} \approx 8 \times 10^6\,\mathrm{m\,s^{-1}}$ ($= 1.3\,V_{\mathrm{A}}$).

2.2.4.2 **Fluxes and Momenta Carried by Accelerated Particles**

Using the estimations from Section 2.2.4.1 above, one can estimate particle densities and the energy and momentum fluxes carried by different populations of accelerated particles.

The energy flux for protons and electrons accelerated by a super-Dreicer field can be calculated as follows:

$$F_{\mathrm{p/e}} = \frac{m}{2} V_{\mathrm{p/e}}^3 n_{\mathrm{p/e}} = \frac{m}{2} V_{\mathrm{p/e}}^2 V_{\mathrm{p/e}} n_{\mathrm{p/e}} = \frac{m}{2} V_{\mathrm{p/e}}^2 V_{\mathrm{in}} n_0 \frac{L}{l} , \tag{2.36}$$

where d is a diameter of the flux tube where the beam precipitates. By the model definition, its size is much smaller than the loop leg thickness since only a small part of the magnetic field lines enters the reconnection region at any given time.

The density of ejected particles can be found using the ratio of the inflow and outflow velocities:

$$n_{\mathrm{p/e}} = n_0 \frac{V_{\mathrm{in}}}{V_{\mathrm{p/e}}} \frac{L}{d} . \tag{2.37}$$

Similar estimations can be made for the separatrix jet particles by substituting the outer region size A instead of L and the jet outflow velocity V_{sj} instead of $V_{\mathrm{e/p}}$.

Finally, the momentum flux carried by different populations of accelerated particles can be found as follows:

$$M(x) = m_{\mathrm{p/e}} n_{\mathrm{p/e/sj}} V_{\mathrm{p/e/sj}} . \tag{2.38}$$

The results of particle density, energy, and momentum flux calculations are summarized in Table 2.1.

Table 2.1 Energy fluxes, particle densities, and momentum fluxes carried out by different populations of accelerated particles upon ejection. From Gordovskyy *et al.* (2005).

	"Fast beam"		"Slow beam"	
	Electrons	Protons	Electrons	Protons
Energy flux, $\mathrm{erg\,cm^{-2}\,s^{-1}}$	1.4×10^8	3.6×10^{10}	3.4×10^7	6.2×10^{11}
Particle density, $\mathrm{cm^{-3}}$	3.0×10^5	2.5×10^6	6×10^8	
Momentum flux, $\mathrm{g\,cm^{-1}\,s^{-2}}$	3×10^2		9×10^2	

2.2.5
Comparison of the Parameters of Accelerated Particles

Summarizing the results of particle acceleration in a 3-D current sheet, one can conclude that protons and electrons are ejected together into both semispaces ($x > 0$ and $x < 0$) if the relative strength of the guiding field $\beta = B_y/B_0$ is lower than 10^{-4}. In contrast, protons and electrons are completely separated if the guiding field $\beta \succeq 10^{-1}$. In the case of intermediate guiding field magnitudes $10^{-4} \preceq \beta \preceq 10^{-1}$, particles are partially separated: protons dominate in one of the semispaces, while electrons dominate in the opposite one.

The energy spectra of particles accelerated by a super-Dreicer electric field appear to be power law, similar to those estimated in the $n = \text{const}$ approach. However, if one takes a more realistic distribution of the density and transversal field (Somov and Oreshina, 2000), the spectra become softer. Apparently, this happens because of a low plasma density near the X point, where the transversal field is low and where high-energy particles are accelerated. Indices of proton energy spectra are 1.8 for $\beta = 10^{-4}$, 1.7 for $\beta = 10^{-2}$, and 1.5 for $\beta = 1$. At the same time, electron energy spectra appear to be softer than those of protons: the spectral indices are 2.2 for $\beta = 10^{-4}$, 2.0 for $\beta = 10^{-2}$, and 1.8 for $\beta = 1$. The spectra deduced from numerical simulations show the maxima, which appear to be at 10–12 keV for protons and at 1–5 keV for electrons.

The calculations made for the simple self-consistent model of RCS show that the deduced RCS parameters are very close to those accepted in our models of acceleration by a super-Dreicer field; therefore, the model is rather realistic. In addition, these calculations provide us with energies of protons and electrons accelerated at slow MHD shocks and ejected from RCS with separatrix jets, which appear to be $\sim$ 100 keV–1 MeV for protons and about 1 keV for electrons. Therefore, one can conclude that acceleration by stationary MHD shocks is a very efficient mechanism for proton and ion acceleration while electrons do not gain energies higher than the thermal ones.

A comparison of the integral characteristics of the different populations of accelerated particles show that the bulk of energy flux ($\sim 6 \times 10^{11}\,\text{erg}\,\text{cm}^{-2}\,\text{s}^{-1}$) is carried by the separatrix jet protons ("slow protons" hereafter). At the same time, protons accelerated by electric field (or "fast protons") carry about 30 times less energy flux ($\sim 2 \times 10^{10}\,\text{erg}\,\text{cm}^{-2}\,\text{s}^{-1}$). Electrons accelerated by electric field ("fast electrons") carry much less energy than protons do in general ($F_e \simeq 1.5 \times 10^8\,\text{erg}\,\text{cm}^{-2}\,\text{s}^{-1}$), but this population of accelerated particles is important because of its high propagation velocity. Finally, electrons of a separatrix jet (or "slow electrons") carry too little energy flux with very low velocities and can be easily neglected.

Hence there are three populations of high-energy particles left to be considered: "fast beam" and "slow jet" protons and "fast beam" electrons. However, one should note that these characteristics of the beam populations (their energy fluxes and initial densities) are very sensitive to the model characteristics, and the presented magnitudes are valid only for the accepted parameters shown above. For instance, if the accepted cross-section of the flux tube d decreases by an order of magnitude, then all the fluxes increase by one order of magnitude (Eqs. (2.36)–(2.38)).

2.2.6
Particle Acceleration in 3-D MHD Models with Fan and Spine Reconnection

A further development in modeling particle acceleration in an RCS involves more advanced MHD models of fan and spine reconnection (Priest and Titov, 1996). The axis of symmetry of the magnetic field, called the *spine* and designated here as the x-axis, is a critical field line connecting to the null point. In the plane $x = 0$, the magnetic field lines are straight lines through the null point describing a fan; hence the (y, z)-plane is termed the *fan plane*. The spine and the fan are the 3-D analogs of the separatrix planes in the classic 2-D X-point geometry. The field configuration near the potential magnetic nulls is given by formulas (2.28)–(2.30).

Two regimes of reconnection at a 3-D magnetic null are possible: *spine reconnection* and *fan reconnection* (Lau and Finn, 1990; Priest and Titov, 1996). While the magnetic field configuration is the same in the two regimes, they are characterized by different plasma flow patterns and different electric fields. Spine reconnection has a current concentration along the critical spine field line, while fan reconnection has a current sheet in the fan plane.

For a potential null, the spine reconnection regime is characterized by plasma flows lying in planes containing the spine, while fan reconnection has nonplanar flows that carry the magnetic field lines in a "swirling"-like motion symmetric about the fan plane. It should be noted that the electric field and flow pattern are nonaxisymmetric in all cases. The model of Priest and Titov (1996) provides a useful analytical model for initial studies; it allows some of the key features of topology and geometry to be identified, and it is the closest 3-D analog of the widely studied constant out-of-plane electric field used in 2-D studies. It describes only the outer ideal reconnection region and does not include acceleration due to the parallel electric field in the inner dissipation or resistive region; indeed, it breaks down very close to the spine/fan as the electric fields formally diverge. However, since few particles will enter the small-volume dissipation regions, the results are likely to be appropriate for the bulk of the energy spectrum. This model has been used as a basis for particle acceleration at 3-D nulls in the regimes of both spine reconnection (Dalla and Browning, 2005, 2006) and fan reconnection (Dalla and Browning, 2008). Particle trajectories are obtained numerically by solving the relativistic equations of motion (2.32). Again, a TP approach is used (for limitations, see Section 2.3).

The strength of the driving electric field is quantified by the parameter

$$\tilde{\mu} = \frac{v_\perp^2}{v_E^2(L)}, \tag{2.39}$$

where $v_\perp$ is the perpendicular speed associated with the gyro motion, which may be equated with the thermal speed of the plasma. The nomenclature indicates that this can also be expressed as the dimensionless magnetic moment when speeds are normalized to the drift speed (Vekstein and Browning, 1997):

$$\tilde{\mu} = \frac{mv_{\perp}^2/2B_0}{mv_E^2(L)/2B_0} , \tag{2.40}$$

where B_0 and E_0 are the amplitudes of the magnetic and electric fields. If $\tilde{\mu} \ll 1$, the electric drift speed at global length scales is strong compared with the thermal gyro motion and we are in the *strong drift regime* corresponding to fast reconnection; in practice, this regime is reached when $\tilde{\mu} \approx 1$. In this case, particles may undergo significant acceleration (Dalla and Browning, 2005) even in the outer "ideal" reconnection region, similar to those discussed in Section 2.2.3.

Typical individual trajectories in fan reconnection are shown in Figure 2.12, while similar plots for spine reconnection can be found in Dalla and Browning (2005). Note the strong dependence of the energy gain on the injection location, similar to those found for a single RCS in Section 2.2.3: particles injected closer to the fan plane (at lower latitudes β) are accelerated the most as they approach the strong-electric-field region near the fan plane.

Energy spectra for spine and fan reconnection (Figures 2.13 and 2.14) show that a significant fraction of protons can be accelerated to energies of up to 10 MeV, comparable to the values obtained in the simple 3-D topology presented in Section 2.2.3. As appears to be natural during any reconnection process, a generally

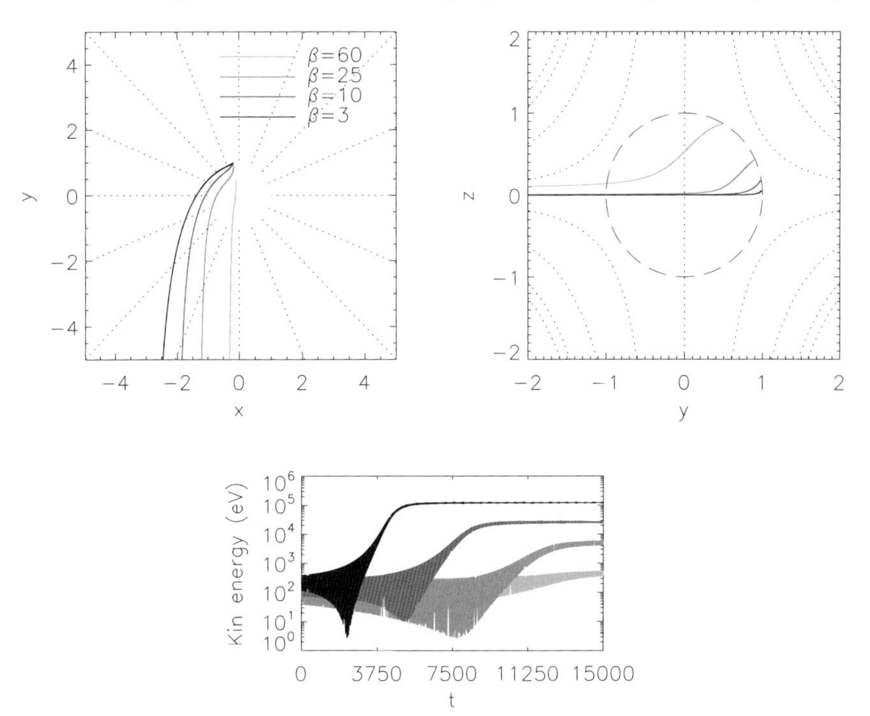

Figure 2.12 Trajectories and energy gains in fan reconnection, for particles injected at a boundary at different latitudes β. From Dalla and Browning (2008).

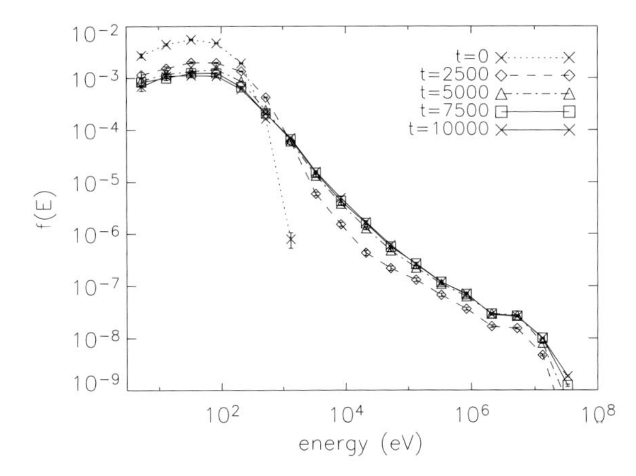

Figure 2.13 Energy spectrum of particles in *spine* reconnection at successive times, showing evolution toward steady state. From Dalla and Browning (2006).

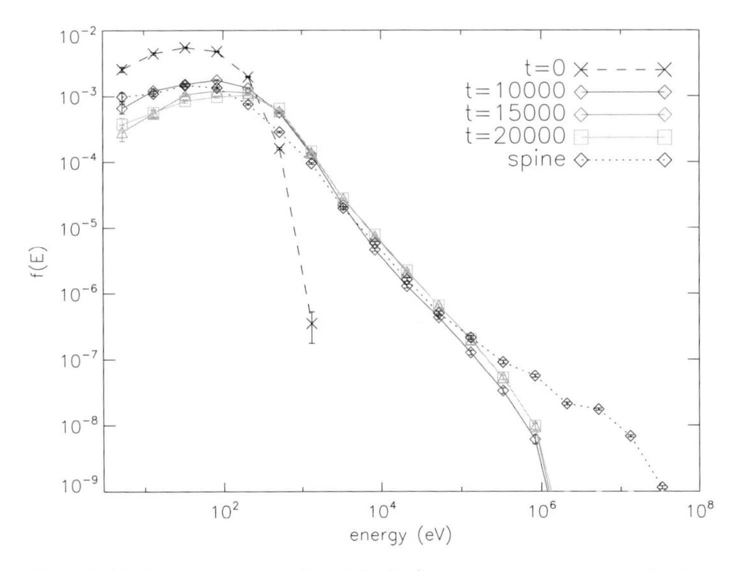

Figure 2.14 Energy spectrum of particles in *fan* reconnection at successive times, showing evolution toward steady state. From Dalla and Browning (2008).

power-law spectrum is obtained for the accelerated particles, but with a significant "bump" at higher energies, particularly for the spine reconnection, for which the bump arises from particles approaching close to the null itself.

This "bump-in-tail" distribution is similar to those reported for a single 3-D current sheet in Figure 2.15 in Section 2.2.3 (see also Zharkova and Agapitov, 2009; Zharkova and Gordovskyy, 2005a) and is formed by the two populations of protons: "transit" and bounced ones, respectively. Note that the slopes of the spectra are similar for the spine and fan reconnection in the intermediate energy range,

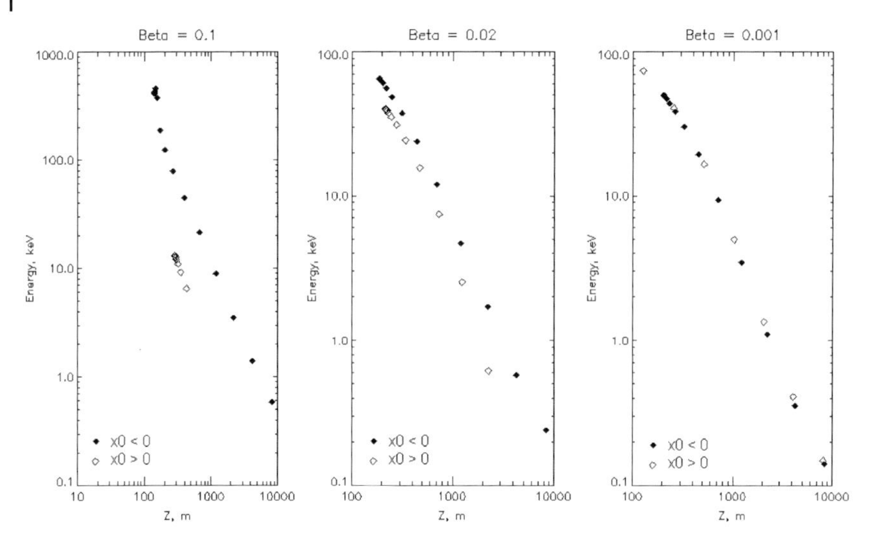

Figure 2.15 Energy spectra resulting from protons entering current sheet from opposite sides, for different ratios $\beta = B_y/B_0$ and different magnitudes of parameter α. From Zharkova and Agapitov (2009).

although fan reconnection is significantly less effective at generating high-energy particles.

The spatial distributions of accelerated of particles and where these particles originate are shown in Figures 2.16 and 2.17 for spine and fan reconnection, respectively. Each location is identified by its latitude β, that is, the elevation angle from the x–y plane, and longitude ϕ. Inflow regions are the top right and bottom left quadrants in the ϕ, β representation, but for clarity only particles originating in the latter are shown (the other quadrant can be added by symmetry). Each point represents one particle and is color-coded according to the final particle energy.

In the case of spine reconnection, particles injected in the plane in which the electric field is weakest (longitude $\phi = 0$) actually have the greatest energy gain, although their acceleration times are longest (Dalla and Browning, 2005). The high-energy particles that escape mostly emerge in jets along the spine. However, most of the highest-energy particles actually remain trapped in the magnetic field (Dalla and Browning, 2006). The occurrence of these trapped high-energy particles may be an explanation for coronal HXR sources (Krucker *et al.*, 2008a).

In fan reconnection, particles starting near the fan plane are those that gain the largest energy, as would be expected since the fan plane is where the electric field is largest. As one moves toward increasingly greater latitudes, particles will reach the fan plane with greater difficulty and so gain less energy. Figure 2.17 shows that in the end, particles are confined to a range of longitudes $[-130°, -50°]$. This is a result of the tendency of the electric field drift in fan reconnection to push particles toward the $x = 0$ plane after they have passed the region near the null. Figure 2.17 also shows that the initial azimuthal locations near $\phi = 90°$ are the most favorable for particle acceleration since near this plane the transverse component of the

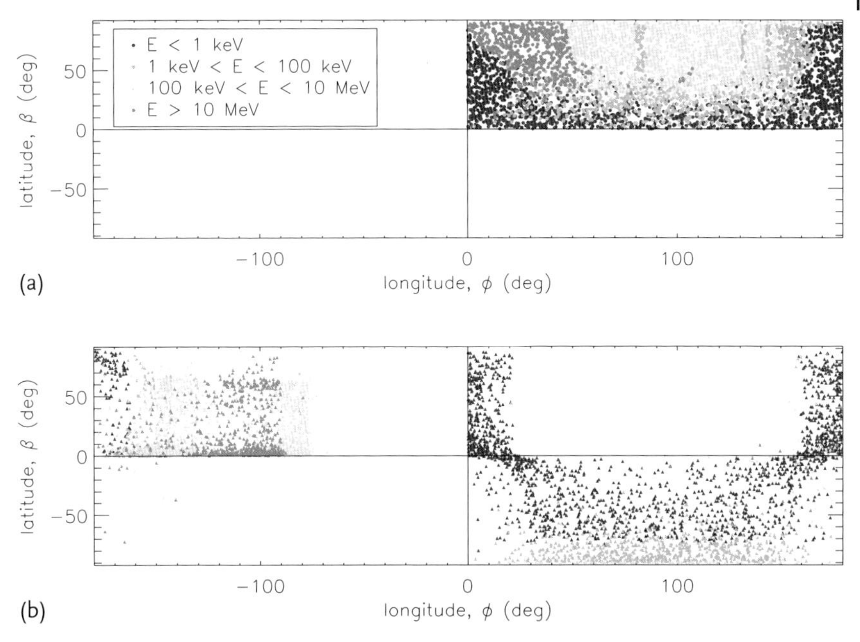

Figure 2.16 Positions of particles at initial (a) and final (b) times for *spine* reconnection, color-coded according to their final energies. From Dalla and Browning (2006). For a color version of this figure, please see the color plates at the beginning of the book.

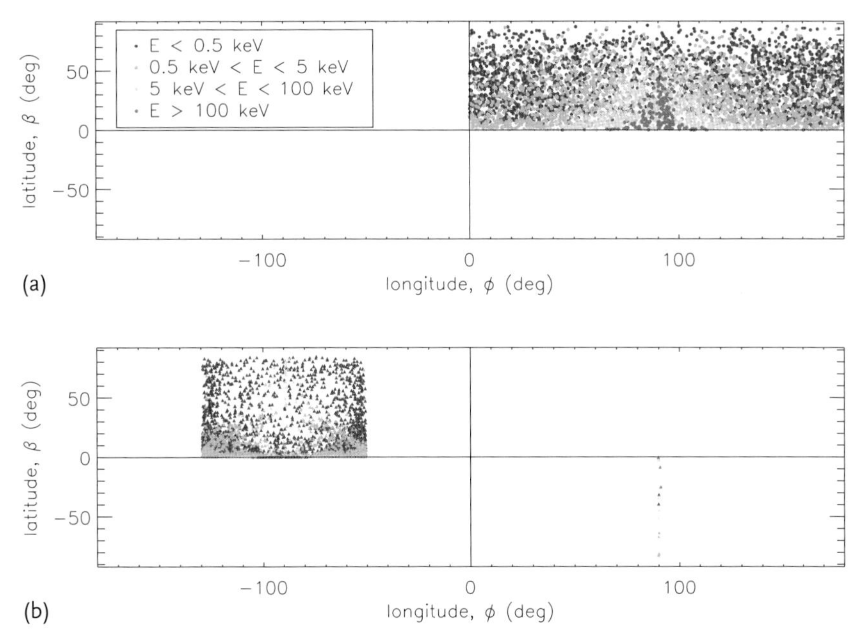

Figure 2.17 Positions of particles at initial (a) and final (b) times for *fan* reconnection, color-coded according to their final energies. From Dalla and Browning (2008). For a color version of this figure, please see the color plates at the beginning of the book.

electric field drift, pushing particles toward the fan plane, is largest. The fan reconnection regime is less efficient in seeding TPs to the region of a strong electric field, compared with spine reconnection, and acceleration times are longer.

These results of proton acceleration, obtained for a single-current-sheet reconnection model with a more complicated magnetic and electric field topology simulated by Priest and Titov (1996), advances further the conclusions derived from the basic 3-D reconnection models discussed in Section 2.2.3. This approach confirms that acceleration by a super-Dreicer electric field generated during a magnetic reconnection is a viable mechanism, which naturally explains the energy input from magnetic field dissipation and its output in the form of power-law energy spectra of accelerated particles with slopes related to a magnetic field geometry occurring during reconnection. Apparently this approach offers a good diagnostic tool for the magnetic field topology from the observed signatures of accelerated particles. Again one can note that the limitations of the TP approach (Section 2.3) can restrict this diagnostic capability, and that a full kinetic approach to particle acceleration in such geometries is required.

2.3
Limitations of the Test-Particle Approach

As discussed above in Sections 2.1.2 (basic reconnection in a single 2-D current sheet), 2.1.4 (acceleration by shocks or plasma turbulence), 2.2.3, and 2.2.6 (reconnection in 3-D geometries), a considerable amount of simulation of the particle acceleration process involves the use of "test particles" moving in a prescribed electromagnetic field configuration. Such a TP approach is certainly useful at some level – for example, it allows us to investigate the individual trajectories of electrons and protons inside 3-D current sheets (Dalla and Browning, 2005; Wood and Neukirch, 2005; Zharkova and Gordovskyy, 2004) and to obtain the aggregated energy distributions for a large volume of particles (Browning and Dalla, 2007; Dalla and Browning, 2005; Zharkova and Gordovskyy, 2005b,c), which can be compared with those inferred from observations. The approach can be used iteratively to extend the results of exact PIC simulations to a much larger region (Section 2.4).

However, considerable caution must be exercised when applying a TP approach to situations in which the number of accelerated particles is sufficiently large. In such cases, the electric and magnetic fields associated with the accelerated particles themselves rapidly become comparable to the fields presupposed at the outset. Observations show that a major flare can accelerate some 10^{37} electrons s^{-1} and that these electrons propagate within a magnetic flux tube of order 10^9 cm in radius. Attempting to model this as a single large-scale ejection leads to the well-known (e.g., Emslie and Hénoux, 1995; Holman, 1985) paradoxes that (i) the magnetic fields induced by the resulting current vastly exceed the presupposed field and (ii) the electric fields resulting from both charge separation and induction effects associated with the accelerated particle stream (Sections 2.2.3 and 2.4) exceed plausible values as per estimates:

- Ampère's law, applied to a steady flow of 10^{37} electrons s^{-1} (current $I \sim 10^{18}$ A), in a cylinder of cross-sectional radius $r \sim 10^7$ m gives a corresponding magnetic field strength of $B \sim \mu_0 I/2\pi r \sim 3 \times 10^4$ T $\sim 3 \times 10^8$ G. Using a propagation volume $V \sim 10^{21}$ m^3, the associated energy content is $(B^2/2\mu_0)V \sim 10^{36}$ J $\sim$ 10^{43} erg, at least ten orders of magnitude greater than the energy in the electron beam itself.
- The typical rise time for an HXR burst in a solar flare is $\tau \sim 1$–10 s. Initiating a current I over a time scale τ in a volume of inductance $L \sim \mu_0 \ell$ (where ℓ is the characteristic scale) requires a voltage $V \sim \mu_0 \ell I/\tau$. Alternatively, the largest current that can appear when a voltage V is applied is $I \sim V\tau/\mu_0 \ell$. With the characteristic voltage associated with the acceleration of HXR-producing electrons being $\sim$ 30 kV, and with solar loop sizes $\ell \sim 10^7$ m, this gives a maximum current of $I \sim 10^4$ A, some *fourteen* orders of magnitude less than the total current involved.

Hence, the self-consistent electromagnetic fields associated with electron acceleration rapidly become comparable to the fields presupposed to be present in the acceleration region; in such an environment, a TP approach breaks down severely.

2.3.1 The Polarization Electric Field

As discussed in Section 2.2.3 (Figure 2.9), reconnection in a 3-D current sheet geometry leads to a separation of protons and electrons across the midplane. As shown in Figure 2.18, such a charge separation results in a substantial (up to 100 V cm^{-1}) Hall-type polarization electric field across the midplane. As pointed out in Section 2.1.2, inclusion of such Hall-type electric fields can significantly *increase* the effectiveness of the whole magnetic reconnection process (Birn *et al.*, 2001; Drake *et al.*, 2005; Huba and Rudakov, 2004; see also Section 2.5).

Electrons, because of their much smaller gyroradii, typically do not show much change in behavior in the presence of a polarization field. On the other hand, protons, with their larger gyroradii, gain slightly more energy when they move in the positive x-direction than when they move in the negative x-direction (Zharkova and Agapitov, 2009). The net (nonlinear) result is that protons gain energy as a result of the polarization field, thus tending to make the proton energy spectrum flatter. A full investigation of the effect of polarization electric fields requires the use of a PIC approach, as described in Section 2.4.

2.3.2 Turbulent Electric Fields

Various turbulent electric fields typically form inside an RCS – these are caused, for example, by the two-stream instability (Bret, 2009) of the two electron beams that enter the current sheet from opposite sides (Zharkova and Agapitov, 2009). Such electric fields, and their associated currents, typically oscillate near the plasma

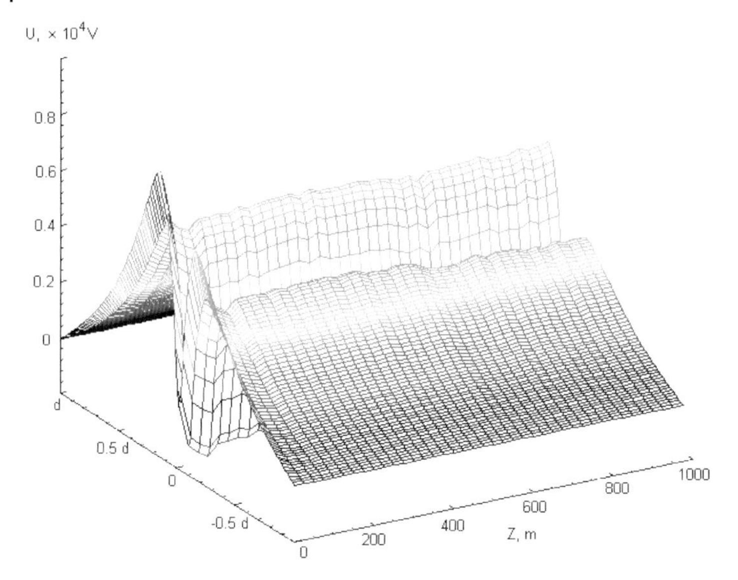

Figure 2.18 Polarization electric field induced by the separation of protons and electrons accelerated in a 3-D current sheet with $B_0 = 10\,\mathrm{G}$, at various distances z from the null point. From Zharkova and Agapitov (2009). For a color version of this figure, please see the color plates at the beginning of the book.

frequency and have an exponential growth rate Γ and an energy saturation level PE given by the analytical relations derived for the cold beam plasma instability (Bret, 2009; Shapiro and Shevchenko, 1988):

$$\Gamma = \sqrt{1/3}\, 2^{-4/3} \Delta^{1/3} \omega_{\mathrm{pe}} \,, \quad \mathrm{PE} = (\Delta/2)^{1/3} E_{\mathrm{beam}} \,, \tag{2.41}$$

where Δ is the ratio of the beam to plasma densities and E_{beam} is the beam kinetic energy. The effect of the electric fields associated with such waves on the accelerated particle distributions was estimated by Zharkova and Agapitov (2009); this work confirmed the estimates of Bret (2009).

Given the ambient plasma feedback discussed above, it is a clear imperative to take into account the self-consistent evolution of the electric and magnetic fields, *including the fields associated with the accelerated particles themselves.* Failure to do so can result in unrealistic numbers of accelerated particles predicted by acceleration models and even paradoxical results, such as the energy in accelerated particles exceeding the free energy in the preflare magnetic field configuration. Hence, further progress requires full kinetic simulations.

2.4 Particle-in-Cell Simulation of Acceleration in a 3-D RCS

In this section, we describe early efforts to accommodate this crucial element of modeling following the approach by Siversky and Zharkova (2009a).

2.4.1 Problem Formulation

2.4.1.1 Magnetic Field Topology

TP simulations have shown that acceleration time is on the order of 10^{-5} s for electrons and 10^{-3} s for protons. Since this time is much shorter than the time of the reconnecting magnetic field variation (Priest and Forbes, 2000; Somov, 2000), we can assume that during simulation the background magnetic field is stationary. Also, from the TP simulations we conclude that travel distances of accelerating particles along the RCS are on the order of 10 km at most (for the protons) (Zharkova and Agapitov, 2009). Thus we can assume that this length is much shorter than the length scale of the magnetic field variation along the current sheet. In addition, as is generally accepted, we suppose that the magnetic field variations across the current sheet have a much shorter length scale than the variation along the current sheet:

$$L_x \ll L_z, L_y \,. \tag{2.42}$$

Thus, our simulation domain is a small part of the RCS (Figure 2.19), large enough to contain the full trajectories of accelerated particles. The background magnetic field is stationary and varies inside this domain only in the x-direction, which is

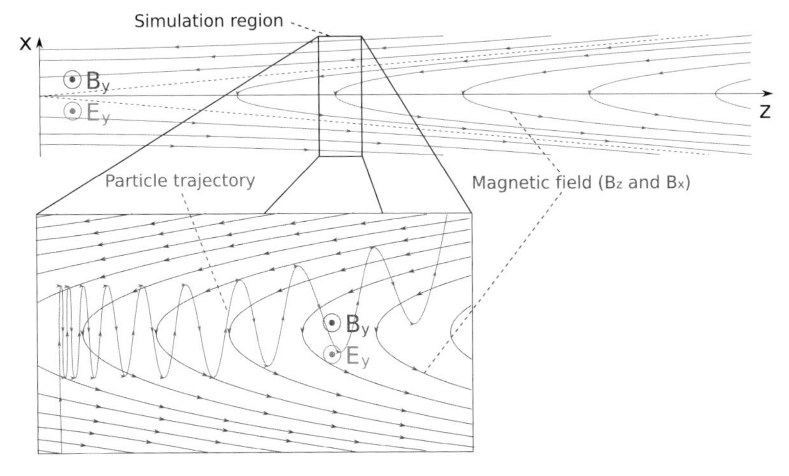

Figure 2.19 Magnetic field topology and electric field used in the PIC simulation model. The vertical rectangle shows the size of the simulation box, which is shifted along the z-axis to account for different magnitudes of tangential magnetic field component B_x. For a color version of this figure, please see the color plates at the beginning of the book.

perpendicular to the RCS. We take into account all three components of the background magnetic field. The main component B_z depends on x as follows:

$$B_z(x) = -B_{z0} \tanh\left(\frac{x}{L_x}\right) . \tag{2.43}$$

The B_x component is assumed (as in Zharkova and Gordovskyy, 2004) to be constant inside the simulation domain:

$$B_x = -B_{x0} . \tag{2.44}$$

The guiding (out-of-plane) magnetic field B_y is maximal in the midplane and vanishes outside the RCS:

$$B_y(x) = B_{y0} \operatorname{sech}\left(\frac{x}{L_x}\right) . \tag{2.45}$$

Note that if $B_{y0} = 0$, then the configuration corresponds to the Harris sheet equilibrium, and if $B_{y0} = B_{z0}$, then the equilibrium is force-free.

The inflow of plasma into the RCS combined with the condition of the frozen-in magnetic field leads to the induction of the drifted (out-of-plane) electric field E_y. In order to provide the inflow of plasma in our simulation domain we set up a background electric field, as those that drifted in with velocity V_{in} by a magnetic diffusion process (Priest and Forbes, 2000; Somov, 2000).

$$E_y = E_{y0} = B_{z0} V_{\text{in}} , \tag{2.46}$$

where V_{in} is the inflow velocity.

In our study we use the following values for the current sheet parameters: the main component of the magnetic field $B_{z0} = 10^{-3}$ T, the current sheet half-thickness $L_x = 1$ m, the drifted electric field $E_{y0} = 250\,\text{V}\,\text{m}^{-1}$, and the guiding, B_{y0}, and transverse, B_{x0}, components of the magnetic field are varied to study how they influence particle acceleration.

2.4.1.2 **Governing Equations and Method of Solution**

Although TP simulations can provide some valuable results of particle motion inside an RCS, it does not take into account the electric and magnetic fields generated by accelerated particles in the simulation region. In order to include these fields, we used 2D3V PIC simulation code developed by Verboncoeur and Gladd (1995). The PIC method, similar to TP, is based on the equation of motion for plasma particles:

$$\frac{d\boldsymbol{V}_{\text{si}}}{dt} = \frac{q_{\text{s}}}{m_{\text{s}}} \left[\boldsymbol{E} + \tilde{\boldsymbol{E}} + \boldsymbol{V}_{\text{si}} \times \left(\boldsymbol{B} + \tilde{\boldsymbol{B}}\right)\right] , \tag{2.47}$$

where besides the background fields $\boldsymbol{E}$ and $\boldsymbol{B}$ given by Eqs. (2.43)–(2.46), the local self-consistent fields $\tilde{\boldsymbol{E}}$ and $\tilde{\boldsymbol{B}}$ induced by the accelerated particles are taken into account. These fields are calculated from the Maxwell equations:

$$\frac{\partial \tilde{E}}{\partial t} = c^2 \nabla \times \tilde{B} - \frac{1}{\epsilon_0}\left(\tilde{j}_e + \tilde{j}_p\right) , \tag{2.48}$$

$$\frac{\partial \tilde{B}}{\partial t} = -\nabla \times \tilde{E} , \tag{2.49}$$

where $\tilde{j}_e$ and $\tilde{j}_p$ are the current densities of electrons and protons.

In this 2D3V PIC simulation the y-dimension is chosen to be invariant. The system is periodic in the z-direction, so that a particle that leaves the system through the right or left boundary (Figure 2.19) appears on the opposite boundary. Thus, it is not necessary to make the system very long in order to handle the whole particle trajectory from entry to ejection. The current sheet half-thickness, L_x, is equal to 1 m, while the width of the whole simulation region along x is chosen to be 20 m, in order to avoid any influence of the boundaries on the particles inside the RCS. Plasma is continuously injected from the $x = \pm 10$ m sides of the simulation region at a rate of $n E_{y0}/B_{z0}$.

In order to avoid numerical instabilities in the PIC method, the following constraints need to be satisfied:

$$c\Delta t < \Delta\xi , \tag{2.50}$$

$$\Delta t < 0.2\omega_{\text{pe}}^{-1} , \tag{2.51}$$

$$\Delta\xi < \lambda_{\text{D}} , \tag{2.52}$$

where Δt is the time step, $\Delta\xi$ is the grid step in any direction, c is the speed of light, $\omega_{\text{pe}} = (4\pi n e^2/m_{\text{e}})^{1/2}$ is the electron plasma frequency, and $\lambda_{\text{D}} = (kT/4\pi n e^2)^{1/2}$ is the Debye length. To satisfy these conditions, while keeping the code running time at a reasonable level, we use a reduced plasma number density $n = 10^{10}\,\text{m}^{-3}$ in our simulations. Also the proton-to-electron mass ratio is reduced to $m_{\text{p}}/m_{\text{e}} = 100$ in order to keep the proton acceleration time within reasonable computational limits. The spatial simulation grid has from 10 to 100 cells in the z-direction and 100 cells in the x-direction with $\Delta z = \lambda_{\text{D}}$, $\Delta x = \lambda_{\text{D}}/5$ and 100 superparticles per cell on average, while each superparticle represents 2×10^7 real particles. The time step is 6×10^{-10} s.

We assume the existence of a background plasma with a real coronal density of $10^{16}\,\text{m}^{-3}$, which is not a part of the PIC simulation due to the limits of current power even in large computer clusters. This assumption is required to provide the background current $j_y = -1/\mu_0 \cdot dB_z/dx$ to sustain the equilibrium with the background magnetic configuration given by Eqs. (2.43)–(2.45) that will produce electron velocities of $V_y = 1/(\mu_0 n e) \cdot dB_z/dx$. For the real coronal plasma density the electron velocity that forms the equilibrium background current is $5 \times 10^5\,\text{m}\,\text{s}^{-1}$. This velocity is a few orders of magnitude lower than the velocities gained by the electrons at acceleration in the drifted electric field; thus it will not change significantly the electron distributions and is neglected in the current approach. In doing so, we possibly eliminate only some low-frequency instabilities (like ion-sound wave) that are not a part of our present study. However, this approach allows us to investigate the acceleration of a smaller number of particles ($n = 10^{10}\ \mathrm{m}^{-3}$) accelerated by a drifted electric field by taking into account the plasma feedback to these particles motion.

The simulations are supposed to run until a quasistationary state is reached. This running time corresponds to the acceleration time of the slowest particles (protons) and is usually less than 10^{-3} s. Also, since we are interested in the stationary state, and the background magnetic field is assumed to be in equilibrium with high-density plasma, the initial state of the simulated plasma particles is not important. In practice, all particles that are initially present in the simulation region would be accelerated, ejected, and replaced by injected particles before the stationary state is reached. Thus, to speed up the simulation we do not initially have any particles in the simulation region. The injected particles are assumed to have a thermal distribution with a typical coronal temperature of 10^6 K.

2.4.2
Test-Particle Simulations

For a better understanding of particle trajectories inside an RCS during PIC simulation let us first investigate the results obtained in the TP approach (similar to those by Zharkova and Gordovskyy, 2004).

A typical trajectory of a plasma particle (Figure 2.20) in electromagnetic configuration, shown in Figure 2.19, consists of three parts: $\boldsymbol{E} \times \boldsymbol{B}$ drift, acceleration by the electric field, and ejection. (i) Outside the current sheet, where the magnetic field is strong, particle motion is adiabatic and can be described as the superposition of a magnetic gyration and a drift in the orthogonal electric and magnetic fields toward the midplane of the RCS. (ii) Inside the current sheet the particle moves along the y-axis and is accelerated by the electric field E_y. The exact trajectory of the particle depends on the value of the guiding field B_y (Litvinenko and Somov, 1993). For a large B_y the particle remains magnetized and reaches higher energy than for a small B_y, when the particle becomes unmagnetized inside the current sheet. (iii) When the particle gains enough energy, the magnetic configuration can no longer confine the particle motion, and it is ejected from the RCS. Outside the current sheet the particle motion is again adiabatic. However, due to the high velocity of the particle along the magnetic field line and nonzero B_x the x-component of the particle velocity is larger than the $\boldsymbol{E} \times \boldsymbol{B}$ drift velocity and the particle moves away from the RCS.

Figure 2.20 shows the trajectories in the x–V_z phase plane of two protons and two electrons (one proton and one electron inflow from the $x > 0$ semispace, another pair from the $x < 0$ semispace). The mass ratio is chosen to be $m_\mathrm{p}/m_\mathrm{e} = 100$ for the comparison with the PIC simulations in later sections. The magnitudes of the magnetic field components B_y and B_x are chosen such that electrons remain magnetized in the vicinity of the midplane while protons are unmagnetized during the acceleration phase. In this case the energy of the unmagnetized protons upon

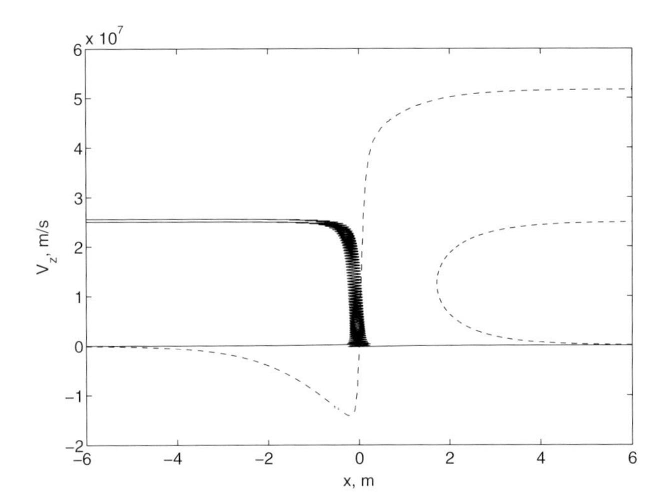

Figure 2.20 Trajectories of two protons (solid) and two electrons (dashed) on the x–V_z phase plane during acceleration inside a current sheet (TP simulations). $B_{y0} = 10^{-4}$ T, $B_{x0} = 2 \times 10^{-5}$ T, $m_p/m_e = 100$.

ejection can be estimated (see Litvinenko, 1996b) as

$$\varepsilon = 2m_{\alpha}\left(\frac{E_{y0}}{B_{x0}}\right)^2, \tag{2.53}$$

where m_{α} is the particle mass. As was shown by Zharkova and Gordovskyy (2004), if B_{y0} is strong enough ($> 1.5 \times 10^{-2} B_{z0}$), then all protons (regardless of the side they entered from) are ejected into one semispace while all electrons are ejected into the opposite semispace with respect to the midplane. Indeed, in Figure 2.20 both protons are ejected into the semispace with negative x and both electrons to the $x > 0$ semispace. In order to distinguish two types of trajectories, the particles that flow in from and are ejected into the same semispace will be referred to as "bounced" particles (this corresponds to the proton coming from $x < 0$ and the electron coming from $x > 0$ in Figure 2.20). Particles that are ejected into the opposite semispace from the one they came from will be referred to as "transit" particles (this corresponds to the proton coming from $x > 0$ and the electron coming from $x < 0$ in Figure 2.20).

As can be seen in Figure 2.20, transit and bounced particles gain different energy, and this effect is stronger for electrons. Note that the difference in motion of transit and bounced particles vanishes when $B_y \to 0$ (see also Figures 6 and 7 in Zharkova and Agapitov, 2009). On the other hand, if $B_x = 0$, then particles cannot escape from the RCS and their acceleration is only limited by the size of the current sheet in the y direction.

Since for the magnitude of B_y and B_x (used in Figure 2.20) the electrons are magnetized, then, as was shown by Litvinenko (1996b), they follow magnetic field lines and gain energy:

$$\varepsilon = 2L_x \left| e E_{y0} \frac{B_{y0}}{B_{x0}} \right| . \tag{2.54}$$

However, this relation is only valid for transit magnetized electrons. Bounced electrons, on the other hand, experience a repelling force caused by E_y while traveling along the magnetic field line toward the midplane. When this force overcomes the attraction due to the $\boldsymbol{E} \times \boldsymbol{B}$ drift, bounced electrons are getting ejected. Since they cannot reach the midplane, "bounced" electrons gain less energy than "transit" ones.

The ejection energies of electrons are plotted in Figure 2.21 as functions of B_{x0} and B_{y0}, respectively. The lines correspond to Eq. (2.53) (dashed) and Eq. (2.54) (solid) obtained analytically by Litvinenko (1996b) in the limits of weak and strong B_y. The energies of transit electrons are in good agreement with the analytical estimations, that is, with Eq. (2.53) for weak B_y and B_x when electrons are unmagnetized, and with Eq. (2.54) for strong B_y and B_x when electrons are magnetized. On the other hand, energies of bounced electrons for any B_y and B_x coincide with those of unmagnetized ones given by Eq. (2.53). This means that B_y enhances the ejection energy of transit electrons only.

2.4.3
PIC Simulation Results

The results of a simulation carried out with an extremely low number density of particles $n = 100\,\text{cm}^{-3}$ are shown in Figure 2.22 for both protons (Figure 2.22a) and electrons (Figure 2.22b). This low-density simulation was performed mainly to verify that the PIC code can reproduce the results obtained in TP simulations (Section 2.2.3, Figure 2.9). Consistent with the TP results, the formation of separate beams of accelerated protons and electrons is clearly observed, and electrons and protons are ejected into opposite semispaces with respect to the $x = 0$ midplane. The protons are largely unmagnetized inside the current sheet, so that the ejection velocity of the bounced and transit protons are almost the same. On the other hand, the electrons are much more magnetized, so that bounced electrons cannot reach the midplane and thus gain much less energy than transit electrons. In this particular case, however, the polarization electric field $\tilde{E}_x$ caused by the charge separation is rather weak ($\sim 20\,\text{V}\,\text{m}^{-1}$).

Simulations for a higher density $n = 10^4\,\text{cm}^{-3}$ are shown in Figure 2.23, for protons (Figure 2.23a) and electrons (Figure 2.23b). Although the assumed density is several orders of magnitude lower than in the solar corona, already several differences between the TP and PIC approaches are revealed. The dynamics of the relatively unmagnetized protons are not substantially different: their trajectories are close to those obtained from the TP simulations (Figure 2.9), they are still ejected mainly into one semispace ($x < 0$) with respect to the midplane, and their acceleration rate coincides with the theoretical magnitude given in Section 2.1.3). In addition, however, a small number of protons are ejected into the

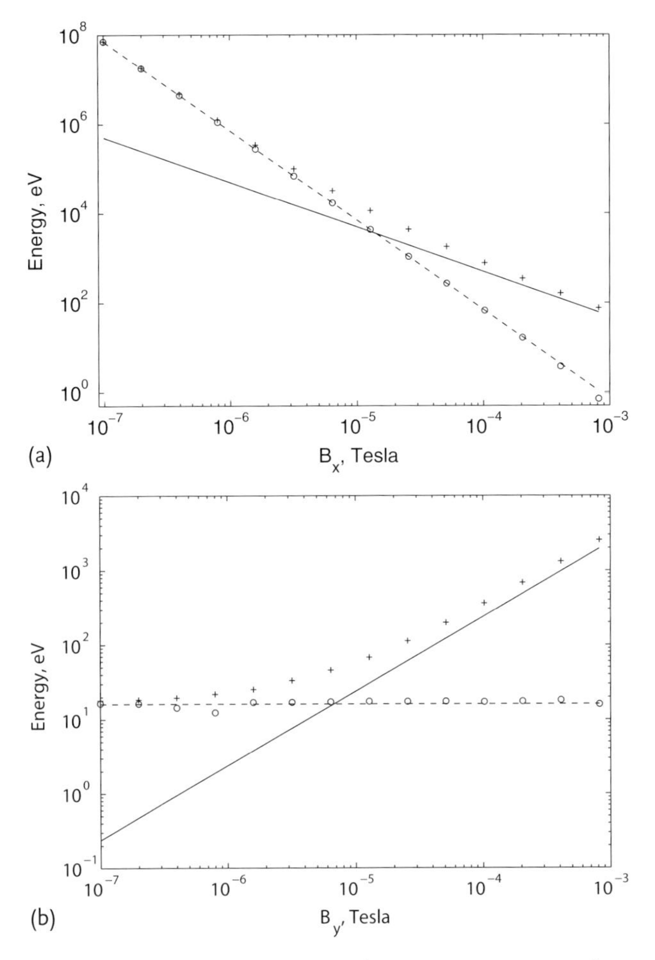

Figure 2.21 Electron energy in TP simulations upon ejection as a function of (a) B_{x0} for $B_{y0} = 10^{-4}$ T and (b) B_{y0} for $B_{x0} = 2 \times 10^{-4}$ T. Crosses: transit electrons, circles: bounced electrons, dashed line: Eq. (2.53), solid line: Eq. (2.54). $B_{z0} = 10^{-3}$ T, $E_{y0} = 250\,\text{V}\,\text{m}^{-1}$, $L_x = 1$ m.

$x > 0$ semispace, where electrons are normally ejected in TP simulations (Figure 2.23a). Figure 2.24 shows the fraction of the protons ejected into the $x > 0$ semispace as a function of B_{x0} and B_{y0}; the dependence on B_{y0} qualitatively coincides with that obtained by Zharkova and Gordovskyy (2004). For small B_{y0}, both PIC and TP simulations show that the protons are ejected symmetrically with respect to the midplane. However, in the PIC simulation all the protons are ejected into the one semispace ($x < 0$) when $B_{y0} > 7$ G ($\approx B_{z0}$), while in the TP simulation (Zharkova and Gordovskyy, 2004) accelerated particles are fully separated if $B_{y0} > 1.5 \times 10^{-2} B_{z0}$. Thus, in PIC simulations particle trajectories have less asymmetry than in the TP approach.

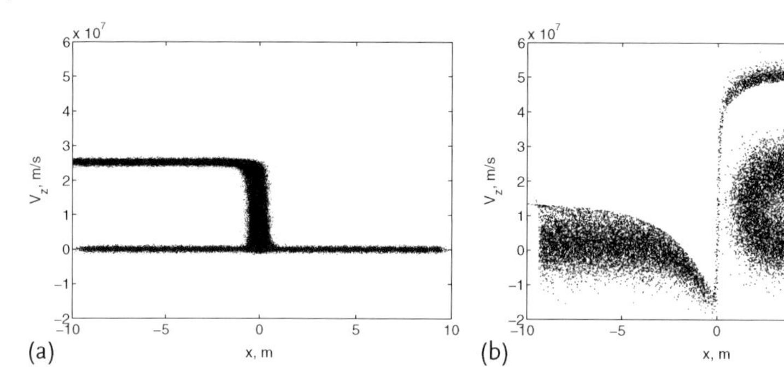

Figure 2.22 PIC simulation snapshots of low-density ($n = 10^2\ \text{cm}^{-3}$) plasma particles on the x–V_z phase plane for protons (a) and electrons (b). The protons and the electrons enter the RCS from both sides and are ejected into opposite semiplanes either as a beam (transit particles) or as a flow (bounced particles); see text for details. Current sheet parameters are the same as in Section 2.2.3. From Siversky and Zharkova (2009a).

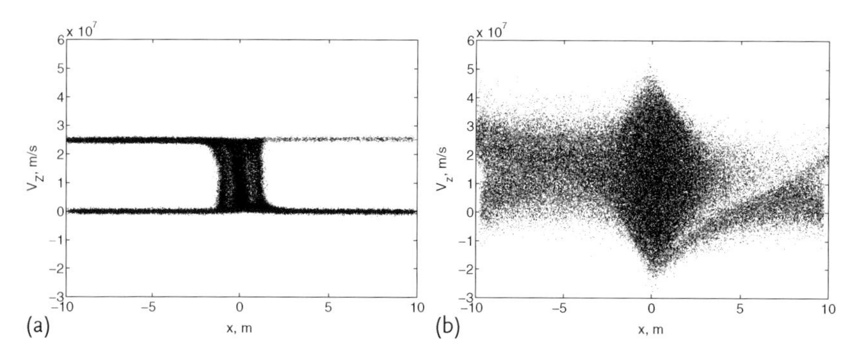

Figure 2.23 Snapshots of high-density plasma particles ($10^4\ \text{cm}^{-3}$) on the x–V_z phase plane (PIC simulations) for protons (a) and electrons (b) entering the RCS from both sides. Protons are still ejected into the same semiplane while the electrons form a cloud that experiences multiple ejections and returns to the acceleration region. As a result, the majority of electrons end up being ejected into the same semiplane as the protons, with only a small fraction of electrons being ejected into the opposite semiplane (see the text for details). Current sheet parameters are the same as in Section 2.2.3. From Siversky and Zharkova (2009a).

2.4.3.1 **Polarization Electric Field Induced by Accelerated Particles**

The differences between the TP and PIC simulations could only occur because of the additional electric and magnetic fields induced locally by the accelerated particles. The PIC simulations have shown that the induced magnetic field $\tilde{\boldsymbol{B}}$ is much smaller than the background B. On the other hand, the induced electric field $\tilde{\boldsymbol{E}}$ is essential. Its absolute value is even larger than the background (drifted) field E_y induced by a magnetic reconnection (see Eq. (2.46)).

Since our 2-D system is invariant in the y-dimension, there is no charge separation in this direction. Thus the induced $\tilde{E}_y$ can be caused only by the time variation of the magnetic field and is found to be small. In this subsection we consider

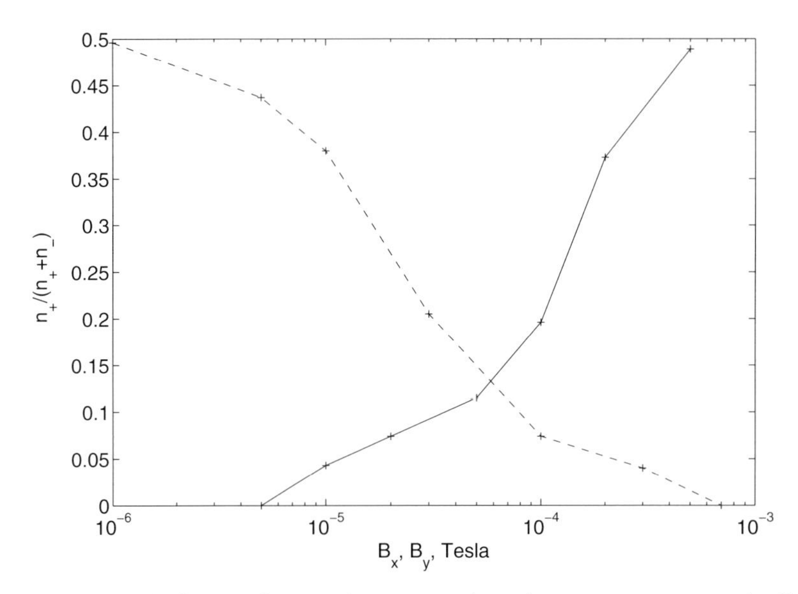

Figure 2.24 The fraction of protons that are ejected into the $x > 0$ semispace. Dashed line: fraction vs. B_{y0} ($B_{x0} = 0.2\,\mathrm{G}$). Solid line: fraction vs. B_{x0} ($B_{y0} = 1\,\mathrm{G}$). All other parameters of the current sheet are the same as in Figure 2.20.

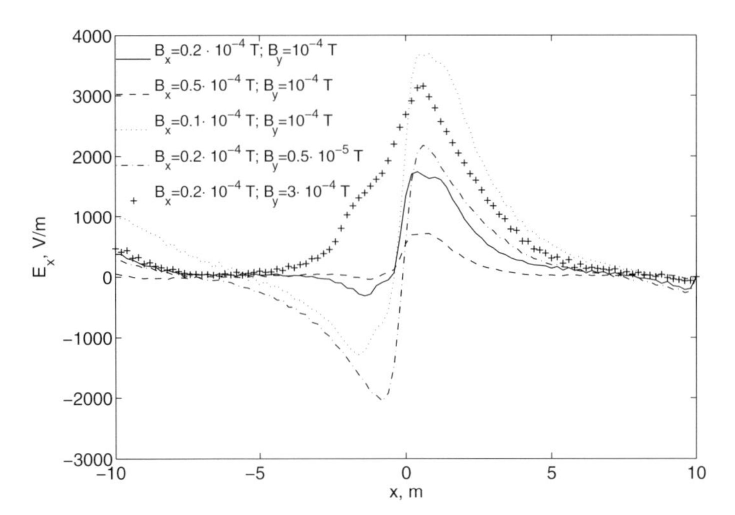

Figure 2.25 Electric field $\tilde{E}_x$ induced by particles in PIC simulations for different values of B_{x0} and B_{y0}. All other parameters of the current sheet are the same as in Figure 2.20.

the electric field $\tilde{E}_x$, which is perpendicular to the current sheet. Figure 2.25 is a plot of $\tilde{E}_x$ as a function of x averaged over the z-coordinate for different values of B_{x0} and B_{y0}. This field appears due to separation of the electrons and protons across the current sheet, which leads to a local nonneutrality of the plasma. The field becomes stronger when B_{x0} decreases or B_{y0} increases. Also, when $B_{y0} \to 0$,

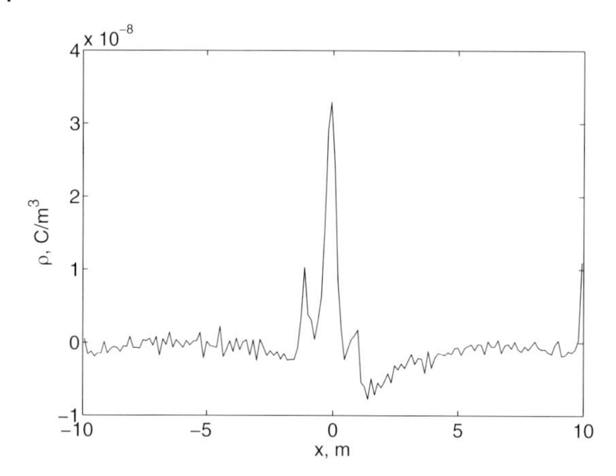

Figure 2.26 Charge density ρ as a function of x deduced from PIC simulation. Current sheet parameters are the same as in Figure 2.20.

the electric field $\tilde{E}_x$ and the system itself become symmetric. The distribution of a charge density $\rho(x)$, which generates the polarization field $\tilde{E}_x(x)$, is shown in Figure 2.26 (Siversky and Zharkova, 2009a). Note that the average charge density over the simulation region is close to zero, which means that the current sheet as a whole remains electrically neutral.

2.4.3.2 Particle Trajectories

In order to reconstruct the particle trajectories, we use the TP code, where the induced electric field $\tilde{E}_x$ obtained from PIC is added to the background electromagnetic configuration given by Eqs. (2.43)–(2.46). The trajectories of the two protons on the x–V_z phase plane are shown in Figure 2.27a. These trajectories are very similar to those obtained in the TP simulations without $\tilde{E}_x$ (Figure 2.20). The only difference is that during the acceleration phase the bounced proton has a wider orbit. It is clear now that this wide orbit is responsible for the two smaller peaks at about ± 1 m in the charge density plot (Figure 2.26). On the other hand, the narrow orbit of the transit proton forms the central peak in this plot.

The trajectories of electrons are much more complicated. First, let us consider an electron that enters from the $x < 0$ semispace (Figure 2.27b). The dynamics of this electron is similar to the dynamics of the transit electron in Figure 2.20 – the electron drifts toward the midplane, becomes accelerated, and is ejected into the $x > 0$ semispace. However, the polarization field $\tilde{E}_x(x)$, which extends beyond the current sheet and has a component parallel to the magnetic field, decelerates the ejected electron. For the chosen magnitudes of B_x and B_y the majority of electrons cannot escape to the $x > 0$ semispace; instead, they are dragged back toward the current sheet and become indistinguishable from the electrons entering from the $x > 0$ semispace.

Electrons that come from the positive x side demonstrate a rather different dynamic in comparison with the case where $\tilde{E}_x = 0$. It turns out that the electron,

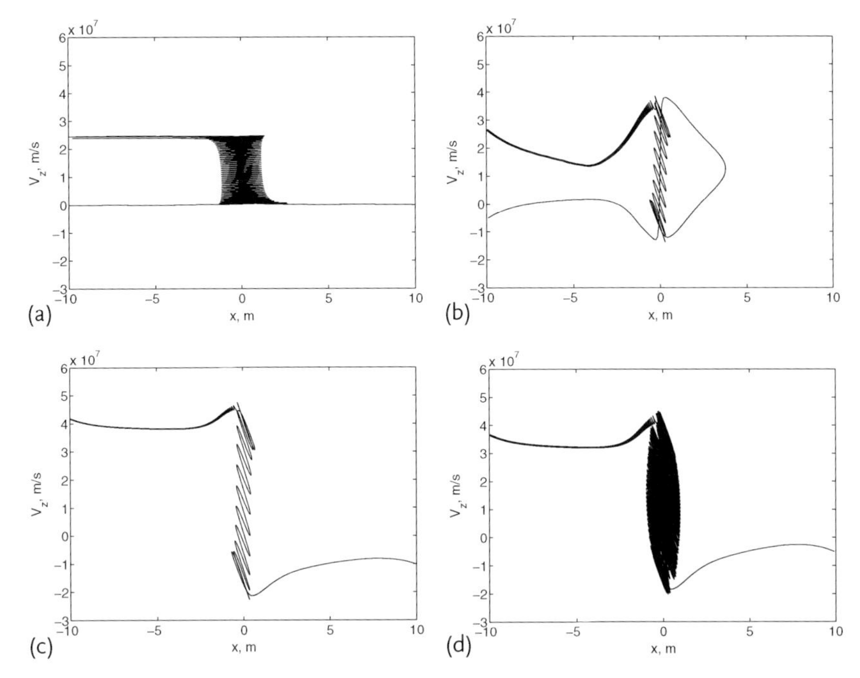

Figure 2.27 Trajectories of particles on V_z–x phase plane during acceleration inside a current sheet: (a) two protons injected at $x = \pm 10$ m with zero initial velocities; (b) electron injected at $x = -10$ m with initial velocity $V_z = -5 \times 10^6$ m s^{-1}; (c) electron injected at $x = 10$ m with initial velocity $V_z = -1 \times 10^7$ m s^{-1}; (d) electron injected at $x = -10$ m with initial velocity $V_z = -5 \times 10^6$ m s^{-1}. The trajectories were obtained in the TP simulations, with the electric field $\tilde{E}_x(x)$ taken from the PIC simulation (Figure 2.25, solid line). Current sheet parameters are the same as in Figure 2.20.

which is "bounced" from the RCS in the absence of $\tilde{E}_x$, can now reach its midplane. In the vicinity of the midplane, the electron becomes unmagnetized and oscillates with a gyrofrequency determined by B_y (Figure 2.27c) and, after some time of oscillation, the electron is ejected. If the electron initial velocity is small, it can be quasitrapped inside the RCS (Figure 2.27d). Such an electron is accelerated on the midplane, ejected from it, then decelerated outside the RCS, and returns back to the midplane. This cycle is repeated several times until it finally gains enough energy to escape the RCS. Since the magnitude of the polarization field $\tilde{E}_x(x)$ is smaller at $x < 0$ than at $x > 0$ (Figure 2.25), it is easier for the electron to escape to the $x < 0$ semispace. Thus, most of the electrons are ejected into the same semispace as protons, contrary to the results of the TP simulations. The asymmetry condition is shifted to much higher magnitudes of the guiding field than those found in the TP approach ($> 0.01 B_{z0}$, Zharkova and Gordovskyy, 2004).

2.4.3.3 **Energy Spectra**

The energy distributions of accelerated electrons for different magnitudes of transverse, B_{x0}, and guiding, B_{y0}, magnetic components are calculated for those elec-

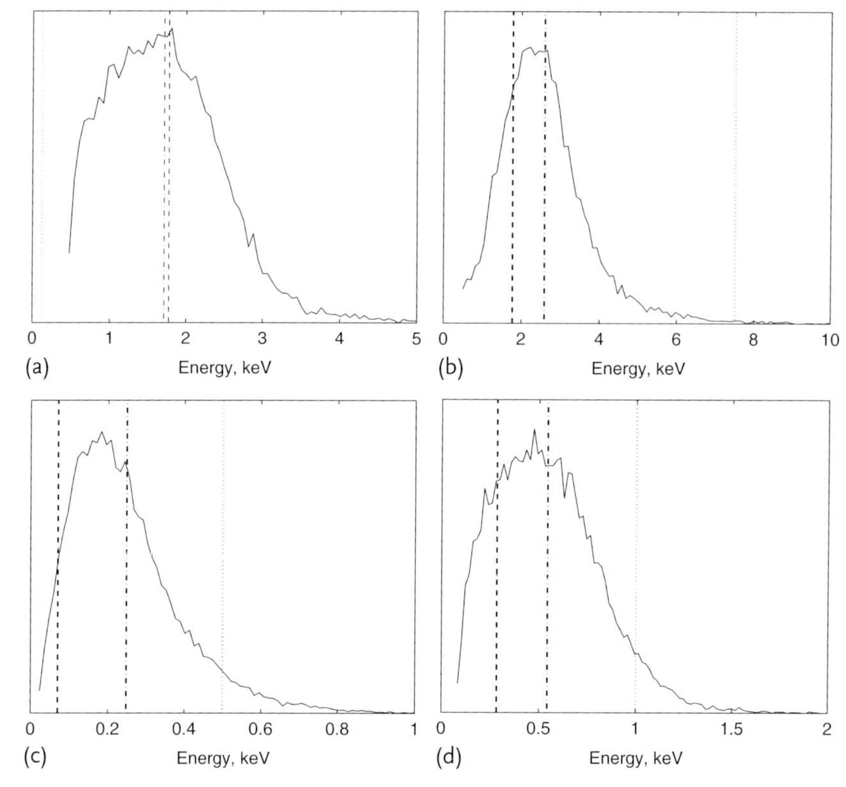

Figure 2.28 Energy spectra of ejected electrons. (a) $B_{x0} = 2 \times 10^{-5}$ T, $B_{y0} = 5 \times 10^{-6}$ T; (b) $B_{x0} = 2 \times 10^{-5}$ T, $B_{y0} = 3 \times 10^{-4}$ T; (c) $B_{x0} = 10^{-4}$ T, $B_{y0} = 10^{-4}$ T; (d) $B_{x0} = 5 \times 10^{-5}$ T, $B_{y0} = 10^{-4}$ T. Dashed line: Eq. (2.53), dotted line: Eq. (2.54), dashed–dotted line: mean energy of ejected electrons. The current sheet parameters are the same as in Figure 2.20.

trons outside an RCS in the negative semispace ($-10\,\mathrm{m} < x < -5\,\mathrm{m}$) and plotted in Figure 2.28. Only escaping electrons are taken into account, that is, the electrons for which the velocity component along the magnetic field projected on the x-axis is larger than the $\boldsymbol{E} \times \boldsymbol{B}$ drift velocity projected on the x-axis. For the accepted configuration and in the limit of a small B_x, this corresponds to electrons with $V_z > E_{y0}/B_{x0}$.

It can be seen from Figure 2.28 that for any B_{x0} and B_{y0} the ejected electrons form a wide single-peak energy distribution with a width on the order of the mean energy. This is contrary to TP simulations and lower-density PIC simulations, where the two narrow energy electron beams are formed (see Figure 2.20 for the TP simulation and Figure 2.22 for the lower-density PIC simulation). The presence of the polarization electric field shifts the high-energy peak toward the lower energy and makes it wider.

If the guiding field, B_y, is negligibly small (Figure 2.28a), then the mean energy of the ejected electrons coincides with the analytical value given by Eq. (2.53).

For the stronger guiding field (Figure 2.28b) the mean energy is somewhat higher than the energy of the slow beam in the TP simulation given by Eq. (2.53) and substantially lower than the energy of the fast beam given by Eq. (2.54).

The energy spectra dependence on B_x is shown in Figure 2.28c,d. They indicate that the mean energy strongly depends on B_x as is predicted by Eqs. (2.53) and (2.54). The lower B_x, the higher their mean energy and the wider their energy distribution.

2.4.3.4 Turbulent Electric Field

Another interesting effect observed in PIC simulations for the selected physical conditions and proton-to-ion mass ratio is the excitation of plasma waves.

As can be seen from Figure 2.29, the $\tilde{E}_z$ component of the induced electric field is structured with a characteristic length scale of about $\lambda_{\text{wave}} \approx 2\,\text{m}$. This structure propagates in time in the positive z-direction at a speed of $V_{\text{wave}} \approx 1.3 \times 10^7\,\text{m}\,\text{s}^{-1}$. This corresponds to an oscillation period of $T_{\text{wave}} \approx 1.5 \times 10^{-7}\,\text{s}$, consistent with the plasma frequency for a density of $n = 10^6\,\text{cm}^{-3}$. Note also that the oscillating component of the excited wave $\tilde{E}_z$ is parallel to the direction of propagation, which corresponds to the polarization of the Langmuir wave. The instability mechanism is evident from an examination of the electron velocity V_z. As shown in Figure 2.30, the electrons have an unstable "bump-on-tail" distribution. The range of velocities V_z for which the derivative $\partial f/\partial V_z$ is positive vary from 1.3×10^9 to $2 \times 10^9\,\text{cm}\,\text{s}^{-1}$, corresponding to the phase velocity of the associated Langmuir wave V_{wave}.

Although it must be noted that the plasma instabilities in the reconnection region have been used by many other authors to produce anomalously high values of resistivity. For different physical models, the 2-D and 2.5-D PIC simulations

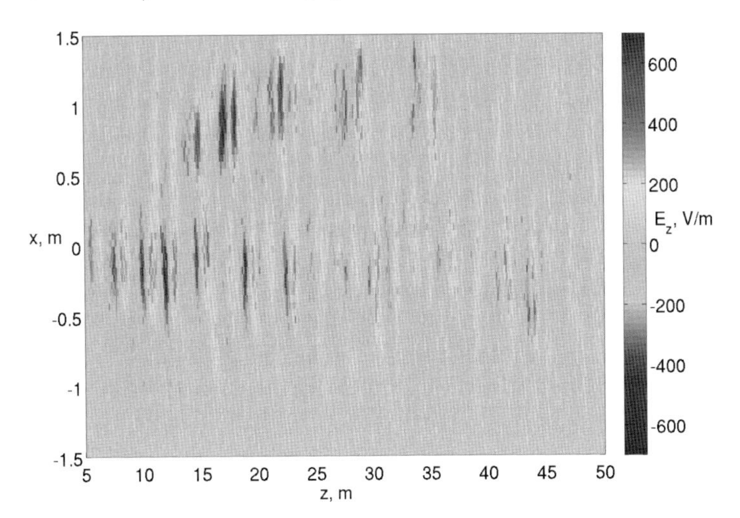

Figure 2.29 Electric field $\tilde{E}_z$ induced by particles in PIC simulation ($B_{z0} = 10^{-3}\,\text{T}$, $B_{y0} = 10^{-4}\,\text{T}$, $B_{x0} = 4 \times 10^{-5}\,\text{T}$, $E_{y0} = 250\,\text{V}\,\text{m}^{-1}$, $m_p/m_e = 10$, $n = 10^6\,\text{cm}^{-3}$). For a color version of this figure, please see the color plates at the beginning of the book.

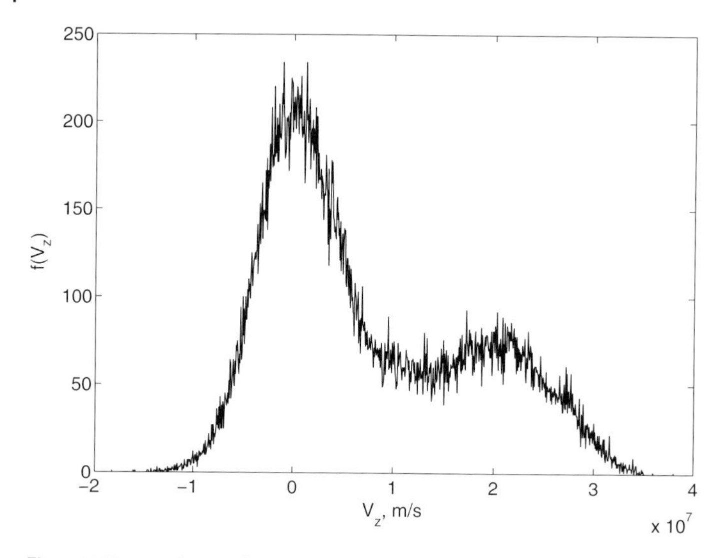

Figure 2.30 Distribution function of electrons calculated for the region with turbulence plotted in Figure 2.29.

showed various kinds of instabilities from the lower hybrid drift instability (LH-DI) leading to the Kelvin–Helmholtz instability (Chen *et al.*, 1997; Landau and Lifshitz, 1960; Lapenta *et al.*, 2003) to kink (Daughton, 1999; Pritchett *et al.*, 1996; Zhu and Winglee, 1996) and sausage (Büchner, 1996) instabilities, or magnetoacoustic waves (Schekochihin and Cowley, 2007) for ions. However, 3-D PIC simulations with a reduced speed of light, $c = 15 \times V_{\text{Alfvén}}$, and proton-to-ion ratio of 277.8 Zeiler *et al.* (2002) managed to get rid of the kink or sausage instabilities above while retaining a lower hybrid drift instability similar to those reported above. This indicates that a change of physical conditions in PIC simulations and mass ratio in a current sheet can cause different types of instabilities to appear if the conditions match exactly those in solar flares.

2.5 Particle Acceleration in Collapsing Magnetic Islands

2.5.1 Tearing-Mode Instability in Current Sheets

As mentioned in Section 2.1.2, in RCSs diffusion converts magnetic energy into ohmic heat, with a concomitant slow expansion of the current sheet at a rate proportional to the resistivity. Consequently, one way of enhancing the reconnection rate is to consider enhancement of the resistivity through resistive instabilities such as the tearing instability (Furth *et al.*, 1963).

In a current sheet wide enough to meet the condition $\tau_{\rm dif} \gg \tau_{\rm A}$, where $\tau_{\rm A} = d/V_{\rm A}$ is the Alfvén transverse time, the tearing instability occurs on a time scale of $\tau_{\rm dif}(\tau_{\rm A}/\tau_{\rm dif})^{\lambda}$, with the parameter λ in the range $0 < \lambda < 1$. The tearing instability occurs on a wavelength greater than the width of the current sheet, $(ka < 1)$, with a growth rate of (Priest and Forbes, 2000)

$$\omega = \left[\tau_{\rm dif}^{3}\tau_{\rm A}^{2}(ka)^{2}\right]^{-1/5} , \tag{2.55}$$

where the wavenumber k satisfies the condition $(\tau_{\rm dif}/\tau_{\rm A})^{1/4} < ka < 1$. The fastest growth is achieved for the longest wavelength (i.e., at $ka = 1$), while the smallest wavelength of the instability grows more slowly, on a time scale of $\tau_{\rm dif}^{3/5}\tau_{\rm A}^{2/5}$. As first demonstrated by Furth *et al.* (1963) (see also Spicer, 1977), tearing-mode instabilities generated during reconnection in a current sheet can produce closed magnetic "islands" with an elliptical O-type topology. Accordingly, large current sheets often observed in solar flares (see, e.g., Sui and Holman, 2003, and references therein) can, depending on their sizes and conditions, be subject to the formation of magnetic islands. For example, for a current sheet of size $\sim$ 10 000 km (as reported by Sui and Holman, 2003), embedded in a (coronal) medium with magnetic field $B \sim 100$ G and density $n = 10^{10}\,{\rm cm}^{-3}$, the Alfvén velocity is about 2000 km s^{-1} and $\tau_{\rm A} \simeq 5$ s. With a rather conservative collision frequency estimate $10^4\,{\rm s}^{-1}$, $\tau_{\rm dif} \simeq 10^{15}$ s. Substituting these values into Eq. (2.55) gives an island formation time in the range > 1000 s.

2.5.2 Particle Acceleration in Magnetic Islands – PIC Approach

In the past decade the tearing-mode instability has been extensively investigated, both in 2.5-D current sheets (Drake *et al.*, 1997, 2005, 2006a,b; Shay and Drake, 1998) and in a full 3-D kinetic PIC approach (Daughton *et al.*, 2009; Drake *et al.*, 2010; Karimabadi *et al.*, 2007; Pritchett and Coroniti, 2004; Shay *et al.*, 2007; Zeiler *et al.*, 2002).

Shay and Drake (1998) studied the dynamics of whistler waves generated inside magnetic islands using a code in which the electric and magnetic fields were calculated at each time step using the Poisson condition $\nabla \cdot E = 4\pi\rho$. The time step was normalized to the electron cyclotron time $\Omega_{\rm e}^{-1}$, calculated for a maximum initial magnetic field B_0, the spatial step normalized to the skin depth $c/\omega_{\rm pe}$, and the velocity normalized to the Alfvén velocity for electrons $V_{\rm Ae} = B_0/\sqrt{4\pi n_0 m_{\rm e}}$. The speed of light was artificially reduced to $5\,V_{\rm Ae}$ in order to marginally resolve Debye radii. The simulation grid was 512×512 with 7 million particles involved. As in previous simulation the authors added small magnetic perturbations and associated currents as the initial conditions in order to force a formation of seed magnetic islands centered around the two current layers. Using such a formalism, Shay and Drake (1998) were able to resolve the electron dissipation region, despite the use of an artificially low ion-to-electron mass ratio of 200. A further modification of the PIC code was used to extend it to a 3-D model with 64 grid cells in the third direction while increasing the proton-to-electron mass ratio to 277.8 and the speed of light to $c = 15 V_{Ae}$ (Zeiler *et al.*, 2002). Small magnetic perturbations and associated currents were added as the initial conditions to form seed magnetic islands centered around the two current layers, as shown in Figure 2.31. Unlike a conventional Harris equilibrium in which the variation of the density produces the equilibrium current, the density is taken to be nearly constant, the current being produced by the $\boldsymbol{E} \times \boldsymbol{B}$ drift of electrons, resulting in the occurrence of the "parallel" electric field in the plane of the island current sheet (Drake *et al.*, 1997; Shay and Drake, 1998; Zeiler *et al.*, 2002) (Figure 2.32).

Pritchett and Coroniti (2004) have further shown that in 3-D current sheets with a guiding magnetic field, the "parallel" electric field is formed only in the two quadrants $(+X, +Y)$ and $(-X, -Y)$. The other electrons entering this current sheet are assumed to be accelerated by this "parallel" electric field. The other quadrants are characterized by an excess of ions, strongly reminiscent of the drift separation of electrons and protons in a 3-D single current sheet (Sections 2.2.3 and 2.4).

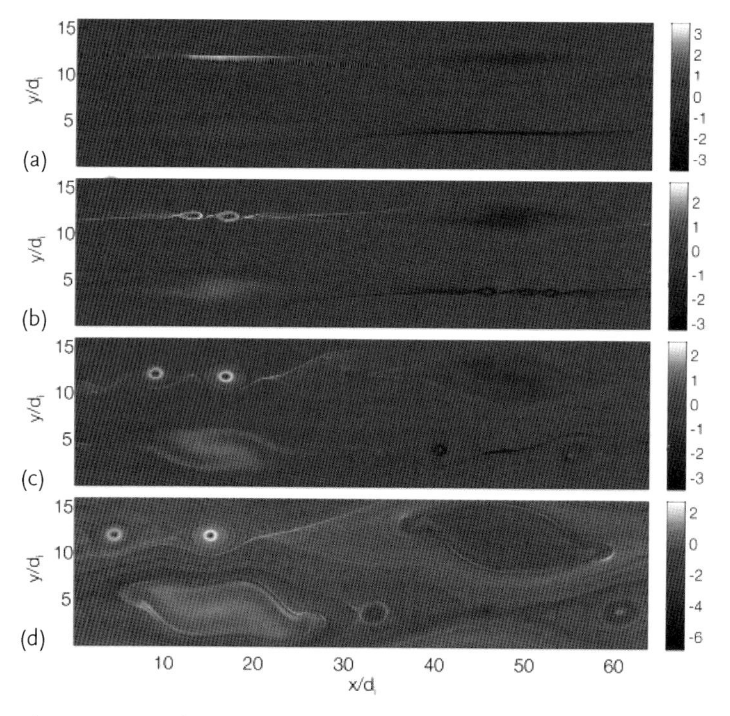

Figure 2.31 Out-of-plane current density in *2.5-D PIC* simulations of reconnection with a guiding field, for four consecutive time intervals (a–d). The environment contains two current sheets and is characterized by the appearance of islands framed by tearing instability. At the start there are only (a) large islands followed at later times by (b–d) the formation of smaller magnetic islands. Note that such islands do not appear in 2-D PIC simulations of the same environment but without a guiding field. From Drake *et al.* (2006b). For a color version of this figure, please see the color plates at the beginning of the book.

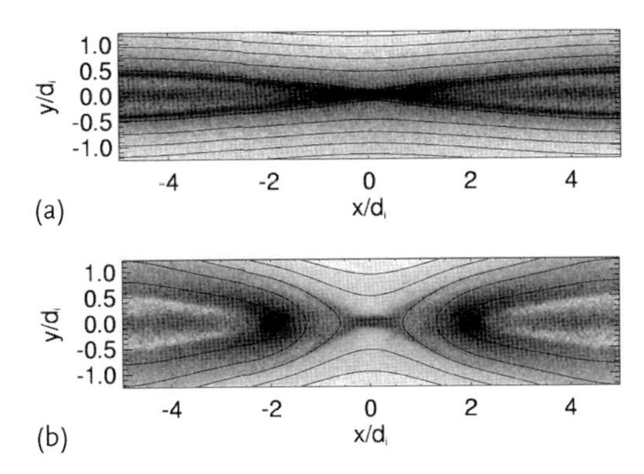

Figure 2.32 Current density in PIC simulations for times (a) $t = 7.7/\omega_{ci}$ and (b) $t = 12.9/\omega_{ci}$. The solid lines correspond to magnetic flux lines. Adopted from Zeiler *et al.* (2002).

Once formed, magnetic islands tend to shrink due to magnetic tension forces. Conservation of the longitudinal invariant leads to efficient first-order Fermi acceleration of particles tied to the islands' field lines. As shown by Drake *et al.* (2006a), the process of island shrinking halts when the kinetic energy of the accelerated particles is sufficient to halt further collapse due to magnetic tension forces; the electron energy gain is naturally a large fraction of the released magnetic energy. A sufficient number of electrons can be produced to account for a moderately sized solar flare. The resultant energy spectra of electrons take the form of power laws, with harder spectra corresponding to lower values of the plasma β. For low-β conditions appropriate to the solar corona, the model predicts accelerated spectra that are much harder ($\delta \sim 1.5$) than observed.

Further progress with 3-D PIC simulations of magnetic islands, using a larger 2560×2560 grid, revealed that the electron diffusion region formed by electron flows occurs over a width of $5d_e$ (where d_e is the electron inertial length – Birn *et al.*, 2001), while other simulations show an extent up to $20d_i$ (where d_i is the ion inertial length – Karimabadi *et al.*, 2007; Shay *et al.*, 2007, see Figure 2.33). The

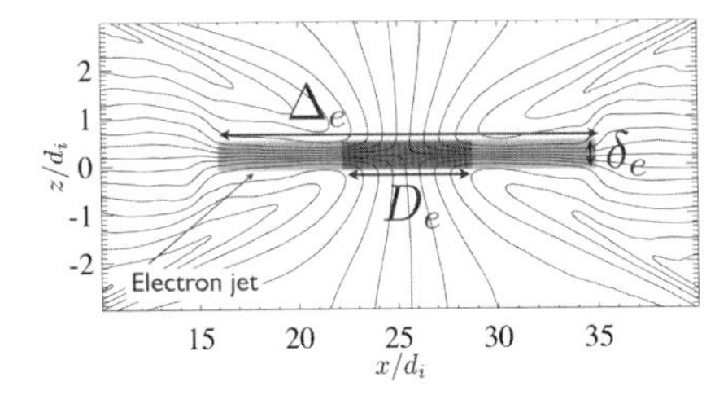

Figure 2.33 Multiscale structure of electron diffusion region around an X-type null point of total length Δ_e and thickness δ_e, as derived from 3-D PIC simulations. The solid lines show the electron trajectories for the time $t\omega_{ci} = 80$ (ω_{ci} is the ion cyclotron frequency). In the inner region of length D_e, there is a steady inflow of electrons and a strong out-of-plane current; the outer region is characterized by electron outflow jets. From Karimabadi *et al.* (2007). For a color version of this figure, please see the color plates at the beginning of the book.

electron diffusion region becomes elongated with a two-scale structure: an inner region (size D_e in Figure 2.33) with a strong electron-drift-associated electric field (Eq. (2.13)) and a wider diffusion region characterized by electron jets (Karimabadi *et al.*, 2007; Shay *et al.*, 2007). Shay *et al.* (2007) found that electrons form super-Alfvénic outflow jets, which become decoupled from the magnetic field and extend to a large distance from the X-type null point, similar to the manner in which electron jets decouple from protons in a TP approach to reconnection in a single 3-D current sheet (Section 2.2.3).

Recently, Daughton *et al.* (2009) showed that for collisional Sweet–Parker reconnection with Lundquist numbers R_m (Eq. (2.5)) higher than 1000, the reconnection region becomes susceptible to the tearing instability, leading to the formation of many magnetic islands with half thickness approaching the ion inertial length d_i (Figure 2.34). This result shows that the current sheet size is a key factor in defining time scales and the role of different physical processes at different stages of the reconnection process. For example, the shrinking of a large current sheet during MHD reconnection is halted when it approaches a size conducive to the formation of magnetic islands.

2.6
Limitations of the PIC Approach

Many PIC simulations (Birn *et al.*, 2001; Drake *et al.*, 2006b; Karimabadi *et al.*, 2007; Shay *et al.*, 2007; Tsiklauri and Haruki, 2007) have been concerned with the study of reconnection rates in the vicinity of a magnetic X-type null point. Even for the very low densities used in these simulations, it is apparent that the feedback of the fields induced by accelerated particles can be substantial (Section 2.3): it can significantly modify the particle trajectories and the energies gained during their acceleration.

However, because of computer power limitations, the size of the region in which the induced electric and magnetic fields are simulated and coupled to the particle motion is necessarily rather small: $\sim 100 \times 50 \times 1000$ cm for 2-D simulations (Birn *et al.*, 2001; Tsiklauri and Haruki, 2007), or $4000 \times 40\,000 \times 86$ cm for the most recent 3-D simulations (Drake *et al.*, 2010; Shay *et al.*, 2007). These are orders of magnitude smaller than the 10^8–10^{10} cm sizes associated with flares (Jiang *et al.*, 2006; Masuda *et al.*, 1994; Sui and Holman, 2003). Only in the PIC simulations discussed in Section 2.4 is the computation region extended well beyond the null point in order to explore the variations of acceleration processes at different magnitudes of the transverse magnetic field. Thus, the *narrowness of the simulation region* still imposes strong limitations on the number of particles being accelerated in PIC simulations.

Other puzzling features of the simulations relate to the origin of the electric fields in the models. One example is the "parallel" electric field reported in many 3-D simulations (Drake *et al.*, 2010; Pritchett and Coroniti, 2004; Zeiler *et al.*, 2002) and produced either as a result of an advective ("drifted") electric field (see the first

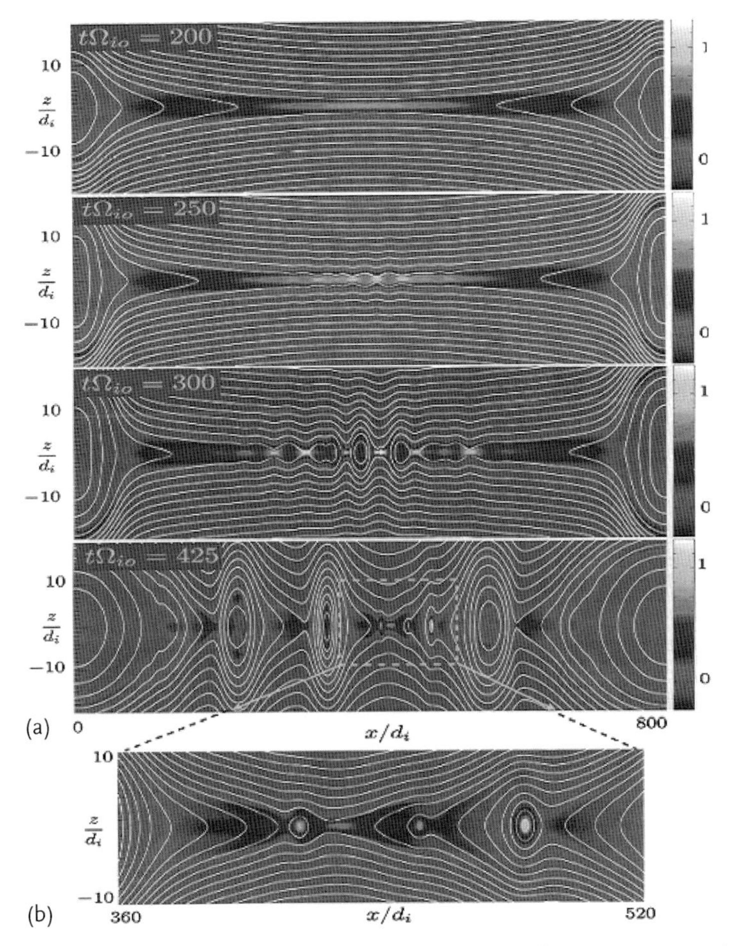

Figure 2.34 Evolution of current density over a region of extent $800d_i$ (where d_i is the ion inertial length) for different times t. The white lines are the magnetic flux surfaces. A close-up of a region in which the formation of new islands occurs is presented in (b). From Daughton *et al.* (2009). For a color version of this figure, please see the color plates at the beginning of the book.

term in formula (2.13)) or by a gradient of the magnetic field component across the current sheet (see the second term in formula (2.13)). This parallel electric field can exceed the local Dreicer field by a factor of up to $\sim$ 30 for very intense beams (Zharkova and Gordovskyy, 2006). Another example is the "return current" electric field, which can be formed from those accelerated electrons that are scattered to pitch angles larger than 90°, thus returning back to the acceleration region. Although both the parallel electric field and the return current electric field associated with the propagation of accelerated electrons can cause various interesting effects (e.g., decelerating electrons and even returning them back to the acceleration region), causality arguments would strongly suggest that neither field can be considered as the main driver of particle acceleration since the parallel electric field is *formed by the accelerated particles themselves.*

A related problem is the difference in electron and ion gyration radii/periods (Birn *et al.*, 2001). Because of this difference, within a time scale of $\lesssim 10^{-7}$ s electrons can move toward the midplane and be accelerated and ejected, while the protons have only just started their gyrations in the RCS environment. In simulations to date, the proton-to-electron mass ratio has been arbitrarily lowered to 100–200, an order of magnitude smaller than the actual value. Even the speed of light is conveniently reduced by some one or two orders of magnitude to allow a reduction in simulation times (Birn *et al.*, 2001; Drake *et al.*, 2006a). Such simulations, then, serve to highlight that by artificially adjusting fundamental physical constants (such as the electron mass or the speed of light) in order to achieve results with a reasonable amount of computational effort, the results themselves can be changed significantly. Such issues can be resolved only by increasing the reconnection region size in the PIC simulations up to sizes comparable with observations and by setting physical constants, such as the proton-to-electron mass ratio, to their actual values. However, even with exponential increases in computing power, there are substantial doubts that one can achieve this goal within the next few decades. Possibly a hybrid approach, combining PIC simulations convoluted with TP extensions to greater source volumes, will allow us to make substantial progress in this area.

2.7
Probing Theories versus Observations

2.7.1
Interrelation between Acceleration and Transport

Before we address the relative merits of the various particle acceleration models, it is important to realize that many flare observations consist of radiation integrated over the source volume; the quantity observed involves integration over not only space but also quantities such as particle pitch angles. Therefore, in order to relate to observed features (Chapter 1), one must incorporate the effects of particle acceleration with the *transport* of the accelerated particles throughout the remainder of the flaring source. Although such an inclusion will in all probability obscure information on the acceleration process, it can also, more profitably, be used as a diagnostic of particle transport processes themselves.

As we have seen (Sections 2.2.3 and 2.2.6), accelerated electrons can attain quasirelativistic energies within an extremely small time scale and can thus propagate the extent of a solar flare of size $\sim 10^9$ cm within $\sim 0.01-0.1$ s. The proton acceleration time is longer than that for electrons by up to two orders of magnitude (Section 2.2.3), so that protons still remain within the acceleration region while the electrons are ejected from it. Thus, particle transport mechanisms are essential for completing the electric circuit for particles escaping the acceleration region (Emslie and Hénoux, 1995) or in establishing cospatial return currents (Emslie and Hénoux, 1995; Knight and Sturrock, 1977; Zharkova and Gordovskyy, 2006). Such return currents can be formed both by ambient electrons (Knight and Sturrock, 1977; Nocera *et al.*, 1985) and by relatively high-energy-beam electrons (Siversky and Zharkova, 2009a; Zharkova and Gordovskyy, 2006; Zharkova *et al.*, 2010) that return to the corona with nearly the same velocities as the injected electrons, which allow one to recycle precipitating electrons 10–100 times and account for the number of HXR photons with a much smaller (10–100) number of emitting electrons. This approach *resolves the particle number problem for high-energy electrons to emit HXR emission in flares,* which was unsolved for decades.

Also, the self-induced electric fields created by beam electrons or the return current have a structure that is very favorable for the generation of plasma turbulence (Karlický and Bárta, 2008), which naturally creates the conditions required for stochastic acceleration of electrons to subrelativistic energies discussed in Section 2.1.4 (see also Miller *et al.*, 1996; Pryadko and Petrosian, 1997). Although the self-induced electric field of precipitating electrons is found to suppress Langmuir turbulence at the chromospheric level and to reduce it at the coronal level (Zharkova and Siversky, 2011), that needs to be taken into account when considering stochastic acceleration.

Another issue that can arise is the consideration of *radiative* transport effects, such as radiative transfer of optically thick lines and continua, and reflection (or albedo effects) emission of HXRs from the photosphere.

2.7.2 Testing Acceleration Models against Observational Constraints

Here we briefly summarize the extent to which the various models discussed in Sections 2.2–2.5 have increased our ability to account for the number and spectrum of accelerated particles in a solar flare. In other words, can *any* hitherto-identified acceleration process reproduce the rate, number, and energy spectrum of accelerated particles inferred from observations?

2.7.2.1 Stochastic Acceleration

Stochastic acceleration remains one of the most promising candidates to account for most of the characteristics deduced from photon spectra in HXR, radio, and γ-ray emission.

The basic transit-time-damping stochastic acceleration model was improved by adding low-frequency fluctuating electric fields parallel to the magnetic field (Grigis and Benz, 2004, 2005). Such fields may originate from low-frequency and high-amplitude turbulence, such as kinetic Alfvén waves as discussed in Section 2.7.1 (see also Miller *et al.*, 1996; Pryadko and Petrosian, 1997). They can accelerate and decelerate electrons (and protons), leading to a net diffusion in the energy space (Arzner and Vlahos, 2004), thereby linking this approach to the complex acceleration models (Zharkova *et al.*, 2011a).

Hence, the recent RHESSI observations have uncovered a few issues with stochastic acceleration that need to be addressed:

1. High-frequency waves near the plasma frequency can be excluded as drivers for stochastic acceleration. They would couple into decimeter radio waves and so would be present in every flare, which is not the case (Benz *et al.*, 2005). On the other hand, acceleration by transit-time damping *is* compatible with a frequent lack of radio emission because the frequency of the postulated turbulence is far below the plasma frequency.
2. A low value of the low-energy rollover of electron energy distribution suggested by RHESSI observations (Kontar *et al.*, 2005; Saint-Hilaire and Benz, 2005) makes the number of flare electrons substantially larger than before. The number of particles (both electrons and protons) that can be accelerated in a current sheet is limited by the current, and the new values are difficult to match with a classic current sheet (Holman, 1985; Litvinenko, 1996a) (Section 2.1.3). However, more complex 3-D current sheets may be reconcilable with this (Sections 2.2.3 and 2.2.6, as discussed below).
3. Nonthermal electron distribution in coronal sources is observed to "grow" out of the thermal population: the coronal source is initially purely thermal in the $< 10\,\text{keV}$ range, then a soft nonthermal population develops, getting harder at the HXR peak and softening toward the end of the emission (Battaglia and Benz, 2007; Grigis and Benz, 2004). This soft–hard–soft behavior of HXR spectra in coronal sources is therefore considered to be inherent in the stochastic acceleration process and not in any transport effect (Grigis and Benz, 2004), although this approach to account for soft–hard–soft HXR spectra is very similar in nature to that caused by an electric field induced by electrons themselves if they move as a beam (Siversky and Zharkova, 2009a; Zharkova and Gordovskyy, 2006).

However, all models to date simply assume an *ad hoc* distribution of turbulence in order to account for the observed energy spectra of accelerated particles.

2.7.2.2 **Collapsing Current Sheets**

Estimates of a particle energy gained in the helmet-type reconnection as collapsing magnetic traps, with and without a termination shock, reveal that there is sufficient energy released in accelerated particles. For example, by approximating the termination shock by a circle and by adopting the parameters deduced for the flare of 28 October 2003, a flux of accelerated electrons $F_e = 2 \times 10^{36}\,\text{s}^{-1}$ above $20\,\text{keV}$ is produced; this value agrees rather well with other RHESSI results (Warmuth *et al.*, 2007). This suggests that the dissipation of magnetic energy in current sheets can indeed provide the required rate and number of accelerated electrons; no estimations have been produced for protons yet.

However, although simple RCSs (discussed in Section 2.1.3) can naturally reproduce power-law energy spectra, they are unable to account for the inferred number of accelerated particles because of the small size of the diffusion region involved. Also, this mechanism has not yet been shown to work for the acceleration of protons.

2.7.2.3 Acceleration in 3-D Reconnecting Current Sheets

Estimates of particle acceleration rates and spectra from MHD simulations of collapsing current sheets have revised our views on the efficiency of particle acceleration inside a diffusion region and energies that both electrons and protons can gain there. Further, recent advances in particle acceleration in a single 3-D RCS with a guiding field and in more complicated reconnection topologies (e.g., fan and spine reconnection) discussed in Sections 2.2.3 and 2.2.6 have established the following points:

1. Drift velocities of electrons and protons in a 3-D RCS have been found to be strongly dependent on the magnetic field geometry and on the locations where particles enter the RCS.
2. The particle orbits in an RCS lead to a separation of electrons and protons into opposite semiplanes, corresponding to ejection in opposite directions.
3. The acceleration of electrons in a 3-D RCS occurs mostly around the RCS midplane, over a very short time scale comparable with the growth rate of HXR emission in solar flares.
4. The acceleration of protons has a much longer time scale and results in a spread of protons in a location outside the midplane.
5. It has been found that, using both TP and PIC approaches, single 3-D current sheets with a guiding field can supply sufficient energy to both electrons and protons in order to account for RHESSI observations of both HXRs and γ-rays.
6. A single 3-D RCS can naturally account for power-law energy spectra for both electrons and protons and for the differences in spectral indices between protons and electrons.
7. 3-D acceleration in an RCS can account for the magnitude and range of lower-energy cutoffs inferred in accelerated electron and proton spectra in flares.
8. The separation of electrons and protons at acceleration in an RCS, and the associated formation of the polarization electric field across the current sheet, leads to the formation of "clouds of accelerated electrons" circling around the current sheet while protons are being accelerated. These *"electron clouds"* can be seen by RHESSI as coronal sources.
9. A polarization electric field also sets the frame for generating strong plasma turbulence and further particle acceleration by various stochastic processes.
10. Spine reconnection naturally produces two symmetric jets of energetic particles escaping along the spine. In contrast, fan reconnection produces azimuthally localized ribbons of high-energy particles in the fan plane. In principle, both features are often observed in flares (e.g., magnetic jets in the flares of 14 July 2000 and 23 July 2002) and two-ribbon flares as discussed in ?. This allows the

underlying mode of reconnection to be determined from the observations of high-energy emission.

The strongly stressed 3-D current sheets discussed in Sections 2.2.3 and 2.2.6 are characterized by much larger diffusion regions than in simple reconnection models (Section 2.1.2), approaching 10^9 cm $\times$ 10^9 cm, with a diffusion region thickness of 10^2–10^5 cm. The resulting volume involved in acceleration is $\sim 10^{20}$–10^{23} cm^3, which, for a coronal plasma of density 10^{10} cm^{-3}, gives particle numbers of 10^{30}–10^{33}. During reconnection electrons are accelerated and ejected within a very short time scale $\sim 10^{-2}$–10^{-3} s. With the replenishment of the coronal plasma provided by a return current, this system can exist (and keep accelerating particles) as long as the reconnection process continues. This allows a production rate of energetic particles of $\sim 10^{33}$–10^{36} particles s^{-1}, comparable to the values inferred from RHESSI observations. And stochastic acceleration or acceleration in collapsing current sheets with termination shocks can finalize the shape of energy distributions of accelerated particles gained in these secondary acceleration processes.

Hence calculations are still required to understand the role of the magnetic field topology of reconnecting loops on the parameters of accelerated particles in different scenarios of a magnetic reconnection process, the effects of plasma feedback on the presence of accelerated particles via polarization and turbulent electric fields formed inside the diffusion region, and the role of converging magnetic fields in the acceleration of electrons and protons in collapsing RCSs.

3
Electron-Beam Precipitation – Continuity Equation Approach

3.1
Introduction

As discussed in Chapter 2, electrons and protons can be effectively accelerated in reconnecting current sheets (RCSs) formed somewhere on the top of interacting magnetic loops producing a flare and injected as beams into the loop legs filled with the ambient plasma particles. The spatial configurations of flaring sites have been observed to be not always identical from one event to another, sometimes comprised of one loop with two footpoints and one coronal source (Masuda *et al.*, 1994; Sui *et al.*, 2002) or several sets of loops with a number of footpoints and coronal sources (Battaglia and Benz, 2006; Krucker *et al.*, 2008a). This results in a number of different models of particle acceleration and transport developed for different types of flaring events (see the reviews by Holman *et al.*, 2011; Zharkova *et al.*, 2011a, in the forthcoming RHESSI book).

Observations of solar flares in HXRs provide important information about these scenarios, in which accelerated electrons gain and deposit their energy into flaring atmospheres. The particle injection can be steady or impulsive as shown by numerous observations of HXR emission and polarization in solar flares from the balloon (Lin and Hudson, 1971), SMM (Bohlin *et al.*, 1980), BATSE, CORONAS (Zhitnik *et al.*, 1991), and RHESSI payload (Lin *et al.*, 2003).

This precipitation results in various emissions produced by the scattering of high-energy beams on ambient particles: HXR bremsstrahlung and radio emission often observed in solar flares by electrons with energies above 10–20 keV up to hundreds of MeV or γ-ray emission produced by protons and ions with energies above 1 MeV up to tens of GeV. In this energy range HXR emission is optically thin and can be an instantaneous sketch of high-energy particle dynamics in flaring atmospheres.

Substantial progress in the quantitative interpretation of HXR emission has been made in recent years using high temporal and spatial resolution observations carried out by the RHESSI payload (Lin *et al.*, 2003). The RHESSI observations provide the locations and shapes of HXR sources on the solar disk, their temporal variations, and energy spectra evolution during the flare (Holman *et al.*, 2003, 2011; Krucker *et al.*, 2008a). These observations are often accompanied by other observa-

Electron and Proton Kinetics and Dynamics in Flaring Atmospheres, First Edition. Valentina Zharkova.

tions (in microwaves (MW), EUV, and optical ranges) which revealed a very close temporal correlation to HXR emission (see, e.g., Fleishman and Kuznetsov, 2010; White *et al.*, 2011). This highlights a deep link between the electron acceleration and transport models, which must simultaneously account both temporally and spatially for all these types of emission.

The particle precipitation problem is rather complicated and difficult to model. This is why its progress depends on an understanding of the basic energy loss mechanisms and their interaction under different physical and geometrical conditions of flaring loops. The most effective scattering mechanism of beam particles are collisions with ambient plasma particles. This was the first factor considered in the interpretation of bremsstrahlung HXR emission by Brown (1971) for the electron flux conservation approach and by Syrovatskii and Shmeleva (1972a) for the continuity equation approach, which will be reviewed in this chapter.

In addition, energy losses by beams in an electrostatic electric field induced by precipitating electrons can be a significant factor that changes electron distributions and the appearance of HXR emission produced by them (Zharkova and Gordovskyy, 2006). In this chapter we present also the analytical solutions to pure ohmic losses found from a continuity equation similarly to the pure collisional solutions by Syrovatskii and Shmeleva (1972a). In this chapter let us discuss basic physics describing energy losses and pitch-angle changes of beam electrons during scattering on ambient electrons, ions, and neutrons, resulting in mean energy spectra and HXR emission for basic cross-sections produced by beams with given parameters (energy flux and spectral index).

3.2 Particle Energy Losses

3.2.1 Particle Trajectories at Scattering

During the interaction of two charged particles 1 and 2 with masses m and M and the same charges, ze is described by the motion equations d^2r/dt^2 with impulses F_{12} and F_{21} applied to particles 1 and 2, respectively. The motion equation for particle m regarding particle M can be written as follows (Ginzburg and Syrovatsky, 1961):

$$m_0 \frac{d^2\boldsymbol{r}}{dt^2} = \boldsymbol{F}_{12} , \tag{3.1}$$

where m_0 is the reduced mass of the system of interacting masses m and M, $m_0 = mM/(m+M)$, and $\boldsymbol{F}_{12}$ is the force acting on each particle (with masses m and M).

Let us define the impulse $\boldsymbol{P}$ carried by the system of these two particles as $\boldsymbol{P} = m_0\boldsymbol{V} = \boldsymbol{F}_{12}$. Now if one multiplies each member of the equation by vector $\boldsymbol{r}$ from the left-hand side, takes a time derivative from each side of the equation, and considers that the vector multiplication of the same vector $\dot{\boldsymbol{r}}$ is zero, this leads to the following equation:

$$\frac{d}{dt}(\boldsymbol{r} \times \boldsymbol{P}) = \boldsymbol{r} \times \boldsymbol{F}_{12} \,. \tag{3.2}$$

Let us now define an axillary vector defining the vector product as $\boldsymbol{L} = \boldsymbol{r} \times \boldsymbol{F}_{12}$, where the components of this vector can be defined as follows:

$$L_x = m_0(y\dot{z} - z\dot{y}) \,, \quad L_y = m_0(z\dot{x} - x\dot{z}) \,, \quad L_z = m_0(x\dot{y} - y\dot{x}) \,.$$

If the motion is central (in the orbit around a particle), then $\boldsymbol{L} = 0$ and both vectors $\boldsymbol{r}$ and $\boldsymbol{F}_{12}$ are located on the same line but directed in opposite directions. If the motion of both particles is on the same plane, say, the XY plane, then $\boldsymbol{L} =$ const, for example, $L_x = L_y = 0$ and $L_z = L =$ const, where

$$L = L_z = m_0(x\dot{y} - y\dot{x}) \,. \tag{3.3}$$

Let us now move from Cartesian to polar coordinates (r, ϕ), for example, $x = r\cos\phi$ and $y = r\sin\phi$. By substituting the full derivatives of $\dot{x}$ and $\dot{y}$ into the formulas for V and L_z one can easily obtain the following:

$$V^2 = \dot{x}^2 + \dot{y}^2 = \dot{r}^2 + r^2\dot{\phi}^2 \,; \quad x\dot{y} - y\dot{x} = r^2\dot{\phi} = \frac{L}{m_0} \,, \tag{3.4}$$

where V is the velocity of the center of mass. Then the time derivative $\dot{\phi}$ can be expressed as $\dot{\phi} = L/(m_0 r^2)$. Now we can evaluate the potential energy of the mass center as equal to $U(r) = Zze^2/r$ (Z being the charge of the particle in the target and z being the charge of the particle which is scattered– here the electron) and the kinetic center equal to $E_k = m_0 V^2/2$. Then the energy conservation equation of the system can be written as

$$\frac{m_0}{2}(\dot{r}^2 + r^2\dot{\phi}^2) + \frac{Zze^2}{r} = E \,, \tag{3.5}$$

where E is a constant energy. Let us now note that $\dot{r} = dr/d\phi d\phi/dt = l/m_0 r^2$ $dr/d\phi$ and substitute it into the energy conservation equation, obtaining

$$\frac{1}{r^4}\left(\frac{dr}{d\phi}\right)^2 = \frac{2m_0 E}{L^2} - \frac{2m_0 Zze^2}{L^2 r} - \frac{1}{r^2} \,. \tag{3.6}$$

Now we need to move to the polar coordinate system to reflect the fact that particle m, approaching the target at a distance r, interacts with the other particle M in such a way that its propagation direction changes at some angle ϕ. The energy conservation equation of the resulting mass center in the coordinate $\rho = 1/r$ is defined as

$$\frac{d^2\rho}{d^2\phi} = \frac{2m_0 E}{L^2} - \frac{2m_0 Zze^2}{L^2}\rho - \rho^2 \,, \tag{3.7}$$

where $L = m_0 r^2 d\phi/dt$.

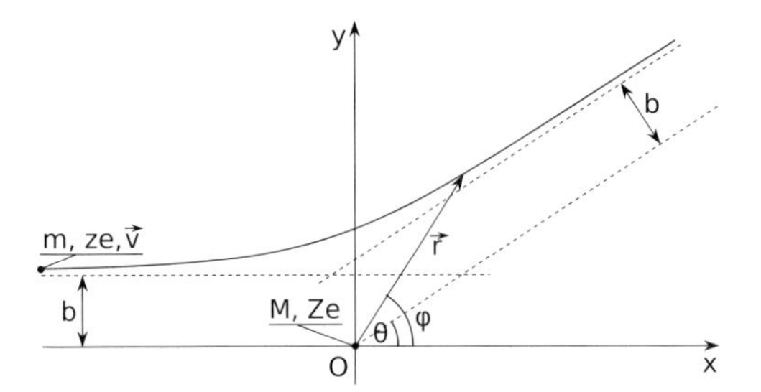

Figure 3.1 Scattering of a moving charged particle with a charge Ze and mass m on another charged particle with the same charge and mass M. Here $\boldsymbol{r}$ is a radius vector, b a target distance, and θ the angle of deviation of a charged particle from its original direction before scattering.

If one differentiates the equation above by $d\phi$ and takes into account the relationships described above, it is easy to obtain the differential equation describing the energy change in the system of m-M particles:

$$\frac{d\rho}{d\phi}\left(\frac{d^2\rho}{d^2\phi}\right) + \rho + \frac{m_0 Zze^2}{L^2} = 0\,. \tag{3.8}$$

This differential equation has the partial solution $\rho_1 = 2m_0 Zze^2/L^2$ and the general solution $\rho_2 = A\cos\phi + B\sin\phi$, which are defined from the initial conditions:

1. When $\phi = \pi$, $r = \infty$ and $\rho = 0$. Hence $A = -m_0 Zze^2/L^2$.
2. When $y = b$ (Figure 3.1), it is easy to show that $B = 1/b$.

Thus the general solution can be written as

$$\rho = -(1+\cos\phi)\frac{m_0 Zze^2}{L^2} + \frac{\sin\phi}{b}\,. \tag{3.9}$$

The deviation of particle m at angle θ is equal to ϕ between the asymptotes of the hyperbolas in Figure 3.1 and for a particle that moves away ($r = \infty$, or $\rho = 0$). By substituting θ and L into the equation above for $\rho = 0$, one can obtain Bohr's (1915) formula:

$$\tan\left(\frac{\theta}{2}\right) = \frac{Zze^e}{m_0 bV^2}\,, \tag{3.10}$$

which allows one to estimate the target parameter b for $\phi = \pi/2$ to be $b = zZe^2/m_0V$.

3.2.2
Energy Loss and Momentum Variations

Let us now evaluate the energy lost by an electron in such scattering by comparing its initial velocity $\boldsymbol{V}_\mathrm{i}$ before and the final velocity $\boldsymbol{V}_\mathrm{f}$ after scattering. In the system coordinate of the mass center particle m has a velocity of $V_i = m/(M+m)(V,0)$ and after its velocity is $V_f = m/(M+m)(V\cos\theta, V\sin\theta)$. In the laboratory system of particle m (rest reference frame) the initial velocity is $V_i = (V,0)$ and the final velocity is

$$V_f = \frac{V}{M+m}(M\cos\theta + m, M\sin\theta) . \tag{3.11}$$

Now one can define the energy loss ΔE during the scattering of particle m on M and the change in its momentum, or the parallel velocity $V_x = V$ from which particle m starts the scattering process. The energy loss ΔE during the scattering of particle m on M is defined by the difference in their kinetic energies, $\Delta E = E_f - E_i$; the change of the parallel velocity V_x is defined as $\Delta V_x = V_{fx} - V$. After a substitution of velocities V_i and V_f above in the laboratory system (the rest reference frame) of particle m gives the following expressions:

$$\Delta E = -\frac{m^2 M}{(m+M)^2} V^2(1-\cos\phi) ; \quad \Delta V_x = -\frac{M}{(M+m)} V(1-\cos\phi) , \tag{3.12}$$

where the minus sign denotes the energy loss. For example, for $\theta = \pi/2$ an electron scattering on a proton gives an energy loss of $\Delta E \sim -2E/(M/m)$, for example, 10.9 eV for an electron with an energy of 1 keV and 326.8 eV for an energy of 300 keV.

3.2.2.1 Energy and Momentum Loss Rates

In order to define the rate of energy loss per second, one needs to apply the operator (Emslie, 1978)

$$\frac{d}{dt} = -2\pi n V \int_b \Delta(b) b\, db \tag{3.13}$$

to both formulas of Eq. (3.12) (In this section the operator is applied to the left-hand side of Eq. (3.12), in the next section to the right-hand side.).

After a substitution of ΔE into the operator above, the energy variations can be written as

$$\frac{d}{dt}\Delta E = -2\pi n V \int_b b\, db \frac{m^2 M}{(m+M)^2} V^2(1-\cos\phi) , \tag{3.14}$$

and after the elementary trigonometric transformation using Eq. (3.10) for the angle dependence on parameter b, the integration gives

$$\frac{dE}{dt} = 2\pi n_e V \frac{m}{M} \frac{Z^2 z^2 e^4}{E} \ln \frac{m_0 b_0 V^2}{Z z e^2} . \tag{3.15}$$

This equation allows one to calculate the rate of energy loss for electron scattering on electrons ($M = m = m_e$) and protons ($M = m_p$) of ambient plasma

(Syrovatskii and Shmeleva, 1972a) as follows:

$$\left(\frac{dE}{dt}\right)_{e,p} = -\frac{2\pi n_e V e^4}{E}\left(\frac{m_e}{m_p}+1\right)\ln\left(\frac{m_0 V^2 b_0}{e^2}\right), \tag{3.16}$$

and on neutral hydrogen atoms as follows (Zharkova and Kobylinskij, 1992):

$$\left(\frac{dE}{dt}\right)_{n} = -\frac{2\pi n_H V e^4}{E}\ln\left(\frac{m_e V^2}{1.105\, I_H}\right), \tag{3.17}$$

where I_H is the full ionization potential of a hydrogen atom.

Now let us take into account that $dt = ds/V$, where s is the distance the electron travels, the ionization degree of the ambient plasma is defined as $X = n_e/n$, and $n_H/n_H + n_e = 1 - X$. Hence the energy loss of an electron scattered on the particles of ambient plasma with partial ionization can be calculated as follows:

$$\frac{dE}{ds} = -\frac{2\pi e^4 n}{E}\left[X\ln\frac{m_0 V^2 b_0}{Zze^2} + (1-X)\ln\frac{m_e V^2}{1.105\, I_H} = -\frac{a}{En}\right], \tag{3.18}$$

where

$$a_e = 2\pi e^4\left[X\ln\left(\frac{m_0 V^2 b_0}{Zze^2}\right) + (1-X)\ln\left(\frac{m_e V^2}{1.105\, I_H}\right)\right] \tag{3.19}$$

is the collisional coefficient for electrons. Parameter a is converted for the temperature 1.5×10^6 to the coefficient defined by Syrovatskii and Shmeleva (1972a):

$$a_e = 1.3 \times 10^{-13}\left[\ln\left(\frac{E}{mc^2}\right) - \frac{1}{2}\ln(n) + 38.7\right], \quad \mathrm{eV^2\,cm^2}, \tag{3.20}$$

where b_0 is the target distance, which is assumed to be equal to the Debye length r_D (see, e.g., Krall and Trivelpiece, 1973):

$$b_0 = r_D = \sqrt{\frac{k_B T}{4\pi n e^2}}.$$

It can be seen that the parameter a_e for bombarding electrons depends linearly on the ionization degree and varies logarithmically with the electron energy, the ambient plasma temperature, and density via the Debye length. For the typical electron energy $\sim$ 100 keV and an ambient plasma density of about 10^{11} cm^{-3} this parameter is equal to 2.7×10^{12} eV^2 cm^2 for fully ionized ambient plasma, decreasing to 2.3×10^{12} eV^2 cm^2 for a fully neutral one.

3.2.2.2 **Energy and Pitch-Angle Losses for the Cold Target Approach**

For the other approximation of the parameter $(b_0)_{max}$ (Brown, 1971) used the mean free path $\eta = V/\nu$ (Landau and Lifshitz, 1960), where ν is the plasma oscillation frequency because beam electrons have much higher energies than ambient plasma particles (cold target approach), so Mach's number is extremely high. Then for the parameter b_0 one can use $b_0 = \min(\eta, r_D)$ and define the Coulomb logarithm Λ as follows (Emslie, 1978):

$$\Lambda = \ln\left[\frac{m_0 V^2 \min(\eta, r_D)}{Z z e^2}\right] , \tag{3.21}$$

which is the same as the first member in the collisional coefficient from formula (3.19). Then the energy loss and momentum change rates of a charged particle scattering on electrons, similar to Emslie (1978) can be written as:

$$\begin{aligned} \frac{dE}{dt} &= -\frac{2\pi Z^2 z^2 e^4 \Lambda}{E}\left(\frac{m}{M}\right) n V^2 , \\ \frac{dV_x}{dt} &= -\frac{\pi Z^2 z^2 e^4 \Lambda}{E^2}\left(1 + \frac{m}{M}\right) n V^2 , \end{aligned} \tag{3.22}$$

which agree with Eq. (3.18) by Brown (1972) for the case of high-energy electrons bombarding ambient electrons or protons ($Z^2 = z^2 = 1$).

By repeating the same operations for electron scattering on protons and combining the outcome with those on electrons one can obtain generalized formulas for the energy loss and momentum rates:

$$\begin{aligned} \left(\frac{dE}{dt}\right)_e &= -\frac{2\pi e^4 \Lambda}{E}\left(\frac{m}{m_p} + \frac{m}{m_e}\right) n V ; \\ \left(\frac{dV_x}{dt}\right)_e &= -\frac{\pi e^4 \Lambda}{E^2}\left(2 + \frac{m}{m_p} + \frac{m}{m_e}\right) n V^2 . \end{aligned} \tag{3.23}$$

For electron scattering on neutral hydrogen atoms (Emslie, 1978) provides a formula for energy loss rates similar to those obtained earlier with the additional coefficient τ, which is equal to 1 if the bombarding particles are electrons and 2 if they are protons:

$$\left(\frac{dE}{dt}\right)_n = -\frac{2\pi e^4 \Lambda'}{E}\left(\frac{m}{m_e}\right) n V , \tag{3.24}$$

where the Coulomb logarithm Λ' for neutral atoms is defined as

$$\Lambda' = \ln\left(\frac{\tau m_e V^2}{1.105 I_H}\right) , \tag{3.25}$$

which is, in fact, the same as the second logarithm term in Eq. (3.19).

The momentum, or parallel velocity, variations can be deduced by using the operator (3.13) applied to the variations of ΔV_x from Eq. (3.12) as follows (Emslie, 1978):

$$\left(\frac{dV_x}{dt}\right)_n = -\frac{\pi e^4 \Lambda''}{E^2}\frac{m_p}{m + m_p} n V^2 , \tag{3.26}$$

where the effective collision logarithm is defined as (Emslie, 1978)

$$\Lambda'' = \ln\left[\frac{1}{\alpha}\frac{m}{m_e}\frac{V}{c}\right] , \tag{3.27}$$

where α is the fine structure constant and c is the speed of light.

By combining all the energy rates and velocity changes above into a single formula for different types of bombarding particles (electrons or protons) one obtains (Emslie, 1978) for

1. Electron bombardment ($m = m_e$):

$$\frac{dE}{dt} = -\frac{2\pi e^4}{E}[X\Lambda + (1-X)\Lambda']nV \,, \tag{3.28}$$

$$\frac{dV_x}{dt} = -\frac{\pi e^4}{E^2}[3X\Lambda + (1-X)\Lambda'' e]nV^2 \,; \tag{3.29}$$

2. Proton bombardment ($m = m_p$):

$$\begin{aligned} \frac{dE}{dt} &= -\frac{2\pi e^4}{E}\frac{m_p}{m_e}[X\Lambda + (1-X)\Lambda']nV \,, \\ \frac{dV_x}{dt} &= -\frac{\pi e^4}{E^2}\left[\frac{m_p}{m_e}X\Lambda + \frac{1}{2}(1-X)\Lambda''\right]nV^2 \,. \end{aligned} \tag{3.30}$$

It must be noted that the derived formulas are valid in the cold target approach, for example, for the ratios between the energy E of bombarding particles and the ambient plasma temperature for electrons $k_B T \ll E$ and for protons $k_B T \ll (m_e/m_p)E$. This condition is fully satisfied for subrelativistic electrons observed from HXRs in flares (Holman *et al.*, 2011), but it can be violated for low-energy protons in the corona (Emslie, 1978).

3.2.2.3 Depth Energy Loss Variations for Bombarding Particles

Let us first introduce a new variable ξ defining a column depth as follows:

$$\xi(x) = \int_0^x n(t)dt \,, \tag{3.31}$$

which defines the total number of particles in the line of sight within an area of one square centimeter. Now let us derive the variations of the energy momentum loss with precipitation depth by taking into account the definition of column depth $d\xi = ndx$, the relations $dx = V_x dt = V\mu dt$ between the derivatives from the kinetic particle energy $E = mV^2/2$, and the parallel velocity $V_x = V\mu$ as follows:

$$\frac{d\mu}{\mu} = \frac{dV_x}{V_x} - \frac{dE}{2E} \tag{3.32}$$

one obtains the following expressions for energy loss and momentum change with depth by bombarding electrons (Emslie, 1978, 1980):

$$\begin{aligned} \frac{dE}{d\xi} &= -\frac{2\pi e^4}{\mu E}[X\Lambda + (1-X)\Lambda'] \,, \\ \frac{d\mu}{d\xi} &= -\frac{\pi e^4 V_x}{2\mu E^2}[3X\Lambda + (1-X)\Lambda''] \,, \end{aligned} \tag{3.33}$$

while for bombarding protons the energy and momentum losses are defined by the following expressions (Emslie, 1980):

$$\frac{dE}{d\xi} = -\frac{2\pi e^4}{\mu E}\frac{m_p}{m_e}[X\Lambda + (1-X)\Lambda'] ,$$
$$\frac{d\mu}{d\xi} = -\frac{\pi e^4 V_x}{2\mu E^2}\left[\frac{m_p}{m_e}X\Lambda + \frac{1}{2}(1-X)\Lambda''\right] . \tag{3.34}$$

Then if one defines the parameters β as one of the following (Emslie, 1978):

$$\beta_e = \frac{2X\Lambda + (1-X)(\Lambda'' - \Lambda')}{\Lambda' + X(\Lambda - \Lambda')} ;$$
$$\beta_p = -\frac{(1-X)\Lambda'}{\Lambda' + X(\Lambda - \Lambda')} , \tag{3.35}$$

and divides the equation for E by those for V_x in expression (3.34), then the following differential equation is obtained for energy loss variations over the velocity V_x:

$$\frac{dE}{dV_x} = \frac{2E}{V_x}\frac{1}{1+\beta_{e,p}} . \tag{3.36}$$

The solution to this ordinary differential equation is as follows (Emslie, 1978):

$$E = E_0\left(\frac{V_x}{V_{x0}}\right)^{\frac{2}{1+\beta}} , \tag{3.37}$$

from which can one can derive, using the ratio

$$\frac{V_x}{V_{x0}} = \left(\frac{V_x}{V_0}\right)\left(\frac{V}{V_0}\right)\left(\frac{V_0}{V_{x0}}\right) = \frac{\mu}{\mu_0}\left(\frac{E}{E_0}\right)^{1/2}$$

and remembering that $V_0 = V_{x0}$ as per Figure 3.1, the following solution:

$$\frac{\mu}{\mu_0} = \left(\frac{E}{E_0}\right)^{\beta/2} . \tag{3.38}$$

Then, by substituting this solution into Eq. (3.34) for electron bombardment for energy loss and momentum change, one obtains

$$E = E_0\left[1 - \left(2 + \frac{\beta}{2}\right)\frac{2\pi e^4 \vartheta \xi}{\mu_0 E_0^2}\right]^{\frac{2}{4+\beta}} ,$$
$$\mu = \mu_0\left[1 - \left(2 + \frac{\beta}{2}\right)\frac{2\pi e^4 \vartheta \xi}{\mu_0 E_0^2}\right]^{\frac{\beta}{4+\beta}} , \tag{3.39}$$

where $\vartheta = m/m_e[X\Lambda + (1-X)\Lambda']$ for electron bombardment as first defined by Brown (1972), which was later generalized for both electron and proton bombardment by Emslie (1978, in which see Table 1).

3.2.2.4 **Collisional Energy Losses in Ambient Solar Atmosphere**

Now let us evaluate the energy losses for a model flaring atmosphere representing the temperature and density variations in the solar atmosphere from the corona to the photosphere as plotted in Figure 10.1. The atmosphere model considered is the one obtained for 4 s after the beam injection. Let us find a solution to Eq. (3.18) by taking into account that $d\xi = nds$, where n is the ambient density, and by assuming that at the top boundary, where $\xi = 0$, a particle has energy E and at the column depth ξ the energy is E^*:

$$(E^*)^2 = E^2 - 2a_e\xi \,, \tag{3.40}$$

where a_e is the collisional parameter for electrons described and estimated in Eq. (3.18) of Section 3.2.

This means that at each depth ξ the particle energy $E^*(\xi)$ is reduced by the energy losses in collisions with ambient particles. Now let us explore the electron energy E^* changes over the whole flaring atmosphere, which can be deduced from Eq. (3.40) as follows:

$$E^*(\xi) = \sqrt{E^2 - 2a_e\xi} \,. \tag{3.41}$$

According to Eq. (3.41), the density of a monoenergetic beam decreases parabolically with column depth as shown in Figure 3.2a for different initial energies of beam electrons.

Now we can introduce a "collisional stopping depth", where an electron with an initial energy E loses it completely ($E^* = 0$) while precipitating vertically into a flaring atmosphere, that is,

$$\xi = \frac{E^2}{2a} \,. \tag{3.42}$$

The collisional stopping depths of electrons with different energies are shown in the first column of Table 3.1. Obviously, electrons with an initial energy of $E_{low} = 16\,\text{keV}$ precipitating into a fully ionized plasma lose their energy completely at a depth of $\xi = 4.7 \times 10^{19}\,\text{cm}^{-2}$, while for those with an energy of $E_{upp} = 380\,\text{keV}$ the energy will only be reduced to $E^*_{upp}(\xi) = 372\,\text{keV}$. Electrons with an energy of 380 keV lose their energy completely at a column density of $(2.8–3.3) \times 10^{22}\,\text{cm}^{-2}$, where the first value corresponds to a fully ionized plasma and the second to a partially ionized one with an ionization degree of $X = 0.001$. This means that lower-energy electrons become fully thermalized to the ambient plasma energies below a column density of $\xi = 4.7 \times 10^{19}\,\text{cm}^{-2}$, while higher-energy electrons can reach a photospheric depth of $3.3 \times 10^{22}\,\text{cm}^{-2}$.

3.2.2.5 **Effect of Pitch-Angle Scattering**

Now we can explore the effect of a pitch-angle change in pure Coulomb collisions on beam electron distributions. Let us consider the same Eq. (3.18) but now allow electrons to move in any direction z toward the vertical with the pitch-angle cosine $\mu \geq 0$. Then, using the substitution $ds = dz\mu$ and calculating the energy losses

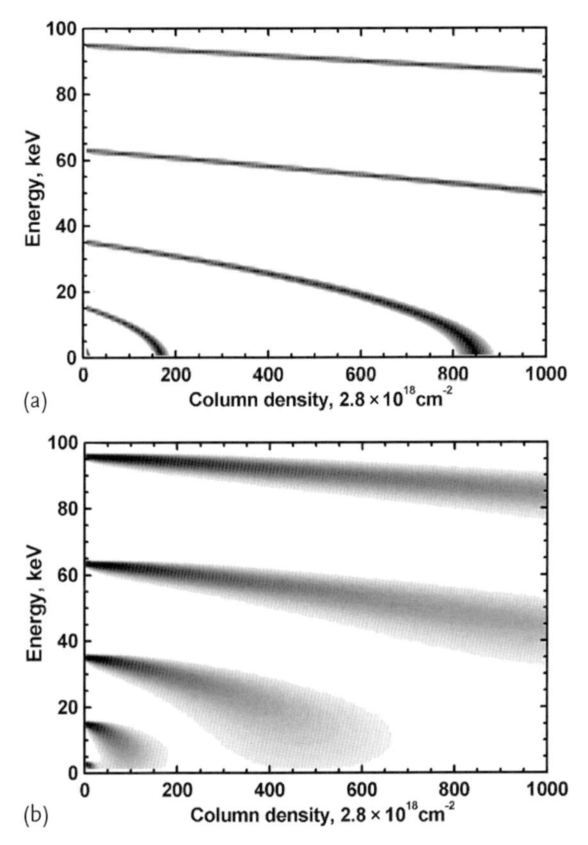

Figure 3.2 Collisional energy loss and pitch-angle scattering: (a) depth variations of initially monoenergetic beams with energies of 16, 36, 64, and 96 keV losing energy in pure Coulomb collisions for $\mu = 1$; (b) the same as in (a) but with pitch-angle scattering, or where μ is arbitrary.

in Eq. (3.18) for all μ from 1 to 0, one can obtain Eq. (3.38) for energy changes dependent on μ, in addition to those energy losses in depth (Eq. (3.41)), as follows:

$$E_\mu = E^* \left(\frac{\mu}{\mu_0}\right)^{\frac{2}{\beta}} . \tag{3.43}$$

The results for such anisotropic pitch-angle scattering are plotted in Figure 3.2b. It can be seen that the monoenergetic beams precipitating with pitch-angle scattering reveal two effects. First, the decrease in electron energy with depth caused by collisions becomes much steeper than the parabolic one (Figure 3.2a) with a wide dispersion around μ_0. Second, the electron energy reaches zero at a much smaller distance (Figure 3.2b). Apparently this happens because the beam electrons, which lose their energy in Coulomb collisions, can be scattered many times not only forward ($\mu \succ 0$) but backward ($\mu \prec 0$) as well. This results in the electrons not reaching a given depth as fast as those without pitch-angle scattering.

Table 3.1 Comparison of "stopping" column depths for electrons with different energies (column 1), for pure collisional losses (column 2), and for pure electric field losses (last three columns).

	a, eV^2 cm^2		$\mathcal{E}$, V cm^{-1}	
	2.7×10^{-12}	10^{-5}	10^{-4}	10^{-3}
E, keV			ξ_{max}, cm^{-2}	
10	1.8×10^{19}	3.2×10^{19}	3.5×10^{17}	2.2×10^{17}
40	2.9×10^{20}	$> 10^{25}$	1.6×10^{18}	2.6×10^{17}
100	1.8×10^{21}	$> 10^{25}$	3.2×10^{19}	3.5×10^{17}
400	2.9×10^{22}	$> 10^{25}$	$> 10^{25}$	1.6×10^{18}

Thus, electrons will lose all their energy at much higher atmospheric depths, or at much lower column densities in the upper atmosphere, and this energy loss depends on the pitch-angles gained during the electrons' precipitation. The pitch-angle scattering is effective for electrons at any energy, while it is more noticeable for low-energy electrons (lower curves) than for the higher-energy ones (compare gray areas in upper and lower curves in Figure 3.2b).

3.3 Continuity Equation Approach for Electrons: Pure Collisions

The main features of electron-beam kinetics in a flaring atmosphere can be described analytically by solving the continuity equation for collisions only first suggested by Syrovatskii and Shmeleva (1972a) and neglecting the electrostatic electric field carried by these beam electrons and their pitch-angle scattering. In such an approach, electron precipitation is described by the continuity equation (Syrovatskii and Shmeleva, 1972a)

$$\frac{\partial}{\partial z}\left[V \cdot N(z,E)\right] + \frac{\partial}{\partial E}\left[\left(\frac{dE}{dz}\right) V \cdot N(z,E)\right] = 0\,, \tag{3.44}$$

where z is the direction of electron precipitation (which is the x-direction in Figure 3.1) and dE/dz are the collisional energy losses as per Eq. (3.18).

Let us rewrite the equation for energy dependence as follows:

$$\frac{\partial}{\partial z}\left[\sqrt{\frac{2E}{m_e}} N(E,z)\right] + \frac{\partial}{\partial E}\left[\left(\frac{dE}{dz}\right)\sqrt{\frac{2E}{m_e}} N(E,z)\right] = 0\,, \tag{3.45}$$

where we make a substitution $\varphi(E,z) = \sqrt{2E/m_e} N(E,z)$ and use Eq. (3.18) for electron energy losses, for example, $dE/dz = an/E$. Then one can rewrite Eq. (3.45) as follows:

$$\frac{\partial}{n\partial z}\varphi(E,z) + \frac{\partial}{\partial E}\left[\frac{a_e}{E}\varphi(E,z)\right] = 0\,.$$

The latter can be easily integrated using a substitution of the column depth ξ instead of a linear depth z, $d\xi = n dz$, that gives the general solution in the form of $\varphi(E, s)$:

$$\varphi(E, \xi) = \Psi\left[E^2 + 2a_e(\xi - \xi_0)\right] ,$$

where Ψ is any function of this argument.

3.3.1 Solutions of Continuity Equation for Power-Law Beam Electrons

If the initial energy spectrum of injected electrons at $\xi_0 = 0$ is power law with some lower E_{low} and upper E_{upp} energy cutoffs and spectral index γ, that is:

$$N(\xi = 0, E) = K^* E^{-(\gamma+0.5)}\Theta(E - E_{low})\Theta(E_{upp} - E) , \tag{3.46}$$

where Θ is the Heavyside function that is equal to 1 for positive arguments and 0 otherwise, the index at the energy E is increased by 0.5 to account for the fact that the spectral index γ refers to the electron flux $(V N(\xi, E))$.

Then the solution of the continuity equation for pure collisions is as follows (Syrovatskii and Shmeleva, 1972a):

$$N(\xi, E) = K\sqrt{E}(E^2 + 2a\xi)^{-\frac{\gamma+1}{2}}\Theta\left(\sqrt{E^2 + 2a\xi} - E_{low}\right) \times \Theta(E_{upp} - \sqrt{E^2 + 2a\xi}) , \tag{3.47}$$

where $N(\xi, E)$ is the electron-beam differential density (per energy unit per column density unit), a is the collisional parameter from Eq. (3.18), and K is the scaling factor found from the normalization of the distribution function $(f(\xi, E) = N(\xi, E))$ on the initial energy flux as follows (Syrovatskii and Shmeleva, 1972a):

$$K = F_0\left(\frac{m}{2}\right)^{1/2} \times \begin{cases} (\gamma - 2)\left(\frac{(E_{low}E_{upp})^{\gamma-2}}{E_{upp}^{\gamma-2} - E_{low}^{\gamma-2}}\right) , & \gamma \neq 2 , \\ \frac{1}{\ln\left(\frac{E_{upp}}{E_{low}}\right)} , & \gamma = 2 . \end{cases} \tag{3.48}$$

Note from Eq. (3.47) that the energy range of beam electrons has the new lower E_1 and upper E_2 energy limits, which vary as the depth of the electron energy distribution between these limits shifts and shrinks:

$$E_1(\xi) = \mathrm{Re}\sqrt{E_{low}^2 - 2a\xi} \le E \le \mathrm{Re}\sqrt{E_{upp}^2 - 2a\xi} , \tag{3.49}$$

where Re means a real part of the square roots. This means there is a continuing shift with depth to lower energies for all beam electrons, which, thus, have their energy distribution changed at every precipitation depth, as discussed in the next section. Those lower-energy electrons reaching energies smaller than the lower-energy cutoff E_{low} required to produce HXR emission will leave the beam and join the ambient electrons, for example, become thermalized.

Differential spectra of electron beams. For a beam injected with a power-law energy spectrum that loses its energy in pure Coulomb collisions the distribution functions with pitch-angle diffusion in the $E-\mu$ coordinates is shown in Figure 3.3 for an initial flux 10^{12} erg cm^{-2} s^{-1}, $\gamma = 3$, lowest and highest energies 16 and

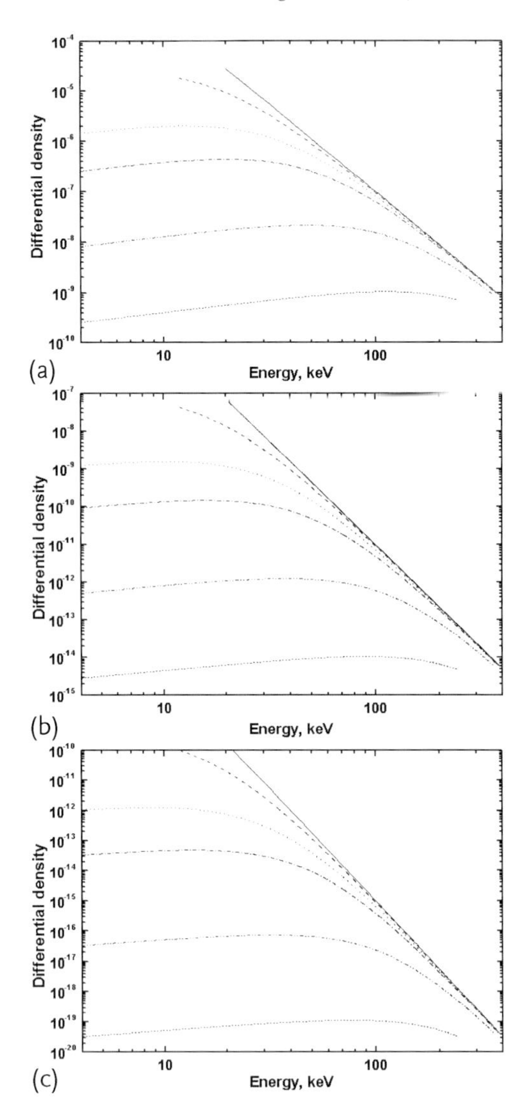

Figure 3.3 Depth variations of differential energy spectra $dN/dE(\xi, E)$ vs. energy of an electron beam injected with an initial power-law energy spectrum of $E_{low} = 16$ keV, $E_{upp} = 384$ keV, $\gamma = 3$ with the initial energy flux of 10^{12} erg cm^{-2} s^{-1}. The differential spectra plotted for a column depth of 2×10^{18} cm^{-2} (corona) (dashed and dotted lines), of 2×10^{19} cm^{-2} (transition region) (dashed–dotted line), and of 2×10^{20} cm^{-2} and 2×10^{21} cm^{-2} (chromosphere) (dashed–dot–dotted and short–dashed lines, respectively).

386 keV, respectively. Due to pitch-angle scattering the number of lower-energy electrons increases with depth and changes from a narrow angular distribution at the injection point to a much wider one in deeper layers, becoming nearly isotropic at larger atmospheric depths below $\xi = 2 \times 10^{22}\,\mathrm{cm}^{-2}$.

The differential spectra $N(\xi, E)$ are shown in Figure 3.3 for $\gamma = 3$ and 7 and initial energy fluxes of 10^8 and $10^{12}\,\mathrm{erg\,cm^{-2}\,s^{-1}}$, respectively. They show the two main features. At first, both the upper and lower energy cutoff magnitudes decrease with depth, which simply demonstrates the process of energy loss by particles with the highest and lowest energies in the beam.

Then, the differential spectra are shown to become harder with depth and the index of the low-energy part of the spectra decreases faster than those of the high-energy part. Starting at some column depth a maximum appears in the energy spectrum that can be easily explained by nonuniform energy losses ($dE/d\xi \sim E^{-1}$) and reaching the maximum at some depth with an energy of $(2\gamma + 1)\,E_{\mathrm{low}}$.

3.3.2
Beam Electron Densities

Using the solutions for differential energy spectra provided by Eq. (3.47), let us calculate variations of the beam density N with depth by integrating it into the energy within the lower upper cutoff limits as follows (Zharkova and Kobylinskij, 1992):

$$N(\xi) = K(2a\xi)^{\frac{1-2\gamma}{4}} \times \begin{cases} \dfrac{2}{2\gamma-1}\dfrac{\Gamma\left(\frac{2\gamma+3}{4}\right)}{\Gamma(0.25)}\left\{\sum_{n=0}^{\infty}(-1)^n \dfrac{\Gamma(0.25+n)}{\Gamma\left(\frac{2\gamma+3}{4}+n\right)} \right. \\ \left. \times \left[t_1^{-(n+0.25)}(1+t_1)^{\frac{1-\gamma}{2}} - t_2^{-(n+0.25)}(1+t_2)^{\frac{1-\gamma}{2}}\right]\right\}, \\ 0 \le \frac{1}{t_1} \le 1 \\ \dfrac{\Gamma\left(\frac{2\gamma-1}{4}\right)\Gamma(0.75)}{2\Gamma\left(\frac{\gamma+1}{2}\right)} - \dfrac{2}{2\gamma-1}(1+t_2)^{\frac{1-\gamma}{2}}\dfrac{\Gamma\left(\frac{2\gamma+3}{4}\right)}{\Gamma(0.25)} \\ \times \sum_{n=0}^{\infty}(-1)^n \dfrac{\Gamma(0.25+n)}{\Gamma\left(\frac{2\gamma+3}{4}+n\right)} t_2^{-(n+0.25)}, \\ 1 < \frac{1}{t_1} \le \frac{1}{t_2}, \end{cases} \tag{3.50}$$

where F_0 is the initial energy flux on the top boundary, m is the electron mass, K is the same normalization coefficient as in Eq. (3.48), $\Gamma(\gamma)$ is the gamma function, and parameters t_1 and t_2 are described by the following expression (Zharkova and Kobylinskij, 1992):

$$t_{1,2} = \max\left[0, \left(\frac{E^2_{\mathrm{low,upp}}}{2a\xi} - 1\right)\right], \tag{3.51}$$

with index 1 referring to E_{low} and index 2 to E_{upp}. The examples of electron density variations with depth for pure collisional losses by beams with different parameters of the initial energy flux and spectral index γ are plotted in Figure 3.4. It can

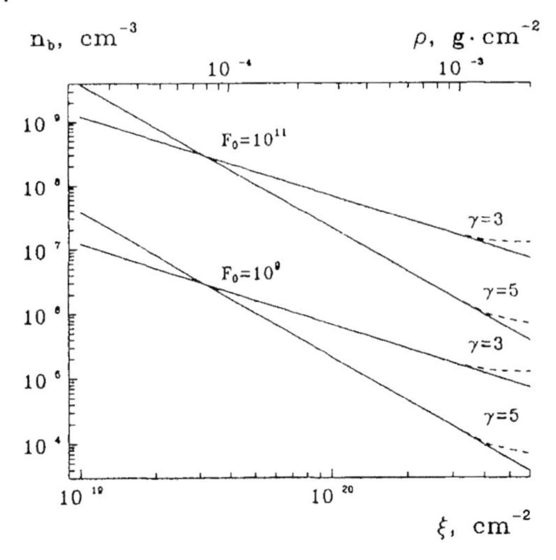

Figure 3.4 Beam density variations with a column depth for precipitating electrons derived from the continuity equation for a beam with $\gamma = 3, 5$ and an initial energy flux of $(1-100) \times 10^9$ erg cm^{-2} s^{-1}. The solid lines correspond to the fully ionized ambient plasma, the dashed lines to the ionization degree of hydrogen plasma defined from a full non-LTE (local thermodynamic equilibrium) consideration for the five-level-plus-continuum hydrogen atom (Zharkova and Kobylinskii, 1991).

be seen that the density of beams is rather high and that in some beams with higher energy fluxes it can be rather comparable with the ambient plasma density. The electron-beam density goes even higher at lower chromosphere depths of the ambient plasma with a partial ionization degree (dashed lines) than in the fully ionized plasma (solid lines). This means that beam electrons lose less energy in Coulomb collisions with charged particles since the plasma becomes almost neutral at these depths.

The beam energy flux for different precipitating depths is defined as follows:

$$F(\xi) = K\sqrt{\frac{2}{m_e}}\int_0^{\infty}\int_{-1}^{+1} N(\xi, E, \mu) E^{1.5} \mu \, d\mu \, dE = K\int_0^{\infty} N(E, \xi) E^{1.5} dE \, . \tag{3.52}$$

3.3.3 Mean Electron Spectra

Let us calculate analytically the mean electron energy flux.

For energies higher than E_{low} the mean electron flux is as follows:

$$\bar{\mathcal{F}}(E) = \sqrt{\frac{2}{m}} K^* E \int_0^{\infty} (E^2 + 2a\xi)^{-\frac{\gamma+1}{2}} d\xi \, . \tag{3.53}$$

For integration let us make the following substitution: $p = 2a\xi + E^2$, thus leading to $\xi = (p - E^2)/2a$ and $d\xi = 1/2a\,dp$.

Due to collisional energy losses, the electron energy distributions are steadily shifted to the energies $E \prec E_{\text{low}}$, so that for an energy range restricted only by the lower cutoff E_{low} the mean electron fluxes can be calculated as a function of energy as follows:

$$\bar{\mathcal{F}}(E) = \frac{1}{a(\gamma-1)}\sqrt{\frac{2}{m}}K * \left[\frac{1}{2a}E_{\text{low}}^2 + \left(1 + \frac{1}{2a}\right)E^2\right]^{-\frac{\gamma-1}{2}} . \tag{3.54}$$

As a result, one obtains the following mean flux spectrum, split by the energy E_{low}:

$$\bar{\mathcal{F}}_{\text{coll}} = \frac{1}{2a_{\text{e}}(\gamma-1)}\sqrt{\frac{2}{m_{\text{e}}}}K\left[E_{\text{low}}^2 + (1+2a)E^2\right]^{-\frac{\gamma-1}{2}} , \quad \text{if} \quad E < E_{\text{low}} ,$$

$$\bar{\mathcal{F}}_{\text{coll}} = \frac{1}{a_{\text{e}}(\gamma-1)}\sqrt{\frac{2}{m_{\text{e}}}}KE^{-\gamma+2} , \quad \text{if} \quad E > E_{\text{low}} . \tag{3.55}$$

It can be noted that for energies of beam electrons lower than E_{low} the mean energy flux is increasing with energy approaching maximum at E_{low} and for higher energies it becomes sharply decreasing the as follows:

$$\bar{\mathcal{F}}(E) \propto E^{-\kappa} .$$

Hence by comparing the initial energy spectrum of beam electrons with the mean electron flux $\bar{\mathcal{F}}(E)$ calculated above, one concludes that the mean flux spectral index κ is related to the initial electron flux spectral index γ as follows:

$$\kappa \simeq \gamma - 2 . \tag{3.56}$$

In other words, the mean spectral index κ of the electron flux observed from the target is two units lower than the spectral index of beam electrons γ injected at the top boundary (Brown, 1971; Syrovatskii and Shmeleva, 1972a).

Therefore, for energies higher than the initial lower-energy cutoff, the mean flux spectral index obtained from the continuity equation using the approach of Syrovatskii and Shmeleva (1972a) is similar to those obtained from the flux conservation approach of Brown (1971).

3.3.4 Hard X-Ray Bremsstrahlung Emission by Beam Electrons

To derive the photon spectra of HXR bremsstrahlung emission produced by beam electrons, we use the approach suggested by Syrovatskii and Shmeleva (1972a) and calculate the spectral density of the radiation flux at the Earth's orbit integrated over the volume V of beam electrons ($dV = dS d\xi/n$, where S is the beam cross-section) as follows:

$$\frac{dI(h\nu)}{h\nu} \sim \frac{S}{2\pi R} \int_{h\nu}^{\infty} dE \frac{d\sigma}{d(h\nu)} \left(\frac{dI_{\mathrm{e}}}{dE}\right)_{\Sigma} , \tag{3.57}$$

where R is the distance from the Sun to the Earth, $d\sigma/d(h\nu)$ is the differential cross-section of bremsstrahlung emission, and $(dI_{\mathrm{e}}/dE)_{\Sigma}$ is the resulting electron flux (mean electron flux $\bar{\mathcal{F}}(E)$) discussed in the section above.

Now let us consider the differential cross-sections by Elwert and Haug (1972), which can be written as follows:

$$\frac{d^2\sigma_{\|}}{d\Omega\, d(h\nu)} = C(A + B(1 - \mu^2))\sigma_0 , \tag{3.58}$$

$$\frac{d^2\sigma_{\perp}}{d\Omega\, d(h\nu)} = CA\sigma_0 , \tag{3.59}$$

where parameters A, B, and C are defined as follows:

$$A = \frac{\eta - \varepsilon/2}{\eta(\eta - \varepsilon)} \ln \frac{\sqrt{\eta} + \sqrt{\eta - \varepsilon}}{\sqrt{\eta} - \sqrt{\eta - \varepsilon}} - 1 , \tag{3.60}$$

$$B = \frac{3/2\varepsilon - \eta}{\eta(\eta - \varepsilon)} \ln \frac{\sqrt{\eta} + \sqrt{\eta - \varepsilon}}{\sqrt{\eta} - \sqrt{\eta - \varepsilon}} + 3 , \tag{3.61}$$

$$C = \frac{1}{\eta\varepsilon} \frac{1 - \exp\left(-\frac{2\pi\alpha c\sqrt{m_{\mathrm{e}}}}{\sqrt{2\eta E_{\mathrm{low}}}}\right)}{1 - \exp\left(-\frac{2\pi\alpha c\sqrt{m_{\mathrm{e}}}}{\sqrt{2(\eta-\varepsilon) E_{\mathrm{low}}}}\right)} . \tag{3.62}$$

ε is a dimensionless photon energy $\varepsilon = h\nu/E_{\mathrm{low}}$, and σ_0 is the elementary cross-section defined as follows:

$$\sigma_0 = \frac{1}{137} \frac{1}{2\pi} \frac{mc^2}{E_{\mathrm{low}}^2} r_0^2 ,$$

where r_0 is the classical electron radius. The resulting cross-section for integral polarization as a function of electron and photon energy $d\sigma/(dh\nu)(E, h\nu)$ is shown in Figure 3.5.

Assuming that the electrons emitting into the target are distributed over pitch angles isotropically, Eq. (7.17) for the differential cross-sections of bremsstrahlung emission by beam electrons can be rewritten as follows:

$$\frac{d\sigma}{dh\nu} = \frac{16}{3} \alpha r_0^2 Z^2 \frac{1}{h\nu\beta\beta_f} \frac{1 - e^{-2\pi\alpha Z/\beta}}{1 - e^{-2\pi\alpha Z/\beta_f}} \ln\left(\frac{\beta + \beta_f}{\beta \quad \beta_f}\right) , \tag{3.63}$$

and the HXR photon spectrum can be calculated as follows:

$$\frac{dI}{dh\nu} = \frac{S}{2\pi R^2} \int_{\xi} d\xi \int_{h\nu}^{\infty} dE \frac{d\sigma}{dh\nu} V N(E, \xi) = \frac{S}{2\pi R^2} \int_{h\nu}^{\infty} dE \frac{d\sigma}{dh\nu} \bar{\mathcal{F}}(E) , \tag{3.64}$$

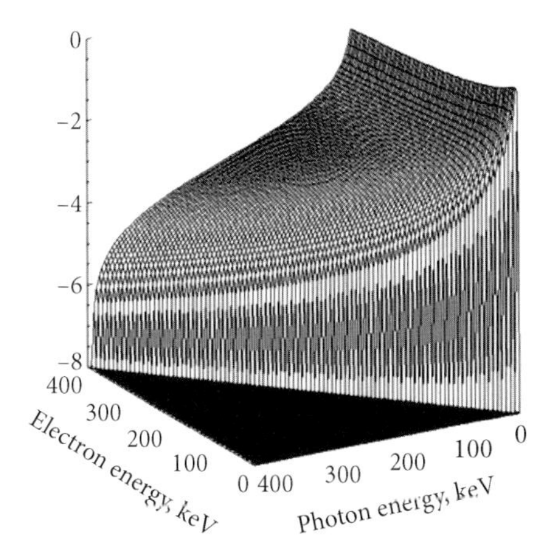

Figure 3.5 Electron bremsstrahlung differential cross-section as function of particle and photon energies (z-axis is in arbitrary units).

where $\bar{\mathcal{F}}(E)$ is the mean electron flux spectrum in the target, $\beta = \sqrt{2E/mc^2}$, and $\beta_f = \sqrt{2(E - h\nu)/mc^2}$.

Let us substitute Eq. (3.63) into Eq. (3.64) and integrate Eq. (3.64) for $\bar{\mathcal{F}}(E) = KE^{-\kappa}$:

$$\frac{dI}{dh\nu} = \frac{S}{2\pi R^2} K \frac{\sqrt{mc^2}}{h\nu} \int_{h\nu}^{\infty} dE\, E^{-\kappa} \frac{1}{\sqrt{E(E - h\nu)}} \frac{1 - e^{-2\pi\alpha Z/\beta}}{1 - e^{-2\pi\alpha Z/\beta_f}} \ln \frac{\beta + \beta_f}{\beta - \beta_f} . \tag{3.65}$$

Assuming that electron velocities are lower than the speed of light (or $\beta < 1$), the exponential functions in the expression above can be neglected, thus

$$\frac{1 - e^{-2\pi\alpha Z/\beta}}{1 - e^{-2\pi\alpha Z/\beta_f}} \approx 1 .$$

At the same time, assuming that the emitted photon has nearly the same energy as the emitting electron (and, hence, $\sqrt{E - h\nu} \ll \sqrt{E}$), the logarithmic function in Eq. (3.65) can be transformed as follows:

$$\ln \frac{\beta + \beta_f}{\beta - \beta_f} \approx \frac{2\beta_f}{\beta - \beta_f} \approx \frac{\sqrt{E - h\nu}}{\sqrt{E}} .$$

Hence Eq. (3.65) can be rewritten as follows:

$$\frac{dI}{dh\nu} = \frac{S}{\pi R^2} K \frac{\sqrt{mc^2}}{h\nu} \int_{h\nu}^{\infty} dE\, E^{-\kappa} \frac{1}{E} .$$

The equation above can be easily integrated yielding the following expression (see, e.g., Syrovatskii and Shmeleva, 1972a,b):

$$\frac{dI}{dh\nu} = \frac{S}{\pi R^2} K \frac{\sqrt{mc^2}}{\delta + 1} h\nu^{-\delta - 1} = \text{const} \cdot (h\nu)^{-\kappa - 1} . \tag{3.66}$$

Hence, if the electron mean flux has a power law with the index κ, then the resulting HXR photon spectrum also has the power-law shape with the photon spectral index δ, for example, $\delta = \kappa + 1 = \gamma - 1$, where γ is the electron flux spectral index from the solution (Eq. (3.47)).

3.3.4.1 Kinetics of a Particle Beam in Pure Converging Magnetic Field

Now let us consider the behavior of particles precipitating in a converging magnetic flux tube without collisions and an electric field following Leach and Petrosian (1981).

The particle precipitation in a converging magnetic flux tube can be investigated using the following equation:

$$\mu \frac{\partial f}{\partial z} = \frac{1-\mu^2}{2} \frac{\partial \ln B(z)}{\partial z} \frac{\partial f}{\partial \mu} . \tag{3.67}$$

In this equation there is no dependence on the particle energy, and therefore one can conclude that the magnetic mirroring equally affects particles of all energies, or, in other words, since there is no energy–pitch-angle dependence in the initial spectrum, the energy spectrum remains the same at all depths.

Let us apply to this equation the standard method of variable separation. The distribution function $f(z,\mu)$ can be represented as the product of two functions, each of which depends only on one of the arguments as follows:

$$f(z,\mu) = Z(z)M(\mu) .$$

Then Eq. (3.27) can be rewritten as follows:

$$\left(\frac{d \ln B}{dz}\right)^{-1} \frac{Z'}{Z} = \frac{1-\mu^2}{2\mu} \frac{M'}{M} ,$$

where $Z' = dZ/dz$ and $M' = dM/d\mu$.

Obviously, both sides of this equation must be equal to a constant, hence

$$\left(\frac{d \ln B}{dz}\right)^{-1} \frac{Z'}{Z} = \lambda , \quad \frac{1-\mu^2}{2\mu} \frac{M'}{M} = -\lambda .$$

Now let us find the solutions to these equations. Assuming that the magnetic field grows with the depth x as $B = B_0 e^{\alpha x}$, where α is a positive parameter, the first equation can be represented as

$$\frac{dZ}{Z} = \alpha\lambda dz .$$

The latter can be easily integrated and gives us the following solution for $X(x)$:

$$Z(z) = C_1 e^{\alpha\lambda z} .$$

At the same time, the equation for $M(\mu)$ can be rewritten as follows:

$$\frac{dM}{M} = -\frac{2\mu}{1-\mu^2} \lambda d\mu ,$$

and its integration gives us the following function $M(\mu)$:

$$M(\mu) = C_2(1-\mu^2)^\lambda .$$

Therefore, the general solution to Eq. (3.27) is as follows:

$$f_{\text{g magn}}(z,\mu) = Ce^{\alpha\lambda z}(1-\mu^2)^\lambda , \tag{3.68}$$

where C is a production of the integration constants C_1 and C_2.

It can be seen that, if the parameter λ is positive, then this general solution cannot be used as a basis for the beamlike initial pitch-angle distribution. However, if the parameter λ is negative, then this general solution can serve as a basis for the wide variety of functions that are symmetrical about $\mu = 0$ (that is, natural: the pitch-angle spectra of the injected and mirrored particles must be the same).

Thus we approximated the normal distributions $f(0,\mu) = e^{-\mu^2-1/\Delta\mu}$ for different $\Delta\mu$ by the series of general solutions (3.21). This allowed us to calculate the functions $f_{\text{magn}}(x,\mu)$ and find out how the beam pitch-angle distribution varies with precipitation depth (Figure 3.6).

Certainly this solution is valid for all pitch angles except for $|\mu| = 1$. This can be easily interpreted from the physical point of view: particles injected with $\mu = 1$ are not affected by the magnetic field convergence. At the same time, particles injected with $\mu < 1$ are mirrored at some level, which is deeper for the particles with higher pitch angles. Therefore, beams with narrower initial pitch-angle distributions are on average mirrored at deeper levels than those with wider distributions, as is demonstrated in Figure 3.6.

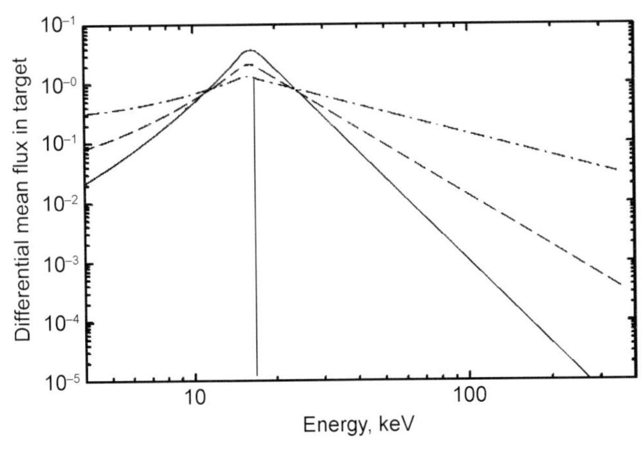

Figure 3.6 Mean flux spectra found from the solution of the continuity equation for an electron beam affected solely by Coulomb collisions. Solid line: $\gamma = 7$; dashed line: $\gamma = 5$; dotted–dashed line: $\gamma = 7$. The vertical line denotes the initial lower-energy cutoff of 16 keV accepted in these calculations.

3.3.5
Heating Functions

3.3.5.1 Heating by Bombarding Particles in the Continuity Equation Approach

The energy deposited by particles at a given column depth, $\wp(\xi, n)$, can be calculated as follows (Syrovatskii and Shmeleva, 1972a):

$$\wp(\xi, n) = \int_0^{\infty} \left(-\frac{dE}{dx}\right)_{\text{e,p}} V N(E, \xi) d\xi = n P(\xi) , \tag{3.69}$$

where $(-dE/dx)_{\text{e,p}} = (-dE/dt)_{\text{e,p}}/V$ defines the energy loss rates by the relevant kind of particles (electrons or protons). Then, using the solutions of the continuity Eq. (3.47), the volume heating function (erg cm^{-3} s^{-1}) can be defined as

$$\begin{aligned} P(\xi) &= aK \left(\frac{2}{m}\right)^{1/2} \int_{E_1}^{E_2} (E^2 + 2a\xi)^{-\frac{\gamma+1}{2}} dE \\ &= aK \left(\frac{2}{m}\right)^{1/2} (2a\xi)^{-\gamma/2} \frac{1}{2} \int_{t_1}^{t_2} \frac{t^{-1/2}}{(1+t)^{\frac{\gamma+1}{2}}} dt , \end{aligned} \tag{3.70}$$

where the integration limits are defined by Eq. (3.49) and the integral I_t by parameter t can be expressed as follows (Syrovatskii and Shmeleva, 1972a):

$$I_t(t_1, t_2) = \frac{1}{2} \int_{t_1}^{t_2} \frac{t^{-1/2}}{(1+t)^{\frac{\gamma+1}{2}}} dt = \begin{cases} \frac{1}{\gamma} \left[t_1^{-\gamma/2} F\left(\frac{\gamma+1}{2}, \frac{\gamma}{2}, \frac{\gamma+2}{2}, \frac{-1}{t_1}\right) - t_2^{-\gamma/2} F\left(\frac{\gamma+1}{2}, \frac{\gamma}{2}, \frac{\gamma+2}{2}, \frac{-1}{t_2}\right) \right] , \\ 0 \le 2a\xi < E_{\text{low}}^2 ; \\ \frac{1}{2} B\left(\frac{1}{2}, \frac{\gamma}{2}\right) - \frac{1}{\gamma} t_2^{-\gamma/2} F\left(\frac{\gamma+1}{2}, \frac{\gamma}{2}, \frac{\gamma+2}{2}, \frac{-1}{t_2}\right) , \\ E_{\text{low}}^2 \le 2a\xi \le E_{\text{upp}}^2 , \end{cases} \tag{3.71}$$

where $t_{1,2} = \max\{0, [E_{\text{low,upp}}^2/(2a\xi) - 1]\}$, where F is the Gauss hypergeometric function and $B(x, y)$ is the beta function expressed by the gamma functions as follows: $B(x, y) = \Gamma(x)\Gamma(y)/\Gamma(x + y)$.

The integral above can be calculated analytically for separate values of the spectral index γ, which for an isotropic distribution of beam electrons with $\gamma = 3$ gives the following expression (Syrovatskii and Shmeleva, 1972a):

$$\wp(\xi) = \frac{1}{2} a F_0 E_{\text{low}} \left(1 - \frac{E_{\text{low}}}{E_{\text{upp}}}\right) (2a\xi)^{-3/2}$$
$$\times \begin{cases} \arctan\sqrt{\frac{E_{\text{upp}}^2}{2a\xi} - 1} - \arctan\sqrt{\frac{E_{\text{low}}^2}{2a\xi} - 1} \\ \quad + \left(\frac{2a\xi}{E_{\text{upp}}^2}\right)\arctan\sqrt{\frac{E_{\text{upp}}^2}{2a\xi} - 1} - \left(\frac{2a\xi}{E_{\text{low}}^2}\right)\arctan\sqrt{\frac{E_{\text{low}}^2}{2a\xi} - 1}\,, \\ 0 \le 2a\xi \le E_{\text{low}}^2\,; \\ \arctan\sqrt{\frac{E_{\text{upp}}^2}{2a\xi} - 1} + \left(\frac{2a\xi}{E_{\text{upp}}^2}\right)\arctan\sqrt{\frac{E_{\text{upp}}^2}{2a\xi} - 1}\,, \\ E_{\text{low}}^2 \le 2a\xi \le E_{\text{upp}}^2\,, \end{cases} \tag{3.72}$$

where F_0 is the beam flux measured in units of $10^{10}\,\text{erg}\,\text{cm}^{-2}\,\text{s}^{-1}$ and ξ_{low} is the stopping depth of electrons with a lower cutoff energy of E_{low}.

For an isotropic beam with $\gamma = 6$ (with infinite upper energy) the heating function is modified as follows:

$$\wp(\xi) = \frac{1}{2} a F_0 E_{\text{low}} \left(\frac{\xi}{\xi_{\text{low}}}\right)^{-3}$$
$$\times \begin{cases} \frac{16}{15} - \frac{2}{3}\sqrt{\frac{\xi_{\text{low}}}{\xi} - 1}\left[\left(\frac{\xi}{\xi_{\text{low}}}\right)^{3/2} + \frac{4}{3}\left(\frac{\xi}{\xi_{\text{low}}}\right)^{3/2} + \frac{8}{3}\left(\frac{\xi}{\xi_{\text{low}}}\right)\right] \\ 0 \le \xi \le \xi_{\text{low}}\,; \\ \frac{16}{15}\,, \\ \xi_{\text{low}} \le \xi\,, \end{cases} \tag{3.73}$$

where, as before, F_0 is the beam flux measured in units of $10^{10}\,\text{erg}\,\text{cm}^{-2}\,\text{s}^{-1}$ and ξ_{low} is the stopping depth of electrons with a lower cutoff energy of E_{low}.

3.3.5.2 Heating by Particle Beams in the Flux Conservation Approach

In this case one can use the same Eq. (3.69) for heating by beam electrons. However, for the density of beam electrons at given depths one can use the electron flux conservation equation by setting to zero in the continuity Eq. (3.44) the first and second terms separately. This step allows one to separate the energy loss variations with depth from the beam density and connecting the beam density at the top boundary n_{B0} and at a given depth ξ (Emslie, 1978) as follows:

$$n_{\text{B}}(E, \xi, t)\mu(\xi, t) d E = n_{\text{B0}}(E, t)\mu_0(t)\frac{V_0(E, t)}{V(E, \xi, t)} d E_0\,, \tag{3.74}$$

where the index "0" refers to the initial energy distribution of an electron beam.

Then, similarly to the continuity equation, using the electron flux at the top boundary $F_0 = n_{\text{B0}} V_0$ and the expressions for variations of E and μ with a col-

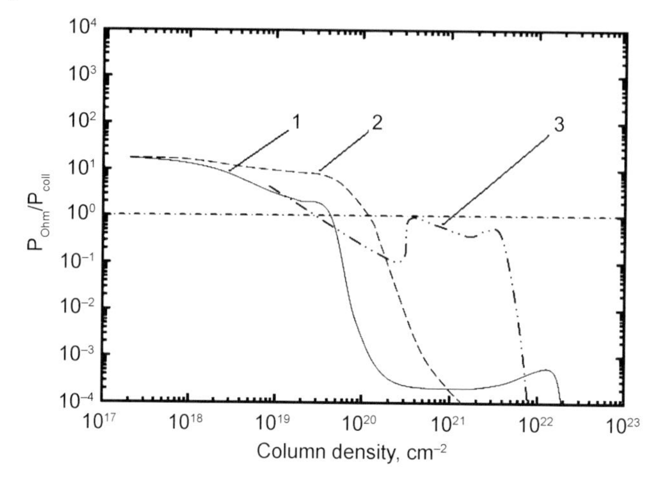

Figure 3.7 Comparison of heating functions by beam electrons with distributions derived for pure collisional energy losses from the continuity equation (solid line 1) and from the conservation equation (dotted–dashed line 3) and for pure ohmic losses (dashed line 2) for an electron beam with an initial energy flux of 10^{11} erg cm^{-2} s^{-1} and spectral index of 6.

umn depth (Eqs. (3.39)), one can obtain (Emslie, 1978)

$$P(\xi, t) = 2\pi e^4 n\vartheta \int_{E_0}^{\infty} \frac{F_0(E_0, t)\, d E_0}{E_0 \left[1 - (2 + \beta/2)\left(\vartheta \pi e^4/\mu_0 E_0^2\right)\right]^{\frac{2+\beta}{4+\beta}}} , \qquad (3.75)$$

where the factors β and ϑ are defined in Section 3.2.2.2 for bombarding electrons and protons.

Strictly speaking, while this approach provides correct variations for energy loss with depth in the corona, it significantly overestimates the number of electrons and, thus, heating at chromospheric depths (Figure 3.7), as can be seen from solutions (3.47) of the continuity equations and, especially, when heating by these electrons is evaluated as discussed above (see, e.g., Mauas and Gomez, 1997).

3.4
Continuity Equation Approach for Electrons – Pure Electric Field

As was established a long time ago (Cox and Bennett, 1970; Hoyng *et al.*, 1978; Knight and Sturrock, 1977), precipitating beams of charged particles (as plotted in Figure 3.4) carry large electric currents throughout an entire flaring atmosphere, which induces an electrostatic electric field $\mathcal{E}$. These currents, obviously, need to be neutralized by the return current formed from the ambient plasma electrons in order to avoid excessive heating in flares caused by large currents. The effect of return currents is strongly dynamic and dependent on the ambient plasma temperature as $\sim T^{-3/2}$, meaning that when the ambient temperature significantly increases, the return current will similarly decrease.

Table 3.2 Magnitudes of electric field induced by beam electrons at the top boundary for different initial energy fluxes of beams and different lower cutoff energies.

	$E_{low} = 30$ keV		$E_{low} = 20$ keV		
	$\mathcal{E}$, V cm^{-1}	$\mathcal{E}/\mathcal{E}_D$	$\mathcal{E}$, V cm^{-1}	F_0, erg cm^{-2} s^{-1}	$\mathcal{E}/\mathcal{E}_D$
10^{12}	1.0×10^{-3}	1.6	1.6×10^{-3}		2.6
10^{10}	1.0×10^{-5}	1.6×10^{-2}	1.6×10^{-5}		2.6×10^{-2}
10^{8}	1.0×10^{-7}	1.6×10^{-4}	1.6×10^{-7}		2.6×10^{-4}

The magnitudes of the electrostatic electric field induced at the top boundary by electron beams with different parameters measured in units of a Dreicer electric field are presented in Table 3.2. It can be seen that the electric field can be rather high for some intense beams, exceeding the local Dreicer field by two orders of magnitude. Hence this electric field must be taken into account in any consideration of electron-beam kinetics.

3.4.1
Estimation of the Ohmic Loss Effect

Let us evaluate the additional energy losses introduced by the electric current carried by charged particles (and consider electrons for certainty). Emslie (1980) has shown that ohmic losses by beam electrons can be defined as

$$\begin{aligned}\left(\frac{dE}{d\xi}\right)_{\mathrm{O}} &= \frac{1}{nV_x}\frac{dE}{dt} = -\frac{e\mathcal{E}}{n}\,,\\ \left(\frac{dV_x}{d\xi}\right)_{\mathrm{O}} &= \frac{1}{nV_x}\frac{dV_x}{dt} = -\frac{1}{nV_x}\frac{e\mathcal{E}}{m_e}\,,\end{aligned} \tag{3.76}$$

where the factor n for the ambient plasma density appears because ohmic losses are dependent on a linear distance, not the column density, and the electric field $\mathcal{E}(\xi)$ carried by beam electrons at a given depth is defined by the expression

$$\mathcal{E}(\xi) = \frac{e}{\sigma}\int_{E_{low}}^{E_{upp}} F(E,\xi)\,dE\,, \tag{3.77}$$

where F is the electron flux at a given depth defined by Eq. (3.52) and σ is the ambient plasma conductivity, which is defined as follows:

$$\eta = \frac{1}{\sigma} = \frac{7.28 \times 10^{-8} X}{T_e^{1.5}} \ln\left(\frac{3}{2e^3}\frac{k_B^3 T_e^3}{\pi n}\right) + \frac{7.6 \times 10^{-18}(1-X)}{X} T_e^{0.5}\,, \tag{3.78}$$

where X is the ionization degree.

3.4.1.1 Comparison of Energy Losses by Electrons in Collisions and Electric Field

Let us consider the energy losses in a pure electric field (Emslie, 1980):

$$\frac{dE}{\mu dx} = -e\mathcal{E} \,, \tag{3.79}$$

where x is the linear precipitation depth.

Obviously, the lower-energy electrons lose their energy at higher atmospheric levels than the more energetic ones. Let us introduce an "electric" stopping depth $x_{\mathcal{E}}$ at which an electron with energy E loses it completely, that is:

$$x_{\mathcal{E}} = \frac{E}{e\mathcal{E}} \quad \text{or}$$
$$\xi_{\mathcal{E}} = \frac{E}{e\mathcal{E}n} \tag{3.80}$$

For example, for an electric field of $10^{-3}\,\mathrm{V\,cm^{-1}}$, electrons with initial energies of 30 and 300 keV lose their energy completely at the associated "electric" column depths of 5×10^{17} and $6 \times 10^{23}\,\mathrm{cm^{-2}}$, respectively, for the atmosphere described in Section 3.2.2.4. Hence, depending on the strength of the induced electric field $\mathcal{E}$, the "electric" stopping column depth can be much shorter than the collisional one. This leads to lower-energy-beam electrons returning to the injection source in the corona from much higher atmospheric levels well before they fully lose their energy in collisions.

In Table 3.1 we present a comparison of the "collisional" and "electric" stopping depths calculated for the electric field magnitudes relevant for the different electron-beam parameters. From Table 3.1 one can conclude that for weak and moderate hard electron beams the collisional stopping depths are always lower than the electric one, thus leading to the domination of collisions above ohmic losses. However, for intense softer beams, which induce higher electrostatic electric fields, their electric stopping depths are located very high in the corona ($\leq 10^{18}\,\mathrm{cm^{-2}}$), much higher than collisional ones, that is, close to the injection source ($10^{17}\,\mathrm{cm^{-2}}$). This means that ohmic losses dominate the electron-beam kinetics at these depths; for example, electrons will return to the source in the corona well before they lose enough energy to become thermalized as happens at the collisional stopping depths.

The measure of the return current effect is defined as the ratio of ohmic energy losses to collisional energy losses: $\mathrm{Re} = P_{\mathrm{Ohm}}/P_{\mathrm{coll}}$. Let us now try to compare the heating rates in collisions and electric fields and calculate the ohmic heating as the square of the current carried by beam electrons, for example, $j^2(\xi)/\sigma$. Ohmic heating rates are strongly dependent on the conductivity, or resistivity, of a flaring atmosphere, which significantly changes with column depth because of its dependence on temperature (Eq. (3.78)), which strongly varies with depth as per the hydrodynamic model presented in Figure 3.8.

Despite the growth in resistivity with increases in column depth, the ohmic losses significantly decrease. This happens because the square of the beam particle flux defines ohmic heating decreases with depth as $j^2 \sim \xi^{-6\ldots-2}$ (depending on

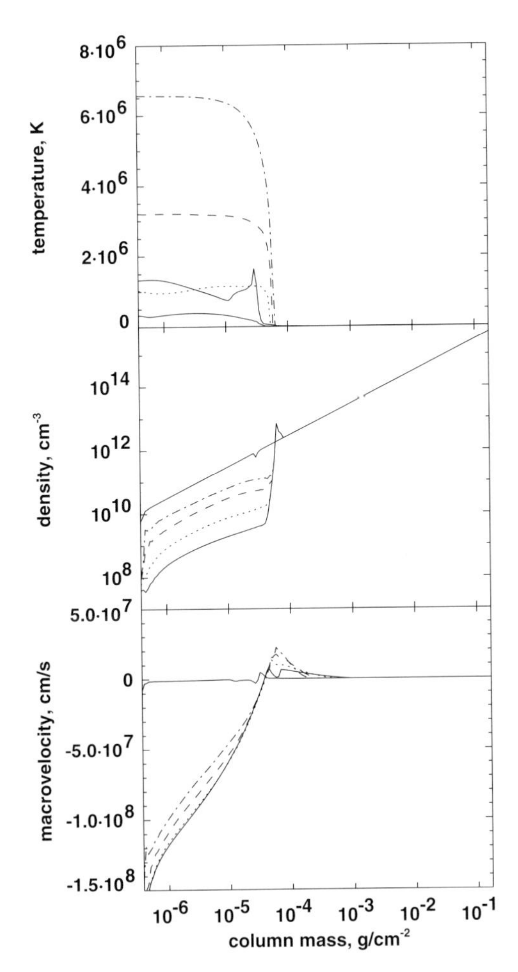

Figure 3.8 Model flaring atmosphere calculated with a hydrodynamic approach to ambient plasma heating by beam electrons with an initial energy flux of 10^{11} erg cm^{-2} s^{-1} and spectral index of 3. The calculations were carried out for the dotted–dashed curve corresponding to a time of 10 s after beam onset (for details see Chapter 6). The mass density is calculated from the column density ξ by multiplying it by the average atmosphere mass per particle, $1.44 m_{\mathrm{H}}$, where m_{H} is the mass of a hydrogen atom.

the initial spectral index γ) (Zharkova and Gordovskyy, 2005a), resistivity decreases much slower as $(\sigma(\xi)n(\xi) \sim \xi^{-1})$. Hence, the ratio $j^2/(\sigma n)$ should reveal a decrease with depth, as shown in Figure 3.9.

It can be seen that the return current effect $\mathrm{Re}(\xi)$ reveals two regions: one where the ohmic losses are nearly constant – in the upper atmosphere at column densities of $\xi \leq 10^{20}$ cm^{-2} (this depth is determined by the collisional stopping depth of the electrons with a lower cutoff energy) – and another in deeper atmospheric regions where the return current effect reveals a sharp decrease, becoming almost insignificant (in comparison with the collisional losses) at depths of $\sim 10^{21}$ cm^{-2}.

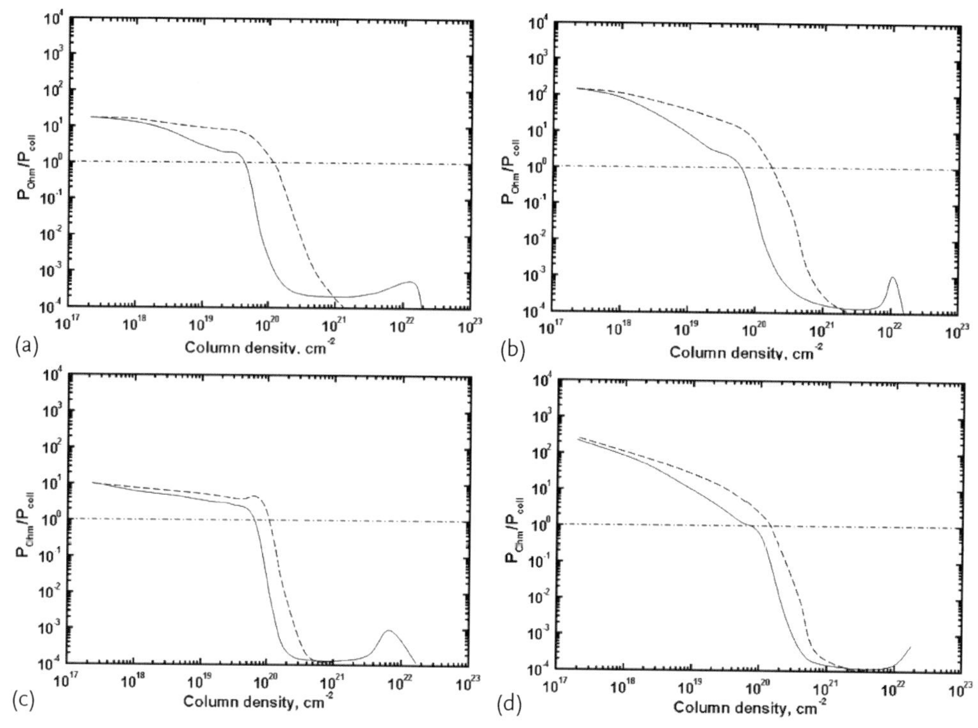

Figure 3.9 Ratios of ohmic to collisional heating rates produced by electron beams with an initial energy flux of $F_0 = 10^{10}\,\mathrm{erg\,cm^{-2}\,s^{-1}}$ (a,c), $F_0 = 10^{12}\,\mathrm{erg\,cm^{-2}\,s^{-1}}$ (b,d), and $\gamma = 3$ (a,b) and 7 (c,d). Adopted from the simulations by Zharkova and Gordovskyy (2005a). The dashed curves represent the heating found from the analytical solutions in the continuity equation approach without pitch-angle scattering, while the solid curves represent those with pitch-angle scattering.

Table 3.3 Comparison of Coulomb and ohmic total energy losses (from whole flaring atmosphere) for a beam with different initial energy fluxes for $\gamma = 3$, $E_{\mathrm{low}} = 16\,\mathrm{keV}$, and $E_{\mathrm{upp}} = 384\,\mathrm{keV}$.

Flux, $\mathrm{erg\,cm^{-2}\,s^{-1}}$	Pure Coulomb heating, $\mathrm{erg\,s^{-1}}$	j^2/σ, $\mathrm{erg\,s^{-1}}$
10^8	1.7×10^{-12}	$\sim 10^{-13}$
10^{10}	1.7×10^{-10}	$\sim 10^{-9}$
10^{12}	1.7×10^{-8}	$\sim 10^{-6}$

A comparison of total heating of the atmosphere in collisions and electric current by beams with different parameters is presented in Table 3.3, which reveals that beams with higher initial energy fluxes and spectral indices produce ohmic heating exceeding that from collisions. Therefore, the effect of an electric field induced by beam electrons is very important to the precipitation of charged particles and needs to be investigated with the same detail as for collisions.

3.4.2
Kinetic Solutions for a Pure Electric Field

Let us consider the kinetics of a beam governed only by the energy losses in a pure electric field $\mathcal{E}$ without any collisions. Based on the analysis in the section above, let us assume that during precipitation of a power-law electron beam there is a constant or variable electric field that is not affected by the precipitating beam (no collisions or pitch-angle scattering). For the sake of simplicity, let us also assume that the beam is injected in the direction $\mu = 1$ with δ-like pitch angular distribution and the particles can either precipitate downward ($\mu = 1$) or, after they fully lose their energy, move backward to the source in the corona ($\mu = -1$).

3.4.2.1 Continuity Equation

The electron-beam differential density in the flaring atmosphere can be found from the continuity equation (Syrovatskii and Shmeleva, 1972a):

$$\frac{\partial}{\partial x}\left[V \cdot N(x, E)\right] + \frac{\partial}{\partial E}\left[\left(\frac{dE}{dx}\right) V \cdot N(x, E)\right] = 0 . \tag{3.81}$$

Let us assume that the initial beam spectrum is power law with the spectral index γ (for the flux) giving the initial electron density as

$$N(E,0) = KE^{-\gamma-0.5}\Theta(E - E_{\text{low}}) \tag{3.82}$$

and find the analytical solutions of the continuity equation for constant and variable electric fields.

3.4.2.2 Constant Electric Field

For a constant electric field ($\mathcal{E} = \mathcal{E}_I$) the beam differential densities can be found analytically from the continuity equation using the same method as in the paper by Syrovatskii and Shmeleva (1972a) with the energy losses taken from Eq. (3.80):

$$N(E, x) = K\mathcal{E}_I^{-0.5} \cdot (E + e\mathcal{E}_I x)^{-\gamma} \times \Theta(E - E_{\text{low}} + e\mathcal{E}_I x)\Theta(E_{\text{upp}} - e\mathcal{E}_I x - E) , \quad \text{for} \quad \mu = +1 ; \tag{3.83}$$

$$N(E, x) = K \cdot \mathcal{E}_I^{-0.5}(e\mathcal{E}_I x)^{-\gamma} \times \Theta(E - E_{\text{low}} - e\mathcal{E}_I x)\Theta(E_{\text{upp}} + e\mathcal{E}_I x - E) , \quad \text{for} \quad \mu = -1 . \tag{3.84}$$

The differential spectra above are asymmetric for the precipitating ($\mu = 1$) and returning ($\mu = -1$) particles, illustrated by Figure 3.10 for electron beams with an initial flux of 10^8 and 10^{12} erg cm^{-2} s^{-1} and an initial index of $\gamma = 3$ (Figure 3.10a – for $\mu = 1$ and 3.10b – for $\mu = -1$, respectively). After completely losing their energy in the electric field, precipitating electrons become accelerated by this field

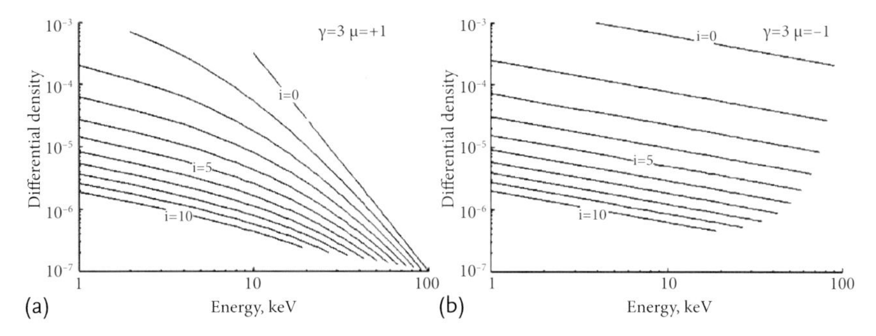

Figure 3.10 Differential densities found in a pure electric field approach for precipitating electrons (a) and returning electrons (b) from the continuity equation for a beam with $\gamma = 3$ and an electric field magnitude of 5×10^{-5} V cm^{-1}. The depths of beam precipitation are calculated according to the numbers i as $x_i = i \times 1.6 \times 10^8$ cm.

in the opposite direction and return to the source; let us call them returning electrons. The energy gained by the latter during acceleration depends on their original energy, or their stopping depth, from which they start their way back to the source. This depth is smaller for lower-energy electrons and larger for higher-energy ones, as demonstrated by Figure 3.10b for $\mu = -1$.

The differential densities of precipitating electrons depend on the linear precipitation depth x, which can be associated with the column depth, and the ambient plasma density as $x \simeq \xi / n$. As expected, for a constant electric field in a flaring atmosphere with exponentially increasing density and a precipitating distance x beam electrons steadily lose the same amount, $e\mathcal{E}x$, of their energy. For lower-energy electrons, with every step of precipitation their loss becomes more noticeable compared to their energy. Therefore, at deeper atmospheric layers the electron differential spectra at lower energies steadily decrease, or become flatter, until all electrons are fully decelerated by the electric field (see Figure 3.10a for $\mu = 1$). Hence, unlike the electron spectra for pure collisional losses, which are dependent only on column depth, the distributions of beam electron densities in an electric field are dependent on the linear depth.

3.4.2.3 **Variable and Combined Electric Field**

As discussed in Section 4.4.2.3, the induced electric field found from the full kinetic solutions is mostly variable in a flaring atmosphere, being a constant only at higher atmospheric levels, after which it falls exponentially (linearly, parabolically, or steeper) (see Figure 3.11).

Let us obtain the pure electric solutions for a few cases of ohmic energy loss in a variable electric field decreasing with depth as

$$\mathcal{E} = \mathcal{E}_0 \left(\frac{x}{x_t} + 1 \right)^{-k} ,$$

where $\mathcal{E}_0$ is a constant electric field (magnitude at injection) and x_t is the turning depth where the electric field starts decreasing and $x \geq x_t$.

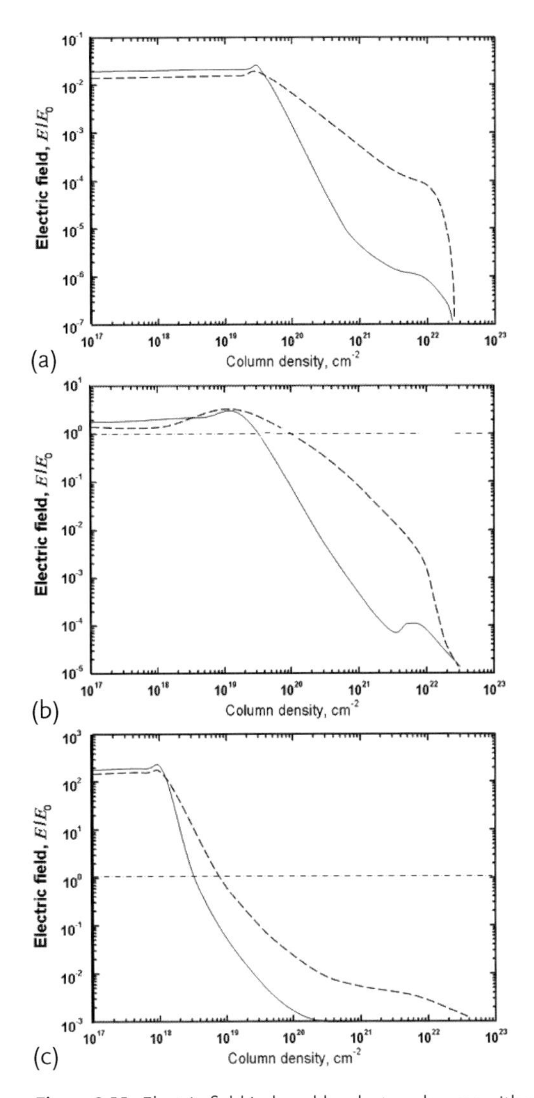

Figure 3.11 Electric field induced by electron beams with γ=3 (dashed lines) and 7 (solid lines) and an initial energy flux of $F_0 = 10^8$ erg cm^{-2} s^{-1} (a), $F_0 = 10^{10}$ erg cm^{-2} s^{-1} (b), and $F_0 = 10^{12}$ erg cm^{-2} s^{-1} (c). From simulations by Zharkova and Gordovskyy (2005a).

For $k = 1$ the differential spectra of precipitating beam electrons can be written as

$$N(E, x) = \frac{K}{\sqrt{E}} \left(E + e\mathcal{E}_0 x_t \ln \frac{x}{x_t} \right)^{-\gamma} \Theta \left(E - E_{low} + e\mathcal{E}_0 x_t \ln \frac{x}{x_t} \right)$$
$$\times \Theta \left(E_{upp} - e\mathcal{E}_0 x_t \ln \frac{x}{x_t} - E \right) ; \tag{3.85}$$

for $k \succeq 1$ they can be defined as follows:

$$N(E, x) = \frac{K}{\sqrt{E}} \left\{ E + \frac{e\mathcal{E}_0 x_t}{k-1} \left[1 - \left(\frac{x}{x_t} \right)^{1-k} \right] \right\}^{-\gamma}$$
$$\times \Theta \left\{ E - E_{low} + \frac{e\mathcal{E}_0 x_t}{k-1} \left[1 - \left(\frac{x}{x_t} \right)^{1-k} \right] \right\}$$
$$\times \Theta \left\{ E_{upp} - \frac{e\mathcal{E}_0 x_t}{k-1} \left[1 - \left(\frac{x}{x_t} \right)^{1-k} \right] - E \right\} . \tag{3.86}$$

Precipitating electrons with lower energies, whose "electric" stopping depth is above the depth of a turning point, will lose less of their energy in deceleration by a decreasing electric field. This will result in less flattening of the differential energy spectra as demonstrated in Figure 3.12 for a beam with an initial flux of 10^{12} erg cm^{-2} s^{-1} and $\gamma = 3$ and for a constant (Figure 3.12a) and decreasing electric field as $\sim x^{-1}$ (Figure 3.12b) and $\sim x^{-2}$ (Figure 3.12c).

In Figure 3.13 differential spectra are plotted for a combined electric field with a turning point occurring at step 4 ($i = 4$) for an electric field decrease of $\sim x^{-1}$ (Figure 3.13a), $\sim x^{-2}$ (Figure 3.13b), and $\sim x^{-5}$ (Figure 3.13c). It can be seen that precipitating electrons with higher energies, whose "electric" stopping depth is deeper than the turning point, suddenly become less decelerated because of the decreasing electric field, and as result they can precipitate much deeper. Hence, a decrease of the electric field leads to a smaller energy loss by electrons and to smaller flattening of their differential spectra in comparison with the constant electric field.

In Figure 3.14 differential spectra are plotted for a combined electric field with a turning point occurring at step 4. The constant part of the electric field $\mathcal{E}$, varies from 5×10^{-5} V cm^{-1} (Figure 3.14a) to 4×10^{-4} V cm^{-1} (Figure 3.14b) and the variable part decreases with depth as $\sim x^{-2}$. Note that for a higher electric field $\mathcal{E}$, there is a much faster decrease in the electron numbers at lower energies and, as a result, greater flattening in their differential spectra.

3.4.2.4 Mean Electron Fluxes

Constant electric field. Let us calculate the mean electron fluxes in a constant electric field by performing in Eq. (3.77) the integration of the differential spectra from Section 3.4.2.2 in a depth range of 0 to infinity as follows:

$$\bar{\mathcal{F}}(E) = \sqrt{\frac{2}{m_e}} K \frac{\gamma - 1}{e\mathcal{E}} E^{-\gamma+1} \sim E^{-\beta_E} . \tag{3.87}$$

The mean electron flux of electron beams with an initial energy flux of 5×10^{10} erg cm^{-2} s^{-1} and spectral index of $\gamma = 3$ is plotted in Figure 3.15a (curve 1) for a constant electric field magnitude of 5×10^{-5} V cm^{-1}.

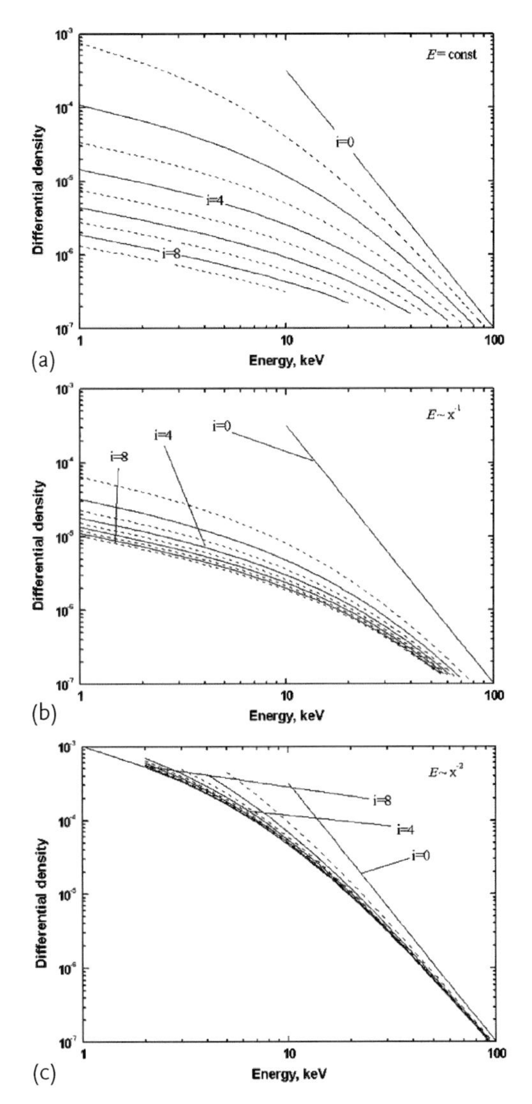

Figure 3.12 Differential spectra of beam electrons with $\gamma = 3$ calculated in a pure electric approach for a constant field (5×10^{-5} V cm^{-1}) (a) and electric field decreasing as $1/x$ (b) and $1/x^2$ (c). The depths of beam precipitation are calculated as in Figure 3.10.

Note that for a constant electric field the spectral index β_{E} is related to the initial spectral index γ of the beam electrons as

$$\beta_{\mathrm{E}} \simeq \gamma - 1 , \tag{3.88}$$

which is different by 1 from $\beta = \gamma - 2$ found for pure collisions (Brown, 1971; Brown *et al.*, 2006; Syrovatskii and Shmeleva, 1972a).

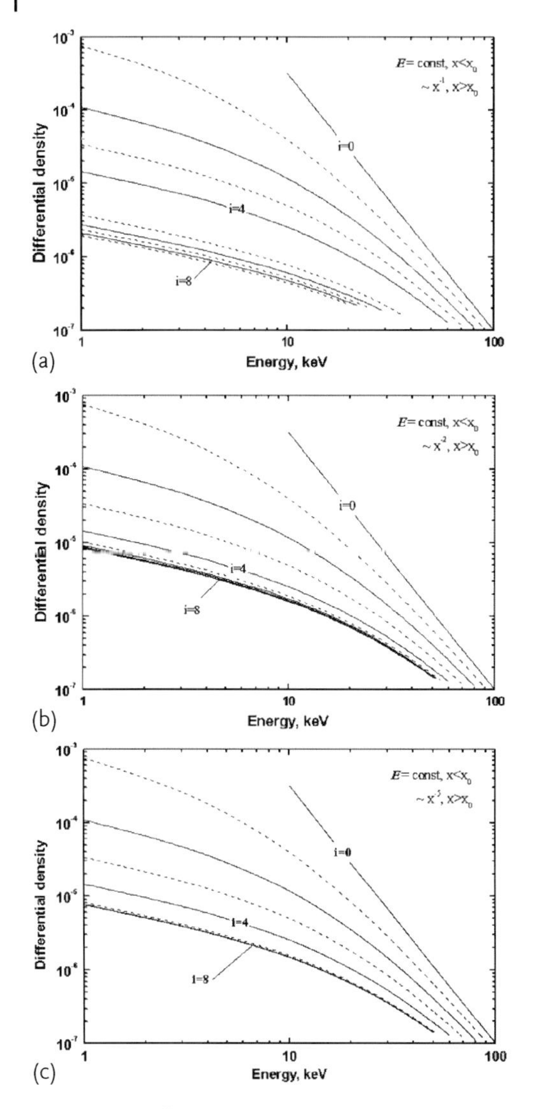

Figure 3.13 Differential spectra of beam electrons with $\gamma = 3$ calculated with a pure electric approach for a combined electric field with a constant part of 5×10^{-5} V cm^{-1}, turning point at $i = 4$, and with a variable field dropping as $1/x$ (a), $1/x^2$ (b), and $1/x^5$ (c). The depths of beam precipitation are calculated as in Figure 3.10.

Variable electric field. By repeating the procedure above for a depth range of between $x_{\min}$ and $x_{\max}$ with the differential densities from Section 3.4.2.3 we can obtain the mean electron fluxes for electrons in a variable electric field presented in Figure 3.15b,c. The mean electron fluxes are calculated for beams with $\gamma = 3$ in a variable electric field decreasing as $\sim x^{-1}$ (curve 2) and as $\sim x^{-2}$ (curve 3). For

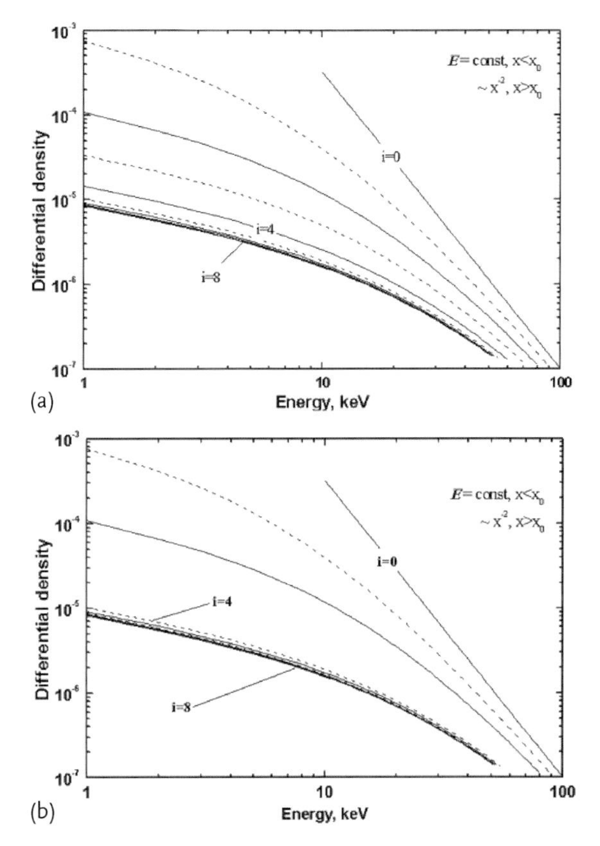

Figure 3.14 Differential spectra of beam electrons with $\gamma = 3$ calculated with a pure electric approach for a combined electric field falling as $1/x^2$ with a turning point at $i = 4$ and with a constant part of 5×10^{-5} V cm^{-1} (a) and 4×10^{-4} V cm^{-1} (b). The depths of beam precipitation are calculated as in Figure 3.10.

a field decreasing as $\sim x^{-1}$, the mean flux index $\beta = \gamma - 0.6 = 2.4$, and for an electric field decreasing as $\sim x^{-2}$, $\beta = \gamma - 0.1 = 2.9$.

Combined electric field. For a combined electric field, that is, constant before the turning depth x_t, or ξ_t, somewhere in the corona (in the fourth depth point, for example) and decreasing afterward, the differential spectra in the constant part are as defined from Section 3.4.2.2, and for an electric field falling with depth as $\sim x^{-k}$ the differential spectra are as defined from Section 3.4.2.3. Then, by performing numerical integration within the limits $x_{\rm min}$ and $x_{\rm max}$, we obtain the mean electron fluxes for a combined electric field.

The results are presented in Figure 3.16a for the same electric field variations as in Figure 3.13: constant at 5×10^{-5} V cm^{-1} until depth 4 and decreasing as $\sim x^{-1}$ (curve 1), $\sim x^{-2}$ (curve 2), or $\sim x^{-5}$ (curve 3). The resulting mean electron flux distributions show a stronger spectra flattening toward lower energies for those

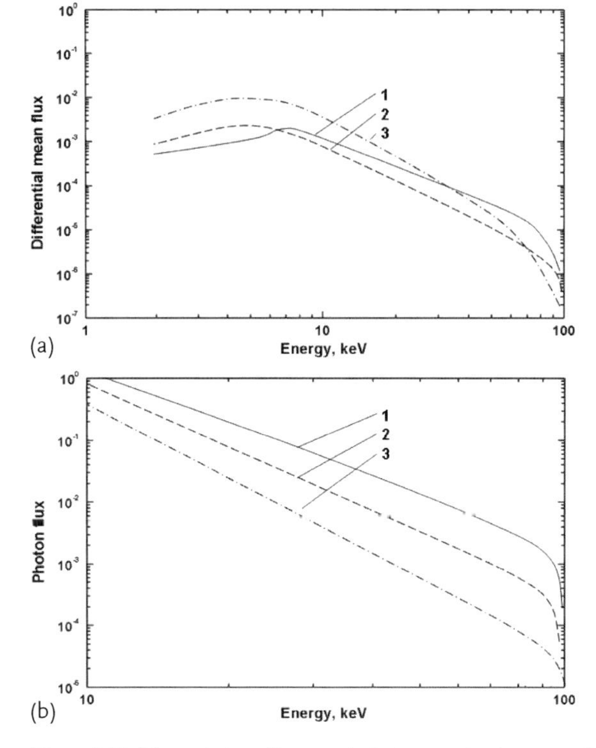

Figure 3.15 Mean electron fluxes and photon spectra in a pure electric field approach for the differential spectra from Figure 3.10.

electric fields that decrease faster with depth (compare curve 1 with curves 2 and 3), which reflects higher energy losses in an electric field for lower-energy particles as discussed in Section 3.4.2.3.

3.4.2.5 Photon Spectra

Constant electric field. If we substitute the solutions for differential spectra from Section 3.4.2.2 into Eq. (3.57), then the photon fluxes for pure electric solutions can be defined as

$$\frac{dI(h\nu)}{d\nu} = \frac{K}{(\gamma - 1) \cdot e\mathcal{E}}(h\nu)^{-\gamma} \sim h\nu^{-\delta}. \tag{3.89}$$

The photon fluxes from beam electrons precipitating in a constant electric field are also power laws with the spectral index δ being equal to the spectral index γ of the initial electron beam (see Figure 3.16b, curve 1). This is different from the pure collisional case where the photon flux spectra index $\delta = \gamma - 1$ (Syrovatskii and Shmeleva, 1972a; Zharkova and Gordovskyy, 2005a).

Combined electric field. The photon spectra produced by electron beams propagating in a variable electric field calculated from the differential spectra in Sec-

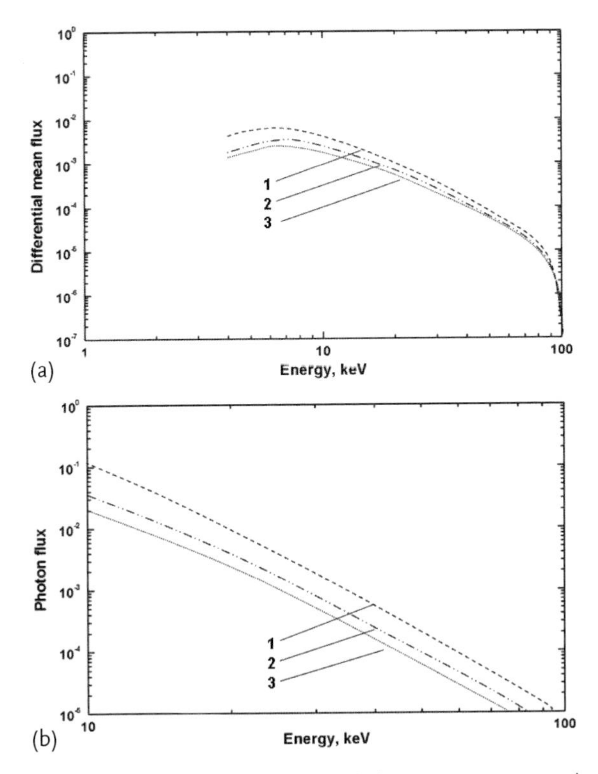

Figure 3.16 Mean electron fluxes and photon spectra in pure electric field approach for the differential spectra from Figure 3.14.

tion 3.4.2.3 for the HXR cross-sections by Haug (1997) are presented in Figure 3.16b. The photon spectra are calculated for the same HXR cross-section as for beam electrons with $\gamma = 3$, with the differential spectra plotted in Figure 3.13 precipitating in depth points 1–4 in a constant electric field of 5×10^{-5} V cm^{-1} and from depth point 5 in a decreasing electric field as $\sim x^{-1}$ (curve 1) or as $\sim x^{-2}$ (curve 2) and $\sim x^{-5}$ (curve 3).

The spectral indices of photon spectra at lower energies are close to $\delta_{\text{low}} = 2.8$ (curves 1–3), which is slightly lower than the electron index of 3. On the other hand, photon spectral indices at higher energies, $\delta_{\text{high}} = 3.5$–$4.3$, increase for electric field decreases (compare curve 1 for $\sim x^{-1}$, curve 2 for $\sim x^{-2}$, and curve 3 for $\sim x^{-5}$). It can be deduced that a decrease of a variable electric field as $\sim x^{-k}$ leads to an increase of photon spectral indices approximately as $\delta \approx \gamma + k/(k+1)$.

Hence the combined (constant plus decreasing) electric field can lead to a noticeable hardening in the photon spectra at lower energies and softening at higher ones. This can be explained by the fact that in a combined electric field, beam electrons are first decelerated by a constant electric field. The magnitude of this deceleration is proportional to the initial energy flux of the precipitating electron beam, leading to a substantial decrease (proportionally to this flux) of the photon flux indices at lower energies. However, starting from the turning point, beam electrons become less and less decelerated by the electric field, as discussed in Section 3.4.2.3. This will lead to a smaller flattening at lower energies of their differential spectra and to the photon spectra indices' being higher than the electron-beam ones, as was established in the current section.

The heating produced by electron beam via ohmic losses in the self-induced electric field shown in Figure 3.7 (dashed line 2 for ohmic losses vs. solid line 1 for collisions) reveals that for softer intense beams ohmic heating dominates collisional heating in the corona and upper chromosphere, while collisional losses dominate in the lower chromosphere.

3.4.3
Estimations of Electron-Beam Stability

Particle beams with such high densities carry a significant electric field, which during the passage through a flaring atmosphere can cause a substantial return current of the ambient electrons neutralizing this electric field (Emslie, 1980; Knight and Sturrock, 1977). As shown by Ivanov and Kocharov (1989), the return current remains stable against excitation of the ion-acoustic and electrostatic ion-cyclotron waves if the drift velocity V_d of the returning electrons of the ambient plasma does not exceed the critical value V_c:

$$V_c = b\frac{T_e}{T_i}\left(\frac{k_b T_e}{m_e}\right)^{1/2}, \tag{3.90}$$

where T_e and T_i are the kinetic temperatures of electrons and ions, respectively. The ratio $b(T_e)/(T_i)$ is evaluated to be equal to 0.46 for $T_e = T_i$ and 0.06 for $T_e = 10T_i$.

Let us check this condition with the solutions found for differential spectra of beam electrons (Eq. (3.47)). The current of the beam is described by the equation

$$J_b(\xi) = e\int_{E_1}^{E_2} N(E,\xi)V(E)dE, \tag{3.91}$$

where E_1 and E_2 are those energies defined by Eq. (3.41), for example, $E_1(\xi) = \mathrm{Re}\sqrt{E_{low}^2 - 2a\xi} \le E \le \mathrm{Re}\sqrt{E_{upp}^2 - 2a\xi}$, where Re means a real part. Thus by conducting the integration and using Eq. (3.91) above for $N(\xi)$ one obtains $J_b(\xi) = eF_0N_1(\xi)$, where

$$N_1(\xi) = \frac{\gamma-2}{\gamma-1}\frac{(E_1E_2)^{\gamma-2}}{E_2^{\gamma-2} - E_1^{\gamma-2}}(t_1+1)^{\frac{1-\gamma}{2}} - (t_2+1)^{\frac{1-\gamma}{2}}, \tag{3.92}$$

where t_1 and t_2 are defined by Eq. (3.51) with E_{low} replaced by E_1 and E_{upp} by E_2.

Then the return current is defined as

$$J_r(\xi) = en_e(\xi)V_d, \tag{3.93}$$

where n_e is the ambient electron density. By assuming $J_b = J_r$ one can define the critical initial energy flux F_c when the drift velocity of the ambient electrons V_d at the top boundary is equal to V_c:

$$F_c(\xi) = \frac{n_e(\xi)}{N_1(\xi)} b \frac{T_e}{T_i} \left(\frac{k_b T_e}{m_e} \right)^{1/2} , \tag{3.94}$$

which defines the limit above which the return current induced by beam electrons will produce plasma instabilities.

The estimations show a dependence on the ratio of the plasma and beam electrons, which is a very dynamic feature even in the pure collisional case. The density of ambient plasma electrons is affected by a hydrodynamic reponse of the plasma to beam electron precipitation, which is a steep function in depth as discussed in Chapter 5. It has been shown that the beam density (Figures 3.4, 3.11, 3.12) stays rather high in the upper atmospheric levels and remains nonnegligible in the lower chromosphere owing to a partial ionization of the ambient plasma.

The ratio of beam to ambient electron density can be rather high and meet the criteria for generating plasma instabilities, which can further reduce the electron-beam density. Zharkova and Kobylinskij (1992) showed that for the relevant hydrodynamic models (Chapter 6) these fluxes do not exceed 1.3×10^{10} erg cm^{-2} s^{-1} for beams with $\gamma = 3$ and 1.93×10^{10} erg cm^{-2} s^{-1} for beams with $\gamma = 5$. However, these estimations can be different at the very top and in deeper atmospheric levels, which will be discussed later.

4
Electron Beam Precipitation – Fokker–Planck Approach

4.1
General Comments on Particle and Energy Transport

It is well accepted that the agents delivering the energy required to account for observed features in HXR, MW, and other emissions are subrelativistic charged particles (beams of electrons (Brown, 1971; Brown *et al.*, 2006; Syrovatskii and Shmeleva, 1972a), possibly mixed with protons (Simnett, 1995)) accelerated somewhere in the corona. These particles precipitate downward to the lower solar atmosphere while depositing their energy into the ambient plasma in different processes at various atmospheric depths. Obviously, the diagnostics of this precipitation from different types of emission can provide a valuable insight into the complicated processes in which these particles deposit their energy at various depths of flaring loops. The variety of observations have resulted in a number of different models of particle acceleration and transport developed for different types of flaring events (see the reviews by Holman *et al.*, 2011; Zharkova *et al.*, 2011a, in the forthcoming RHESSI book), which can account for the observed high-energy emission.

In recent years good progress has been made in the quantitative interpretation of HXR emission using RHESSI observations (Lin *et al.*, 2003), which provide the locations and shapes of HXR sources on the solar disk, their light curves, and photon and electron spectra evolution during a flare (Brown *et al.*, 2006; Holman *et al.*, 2003, 2011; Krucker *et al.*, 2008a). Flare events are sometimes comprised of one loop with two footpoints and one coronal source (Masuda *et al.*, 1994; Sui *et al.*, 2002) or of several sets of loops with a number of footpoints and coronal sources (Battaglia and Benz, 2006; Krucker *et al.*, 2008a). Observations of solar flares by TRACE and RHESSI have revealed that the areas of flaring loops decreases and, thus, their magnetic field increases with the depth of the solar atmosphere (Kontar *et al.*, 2008).

RHESSI observations of double power-law energy spectra with flattening toward lower photon energies (Holman *et al.*, 2003), which leads to the soft–hard–soft temporal pattern of the photon spectra indices below 35 keV (Grigis and Benz, 2004), have highlighted the role in their formation of the self-induced electric field (Zharkova and Gordovskyy, 2006). This revealed a need for further improvements in electron transport models that could simultaneously account, both temporarily and spatially, for all these types of emission. Furthermore, numerous observations of solar flares by SMM, TRACE, and RHESSI suggest that the areas of flaring loops decrease and, thus, their magnetic fields increase with the depth of the solar atmosphere (Kontar *et al.*, 2008; Lang *et al.*, 1993). This increase in the magnetic field can act as a magnetic mirror for the precipitating electrons, forcing them to return back to the source in the corona, in addition to the self-induced electric field.

In addition, some high-temporal-resolution observations with RHESSI have revealed that the temporal intervals of impulsive increases of HXR emission vary from very short – tens of milliseconds (Charikov *et al.*, 2004; Kiplinger *et al.*, 1983) – to tens of minutes (Holman *et al.*, 2003). This encourages us to revise the electron transport models and to consider solutions of a time-dependent Fokker–Planck equation for different time scales of beam injection (milliseconds, seconds, and minutes). Electron transport, in turn, can slow down also by anisotropic scattering of beam electrons in this self-induced electric field enhanced by their magnetic mirroring in converging magnetic loops. The further delay can be caused by particle diffusion in pitch angles and energy, which can significantly extend the electron transport time into deeper atmospheric layers where they are fully thermalized.

To account for all the observational features above, one needs to consider the time-dependent Fokker–Planck equation and compare the solutions to electron precipitation for stationary and impulsive injection for the different models of a converging magnetic field by taking into account all the mechanisms of energy loss (collisions, ohmic losses) and anisotropic scattering.

4.2 Problem Formulation

4.2.1 The Fokker–Planck Equation

Let us consider precipitation of a beam of high-energy electrons into a flaring atmosphere along the linear depth l. The distribution function f of beam electrons as a function of time t, depth x, velocity V, and pitch angle θ between the direction of the velocity and the magnetic field, can be found from the Fokker–Planck–Landau equation (Landau, 1937):

$$\begin{aligned}\frac{\partial f}{\partial t} &+ V\cos\theta\frac{\partial f}{\partial x} - \frac{e\mathcal{E}}{m_e}\cos\theta\frac{\partial f}{\partial V} - \frac{e\mathcal{E}}{m_e V}\sin^2\theta\frac{\partial f}{\partial\cos\theta}\\ &= \frac{1}{V^2}\frac{\partial}{\partial V}\left(\nu V^3 f\right) + \nu\frac{\partial}{\partial\cos\theta}\left(\sin^2\theta\frac{\partial f}{\partial\cos\theta}\right)\\ &+ \frac{V\sin^2\theta}{2}\frac{\partial\ln B}{\partial x}\frac{\partial f}{\partial\cos\theta}\,,\end{aligned} \tag{4.1}$$

where the second term on the right-hand side defines the variations of f with electron precipitation depth, the third and fourth terms define the effect of ohmic losses on energy and pitch-angle distribution, the first two terms on the right-hand side define the energy losses and pitch-angle change in collisions, and the third term defines the pitch-angle change in the converging magnetic field.

Here $\mathcal{E}$ is the self-induced electrostatic electric field, B is the background magnetic field, e and m_e are the electron charge and mass, respectively, $\nu(E)$ is the frequency of collisions, which can be written as follows (Chapter 3):

$$\nu(E) = \frac{k}{\sqrt{2m_e}} \pi n(x) \lambda e^4 E^{-3/2} , \tag{4.2}$$

where E is a kinetic electron energy, $n(x)$ is the ambient plasma density, and λ is the Coulomb logarithm for beam electron scattering on the ambient electrons. As a first approach let us assume that the Coulomb logarithm is constant ($\Lambda \approx 20$), although later we will consider the effect of Λ varying with depth and its effect on electron distributions. The parameter k is taken in the form discussed in Chapter 3:

$$k = X\frac{(\lambda + \lambda')}{\lambda} + \frac{(1 - X)\lambda''}{\lambda} .$$

Here X is the ionization degree of the ambient plasma and λ' and λ'' are the Coulomb logarithms for protons and neutral particles of the ambient plasma, respectively (see Chapter 3 for details).

The effect of a magnetic field convergence can be taken into account in a form suggested by McClements (1992a)

$$\left(\frac{\partial f}{\partial t}\right)_{\text{magn}} = \sqrt{\frac{2E}{m_e}} \frac{\sin^2\theta}{2} \frac{\partial \ln B(x)}{\partial x} \frac{\partial f}{\partial \cos\theta} , \tag{4.3}$$

where $B(x)$ is the longitudinal component of the magnetic field at the given depth x.

The electric current carried by precipitating beam electrons can be determined as

$$j(x) = 2\pi e \int_0^\infty \int_{-1}^1 f(x, E, \theta) v_x \, dE \, d\cos\theta ,$$

or, since $V = v_x = \sqrt{2E/m_e}\cos\theta$, it can be rewritten as follows:

$$j(x) = 2\sqrt{2}\pi \frac{e}{\sqrt{m_e}} \int_0^\infty \int_{-1}^1 f(x, E, \theta) \sqrt{E} \cos\theta \, dE \, d\cos\theta . \tag{4.4}$$

The electric current carried by precipitating electrons is to be fully compensated by those carried by return current electrons. Hence, the self-induced electrostatic electric field of the beam, compensated by the return current of the ambient electrons, is calculated from Ohm's law (Diakonov and Somov, 1988):

$$\mathcal{E} = \frac{j_{rc}(x)}{\sigma(x)} = \frac{j(x)}{\sigma(x)} = \frac{2\sqrt{2\pi}}{\sigma(x)} \frac{e}{\sqrt{m_e}} F_0 \int_0^{\infty} \int_{-1}^{1} f(x, E, \theta) \sqrt{E} \cos\theta \, d E \, d\cos\theta \, , \tag{4.5}$$

where $\sigma(x)$ is the classic conductivity of the ambient plasma related to the plasma resistivity η as follows:

$$\sigma(s) = \frac{1}{\eta} = 1.97\sqrt{2\pi}\frac{3}{4}\frac{n_0\left[k_B T(s)\right]^{3/2}}{\sqrt{m_e} F_0} \, . \tag{4.6}$$

Here k_B is the Boltzmann constant, T is the kinetic temperature of the ambient plasma, and F_0 is the normalization factor for the beam energy flux of a beam. Since in Chapter 3 we established that electron collisional energy losses change with the column depth, and not with linear depth x, let us convert the linear depth x into a column depth ξ using Eq. (3.31):

$$\xi = \int_0^x n(y) dy \, . \tag{4.7}$$

The ambient plasma is assumed to be preheated by previous precipitation of an electron beam in pure collisions (as discussed in Chapter 3). Hence, its density, n, and temperature, T, defined as functions of the column depth, ξ, are calculated using the hydrodynamical model for pure collisional plasma heating by electrons (Somov *et al.*, 1981; Zharkova and Zharkov, 2007) discussed in Chapter 5. The profiles $n(\xi)$ and $T(\xi)$ for different beam parameters are shown in Figure 4.1. These profiles are assumed to be unchanging during beam precipitation since the thermal conduction time scale is much longer than the precipitation time scales of beam electrons (Somov *et al.*, 1981).

4.2.2
Normalization of a Distribution Function

The distribution function is normalized on a particle density at the top boundary and can be defined as follows:

$$N(\xi_{min}) = K \int_{E_{low}}^{E_{upp}} \int_{-1}^{1} E^{-\gamma-0.5} \exp\left[-\left(\frac{\mu-1}{\Delta\mu}\right)^2\right] d E \, d\mu \, . \tag{4.8}$$

In order to define a scaling parameter K in normalization (4.8) of the distribution function f, one must calculate the initial energy flux $F(\xi_{min})$ carried through the top

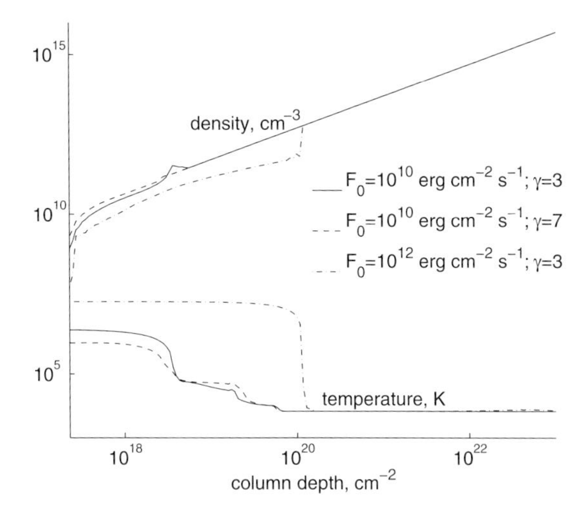

Figure 4.1 Density and temperature of ambient plasma calculated by hydrodynamical model (Zharkova and Zharkov, 2007) heated by collisions with ambient particles of an electron beam with energy flux F_0 and spectral index γ (Eq. (4.30)).

boundary by beam electrons with a given spectral index γ and lower- and upper-energy cutoffs E_{low} and E_{up}:

$$F(\xi_{\min}) = K\sqrt{\frac{2}{m_e}}\int_{E_{\text{low}}}^{E_{\text{upp}}}\int_{-1}^{1} E^{-\gamma+1}\exp\left[-\left(\frac{\mu-1}{\Delta\mu}\right)^2\right]\mu\, d E\, d\mu\,.$$

Thus, if one measures the electron energy flux $F(\xi_{\min})$ at the top boundary, then the scaling factor K is defined from the beam parameters by the following formula:

$$K = \frac{F(\xi_{\min})}{\sqrt{\frac{2}{m_e}}\int_{E_{\text{low}}}^{E_{\text{upp}}}\int_{-1}^{1} E^{-\gamma+1}\exp\left[-\left(\frac{\mu-1}{\Delta\mu}\right)^2\right]\mu\, d E\, d\mu}\,. \tag{4.9}$$

4.2.3
Dimensionless Equations

To simplify Eq. (4.1), let us introduce, similar to Diakonov and Somov (1988), new dimensionless variables. Since the characteristic (average) energy of a beam is the lower-energy cutoff E_{low}, the characteristic column density can be defined as $\xi_0 = E_{\text{low}}^2/2a$. This is the column density where electrons with energy E_{low} completely lose their energy in collisions and its magnitudes presented in Chapter 3, column 1 of Table 3.1. The collisional parameter a takes into account beam electron scattering on particles of the ambient plasma for a given ionization degree X discussed in Section 3.2.2.1. Parameter a, or a_e for electrons as discussed in Chapter 3, depends directly on the ionization degree and varies logarithmically with the electron energy and the ambient plasma temperature and density via the Debye length or mean free path as discussed in Sections 3.2.2.1 and 3.2.2.2.

From the definition of a column density and kinetic energy E one can deduce the characteristic length $x_0 = \xi_0/n(\xi)$ and velocity

$$V_0 = \sqrt{\frac{2E_{\text{low}}}{m_e}},$$

which in turn helps to define the characteristic time

$$t_0 = \frac{x_0}{V_0} = \frac{\sqrt{2m_e}E_0^{\frac{3}{2}}}{2\pi e^4 n\lambda}.$$

The electrostatic electric field ε induced by beam electrons can be measured in units of the Dreicer electric field

$$\varepsilon_0 = \mathcal{E}_D = \frac{2\pi e^3 \lambda n(x)}{k_B T_e}.$$

Therefore, the original variables can be referred to these characteristic ones, which allows one to introduce the dimensionless-variable depth $s = \xi/\xi_0$, energy $z = E/E_{\text{low}}$, electric field $\varepsilon = \mathcal{E}/\mathcal{E}_D$, pitch-angle cosine $\mu = \cos\theta$, and time $\tau = t/t_0$ as follows:

$$\tau = \tau = \frac{t}{t_0} = t\frac{2\pi e^4 n_0 \ln\Lambda}{\sqrt{2m_e}E_0^{3/2}}, \tag{4.10}$$

$$s = s = \frac{\xi}{\xi_0} = \xi\frac{\pi e^4 \ln\Lambda}{E_0^2}, \tag{4.11}$$

$$z = \frac{E}{E_0} = \frac{m_e V^2}{2E_0}, \tag{4.12}$$

$$\mu = \cos\theta, \tag{4.13}$$

$$\varepsilon = \varepsilon = \frac{\mathcal{E}}{\mathcal{E}_D} = \mathcal{E}\frac{E_0}{2\pi e^3 n_0 \ln\Lambda}, \tag{4.14}$$

$$n_p = \frac{n}{n_0}, \tag{4.15}$$

where n_p is the dimensionless plasma density measured in units of n_0, $n_0 = 10^{10}\,\text{cm}^{-3}$ is the ambient density in the corona, $E_0 = E_{\text{low}} = 12\,\text{keV}$ is the lower cutoff energy, and ξ is the column depth defined above.

In dimensionless variables, Eq. (4.1) can be rewritten in the following form:

$$\begin{aligned}\frac{\partial f}{\partial\tau} &+ n\sqrt{z}\mu\frac{\partial f}{\partial s} - 2\varepsilon\mu\sqrt{z}\frac{\partial f}{\partial z} - \varepsilon\frac{1-\mu^2}{\sqrt{z}}\frac{\partial f}{\partial\mu} \\ &= n\frac{1}{\sqrt{z}}\frac{\partial f}{\partial z} + n\frac{1-\mu^2}{2z^{3/2}}\frac{\partial^2 f}{\partial\mu^2} - n\frac{\mu}{z^{3/2}}\frac{\partial f}{\partial\mu} \\ &+ n\frac{(1-\mu^2)\sqrt{z}}{2}\alpha_B\frac{\partial f}{\partial\mu},\end{aligned} \tag{4.16}$$

where α_{B} is the parameter of a magnetic convergence:

$$\alpha_{\mathrm{B}} = \frac{\partial \ln B}{\partial s} \,. \tag{4.17}$$

4.2.4
Integral Characteristics of an Electron Beam

Having solved numerically Eq. (4.16), one can calculate the following integral parameters for the electron beam.

Beam density (in $\mathrm{cm^{-3}}$):

$$n_{\mathrm{b}}(\tau, s) = F_0 \sqrt{\frac{m_{\mathrm{e}}}{2E_0^3}} \int\limits_{z_{\min}}^{z_{\max}} dz \int\limits_{-1}^{1} d\mu \sqrt{z} A(s) f(\tau, s, z, \mu) \,, \tag{4.18}$$

where the coefficient $A(s) = B_0/B(s)$ accounts for the variations of the flaring atmosphere cross-section in a converging magnetic field.

Differential particle energy flux at a given depth (in $\mathrm{erg^{-1}\,cm^{-2}\,s^{-1}}$):

$$\mathcal{F}_{\mathrm{n}}(\tau, s, z) = \frac{F_0}{2E_0^2} \int\limits_{-1}^{1} d\mu\, z A(s) f(\tau, s, z, \mu) \,. \tag{4.19}$$

Mean particle energy flux (in $\mathrm{erg^{-1}\,cm^{-2}\,s^{-1}}$) (Brown *et al.*, 2003):

$$\langle \mathcal{F}_{\mathrm{n}} \rangle (\tau, z) = \frac{F_0}{E_0^2} \frac{\int_{s_{\min}}^{s_{\max}} ds \int_{-1}^{1} d\mu\, n^{-1}(s) A(s) z f(\tau, s, z, \mu)}{2 \int_{s_{\min}}^{s_{\max}} n^{-1}(s) ds} \,. \tag{4.20}$$

Pitch angle distributions of particles (in arbitrary units):

$$\frac{d N_{\mathrm{b}}(\tau, \mu)}{d\mu} = \int\limits_{s_{\min}}^{s_{\max}} ds \int\limits_{z_{\min}}^{z_{\max}} dz\, n^{-1}(s) A(s) \sqrt{z} f(\tau, s, z, \mu) \,. \tag{4.21}$$

Energy deposition (or heating function) by a beam (in $\mathrm{erg\,cm^{-3}\,s^{-1}}$):

$$I(\tau, s) = \frac{F_0 n_0}{E_0} n(s) A(s) \int\limits_{z_{\min}}^{z_{\max}} dz \int\limits_{-1}^{1} d\mu \left(-\frac{dz}{ds}\right) \mu z f(\tau, s, z, \mu) \,, \tag{4.22}$$

where dz/ds is the electron's energy losses with depth, which can be estimated as (Emslie, 1980)

$$\frac{dz}{ds} = \left(\frac{dz}{ds}\right)_{\mathrm{c}} + \left(\frac{dz}{ds}\right)_{\mathrm{r}} = -\frac{a_e}{\mu z} - 2\frac{\varepsilon}{n} \,, \tag{4.23}$$

where the two terms represent the collisional and ohmic energy losses, respectively.

4.3 Simulation Method

For the simulations we have applied the summary approximation method proposed by Samarskii (2001). In order to replace the partial derivatives with finite differences, let us group together the corresponding derivatives and rewrite the Fokker–Planck equation (4.16) in the following form:

$$\begin{aligned}\frac{\partial f}{\partial \tau} &= -n\sqrt{z}\mu\frac{\partial f}{\partial s} + \left(2\varepsilon\mu\sqrt{z} + n\frac{1}{\sqrt{z}}\right)\frac{\partial f}{\partial z} \\ &+ \left(\varepsilon\frac{1-\mu^2}{\sqrt{z}} - n\frac{\mu}{z^{3/2}} + n\frac{(1-\mu^2)\sqrt{z}}{2}\alpha_{\mathrm{B}}\right)\frac{\partial f}{\partial \mu} \\ &+ n\frac{1-\mu^2}{2z^{3/2}}\frac{\partial^2 f}{\partial \mu^2} = \phi_s\frac{\partial f}{\partial s} + \phi_z\frac{\partial f}{\partial z} + \phi_\mu\frac{\partial f}{\partial \mu} + \phi_{2\mu}\frac{\partial^2 f}{\partial \mu^2}\,. \end{aligned} \tag{4.24}$$

Equation (4.24) is solved numerically using the summary approximation method (Samarskii, 2001), where the 4-D problem is reduced to a chain of 2-D problems. This is done by considering the 3-D differential operator on the right-hand side of Eq. (4.16) as a sum of 1-D operators, each acting on the distribution function separately during one third of the time step. For each substep in time, the distribution function is calculated implicitly by solving the system of equations written as follows:

$$f^{\tau+\frac{1}{3}\Delta\tau} - f^{\tau} = \Delta\tau\phi_s L_s f^{\tau+\frac{1}{3}\Delta\tau}\,, \tag{4.25}$$

$$f^{\tau+\frac{2}{3}\Delta\tau} - f^{\tau+\frac{1}{3}\Delta\tau} = \Delta\tau\phi_z L_z f^{\tau+\frac{2}{3}\Delta\tau}\,, \tag{4.26}$$

$$f^{\tau+\Delta\tau} - f^{\tau+\frac{2}{3}\Delta\tau} = \Delta\tau(\phi_\mu L_\mu + \phi_{2\mu} L_{2\mu}) f^{\tau+\Delta\tau}\,, \tag{4.27}$$

where L_α are the finite difference operators that approximate the first-order differential operators $\partial/\partial\alpha$. If the coefficient ϕ_α is positive, then the right difference scheme is used, that is, $L_\alpha f = (f^{\alpha+\Delta\alpha} - f^{\alpha})/\Delta\alpha$, otherwise the left scheme is used, that is, $L_\alpha f = (f^{\alpha} - f^{\alpha-\Delta\alpha})/\Delta\alpha$. The $L_{2\mu} f = (f^{\mu+\Delta\mu} - 2f^{\mu} + f^{\mu-\Delta\mu})/\Delta\alpha^2$ is the central difference that approximates the second-order derivative $\partial^2 f/\partial\mu^2$. The computational grid has 200 nodes in the s dimension, 50 nodes in the z dimension, and 30 nodes in the μ dimension. The nodes are distributed logarithmically in the s and z dimensions and linearly in the μ dimension.

Equation (4.25), together with boundary conditions, forms a set of linear equations, which, after being solved, gives $f^{\tau+1/3\Delta\tau}$ from a known f^{τ}. The distribution function $f^{\tau+1/3\Delta\tau}$ is then used in Eq. (4.26) to obtain $f^{\tau+2/3\Delta\tau}$. Finally, from Eq. (4.27) we obtain $f^{\tau+\Delta\tau}$, which is in turn used in Eq. (4.25) in the next step.

The electric field ε is calculated at each time step according to Eq. (4.5), where the distribution function is taken from the previous step. Thus the numerical scheme is not fully implicit. This means that, in order to avoid numerical instability, the time step, $\Delta\tau$, must be shorter than some critical value $\Delta\tau_c$. In practice, the time step was determined by the trial-and-error method. For example, for the energy flux

10^{10} erg cm^{-2} s^{-1} the time step is 1.7×10^{-4} s. It was found that when the energy flux is increased, the time step needs to be decreased proportionally to keep the numerical scheme stable.

4.4 Stationary Fokker–Planck Approach ($d f/dt = 0$)

Let us first consider the case of precipitation of a stationary particle by fixing the injection of electrons at the top boundary (by setting $\partial d/\partial t = 0$) and then considering their distributions in depths, energies, and pitch angles, similarly to Zharkova and Gordovskyy (2005a). This approach will help us to understand better the physical process of energy losses by beam electrons during their precipitation.

4.4.1 Initial Condition

The distribution function at the top boundary $s = s_{\min}$ is defined by the initial energy–pitch-angle distribution of an injected beam. We assume that the injected beam has a power-law energy spectrum in a range from E_{low} to E_{upp} with a spectral index of γ, a normal distribution in pitch-angle cosines μ of beam electrons with a half-width dispersion of $\Delta\mu = \mu_{\max 0} - \mu_{\min 0}$, where the subscripts max and min define the maximum and minimum pitch-angle cosines at injection. Therefore, the boundary condition at the upper boundary at depth $s = s_{\min}$ is defined by the following formula:

$$f(s_{\min}, z, \mu) = \begin{cases} z^{-\gamma-0.5} \exp\left[-\left(\frac{\mu-1}{\Delta\mu}\right)^2\right] U(\tau)\,, & 1 < \eta < \eta_{\max}, \quad \text{and} \quad \mu > 0 \\ 0, & \text{elsewhere.} \end{cases} \tag{4.28}$$

Here $U(\tau)$ is a temporal profile of the electron beam injection, which denotes the initial beam flux variations. If a steady injection is considered, $\partial f/\partial t = 0$, then $U(\tau) =$ constant.

One can estimate the decay time for nonthermal electrons, when they will be completely thermalized, from the collision frequency Eq. (4.2) as follows:

$$\tau_{\text{dec}} \simeq \frac{1.47 \times 10^8 \; E^{3/2}\,(\text{keV})}{n\,(\text{cm}^{-3})}\,.$$

For an electron energy as high as ~ 300 keV and a density in the lower corona of $\sim 10^{11}$ cm^{-3}, this equation gives a decay time of less than ~ 7 s. However, the bulk of electrons have energies much smaller than 300 keV since they have a power law in energy. Thus those electrons that are not trapped in the magnetic mirror in the corona deposit their energy in the much denser plasma in the chromosphere. Then the real decay time for precipitating electrons is less than $\precsim 0.1$ s.

Table 4.1 Electron densities at injection at top boundary for beams with power-law energy spectra with different sets of initial energy flux and spectral indices.

Flux, erg cm^{-2} s^{-1}	10^8	10^{10}	10^{12}
γ		Particle density, cm^{-3}	
3	2.2×10^5	2.2×10^7	2.2×10^9
5	3.4×10^5	3.4×10^7	3.4×10^9
7	4.8×10^5	4.8×10^7	4.8×10^9

Adopted parameters. The calculations of electron distributions in energy and pitch angles for different depths were carried out for an initial beam of electrons with power-law energy spectra having a lower energy cutoff of $E_{low} = 16$ keV and an upper energy cutoff of $E_{upp} = 384$ keV, or $z_{max} = 24$, which is reasonably large to investigate electron precipitation into deeper atmospheric levels in the first approach. Here we ignore the relativistic effects for precipitation of these higher-energy electrons, since we are concerned with the interpretation of the lower-energy photon spectra affected by ohmic losses. Spectral indices of 3, 5, and 7 and initial energy fluxes of 10^8, 10^{10}, or 10^{12} erg cm^{-2} s^{-1} were adopted. Two types of pitch-angle distribution were considered: a "narrow distribution" ($\Delta\mu = 0.1$) and a "wide distribution" ($\Delta\mu = 0.6$).

The corresponding total beam initial densities at the top boundaries are presented in Table 4.1.

4.4.2 Beam Electron Distribution Functions

4.4.2.1 Effect of Pure Collisional Energy Losses

An electron beam's differential densities, or differential spectra, calculated as an integral over pitch angles of the electron distribution functions at given depths are the same as those presented in Figure 3.3.

It can be seen that the energy loss is stronger for lower-energy electrons since the following proportion is valid: $dz/ds \propto z^{-1}$. With every depth step this leads to a growing deficit of low-energy electrons in comparison with the initial energy spectrum, which in turn appears in the electron differential spectra as a "flattening" at lower energies. At larger column depths the initial power-law energy distribution with a spectral index $\gamma = 3$ is transformed into distributions with a flat lower-energy part and becomes flatter with every step in depth until it approaches a maximum. This flattening in the electron differential spectra can result in the flattening of the integrated (over depth) electron spectrum, that is, of the total number of electrons with a given energy at a given instance.

Any additional electron energy sinks, such as pitch-angle diffusion or ohmic heating, can further speed up the electron shift to a lower energy and results in a noticeable spectrum flattening (Sections 4.4.2.2 and 4.4.2.4). These additional energy losses during beam precipitation can strongly affect the HXR bremsstrahlung emission emitted from a whole flaring atmosphere, discussed in Chapter 5.

4.4.2.2 **Effect of Pitch-Angle Scattering**

The monoenergetic beams precipitating with pitch-angular scattering reveal, first, a much steeper than parabolic decrease of energy with depth caused by collisional energy losses (Figure 3.2a) and, second, lost energy at much greater depths because of pitch-angle scattering (Figure 3.2b). Apparently, the latter happens because the beam electrons, which lose their energy in Coulomb scattering, can be scattered many times not only forward ($\mu \succ 0$) but backward ($\mu \prec 0$). Pitch-angle scattering is more effective for low-energy electrons (lower curves) than for higher-energy ones (compare gray areas in upper and lower curves in Figure 3.2b).

As a result, for a beam injected with a power-law energy spectrum and losing its energy in pure Coulomb collisions, distribution functions with pitch-angle diffusion in the $E - \mu$ coordinates is shown in Figure 4.2 for the initial flux 10^{12} erg cm^{-2} s^{-1}, $\gamma = 3$, with lowest and highest energies of 16 and 386 keV, respectively. Because of pitch-angle diffusion the number of lower-energy elec-

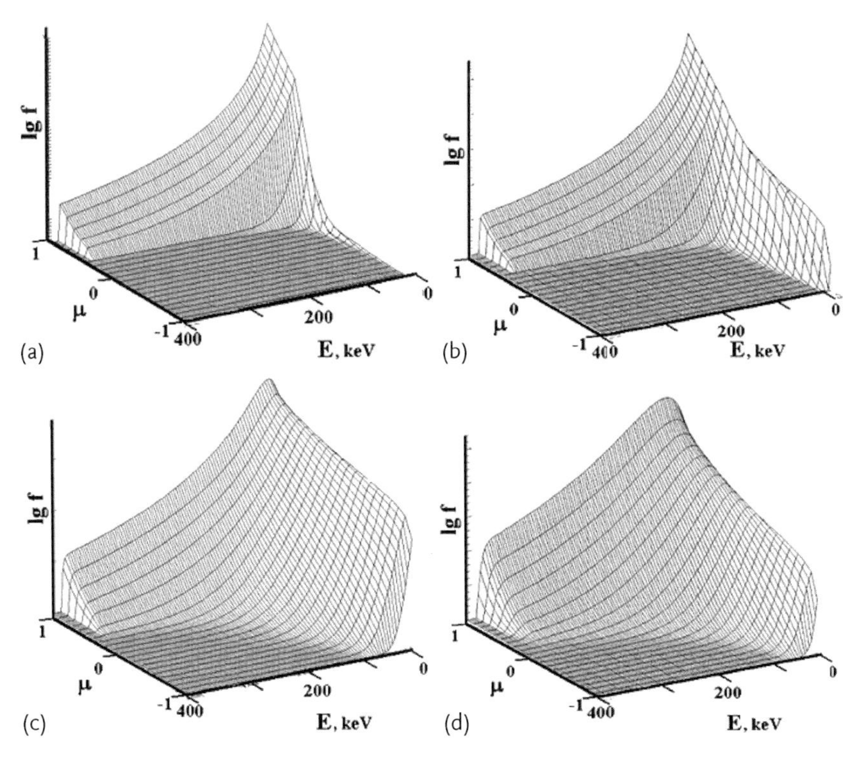

Figure 4.2 Electron distribution function (vertical axis) vs. energy E and pitch-angle cosines μ calculated without a return current effect for a beam injected with an energy range of 16–384 keV, a spectral index of $\gamma = 3$, and an initial energy flux of 10^{10} erg cm^{-2} s^{-1}: column density of (a) 1.8×10^{18}, (b) -1.3×10^{19}, (c) -5.1×10^{20}, and (d) -2.9×10^{22} cm^{-2}.

trons increases with depth and changes from a narrow angular distribution at the injection point to much wider ones in deeper layers, becoming nearly isotropic at greater atmospheric depths below $\xi = 2 \times 10^{22}$ cm^{-2}.

4.4.2.3 **Depth Distributions of Self-Induced Electrostatic Electric Field**

The kinetics of very intense electron beams can be strongly affected by a self-induced electric field (Zharkova and Gordovskyy, 2005a; Zharkova *et al.*, 1995). The induced electric field is calculated at every precipitation depth and its magnitude is linearly proportional to the precipitating beam particle flux (Eq. (4.5)). The electric field distributions with depth found from the full kinetic simulations by Zharkova and Gordovskyy (2005a) are plotted in Figure 4.3 for electron beams with $\gamma = 3$ and initial fluxes of 10^8 erg cm^{-2} s^{-1} (Figure 4.3a), 10^{10} (Figure 4.3b), and 10^{12} (Figure 4.3c) taken from Zharkova and Gordovskyy (2005a).

The electric field is measured in units of a local Dreicer field induced by ambient plasma particles, that is, $\mathcal{E}_{\rm D} = (2\pi e^3 n \ln \Lambda)/(k_{\rm B} T)$, where T and n are the local temperature and density and $\ln \Lambda$ is the Coloumb logarithm. For weak hard beams ($F_0 = 10^8$ erg cm^{-2} s^{-1}, $\gamma = 2$–3) the induced field is weaker (10^{-7} V cm^{-1}) than the local Dreicer field, while for intense softer beams (10^{11}–10^{12} erg cm^{-2} s^{-1}), it can reach 10^{-3} V cm^{-1}, which is comparable or even higher by a factor of 10–100.

These full kinetic simulations confirm the estimations carried out for the continuity equation with energy losses in a pure electric field accepted in Section 3.4. A self-induced electric field is constant at upper coronal levels and strongly decreases with depth in the lower corona and transition region (Figure 4.3). For weaker hard beams the induced electric field is nearly constant to a column depth of a few units of 10^{20} cm^{-2} (upper chromosphere) and then smoothly decreases toward the photosphere (Figure 4.3a). For more intense hard beams the magnitude of a constant electric field increases proportionally to the initial flux but remains constant only at levels of about 10^{19} cm^{-2} (for the initial flux of 10^{10} erg cm^{-2} s^{-1}) (Figure 4.3b) or even about 10^{18} cm^{-2} (for 10^{12} erg cm^{-2} s^{-1}) (Figure 4.3c), after which it drops sharply. The level of the decrease depends on the beam spectral indices: it is steeper for softer beams (5–7) than for harder ones (3), as is shown by the solid and dashed lines in Figure 4.3a–c.

Therefore, an induced electric field is constant only at higher atmospheric levels after beam injection. Then it becomes variable throughout most depths of a flaring atmosphere, falling exponentially (linearly, parabolically, or more steeply) at a linear depth of x. As discussed in Chapter 3, the "turning" depth, where the electric field starts decreasing, is very important for understanding the resulting electron-beam distributions in depth and energy and the measurable parameters, such as electron mean energy flux and photon spectrum.

4.4.2.4 **The Effect of a Self-Induced Electric Field**

The effect of an electrostatic electric field induced by a precipitating beam is found to be essential for initial beams with a wide pitch-angle dispersion at injection, $\Delta\mu = 0.6$.

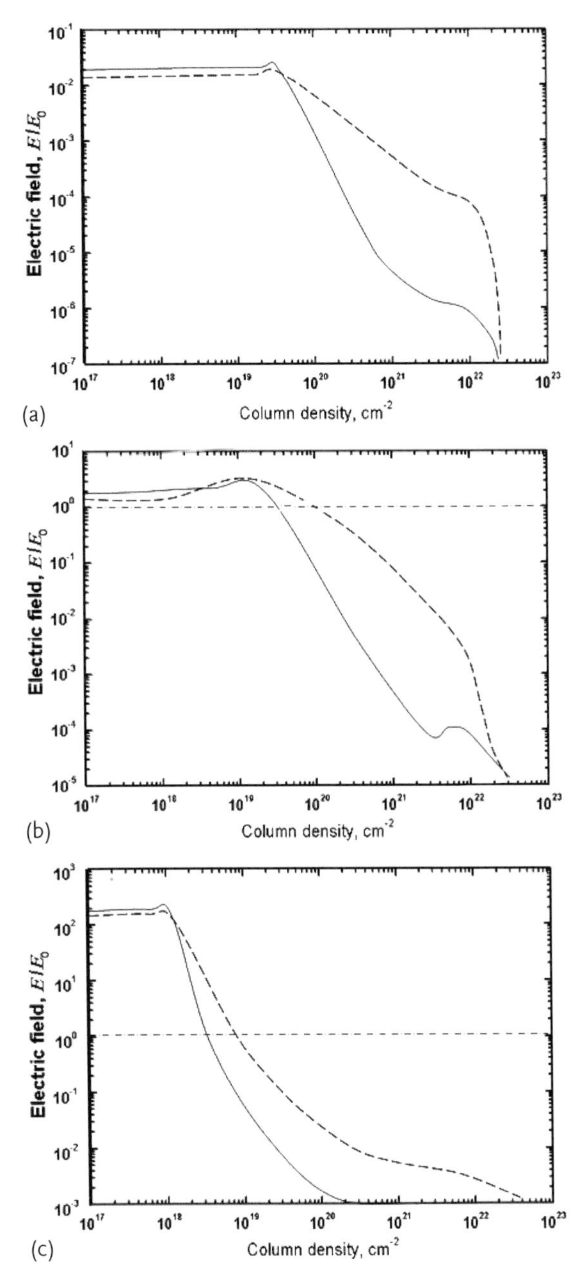

Figure 4.3 Electric field induced by electron beams with $\gamma = 3$ (dashed lines) and 7 (solid lines) and initial energy fluxes of $F_0 = 10^8$ erg cm^{-2} s^{-1} (a), $F_0 = 10^{10}$ erg cm^{-2} s^{-1} (b), and $F_0 = 10^{12}$ erg cm^{-2} s^{-1} (c). Adopted from Zharkova and Gordovskyy (2006).

The return current effect appears at the very start of electron precipitation, at great depths in the corona, as an additional maximum in negative pitch-angle cosines μ in electron distributions with lower energies ($E < 50$–$90\,\text{keV}$). For moderate beams with an initial energy flux of $10^{10}\,\text{erg}\,\text{cm}^{-2}\,\text{s}^{-1}$ the return current appears at a depth of $\xi = 1.8 \times 10^{18}\,\text{cm}^{-2}$ for a harder beam (Figure 4.4) or even higher at $\xi = 1.2 \times 10^{18}\,\text{cm}^{-2}$ for a softer beam (Figure 4.5). The return current is maintained during beam precipitation downward into the transition region, $\xi = 4.0 \times 10^{19}\,\text{cm}^{-2}$ (Figure 4.5), or even into the upper chro-

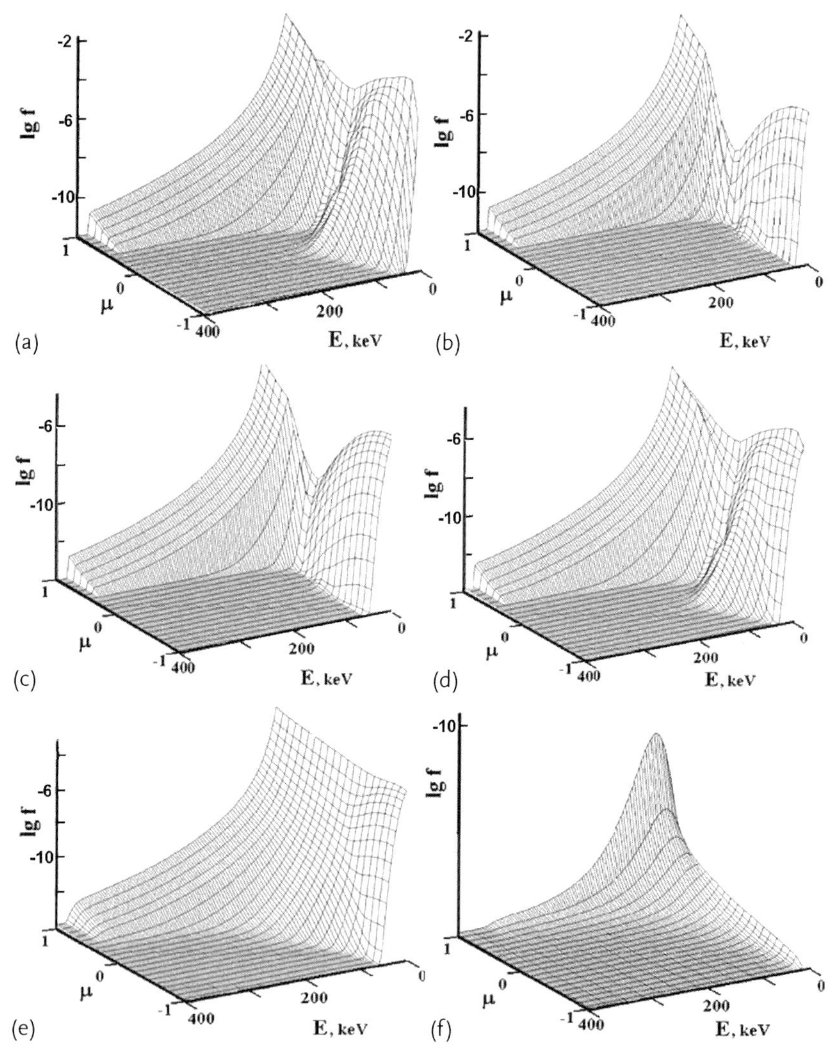

Figure 4.4 Energy pitch-angle (E–μ) distributions (vertical axis) vs. energy E and pitch-angle cosines μ calculated with a return current effect for a beam injected with an energy range of 16–384 keV, spectral index of $\gamma = 3$, and initial energy flux of $10^{10}\,\text{erg}\,\text{cm}^{-2}\,\text{s}^{-1}$. Column densities: (a) $4.2 \times 10^{17}\,\text{cm}^{-2}$, (b) -8.4×10^{17}, (c) -1.8×10^{18}, (d) -1.3×10^{19}, (e) -5.1×10^{20}, and (f) $-2.9 \times 10^{22}\,\text{cm}^{-2}$.

mosphere $\xi = 1.0 \times 10^{20}\,\mathrm{cm}^{-2}$ (Figure 4.4), and then disappears at the lower chromosphere due to a decrease in the ambient plasma conductivity caused by a drop in the degree of plasma ionization. For harder beams the returning electrons become fully isotopic in pitch angles at the upper chromospheric depth, $\xi = 1.0 \times 10^{20}\,\mathrm{cm}^{-2}$ for moderate initial fluxes (Figure 4.4), and even just below the transition region at $\xi = 4.0 \times 10^{19}\,\mathrm{cm}^{-2}$ for higher initial energy fluxes of $10^{12}\,\mathrm{erg\,cm^{-2}\,s^{-1}}$ (Figure 4.6). For more intense beams with an initial energy flux of $10^{12}\,\mathrm{erg\,cm^{-2}\,s^{-1}}$ the return current appears immediately after beam

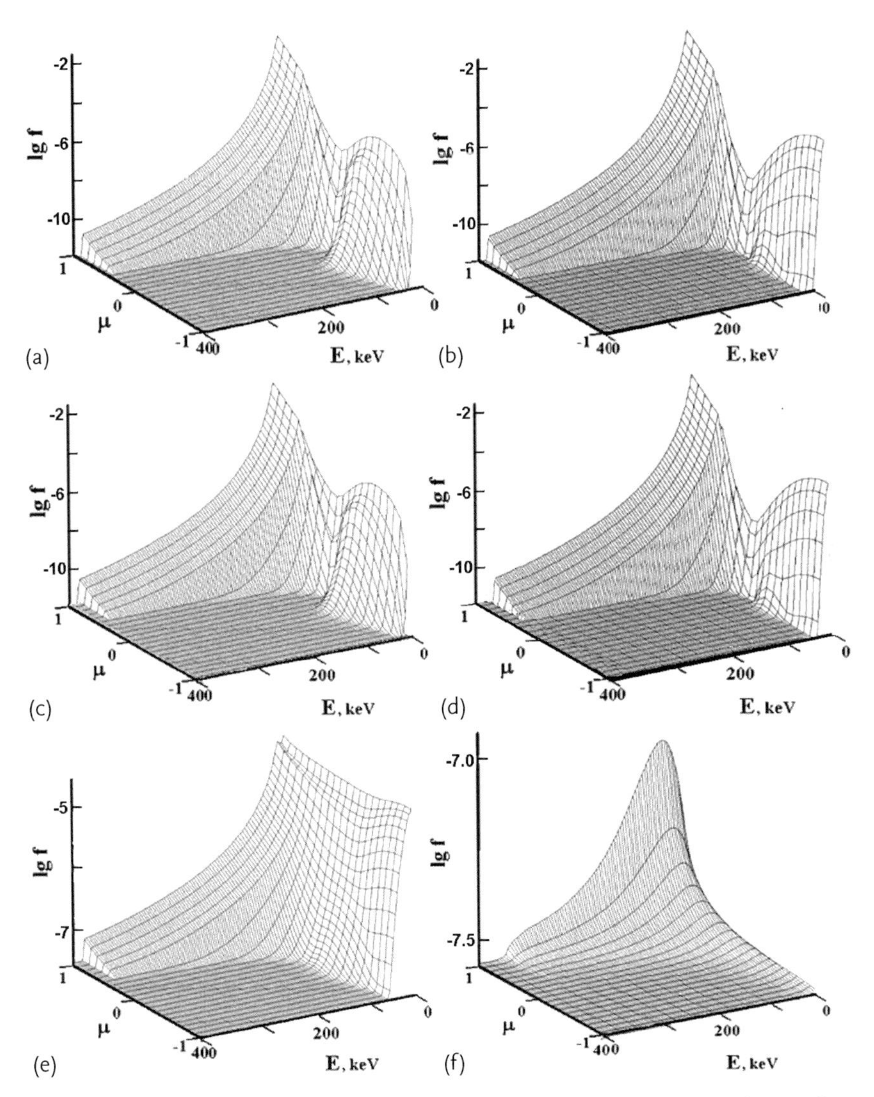

Figure 4.5 The same as in Figure 4.4 but for a spectral index of $\gamma = 7$ and an initial energy flux of 10^{10}.

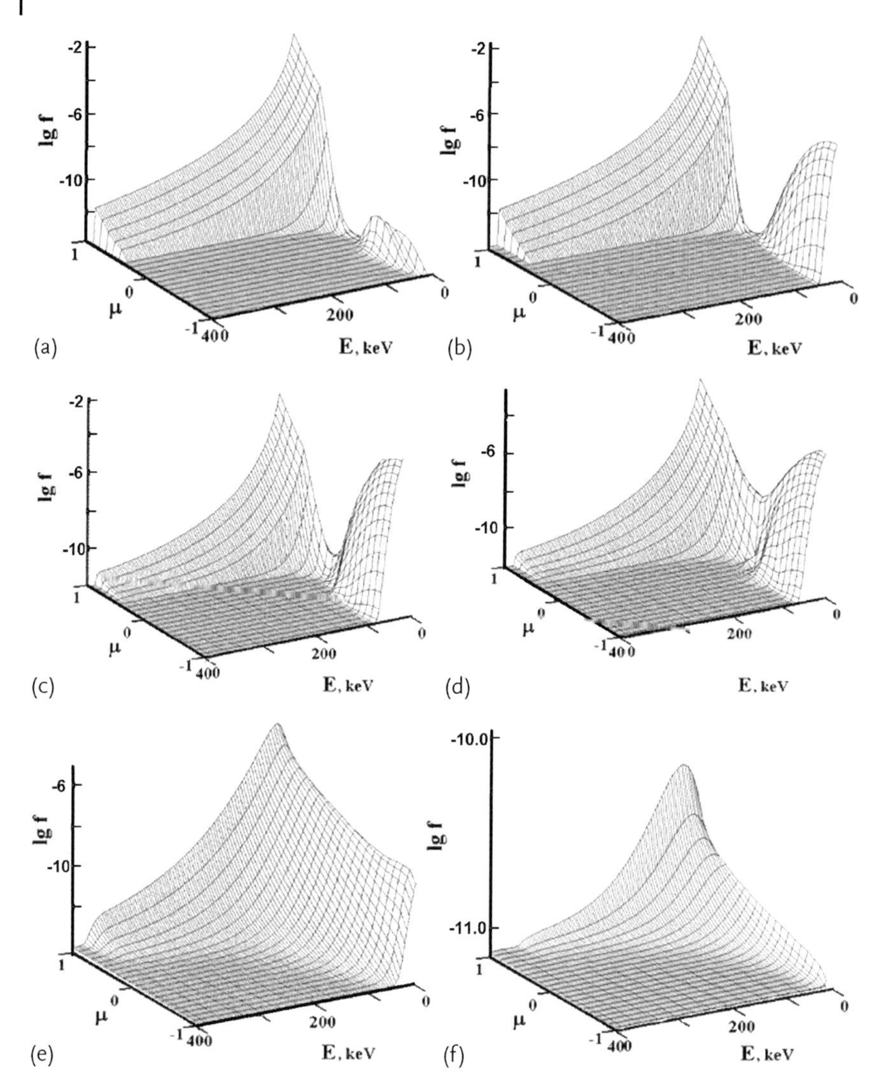

Figure 4.6 The same as in Figure 4.4 but for a spectral index of $\gamma = 3$ and an initial energy flux of 10^{12}.

injection at a depth of $\xi = 1.2 \times 10^{18}\,\text{cm}^{-2}$ for a harder beam (Figure 4.6) or $\xi = 1.0 \times 10^{18}\,\text{cm}^{-2}$ for a softer beam (Figure 4.7); it is maintained during beam precipitation downward to the transition region, $\xi = 2.0 \times 10^{19}\,\text{cm}^{-2}$ (Figure 4.7), or into the upper chromosphere $\xi = 9.0 \times 10^{19}\,\text{cm}^{-2}$ (Figure 4.6).

The density of a returning beam is less than that of a precipitating one of moderate intensity, the difference being by two orders of magnitude for harder beams, and only by a factor of 2–3 for softer ones (cf. Figures 4.4 and 4.5), while for intense beams the returning beam density becomes comparable with the direct one (Figures 4.4 and 4.6 or 4.5 and 4.7). It appears that ambient electrons play an es-

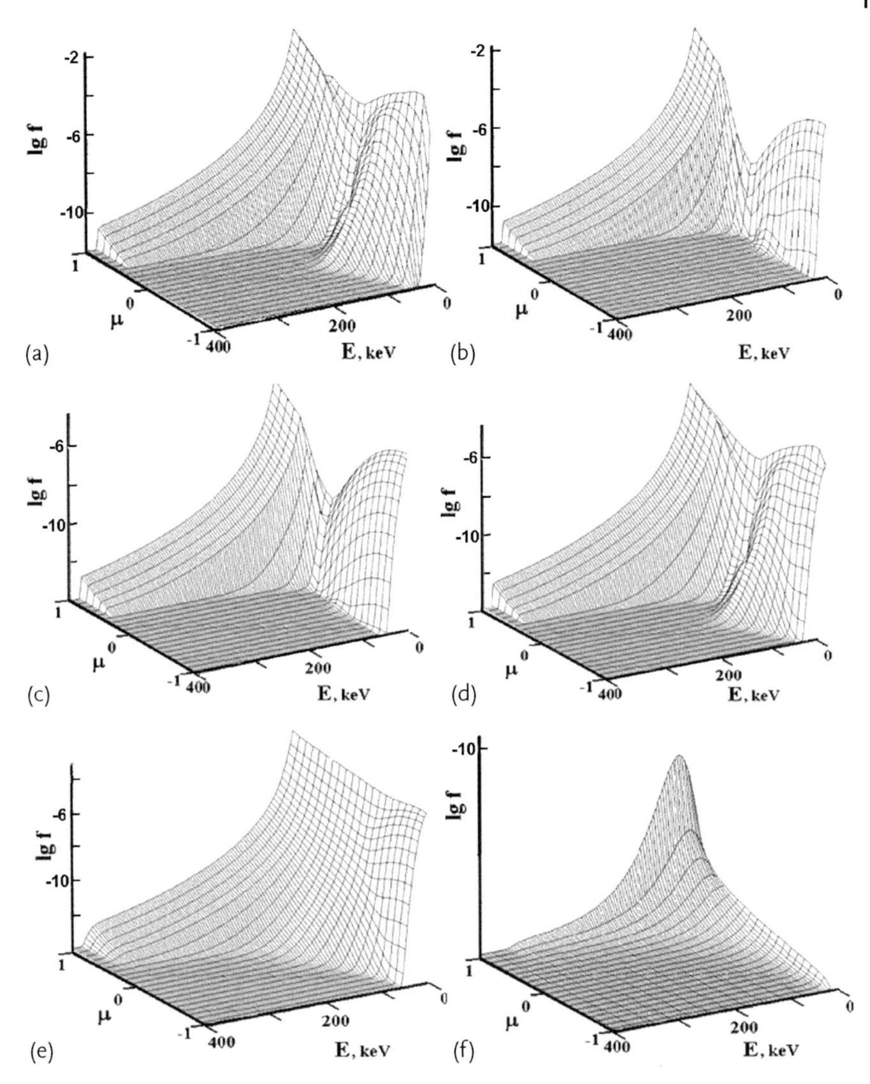

Figure 4.7 The same as in Figure 4.4 but for a spectral index of $\gamma = 7$ and an initial energy flux 10^{12}.

sential role in reducing the electric field induced by precipitating electrons, which needs to be compensated by ambient electrons. As a result, at greater depths the return current is mostly formed by ambient plasma electrons, which is clearly seen in panels (a) and (b) of Figures 4.4–4.7.

The energy distributions of the "bump" from returning beam electrons are limited to energies lower than $\sim$ 45–50 keV for harder beams ($\gamma \approx 3$) (Figures 4.5 and 4.7) then increase to 65–70 keV for softer beams ($\gamma \approx 7$) (Figures 4.4 and 4.6). The energy distributions of returning electrons appear to be exponential, with a maximum of 15–25 keV for weaker hard beams and shifting to higher energies of 30–35 keV for softer and more intense beams. This is likely to be caused by a combination of much more effective pitch-angle scattering and collisional energy losses for lower-energy electrons in softer beams than for the higher-energy electrons discussed above (Sections 3.2.2.5 and 4.4.2.1). This leads to the maximum in the returning electron numbers whose appearance depends on the maximum depth of the self-induced electric field. These returning electrons can significantly contribute to the resulting photon spectra that will be discussed in Chapter 5.

By comparing the precipitating electron distribution functions in Figures 4.4–4.7 at depths after the self-induced electric field becomes negligible, one can deduce that in the lower chromosphere the bulk of lower-energy electrons ($\preceq$ 100 keV) become fully isotropic (Leach and Petrosian, 1981; Zharkova *et al.*, 1995). Although electrons with initial energies higher than 100 keV can reach these levels, their numbers are significantly lower than the number of lower-energy electrons because of their power-law energy distributions.

4.4.2.5 Differential Energy Spectra of Beam Electrons

The differential spectra (calculated at given depths) for the full kinetic equation are plotted in Figure 4.8 for beams with initial energy fluxes of 10^8, 10^{10}, and 10^{12} erg cm^{-2} s^{-1} and initial spectral indices of 3 and 7 (Zharkova and Gordovskyy, 2005a).

In the case of weak beams (10^8 erg cm^{-2} s^{-1}) (Figure 4.8a,b), when the induced electric field is weak and constant at most depths, the collisional stopping depth (2×10^{19} cm^{-2} for $E_{\rm low}$) is shorter than the "electric" one (10^{20} cm^{-2}) found for the electric field from Figure 4.3. Thus, beam precipitation is dominated by collisions, and their resulting differential spectra toward lower energies become flatter with depth because the electrons shift in the energy distribution from higher to lower energies. Since the initial electron spectrum is power law, a number of lower-energy electrons will be repopulated by higher-energy ones whose numbers are reduced by the accepted energy power law. This leads to a flattening of the lower-energy parts of differential spectra (Syrovatskii and Shmeleva, 1972a) and, at some chromospheric depth, to the appearance of maxima at $E_{\rm max} = (\gamma + 1) \cdot E_{\rm low}$ (Zharkova and Gordovskyy, 2006).

The induced electric field of more intense beams (10^{12} erg cm^{-2} s^{-1}) (Figure 4.8c,d) is, first, substantially higher in the corona and, second, constant only at depths of a factor of 10^{18} cm^{-2} (the turning point), after which it decreases either linearly (for harder beams) or parabolically and more steeply (for softer ones). For higher electric fields the electric stopping depths move into upper atmospheric levels above the collisional one. For electrons with $E_{\rm low} = 12$ keV the electric stopping depth moves to 4×10^{17} cm^{-2}, for 40 keV electrons to 2×10^{18} cm^{-2}, and only for 100 keV electrons to a depth lower than 2×10^{19} cm^{-2}, which is the collisional stopping depth for $E_{\rm low} = 10$ keV.

The more intense and softer the beam is, the higher is the magnitude of its constant electric field (Section 4.4.2.3). Thus the energy that electrons lose before the turning point is also higher, and their differential spectra at lower energies are flat-

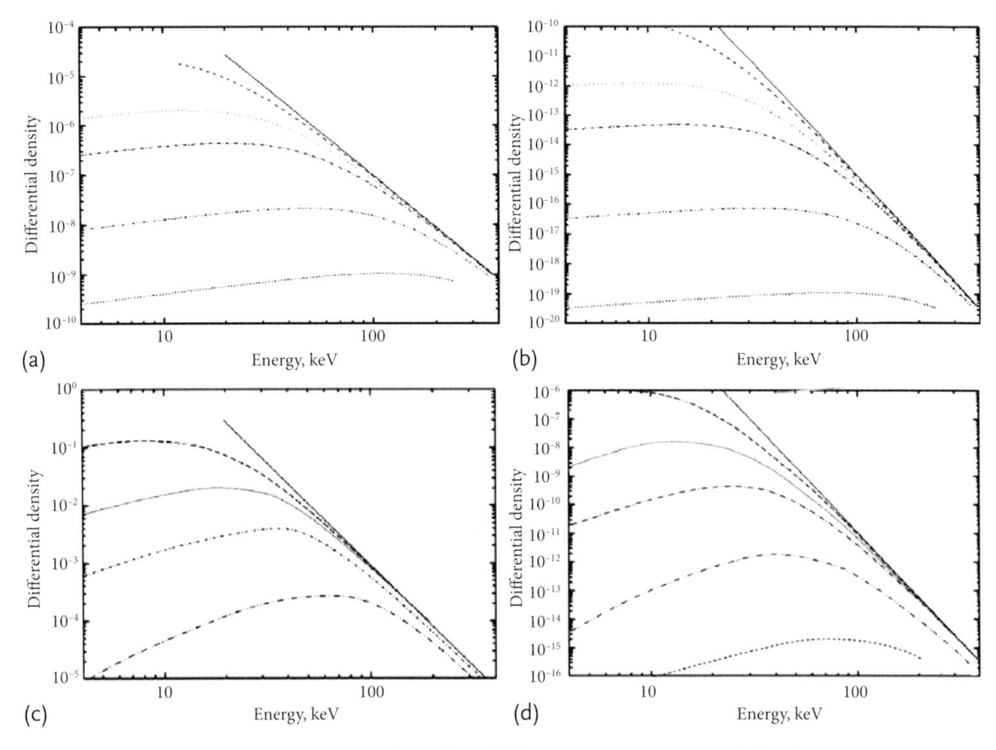

Figure 4.8 Differential energy spectra calculated for a full kinetic stationary approach for electron beams with initial energy fluxes of 10^8 (a,b) and 10^{12} erg cm^{-2} s^{-1} (c,d) for $\gamma = 3$ (a) and $\gamma = 7$ (b).

ter. However, their turning point also shifts upward to the corona, which reduces the distance of a constant electric field and increases the depth range where the self-induced electric field decreases. Hence the energy losses of electrons between 10 and 100 keV are dominated by deceleration in an induced electric field. But electrons with energies of 10–50 keV lose the bulk of their energy, mostly in a constant electric field and only a small part in collisions. Electrons with higher energy (50–100 keV) lose less of their energy in a decreasing electric field and still not enough in collisions because they do not reach the collisional stopping depth for electrons with a lower cutoff energy. As a result, electron distributions at lower energies become flatter.

On the other hand, for a given beam intensity, the softer the beam, the shorter the distance to where the induced electric field is constant and the smaller the energy lost by electrons in this deceleration. As a result, electrons with higher energy can reach depths between the turning point and the collisional stopping depth (cf. Figure 4.8a,b and 4.8c,d). In addition, harder beams induce an electric field that falls smoother than softer ones (Section 4.4.2.3). This also leads to a faster energy loss between the turning point and the collisional stopping depth and to a narrower maximum in their differential spectra than for softer beams (compare Figure 4.8c and d).

The position of maxima in the differential spectra is also dependent on the beam parameters. For weaker beams, the maximum appears at lower energies and for stronger ones at higher energies (cf. Figure 4.8a,b and 4.8c,d). The maximum positions are defined by a greater distance between the electric field turning point and the collisional stopping depth (Section 4.4.2.3). The maxima are higher and wider for softer beams and lower and narrower for harder ones (cf. Figure 4.8a,c and 4.8b,d). The more intense and softer the beams are, the shorter the constant part of the electric field is and the steeper its decrease is. The former means that the maximum energy of an electron affected by the deceleration in the strong constant electric field is reduced (i.e., 60 keV for a beam of 10^{10} but only 30 keV for a beam of 10^{12}). The latter means that electrons with higher energies can precipitate between the electric stopping depth for the constant field and the collisional stopping depth, that is, the energy range of electrons "released" from the electric field deceleration is increased.

4.4.3 Electron-Beam Density Variations with Depth

Using the electron distribution functions above, one can calculate the depth variation of electron-beam densities and energy fluxes according to Eqs. (4.8) with the normalization coefficient K determined by Eq. (4.9).

4.4.3.1 Effect of Nonuniform Ambient Ionization

Energy diffusion in partially ionized plasma is defined by a parameter a and Coulomb logarithms for electrons, ions, and neutral atoms associated with it, as discussed in Chapter 3 (Eq. (3.19)). This parameter is held to be constant at all depths. However, for partially ionized ambient plasma, that is, in the lower chromosphere and photosphere, this is not strictly valid (Kontar *et al.*, 2002; Zharkova and Gordovskyy, 2005a). In the lower chromosphere the ionization degree of ambient plasma strongly decreases to 10^{-3}–10^{-4} (Zharkova and Kobylinskii, 1993), as demonstrated in Figure 4.9a, leading to a sharp decrease in the number of electrons and ions in the lower chromosphere. Keeping in mind that the energy of beam electron changes with depth following Eq. (3.41), the smaller parameter a is, the less will be the particle energy loss. This will result in slower beam dissipation at greater depths and an increase of the maximum penetration depth for electrons with the same energy, in comparison with those precipitating into a fully ionized ambient plasma.

This is demonstrated in Figure 4.9b, which shows the depth variations of electron-beam density for pure collisional energy losses corresponding to a fully ionized atmosphere (solid lines) and in a partially ionized one (dashed lines). In partially ionized plasma the density of beam electrons precipitating into a given

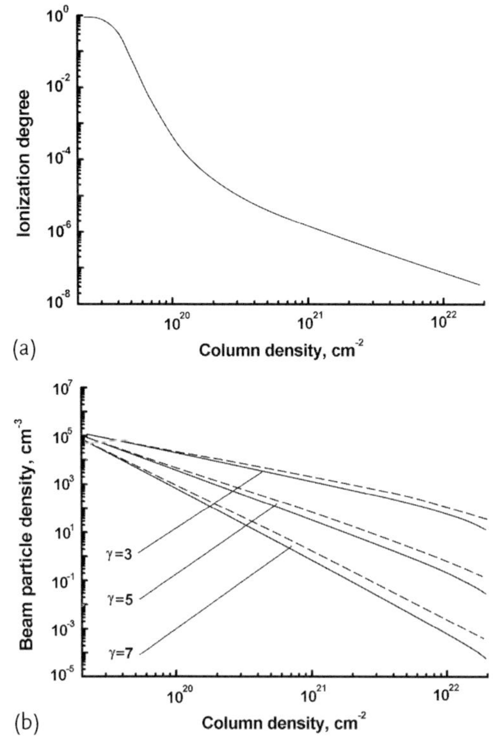

Figure 4.9 (a) Variations in degree of ionization with depth (Zharkova and Kobylinskii, 1993) and (b) beam density variations with depth losing energy in pure collisions for an initial beam flux of $F_0 = 10^8$ erg cm^{-2} s^{-1} and different initial spectral indices. The solid curves correspond to beam precipitation into a fully ionized atmosphere, while the dashed curves correspond to precipitation into a target with the ionization degree decreasing with depth.

depth is higher than in a fully ionized one, allowing more electrons to reach deeper atmospheric layers in the chromosphere and photosphere.

4.4.3.2 **Effect of Electric Field**

The particle densities computed for the precipitation of an electron beam into a partially ionized ambient plasma taking into account energy and pitch-angle scattering for collisional and ohmic energy losses are plotted in Figure 4.10. Comparing Figure 4.10b with Figure 4.10c, one can note that without ohmic losses the electron-beam densities at every depth are much higher than with them, for example, electron beams without electric field losses reach deeper layers below 10^{22} cm^{-2}, in comparison with a depth of 10^{21} cm^{-2} for $\gamma = 3$ when the electric field is taken into account.

In deeper layers, softer beams with higher spectral indices (solid curves) have much lower densities, and thus they reach much lower depths than harder beams with a lower index $\gamma = 3$ (dotted–dashed curves). With the inclusion of an electric

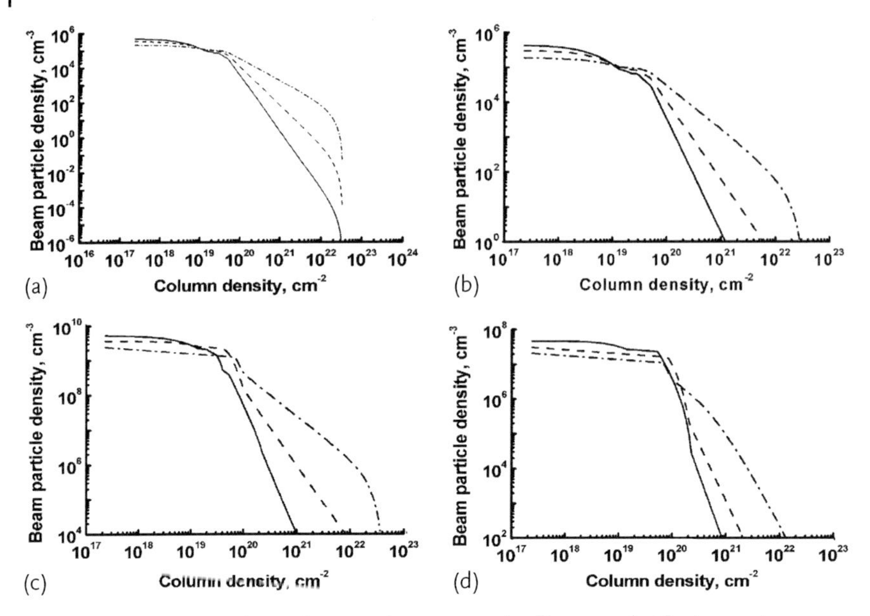

Figure 4.10 Variations in beam density with depth calculated for an initial energy flux of 10^{10} erg cm^{-2} s^{-1} without (a) and with (b) an induced electric field and with an electric field for initial energy fluxes of 10^8 erg cm^{-2} s^{-1} (c) and 10^{12} erg cm^{-2} s^{-1} (d). Initial spectral indices are equal to 7 (solid curves), 5 (dashed curves), and 3 (dotted–dashed curves). Adopted from Zharkova and Gordovskyy (2005a).

field the densities become even lower for beams with $\gamma = 7$, reaching smaller depths of 10^{21} cm^{-2} (Figure 4.10b) instead of 2×10^{22} cm^{-2} without an electric field (Figure 4.10d). This reflects higher energy losses in the denser atmosphere layers owing to energy diffusion by a softer beam that thermalizes much faster than the hard one (Section 4.4.2.1), which becomes even stronger if the electric field is taken into account (Section 4.4.2.4).

Therefore, the most important effect revealed in Figure 4.10 is the dependence of depth density variations on a beam's initial spectral index, which can be followed with the different curves in every plot. Apparently the domination of either collisions or ohmic losses is defined by the stopping depth of beam electrons, which is found as a minimal stopping depth from the following two: collisional and electric stopping depths discussed in Section 3.4 and summarized in Table 3.1.

4.4.4
Mean Electron Fluxes

The mean electron energy fluxes found from the integration in depth and pitch angles of the distribution functions of both precipitating and returning electrons above are presented in Figure 4.11 for beams with two initial energy fluxes of 10^8 (Figure 4.11a) and 10^{12} erg cm^{-2} s^{-1} (Figure 4.11b) for $\gamma = 3, 5$, and 7.

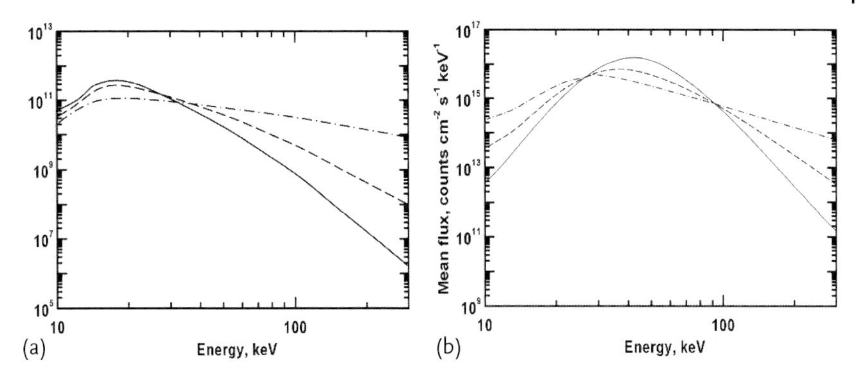

Figure 4.11 Mean electron spectra for a full kinetic solution (collisions + electric field) calculated for initial fluxes of 10^8 (a) and 10^{12} erg cm^{-2} s^{-1} (b) and spectral indices of $\gamma = 3$ (dotted–dashed lines), 5 (dashed lines), and 7 (solid lines).

First, one can notice that for more intense beams the mean flux spectra are noticeably softer, or their indices are higher by about 0.8–1.2, than those for weaker beams (cf. Figure 4.11a and 4.11b). This is likely to be caused by the dominant energy loss mechanisms: collisions for weaker hard beams whose mean spectral index is proportional to those of beam minus 1 (Section 3.4.2.4) and ohmic losses for more intense soft beams with the mean flux spectral index being equal to those of electrons.

Second, the mean electron fluxes calculated for joint collisional and ohmic energy losses have also been found to have well-defined maxima, which for more intense softer beams appear at the energies of 50–65 keV and for harder beams shift toward lower energies and become narrower, similarly to their differential spectra. These maxima are likely to point to an electric field turning point in beam precipitation followed by a slower decrease of the electric field with depth for harder beams and faster for softer ones.

4.5
Time-Dependent Fokker–Planck Equation

In this case the solutions are sought in the same flaring atmosphere as discussed above for the steady injection of an electron beam, so that $\partial f/\partial t$ is taken into consideration at all times. The initial cutoff energies of beam electrons are accepted to be equal: the lower cutoff $E_{\text{low}} = 12$ keV and the upper cutoff $E_{\text{upp}} = 1.2$ MeV. The upper cutoff energy is increased to match the limits sometimes observed in solar flares (Kuznetsov *et al.*, 2006). The initial spectral index of the injected beam for energies above the lower cutoff energy is chosen to be $\gamma = 3$, the energy flux at the top boundary is $F_{\text{top}} = 10^{10}$ erg cm^{-2} s^{-1}, and the initial angle dispersion is $\Delta\mu = 0.2$.

4.5.1
Initial and Boundary Conditions

We assume that before injection onset there are no beam electrons, for example, the initial condition is defined as

$$f(\tau = 0, s, z, \mu) = 0 \,. \tag{4.29}$$

The boundary condition at the top boundary, at $s = s_{\text{min}} = 2.08 \times 10^{-3}$ (or $2.29 \times 10^{17}\,\text{cm}^{-2}$), is selected to have two power-law distributions in energy (before the lower energy cutoff and after it) and a normal distribution in pitch-angle cosines μ:

$$f(\tau, s = s_{\text{min}}, z, \mu > 0) = f_{\text{n}} \psi(\tau) \frac{z^{\delta-1}}{z^{\delta+\gamma} + 1} \exp\left[-\frac{(1-\mu)^2}{\Delta\mu^2} \right] , \tag{4.30}$$

where $\Delta\mu$ is the initial pitch-angle dispersion of the injected electron beam and $\psi(\tau)$ determines the variations of the beam flux with time. If the energy is much larger than the lower cutoff energy, $z \gg 1$, then the distribution is power law with the index $-\gamma - 1$; thus the flux spectrum ($\sim z f$) is power law with the index γ. In the opposite case, $z \ll 1$, the distribution is also power law but with the index $\delta - 1$. The low-energy index δ is chosen to be 10, which accounts for some simulated energy spectra upon ejection from a current sheet (see, e.g., Zharkova and Gordovskyy, 2005b), while for the high-energy index two values, 3 and 7, are considered, which give the limits of spectral indices of particles ejected from a reconnecting current sheet (RCS) (Chapter 2).

f_{n} is the normalization coefficient, which is chosen to account for the energy flux of the injecting electron beam at the top boundary, $s = s_{\text{min}} = 2.08 \times 10^{-3}$ (or $2.29 \times 10^{17}\,\text{cm}^{-2}$):

$$F(s = s_{\text{min}}) = F_0 f_{\text{n}} \int\limits_{z_{\text{min}}}^{z_{\text{max}}} dz \int\limits_{-1}^{1} d\mu z^2 \mu f(s = s_{\text{min}}, z, \mu) \tag{4.31}$$

is equal to its initial value at the top boundary F_{top}, where $F_0 = 10^{10}\,\text{erg}\,\text{cm}^{-2}\,\text{s}^{-1}$ is the normalization factor of the energy flux.

At the bottom boundary, or the largest depth, $s = s_{\text{max}} = 9.17 \times 10^2$ (or $1.01 \times 10^{23}\,\text{cm}^{-2}$), the number of electrons in the beam is assumed to be negligibly small; thus the corresponding boundary condition is written as

$$f(\tau, s = s_{\text{max}}, z, \mu < 0) = 0 \,. \tag{4.32}$$

The distribution function is calculated in the following range of energies: $z_{\text{min}} \leq z \leq z_{\text{max}}$, where $z_{\text{min}} = 0.1$ (or 1.2 keV) and $z_{\text{max}} = 100$ (or 1.2 MeV). The boundary conditions in the energy boundaries are as follows:

$$\frac{\partial f(\tau, s, z = z_{\min}, \mu)}{\partial z} = 0\,, \tag{4.33}$$

$$\frac{\partial f(\tau, s, z = z_{\max}, \mu)}{\partial z} = 0\,. \tag{4.34}$$

The boundary conditions in pitch-angle cosines are as follows:

$$\frac{\partial f(\tau, s, z, \mu = 1)}{\partial \mu} = 0\,, \tag{4.35}$$

$$\frac{\partial f(\tau, s, z, \mu = -1)}{\partial \mu} = 0\,. \tag{4.36}$$

4.5.2
Relaxation to a Steady State

Electron injection starts at $t = 0$ and continues with a simulation until the stationary state is reached. This means that beam electrons keep being injected while the previously injected ones precipitate. In this case, the number of beam electrons at every depth depends on the time of their energy loss in both collisions and an electric field. Thus, at some depths close to the stopping depths of the electrons with lower cutoff energy the bulk of electrons has to lose all their energy and then, under the influence of the electric field of the beam, which is still being injected, they should become accelerated by this field in the opposite direction, and thus, return back to the top site where they were injected and create the close electric circuit with direct and return current.

Now let us evaluate how the system relaxes to the stationary state while losing energy in collisions and ohmic losses. The ratio of a dimensionless electric field to ambient plasma density as a function of column depth is plotted in Figure 4.12, revealing that the electric field is now substantially different from that simulated for the steady precipitation discussed in Section 4.4.

Figure 4.13 shows the profiles of a dimensionless electric field and beam density at different times after beam onset, which shows that the electric field relaxes to the steady state somewhat faster than the beam density. It turns out that the electric field formed by returning electrons is a necessary part of the steady precipitation of beam electrons as they complete the electric circuit formed by the process producing these electrons, for example, by a reconnecting current sheet. The circuit exists as long as the electron injection continues allowing for the recycling of the bulk of beam electrons precipitating into the same flaring atmosphere 10–100 times per second. This in turn reduces the requirements for the particle numbers responsible for the observed HXR emission discussed in Chapter 2.

The relaxation time t_r, after which the precipitating density stops changing, for example becomes stationary, can be estimated as ~ 0.07 s. This relaxation time is longer for a stronger beam, which is caused by a smaller density of the ambient plasma at coronal depths where the bulk of precipitation occurs (Figure 4.1), and,

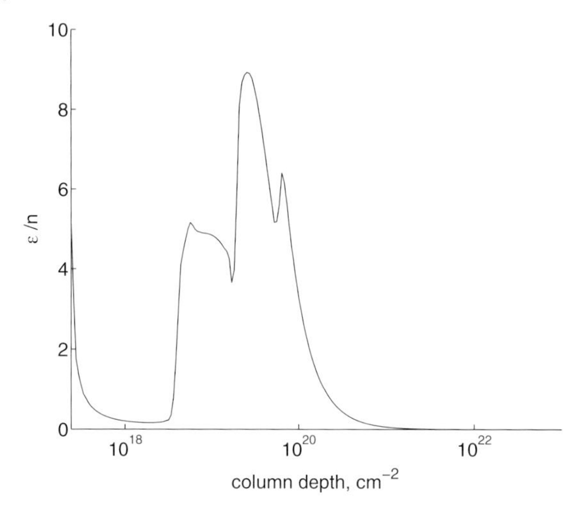

Figure 4.12 Ratio of dimensionless electric field to ambient plasma density as a function of column depth. Collisions and electric field are taken into account.

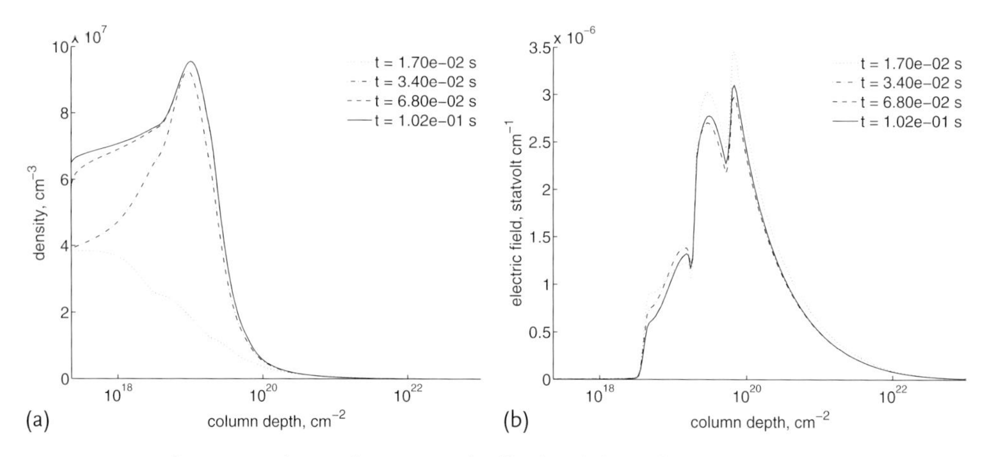

Figure 4.13 Electron density (a) and self-induced electric field ε (b) profiles, where t is time elapsed since injection onset. Collisions and electric field are taken into account. Beam parameters are as in Figure 4.14.

hence, a larger linear depth for the same column depth. For example, the depth $10^{20}\,\text{cm}^{-2}$ corresponds to the linear depth of 2×10^8 and 11×10^8 cm for beam energy fluxes of 10^{10} and $10^{12}\,\text{erg}\,\text{cm}^{-2}\,\text{s}^{-1}$, respectively. This leads to a longer relaxation time for the atmosphere preheated by a stronger beam, for example, for a beam with the energy flux of $10^{12}\,\text{erg}\,\text{cm}^{-2}\,\text{s}^{-1}$ the relaxation time is found to be ~ 0.2 s. The local maximum, which appears in the density profile at a depth of about $\sim 2 \times 10^{19}\,\text{cm}^{-2}$ (Figure 4.13), is caused by beam deceleration, while the flux of electrons remains nearly constant. After this depth the bulk of electrons leave the distribution (thermalize) because their energy drops below $z_{\min}$, and the

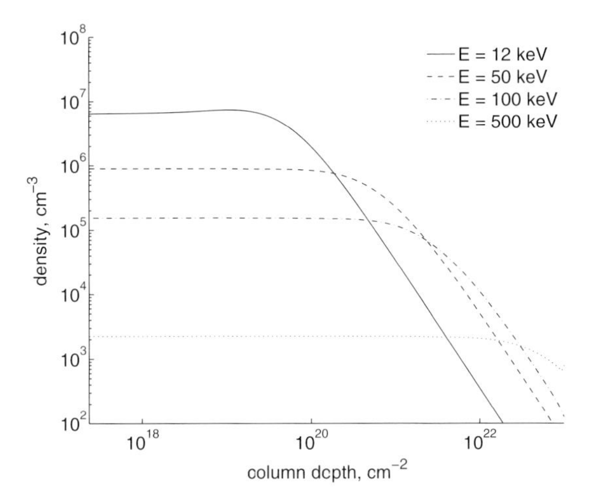

Figure 4.14 Electron beam density in different energy bands calculated for collisional energy losses with pitch-angle scattering. The beam parameters, see Eq. (4.30), are $\gamma = 3$, $F_{\text{top}} = 10^{10}\,\text{erg}\,\text{cm}^{-2}\,\text{s}^{-1}$, and $\Delta\mu = 0.2$.

density rapidly decreases at a depth of about $\sim 5 \times 10^{19}\,\text{cm}^{-2}$ (Figure 4.13). This corresponds to the stopping depth for electrons with energies close to the lower cutoff energy (12 keV).

Electrons with higher energies can travel deeper (Figure 4.14). In particular, it can be seen that electrons with energies of $> 500\,\text{keV}$ can travel down to the photosphere almost without any energy loss. Figure 4.14 is in a good agreement with the analytical results obtained by Zharkova and Gordovskyy (2006).

4.6 Regime of a Stationary Injection

4.6.1 Distributions of Electron Beams with a Lower-Energy Part

Now let us explore the variations of electron beam distributions by adding to the initial power-law energy distribution a lower-energy knee with a positive spectral index below the lower cutoff energy and including a magnetic field convergence in addition to collisional and ohmic losses. The simulations lasted longer than 0.07–0.2 s, until the stationary precipitation was established.

In the simulations the following parameters were adopted: initial energy flux of the accelerated electrons: $F = 10^{10}\,\text{erg}\,\text{cm}^{-2}\,\text{s}^{-1}$; electron energy range: $E_{\text{min}} = 12\,\text{keV}$, $E_{\text{max}} = 1200\,\text{keV}$; initial pitch-angle dispersion of electrons: $\Delta\mu = 0.2$; transition-region boundary at $\xi_c = 10^{20}\,\text{cm}^{-2}$. The initial power-law index of accelerated electrons was taken to be $\gamma = 3$ or $\gamma = 7$. The convergence factor B_c/B_0 was equal to $B_c/B_0 = 3$, and the characteristic column depth was equal to 2×10^{20} cm^{-2} (for the details of the selection rule see Siversky and Zharkova, 2009b).

Examples of electron distribution functions at a few column depths are shown in Figures 4.15–4.17: (i) for a column density of $\xi = 10^{18}$ cm^{-2} (Figure 4.15), (ii) for a column density of $\xi = 10^{20}$ cm^{-2} (Figure 4.16), and (iii) for a column density of $\xi = 10^{22}$ cm^{-2} (Figure 4.17).

One can see that both the self-induced electric field and the converging magnetic field result in the formation of a particle stream propagating upward, or returning back to the coronal source from which they are injected. Such a stream significantly exceeds a random stream of particles being scattered isotropically, and it becomes comparable with a stream of particles propagating downward, or precipitating particles.

Effect of an electric field. The self-induced electric field of a beam produces a stream of returning beam electrons (Figure 4.15a,d), which compensates the charge of precipitating electrons from Eq. (4.5) during the established process of a steady injection. The self-induced electric field at every depth is comprised of the electric field of precipitating electrons minus the electric field of those beam electrons that turned to negative μ at the depths above. This resulting electric field is the one that is then compensated by the ambient electrons known as a return current (van den Oord, 1990).

Note that for electron beams with an extended energy range of up to 1.2 MeV the effect of a self-induced electric field remains more important at upper atmospheric depths for those beams with softer energy spectra (compare graphs a and d in Figures 4.15–4.17), which confirms the conclusions of the steady state approach discussed in Section 4.4. One can also see that a self-induced electric field is able to accelerate the returning beam particles to rather high energies (up to 50–100 keV) while their number drops to zero at higher energies.

The distributions of returning electrons still depart from power laws and resemble more quasithermal distributions like those discussed in Section 4.4. The variations of a power-law index γ in the energy distributions of injected particles lead to a number of precipitating electrons decreasing with energy much faster for $\gamma = 7$ than for $\gamma = 3$. As a result, for $\gamma = 7$, at some atmospheric depths (Figure 4.15), the number of returning particles with energies below 100 keV can even exceed the number of precipitating particles with the same energy that affects dramatically the properties of X-ray emission, as discussed in Chapter 7.

Variations in differential energy spectra. Differential flux spectra are plotted in Figure 4.18 for forward ($\mu > 0$) and backward ($\mu < 0$) moving electrons at different depths. It can be seen that the self-induced electric field does not change the spectra of the downward ($\mu > 0$) moving electrons but essentially affects the spectra of the upward ($\mu < 0$) moving ones. Since the angle diffusion due to the electric field is more effective for lower-energy electrons, the spectra of the returned electrons is enhanced at low and medium energy levels (compare plots b and d in Figure 4.18).

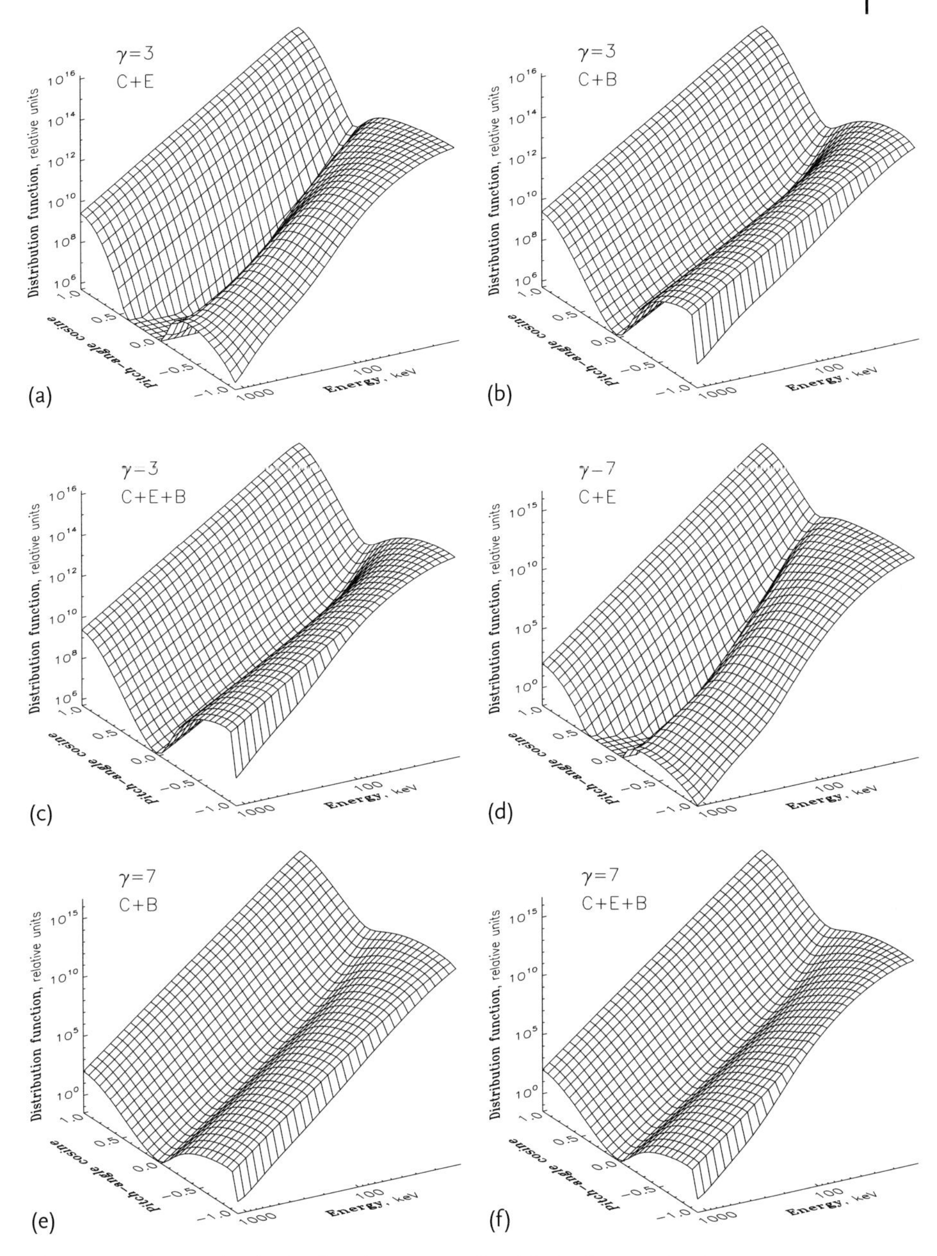

Figure 4.15 Electron distribution functions (in $E-\mu$ space) obtained by solving the Fokker–Planck equation at a column depth of $\xi = 10^{18}\ \text{cm}^{-2}$. The initial power-law indices of the accelerated electrons are equal to 3 and 7. The factors taken into account are as follows: collisions (*C*), self-induced electric field (*E*), and convergence of the magnetic field (*B*).

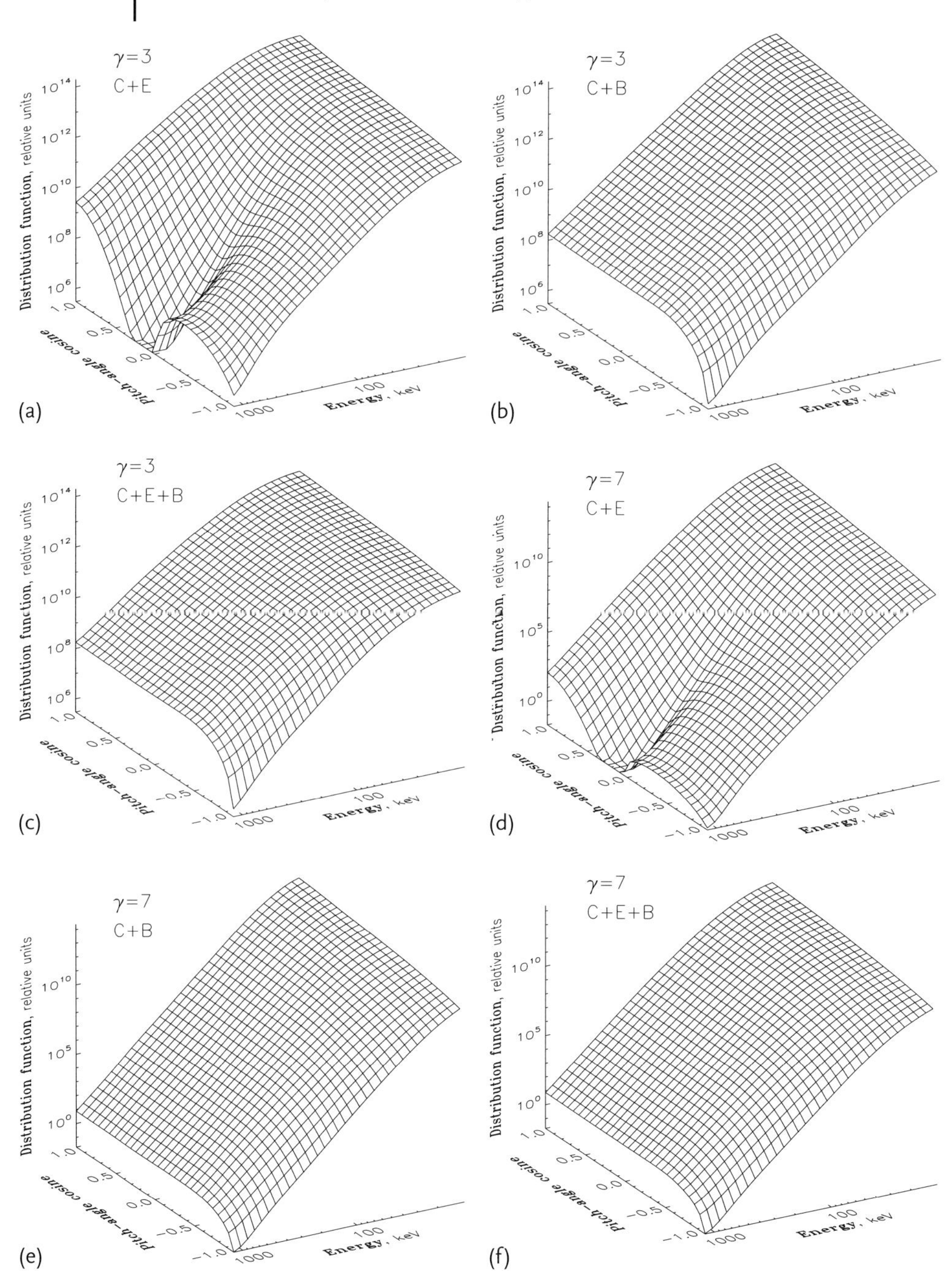

Figure 4.16 Electron distribution functions at a column depth of $\xi = 10^{20}\ \mathrm{cm}^{-2}$. Other simulation parameters are as in Figure 4.15.

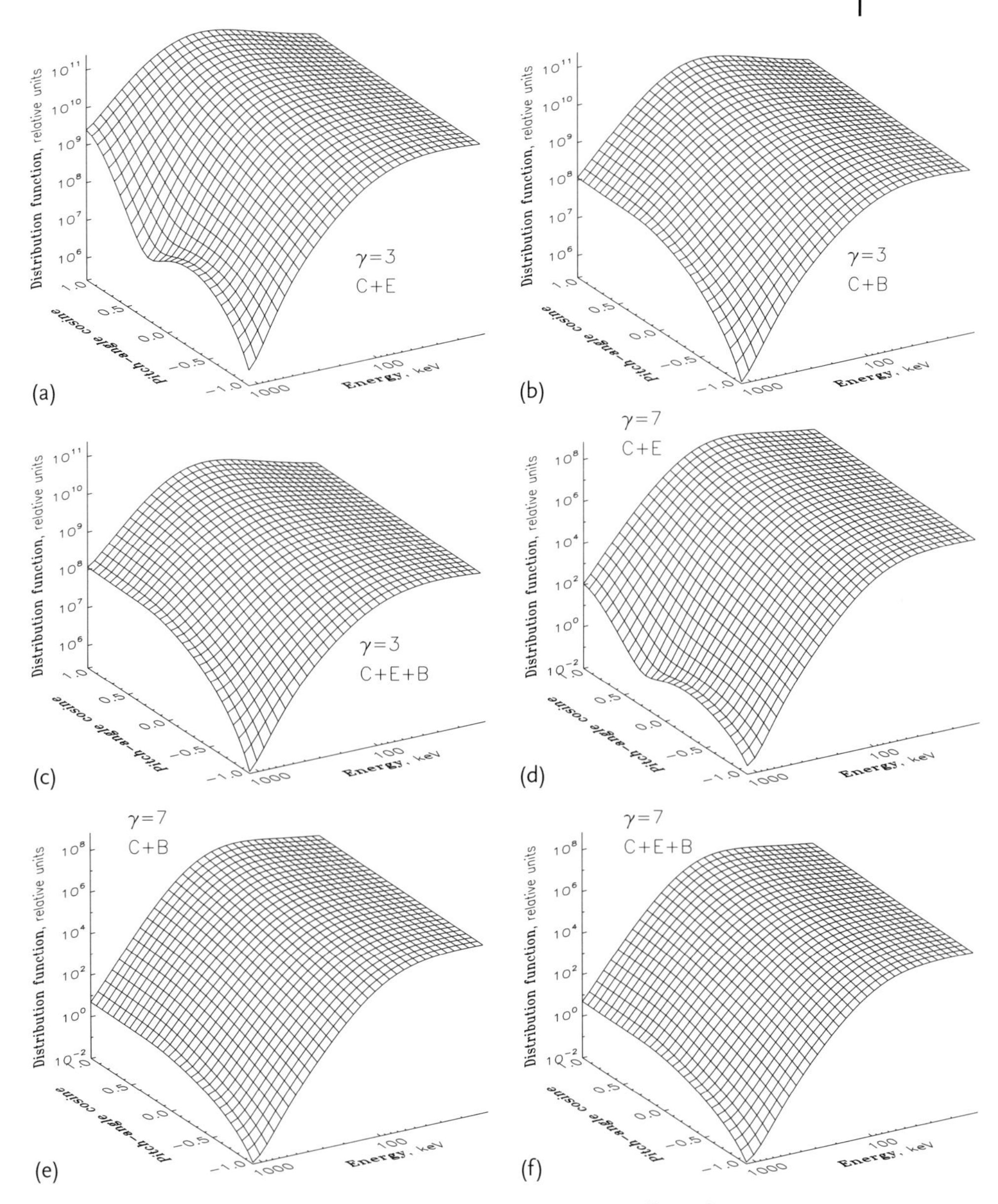

Figure 4.17 Electron distribution functions at a column depth of $\xi = 10^{22}$ cm^{-2}. Other simulation parameters are as in Figure 4.15.

Effect of a converging magnetic field. A converging magnetic field produces the well-known loss-cone distributions (Leach and Petrosian, 1981) reflecting the precipitating electrons' being scattered to pitch angles within the loss cone defined by the magnetic convergence factor (plots b of Figures 4.15–4.17 and plots e of

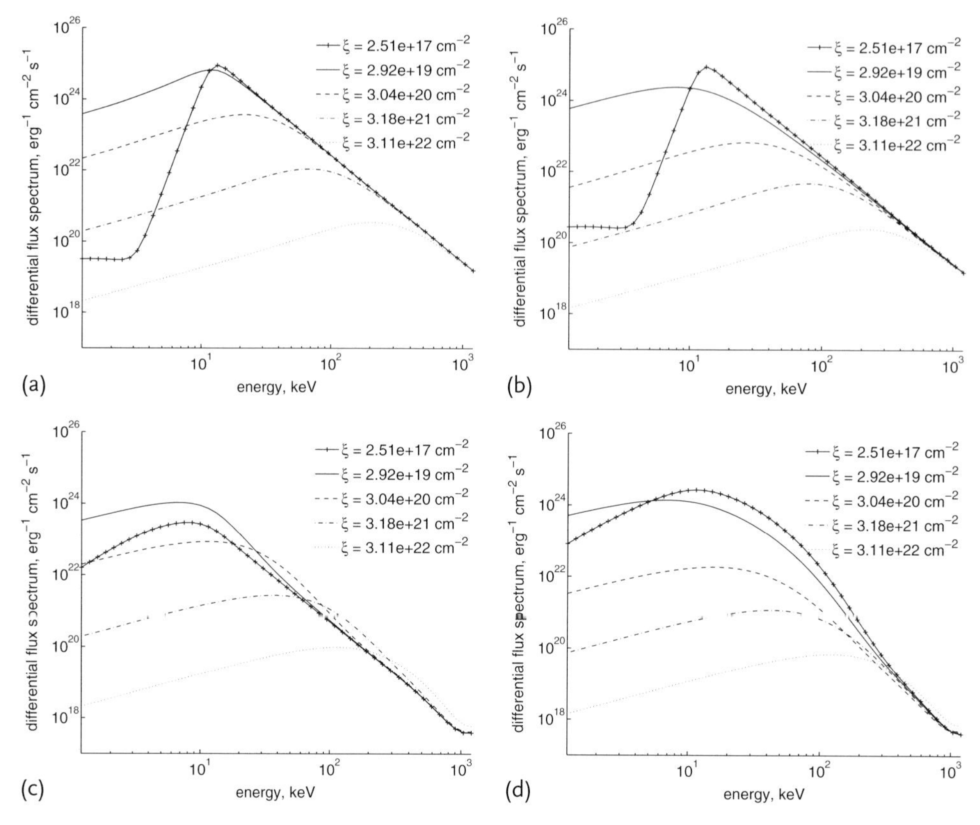

Figure 4.18 Differential energy spectra of beam electrons integrated over positive and negative pitch angles. Beam parameters are as in Figure 4.13: (a) collision, $\mu > 0$; (b) collision and electric field, $\mu > 0$; (c) collisions, $\mu < 0$; (d) collisions and electric field, $\mu < 0$.

Figures 4.15–4.17). The energy spectrum of the returning particles (reflected by a converging magnetic field) does not differ significantly from the initial energy spectrum of precipitating particles. However, the effect of magnetic mirroring is more significant for higher energy electrons compared to other energy losses (collisions or ohmic losses) (compare Figure 4.17a,d and c,f with Figure 4.15a,d and c,f, respectively). This is because the latter inversely depends on the particle energies, thus becoming with depth (and energy loss) much smaller than the particle pitch-angle changes in magnetic mirroring (convergence). In real events (e.g., solar flares), all the energy loss factors work simultaneously, likely producing the beam electron distributions in energy and pitch angles similar to those shown in plots c of Figures 4.15–4.17 calculated for the models C + E + B for the beams with $\gamma = 3$ and in plots f of Figures 4.15–4.17 for the same model and $\gamma = 7$.

At depths of $\xi \simeq 10^{20}$ cm^{-2}, where the bulk of electrons lose their energy, magnetic field convergence, along with collisions, governs the remaining electron distributions (Figure 4.16). The electrons become distributed almost isotropically at all energies (due to collisional scattering), and the presence of a converging magnetic

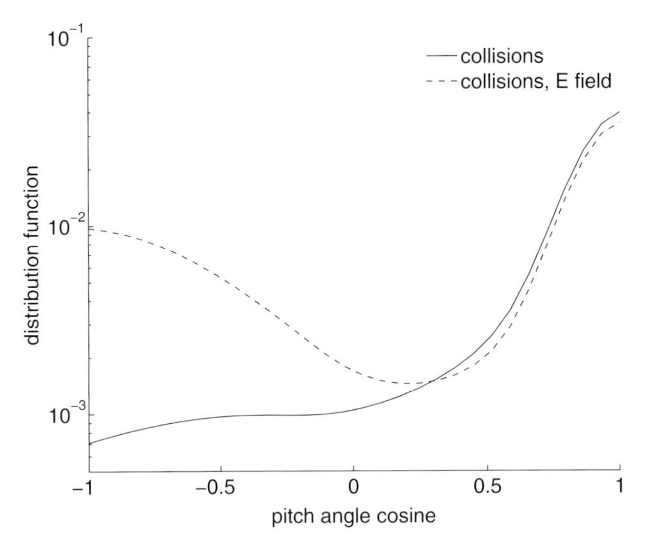

Figure 4.19 Pitch-angle distribution of an electron beam with parameters as in Figure 4.13 and the magnetic convergence (for dotted curve) given by Eq. (4.41).

field speeds up the isotropization process. In the even deeper layers (Figure 4.17), the number of particles decreases rapidly with depth due to collisions. As the collisional losses are more significant for low-energy electrons, the "humped" distribution (with the maximum at about 100 keV) is formed. The effect of magnetic mirroring for C+E+B models only enhances the number of returning electrons in the whole energy range as per Figures 4.15–4.17.

Pitch-angle diffusion. The number of electrons returning back to the source plotted in Figure 4.19 is smaller compared to the case where the self-induced electric field is not taken into account. However, even without the electric field ε, a number of electrons with $\mu < 0$ is essential owing to the pitch-angle scattering (second and third terms on the right-hand side of Eq. (4.16)).

4.6.2 Variations of Electron-Beam Density

The beam densities of precipitating and returning electrons are shown to significantly vary with the variations of beam parameters. Figure 4.20 shows the density variations of energetic electrons with a column density from the injection point for different models of energy losses by beam electrons. Note that if the self-induced electric field is not taken into account (panels labeled "C" and "C+B"), then the electric currents directed upward and downward are not compensated because only a part of the precipitating electrons is scattered to negative pitch-angle cosines as per Eq. (4.5) for a self-induced electric field.

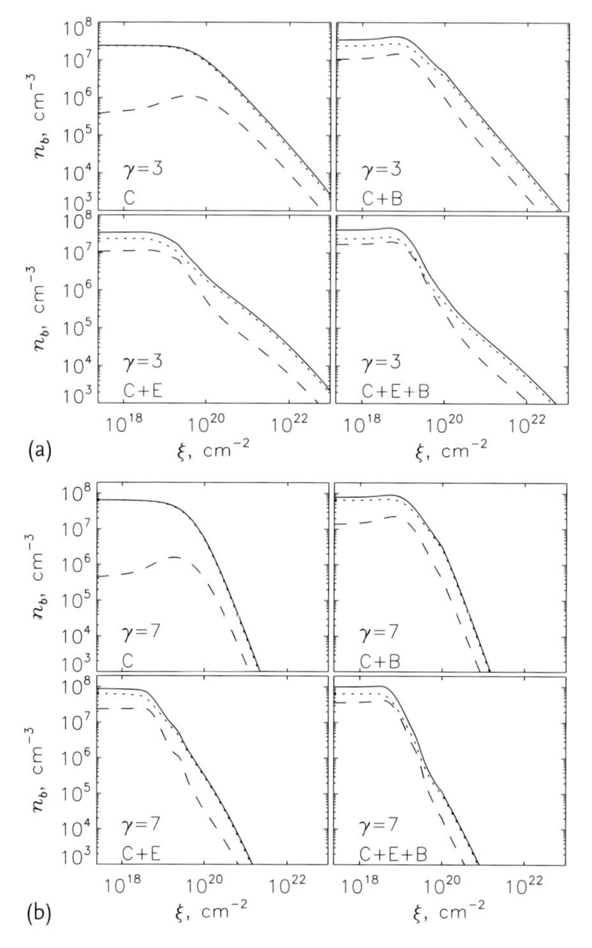

Figure 4.20 Depth variations of density of precipitating electrons and returning electrons for different energy loss models detailed in Figure 4.15. Dotted line: downward propagating particles ($\mu > 0$); dashed line: upward propagating particles ($\mu < 0$); solid line: total concentration (downward + upward).

The beam densities of returning electrons are dependent on the energy loss mechanisms of precipitating electrons. Note that in the corona (at $\xi < \xi_{10} \simeq 2 \times 10^{19}\,\text{cm}^{-2}$), the electron energy losses due to collisions are not very significant, while ohmic losses can play the dominant role in the precipitation (compare the magnitudes of collisional and electric stopping depths defined in Table 1 from Zharkova and Gordovskyy, 2006). Thus, the number of returning particles (integrated over all energies) is always slightly less than that of the total precipitating particles.

With every precipitation depth in the corona, there is a substantial increase in the numbers of particles propagating upward (panels labeled "C+E" and "C+E+B" in Figures 4.15–4.17). This increase is higher for beams with bigger spectral indices (compare Figures 4.20a and 4.20b). In other words, the electric current is not simply proportional to the density of precipitating particles as discussed in Section 4.4.3.2 but depends on the spectral characteristics of beam electrons as per Eq. (4.5).

Our simulations confirm the results obtained in Chapter 3 for a pure collisional continuity equation that electron beams with harder spectra can reach deeper layers of the solar atmosphere (Siversky and Zharkova, 2009b; Zharkova and Gordovskyy, 2005a; Zharkova and Kobylinskii, 1993). This results in higher densities at deeper atmospheric levels for electron beams with $\gamma = 3$ than for those with $\gamma = 7$, while in the upper atmospheric levels the densities of returning electrons are higher for electrons with $\gamma = 7$, which reflects the ohmic-loss dependence of the power-law index (Zharkova and Gordovskyy, 2006).

4.6.3
Effects of Magnetic Field Convergence

Let us now determine how large the magnetic convergence parameter, α_B, must be to produce any noticeable effect on the distributions of beam densities. To do so, we compare the terms in front of $\partial f/\partial \mu$ in Eq. (4.24). Since the collisional pitch-angle diffusion is much smaller than the pitch-angle scattering caused by an electric field (Figure 4.19), we compare the effects of a magnetic convergence with those caused by an electric field.

Magnetic convergence effects are stronger if $\alpha_B > 2\varepsilon/(nz)$. To prove this, let us plot the dimensionless ratio ε/n in Figure 4.12. The minimal value of the ratio ε/n in the interval from $s = s_{min}$ to the stopping depth of low-energy ($z = 1$) electrons, $\sim 5 \times 10^{19}\,cm^{-2}$, is about 0.2. Thus, a magnetic convergence would be more effective than an electric field for electrons with energies higher than their cutoff energy if $\alpha_B \gtrsim 0.4$. High-energy electrons can travel much deeper into the chromosphere (Figure 4.14), where the ratio ε/n can be as low as 2×10^{-4}; thus the magnetic convergence would be more effective for them if $\alpha_B \gtrsim 4 \times 10^{-6}$.

In the following sections we present the results of simulations for different models of a converging magnetic field.

4.6.3.1 Exponential Approximation

Following the approximation proposed by Leach and Petrosian (1981), let us assume that the convergence parameter does not depend on depth:

$$\alpha_B = \alpha_{B0} = \text{const}\,; \tag{4.37}$$

then the magnetic field variation is

$$B(s) = B_0 \exp[\alpha_{B0}(s - s_{min})]\,. \tag{4.38}$$

Suppose that the magnetic field at a depth of s_{max} is 1000 times stronger than at s_{min}; then $\alpha_B \approx \ln(1000)/s_{max} = 7.5 \times 10^{-3}$. As we discussed earlier, the effect of a magnetic convergence with such low α_B would be noticeable only for high-energy (> 100 keV) electrons. While an electric field returns mostly low- and medium-energy electrons and makes the spectrum of the returning electrons softer (in comparison with the purely collisional case), the magnetic mirror returns the high-energy electrons and makes their spectrum harder and similar to the initial power law.

4.6.3.2 Parabolic Approximation

McClements (1992a) suggested the following profile of a magnetic field variation:

$$B(s) = B_0 \left[1 + \frac{(s - s_{\min})^2}{s_0^2} \right] . \tag{4.39}$$

If $B(s_{\max})/B(s_{\min}) = 1000$, then $s_0 \approx s_{\max}/\sqrt{(1000)} = 31.6$. The convergence parameter is

$$\alpha_B(s) = \frac{2(s - s_{\min})}{s_0^2 + (s - s_{\min})^2} . \tag{4.40}$$

The magnetic convergence parameter α_B at maximum is $1/s_0 \approx 3.16 \times 10^{-2}$. This magnitude is not high enough to affect all the beam electrons, but, as is seen in Figure 4.21, such a convergence model, similar to the previous one, increases the number of high-energy electrons with negative μ. Note that McClements (1992a) considered a constant plasma density for which the profile of the magnetic field given by Eq. (4.39) can be more effective. While in our case the plasma density exponentially increases with depth, that significantly reduces the effect of convergence. Also we assumed that the magnetic field changed according to Eq. (4.39) in the whole range of column depths. However, observations show that the magnetic field variation is different in the corona and chromosphere.

4.6.3.3 Hybrid Approximation

In this model we propose that α_B is close to constant at shallow depths (in the corona) and tends to zero after some depth s_0 (in the chromosphere):

$$\alpha_B(s) = \alpha_{B0} \frac{s_0^2}{s_0^2 + (s - s_{\min})^2} ; \tag{4.41}$$

then the magnetic field variation is

$$\begin{aligned} B(s) &= B_0 \exp\left[\int_{s_{\min}}^{s} \alpha(s') ds' \right] \\ &= B_0 \exp\left[\alpha_{B0} s_0 \arctan\left(\frac{s - s_{\min}}{s_0} \right) \right] . \end{aligned} \tag{4.42}$$

At shallow depths, where $s \ll s_0$, the magnetic field varies as $B \approx B_0 \exp[\alpha_{B0}(s - s_{\min})]$, and at greater depths, $s \gg s_0$, the magnetic field is constant, $B \approx B_0 \exp(\alpha_{B0} s_0 \pi/2)$. In most simulations (where it is not stated explicitly) we accept

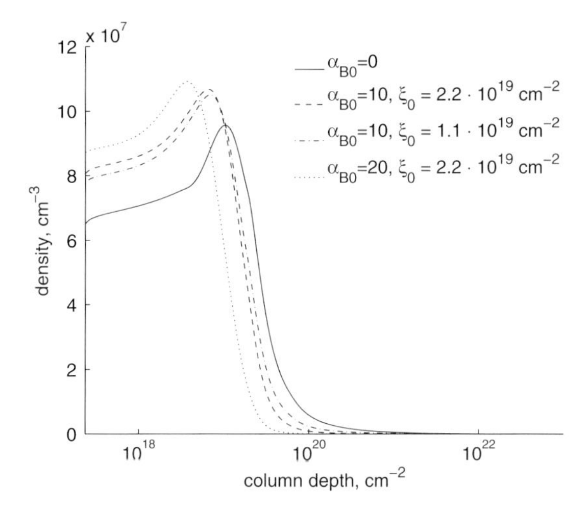

Figure 4.21 Beam density variations as a function of column depth. The magnetic convergence parameter is given by Eq. (4.41) with various α_{B0} and s_0. Beam parameters are the same as in Figure 4.13.

$\alpha_{B0} = 10$ and $s_0 = 0.2$ (or $2.2 \times 10^{19}\,\text{cm}^{-2}$, which corresponds to the transitional region); as a result the ratio $B(s_{max})/B(s_{min})$ equals 23.1.

Electrons with velocities inside the loss cone are not reflected by the magnetic mirror and reach the deep layers. For the current convergence model the critical pitch-angle cosine of the loss cone is $\mu_{lc} = \sqrt{1 - B(s_{min})/B(s_{max})} \approx 0.98$. This means that for the accepted initial angle dispersion of 0.2 about 90% of the electrons are reflected back.

As seen in Figure 4.19, the effect of magnetic convergence on pitch-angle distribution is similar to the effect of an electric field. However, since an electric field is most effective for electrons with $\mu = \pm 1$, the pitch-angle distribution has a maximum at $\mu = 1$ when the convergence cannot affect beam electrons. In contrast, a magnetic field does not affect electrons moving along field lines; thus the angle distribution of the upward moving electrons has a maximum at $\mu_m \approx -0.8$ (Figure 4.19), which is consistent with the conclusions of Zharkova and Gordovskyy (2006).

The profiles of electron density with different energies are plotted in Figure 4.22. If only magnetic convergence is taken into account (Figure 4.22a), it can be seen that magnetic mirroring does not depend on electron energy. Electrons with a pitch angle outside the loss cone are turned back at a depth of $\sim 10^{19}$–$10^{20}\,\text{cm}^{-2}$. The remaining electrons (with a pitch angle inside the loss cone) can travel down to the lower boundary in the atmosphere. When collisions are taken into account, electrons, especially those with low energies, lose their energy due to collisions (Figure 4.22b). It is important to compare collisional beam relaxations with and without magnetic convergence, as plotted in Figures 4.22b and 4.19, respectively. Note that the combination of the effects of collisions and convergence is stronger than a simple sum of the two separate effects. This occurs because electrons with

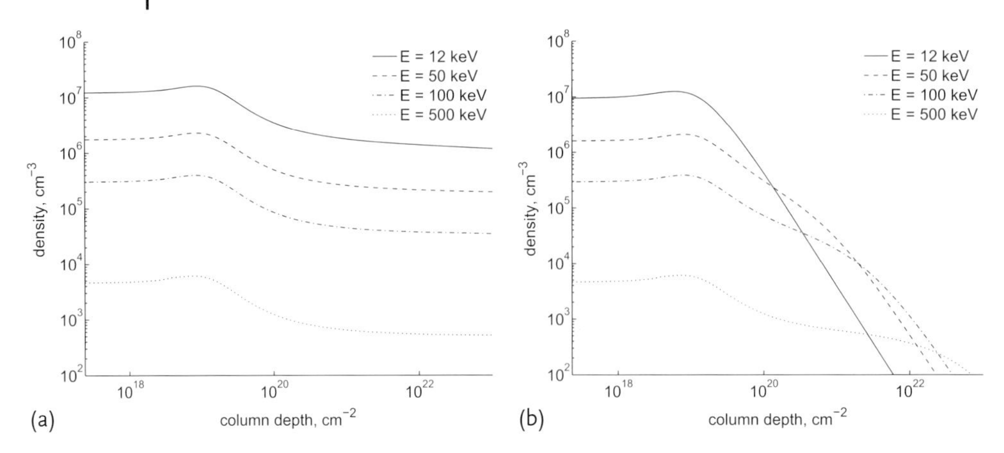

Figure 4.22 Densities of beam electrons in different energy bands calculated for the models including (a) convergence; (b) collisions and onvergence. The beam parameters are the same as in Figure 4.13. The magnetic field convergence parameter is given by Eq. (4.41).

initial pitch angles inside the loss cone are scattered by collisions to pitch angles that fall outside of the loss cone; thus more electrons are returned back to the top by the magnetic mirror.

In Figure 4.21 the results of simulations are presented for different magnitudes for the parameters α_{B0} and s_0 of the magnetic convergence model given by Eq. (4.41). The increase of α_{B0} clearly affects the beam electrons by reducing the depth of their penetration and by increasing the number of returning electrons. It is obvious that the system must be sensitive to the variations of s_0 if it is smaller than the penetration (stopping) depth, which is proven in Figure 4.21.

4.6.3.4 Magnetic Field Model Fitted to Observations

Although a magnetic field cannot be directly measured in the solar atmosphere, there are some indirect techniques that allow one to estimate the magnitude of the magnetic field. The coronal magnetic field can be determined from radio observations of gyroresonance emission (Brosius and White, 2006; Lang *et al.*, 1993). In particular, Brosius and White (2006) suggest that the magnetic scale height above sunspots, $L_{B\ cor} = B/\Delta B$, is ~ 7 Mm. On the other hand, Kontar *et al.* (2008) determined the chromospheric magnetic field by measuring the sizes and heights of HXR sources in different energy bands. They found that the chromospheric magnetic scale height, $L_{B\ chr}$, is ~ 0.3 Mm. Assuming that the magnetic scale height, L_B, changes linearly with depth l, the convergence parameter is

$$\alpha_B(s) = \frac{\alpha_{B0}}{n L_B} , \tag{4.43}$$

where $\alpha_{B0} = E_0^2/(\pi e^4 n_0 \ln \Lambda\, L_{B\ cor}) = 15.7$, and the dimensionless magnetic scale height as a function of column depth is

$$L_B = 1 - \left(1 - \frac{L_{B\ chr}}{L_{B\ cor}}\right) \frac{l(s)}{l(s_{max})} , \tag{4.44}$$

where the linear depth is $l(s) \propto \int n(s)^{-1} ds$. Since the magnetic field is defined as a function of linear depth, the model depends on the density profile $n(s)$ of the background plasma.

4.6.4
Mean Electron Fluxes of a Steady Beam

The mean electron fluxes shown in Figure 4.23 are found to have similar wide bumps at energies close to the turning point of the induced electric field magnitude, after which it steeply drops with depth, as discussed in Section 4.4.4. In addition, the mean fluxes are found to be affected by magnetic field convergence at higher energies.

It can be seen that the exponential magnetic field convergence strongly reduces the number of electrons that are able to reach lower atmospheric levels, while the parabolic and hybrid convergence models reveal much smaller reductions, for example, allowing more electrons to precipitate to the lower atmosphere. Therefore, mean electron fluxes are good indicators of various energy loss mechanisms for beam electrons at various precipitation depths, which is transferred to distributions of photon energy. Since the hybrid model fits most closely the magnetic field convergence deduced from observations, this makes it the most suitable for the interpretation of real flaring events and deduction of magnetic field convergence from observed mean electron spectra.

4.6.5
Plasma Heating by a Stationary Beam in Converging Magnetic Field

Heating by an electron beam losing its energy in collisions or ohmic losses with pitch-angle scattering is plotted in Figure 4.24, where for comparison a heating

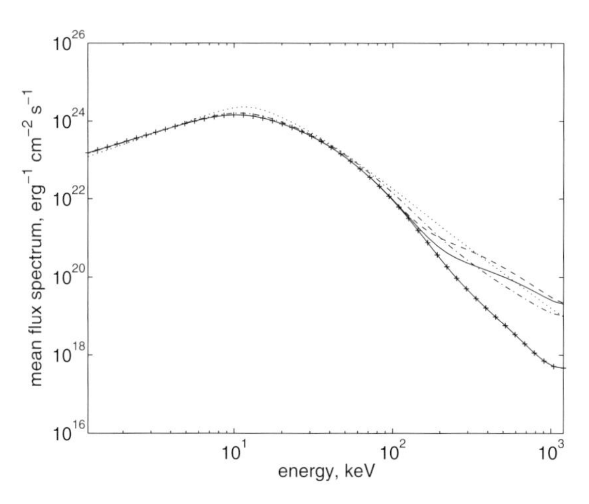

Figure 4.23 Mean flux spectra of electrons propagating upward ($\mu < 0$). Beam parameters are as in Figure 4.13.

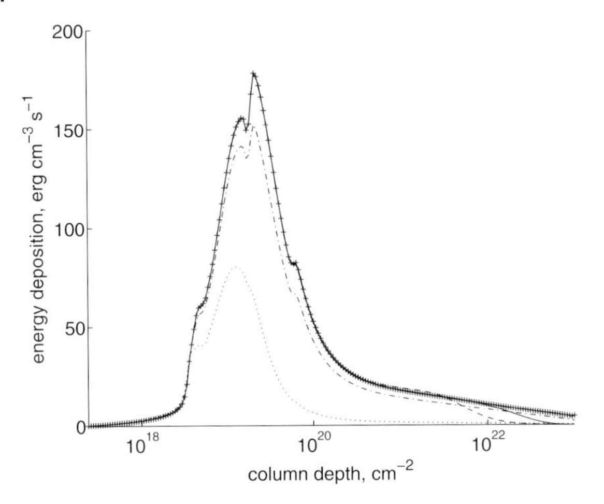

Figure 4.24 Energy deposition rates by beam electrons precipitating into the atmosphere without (crosses) and with magnetic convergence given by Eq. (4.38) (solid curve), Eq. (4.39) (dashed curve), Eq. (4.42) (dotted curve), and Eq. (4.43) (dotted–dashed curve). Beam parameters are as in Figure 4.13.

curve for pure collisional heating in the continuity equation approach (Syrovatskii and Shmeleva, 1972b) is also presented.

It can be seen that heating by softer intense beam electrons caused by their self-induced electric field is most effective at upper column depths ($2 \times 10^{19}\ \mathrm{cm}^{-2}$), dropping sharply with depth, while for harder weaker beams the heating spreads more evenly to the lower atmospheric depths just below the transition region ($10^{20}\ \mathrm{cm}^{-2}$). These heating patterns of beam electrons are similar to the results obtained earlier for the stationary Fokker–Planck equation and can be interpreted by the interplay of collisional and electric stopping depths for beam electrons.

Indeed, the inclusion of an electric field induced by the beam electrons themselves decreases the electric stopping depth for softer intense beams (Zharkova and Gordovskyy, 2006), which in turn increases the number of returning electrons. Thus most electrons will lose their energy at higher atmospheric depths, reducing the number of electrons at greater depths. All these factors lead to an upward shift of the heating function maximum and its magnitude reduction compared to pure collisions when the heating is spread more evenly in depth until the collisional stopping depth for electrons with a lower cutoff energy is reached.

On the other hand, magnetic convergence reduces heating at greater depths (Figure 4.24), where it is caused by high-energy electrons, because they were mirrored back to the corona. As the convergence parameter α_B is relatively high in the hybrid model, the whole spectrum of electron energy is affected (Figure 4.25). The energy deposition profile for this magnetic field approximation (Figure 4.25) indicates that the heating is only about 30% of the heating produced in the case of a constant magnetic field, which is because many electrons are reflected by the magnetic mirror before they reach dense plasma.

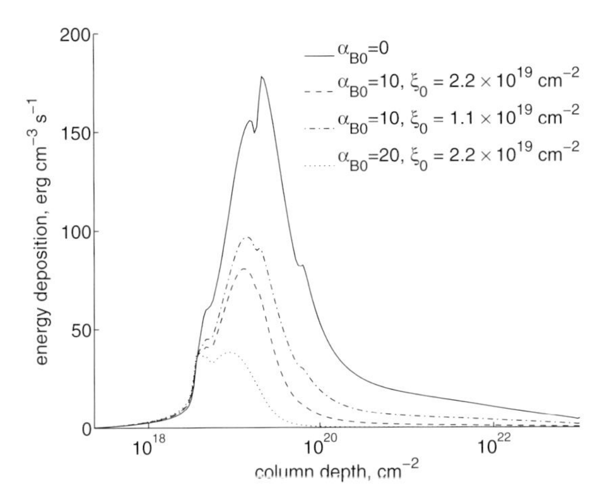

Figure 4.25 Energy deposition by beam electrons plotted as a function of column depth. The magnetic convergence parameter is given by Eq. (4.41) with various α_{B0} and s_0. Beam parameters are as in Figure 4.13.

Thus, a magnetic field convergence also leads to a reduction in the heating by about 20% produced by beam electrons, in comparison with the constant magnetic field profile (Figure 4.25).

4.7 Impulsive Injection

Now we can consider precipitation of an electron beam injected as a short impulse. The early SMM (Kiplinger *et al.*, 1983) and recent CORONAS/IRIS (Charikov *et al.*, 2004) observations have reported millisecond impulses in HXR emissions from solar flares. Furthermore, the electron acceleration time in an RCS was recently shown (Siversky and Zharkova, 2009b) to be as short as 10^{-5} s. These facts suggest that the time scale of a beam of accelerated electrons may be rather short. In this section we study the evolution of such short impulses in the solar atmosphere.

The injection time, δt, is chosen to be 1.7×10^{-3} s, which is much shorter than the relaxation time, $t_r \approx 0.07$ s, that was found in Section 4.5.2 (Figure 4.13). The default parameters of the beam are similar to the case of stationary injection: the initial spectral index of the beam is $\gamma = 3$, the maximal energy flux at the top boundary is $F_{top} = 10^{10}\,\text{erg}\,\text{cm}^{-2}\,\text{s}^{-1}$, and the initial angle dispersion is $\Delta\mu = 0.2$. Also, energy deposition profiles produced by a softer beam ($\gamma = 7$) and by a stronger beam ($F_{top} = 10^{12}\,\text{erg}\,\text{cm}^{-2}\,\text{s}^{-1}$) are obtained.

Obviously, an impulsive injection has to lead to smaller densities of electrons at a given depth, in comparison with a stationary injection (Figures 4.13a and 4.26a). A smaller density, in turn, results in a lower self-induced electric field. Thus, in the

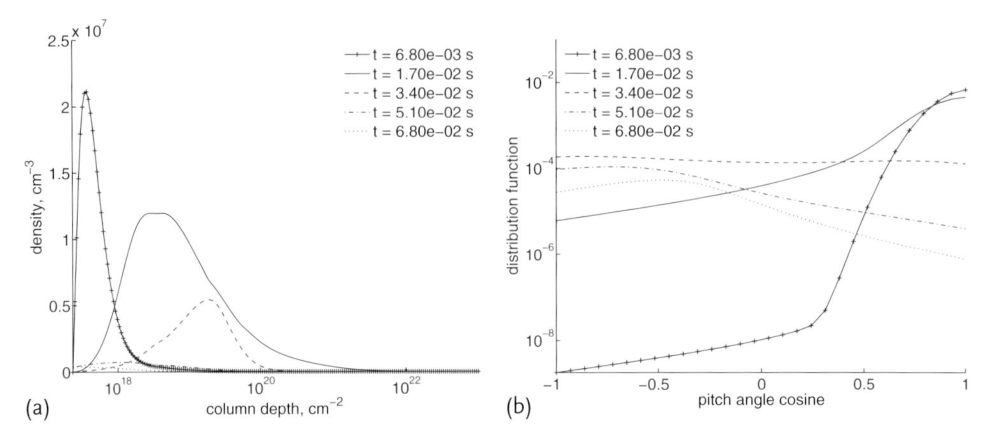

Figure 4.26 Electron density at impulsive injection as a function of column depth (a) and pitch angle (b). Collisions and electric field are taken into account. The beam parameters, see Eq. (4.30), are $\gamma = 3$, $F_{\text{top}} = 10^{10}$ erg cm^{-2} s^{-1}, $\Delta\mu = 0.2$, and the injection time is $\delta t = 1.7 \times 10^{-3}$ s.

case of a short impulsive injection, the electric field does not affect the distributions very much.

As a result, the only mechanism that can essentially increase the number of returning electrons is a magnetic convergence. It should also be noted that in the current study we assume that a beam's current is always compensated by the return current, and thus the self-induced electric field develops immediately. However, as was shown by van den Oord (1990), a beam's current neutralization time is on the order of the collisional time. Thus the effect of an electric field for short impulses can be even smaller than our estimations.

Anisotropic scattering of beam electrons in collisions with ambient plasma makes the pitch-angle distribution more flat with time (Figure 4.26b). Electrons propagating downward reach the lower atmospheric depths where the ambient plasma has a high density, lose the bulk of their energy in collisions, and leave the distribution (become thermalized) when their energy is less than $z_{\min}$.

In contrast, returning electrons move into the less dense plasma of the upper atmosphere almost without losing any energy, but only gaining it during their acceleration in the self-induced electric field. Thus, after some time, the number of upward moving electrons can exceed the number of downward moving ones, which is clearly seen in Figure 4.26b. The angle distributions show that after $\sim 3.4 \times 10^{-2}$ s most of the downward propagating electrons are gone and the majority of electrons have $\mu < 0$, that is, they move back to the source in the corona.

4.7.1
Mean Electron Flux for Beam Impulse

Since the first term on the right-hand side of Eq. (4.16), which is responsible for energy losses due to collisions, is proportional to $z^{-1/2}$ (where z is the dimension-

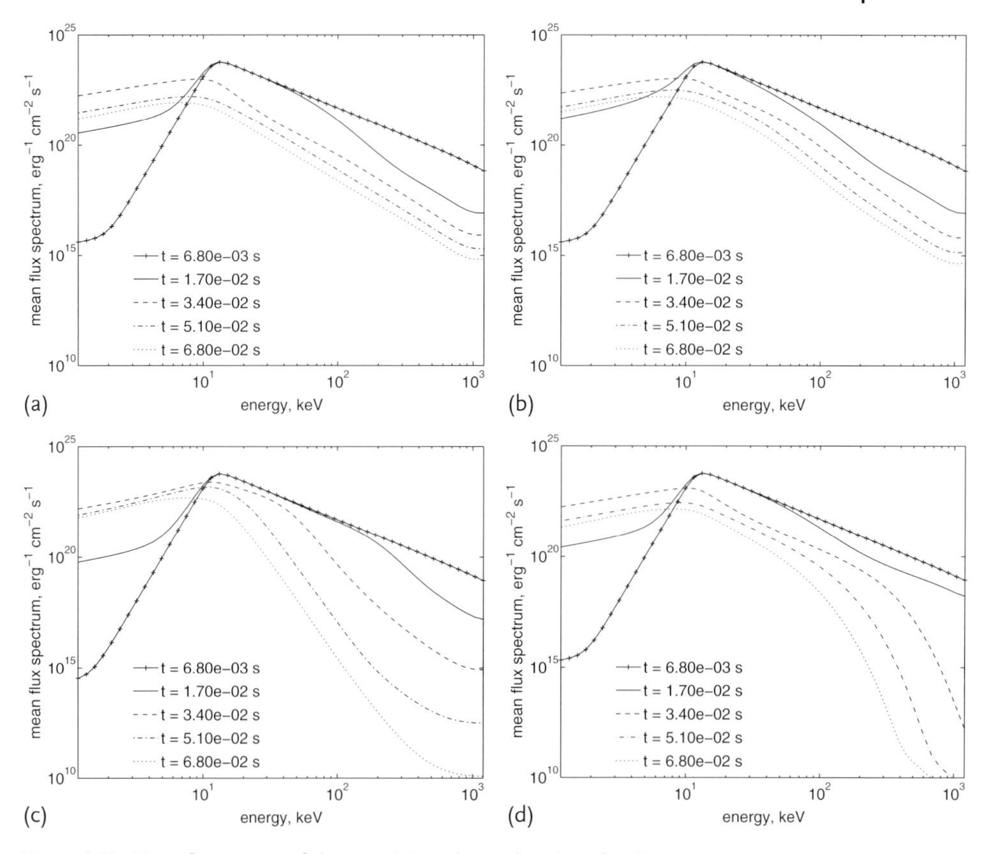

Figure 4.27 Mean flux spectra of electrons injected as a short impulse. Beam parameters are as in Figure 4.26: (a) collisions; (b) collisions and electric field; (c) collisions and convergence according to Eq. (4.41); (d) collisions and convergence according to Eq. (4.43).

less energy), one can expect that the electron spectra should become harder with the precipitation time, or depth. However, downward moving electrons with higher energy reach the dense plasma faster and lose their energy faster than lower-energy electrons, which makes the energy spectra softer with time (Figure 4.27a). The same is true of the spectra of upward moving electrons. In this case, high-energy electrons escape distribution faster by reaching the top boundary ($s = s_{\min}$). During beam precipitation this effect can lead to an increase in the power law index from an initial value of 3 up to 4 (Figure 4.27a).

A magnetic field on its own cannot change the energy of electrons. However, a converging magnetic field acts as a magnetic mirror and returns the essential part of electrons back to the source. As was shown above in Figure 4.23 for high-energy electrons, magnetic convergence is more effective than an electric field and pitch-angle diffusion. Thus high-energy electrons can quickly escape through the $s = s_{\min}$ boundary and the power-law index can reach higher values than in the case with a constant magnetic field. For example, for the magnetic field profile given by Eq. (4.42) the power-law index increases from the initial value 3 up to 8 (Figure 4.27b,c). If the convergence parameter is defined by Eq. (4.43), then the initial power-law distribution converts to some kind of quasithermal distribution with an essential drop in high energies (Figure 4.27d).

4.7.2
Energy Deposition by a Beam Impulse

Figure 4.28 shows the evolution of energy deposition, or heating functions, when different precipitation effects are taken into account. In the purely collisional case (Figure 4.28a) the heating maximum appears at the bottom boundary, moves upward with time, and vanishes near a column depth of $\sim 10^{19}–10^{20}\,\mathrm{cm}^{-2}$. This evolution is consistent with stopping depths obtained for electrons with different energies (Figure 4.22). Indeed, high-energy electrons are the first to reach deep lay-

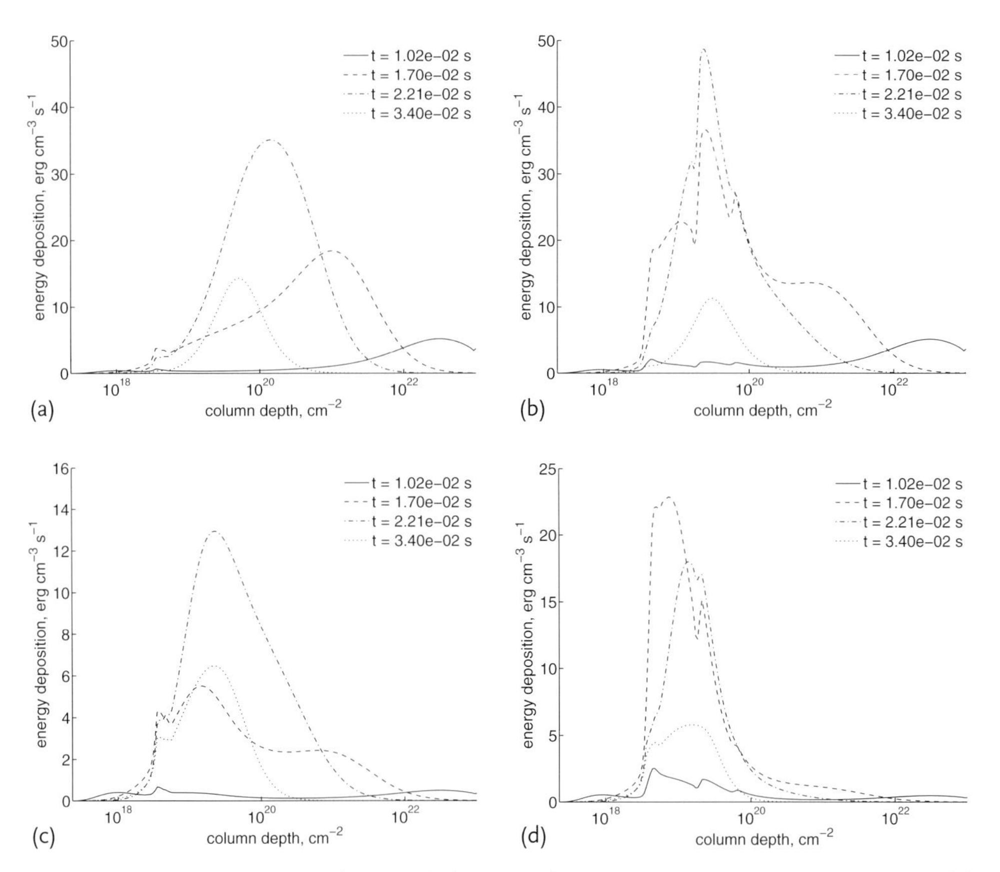

Figure 4.28 Energy deposition by beam impulse. Beam parameters are as in Figure 4.26 and the magnetic convergence is given by Eq. (4.41): (a) collisions; (b) collisions and electric field; (c) collisions and convergence; (d) collisions, convergence, and electric field.

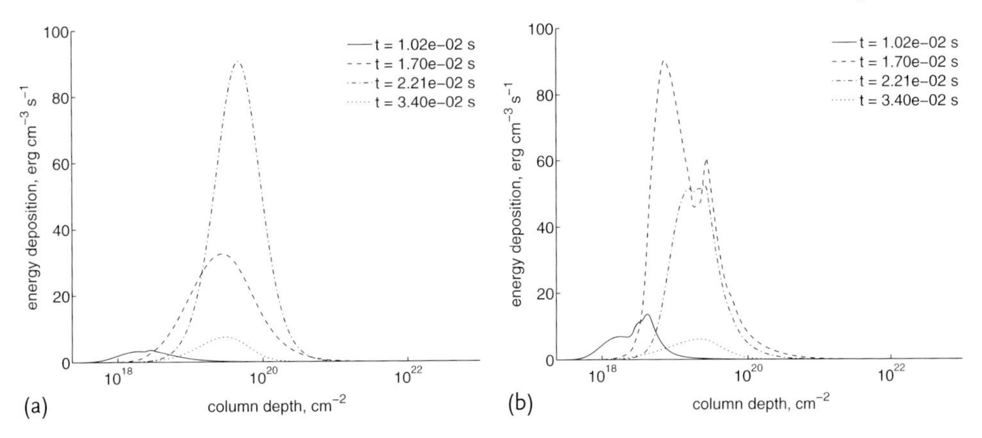

Figure 4.29 Energy deposition by impulse of a beam with $\gamma = 7$. The other beam parameters are as in Figure 4.26: (a) collisions; (b) collisions and electric field.

ers where the density is high enough to thermalize them. Less energetic electrons travel longer because of their lower velocity but lose their energy higher in the atmosphere. Thus the heating function maximum moves with the precipitation time from the photosphere (10 ms) to the lower (17 ms) and then upper (22 ms) chromosphere toward the stopping depth of low-energy electrons, where it vanishes.

In the presence of a self-induced electric field (Figure 4.28b) the heating by collisions becomes smaller than in the purely collisional case because some electrons are reflected by the electric field and do not reach the dense plasma and, thus, do not deposit their energy into the ambient plasma. On the other hand, there is a second maximum on the heating function, which is also caused by the energy losses due to ohmic heating by the self-induced electric field. This maximum does not move but grows in time with more electrons coming to the region with a high electric field (Figure 4.13b).

As was shown in Section 4.6.3.3 for the magnetic convergence given by Eq. (4.42) only about 10% of electrons can escape through the loss cone and heat the deep layers. Thus magnetic convergence substantially reduces energy deposition at lower atmospheric levels by mirroring the electrons back to the top, which shifts the heating maximum upward to the corona (Figure 4.28c,d).

The heating function of a beam with an initial spectral index of $\gamma = 7$ is shown in Figure 4.29. Contrary to the $\gamma = 3$ beam, the heating peak appears at smaller depth and moves downward with time. Apparently this occurs because the number of high-energy electrons is extremely low for a softer beam ($\gamma = 7$), and thus the heating that they can produce at larger depths is too low to be noticeable. Also, the heating profile is narrower but higher and its maximum located higher in the atmosphere in comparison with the $\gamma = 3$ case. When the electric field is taken into account (Figure 4.29b), the heating becomes stronger at smaller depths in the corona, in comparison to the pure collisional case (Figure 4.29a), where it has its maximum in the upper chromosphere.

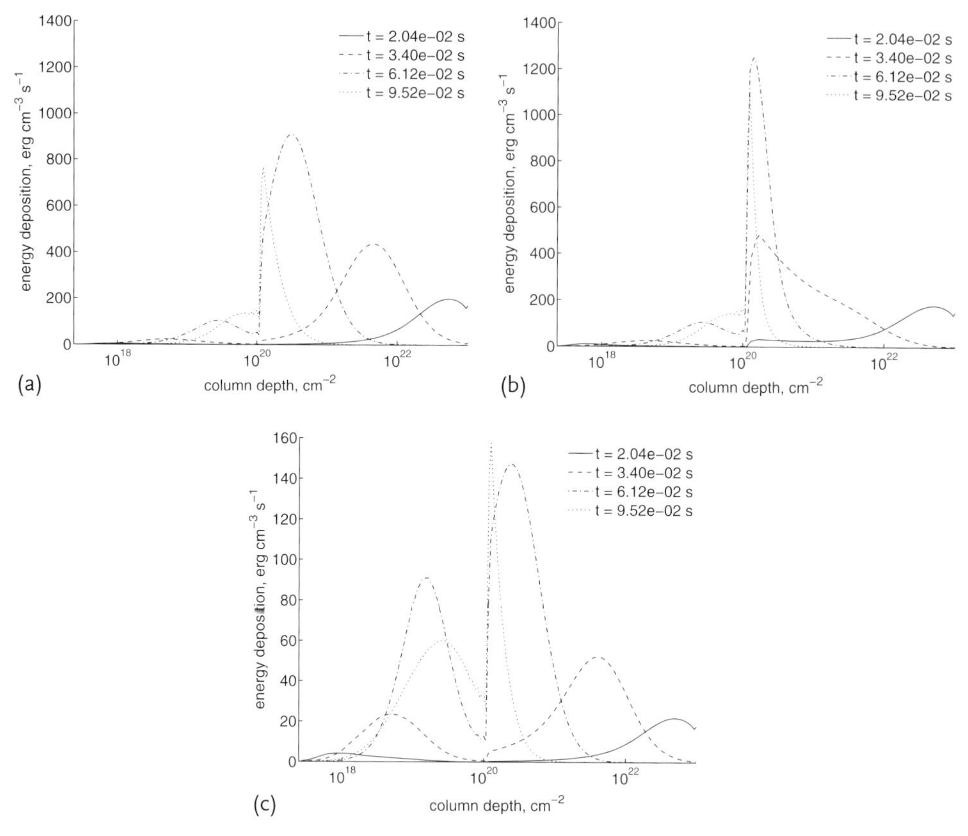

Figure 4.30 Energy deposition of beam impulse with an energy flux of 10^{12} erg cm^{-2} s^{-1}. Other beam parameters are as in Figure 4.26 and the magnetic convergence is given by Eq. (4.41): (a) collisions; (b) collisions and electric field; (c) collisions and convergence.

A more powerful beam, with an energy flux of 10^{12} erg cm^{-2} s^{-1}, obviously deposits more of its energy in the ambient atmosphere (Figure 4.30) than a beam with an energy flux of 10^{10} erg cm^{-2} s^{-1}. Two maxima are clearly seen on the heating function profile – one in the chromosphere, another in the corona. If magnetic convergence is absent, the chromospheric heating is much stronger (Figure 4.28a,b). On the other hand, if convergence is taken into account, then only about 10% of electrons can reach the chromosphere (Figure 4.28c).

Thus, the heating under the transition region is reduced by an order of magnitude, while the coronal heating remains nearly the same as in the case of the constant magnetic field (Figure 4.28c). Note that in this case a different hydrodynamical model is used to estimate the density and temperature of the ambient plasma. The discontinuity at a depth of 10^{20} cm^{-2} (Figure 4.30) is caused by a sharp increase of the ambient plasma density (Figure 4.1), which apparently corresponds to the transition region.

It can be noted that the evolution time of an electron impulse is longer for a stronger beam. This is the result of the smaller density of ambient plasma heated by a stronger beam, which leads to a longer relaxation time, as was shown in Section 4.5.2.

4.8 Conclusions

Simultaneous inclusion of different mechanisms of energy loss utilized in the Fokker–Planck approach to electron precipitation into a flaring atmosphere, for example Coulomb collisions, ohmic losses in an electrostatic electric field induced by the beam itself, pitch-angle diffusion caused by collisions, electric field, and converging magnetic field significantly changes the distributions of beam electrons at their precipitation to every depth.

The domination of collisional or ohmic energy losses is defined by the minimal stopping depth, either collisional or electric ones, as well as by the level of magnetic field convergence. It turns out that the electric field of the beam causes many lower-energy electrons to turn around and return to the source at the top of the atmosphere. These returning electrons reduce the electric field of the beam, and thus the magnitude of the direct current (and electric field) of precipitating electrons, which needs to be neutralized by a return current from the ambient plasma.

The mean electron fluxes integrated from the whole flaring atmosphere are shown to have a distribution with a maximum occurring around the energy corresponding to the midpoint between the turning point of the electric field and the collisional stopping depth. The larger this difference is, the wider is the maximum that is valid for softer intense beams.

By solving numerically the time-dependent Fokker–Planck equation one is able to study the temporal evolution of electron-beam precipitation in the solar atmosphere and to evaluate the relaxation time required for the beam to reach a stationary regime. For beams with an energy flux of 10^{10} erg cm^{-2} s^{-1} this relaxation time is $\sim$ 0.07 s and becomes longer by a factor of about 3 (or $\sim$ 0.2 s) for beams with an energy flux of 10^{12} erg cm^{-2} s^{-1}.

The study of impulsive injections of electrons with an initial power-law energy spectrum and an impulse duration of about 1.7×10^{-3} s, which is much shorter than the relaxation time, has shown that the shape of electron distributions is mainly defined by collisional energy losses, while the effect of a self-induced electric field is considerably smaller than for a steady injection.

It has been shown that during the evolution of an impulse the power-law index increases with time. For example, if the joint effects of collisions and magnetic convergence are taken into account, an initial power-law index of 3 can increase up to 8 and resembles some quasithermal distributions. If the electric field is taken into account, then the maximum of the energy deposition profile is shifted upward, making the coronal heating stronger and the chromospheric heating weaker than in the case of pure collisional precipitation. Different models of a converging magnetic field are shown to affect the effectiveness of a beam's electron refraction by a magnetic mirror. All these effects play an important role in electron precipitation, not only in time but in depth, that must be reflected in their HXR emission and polarization, discussed in Chapter 6.

The energy deposition profile has been shown to depend on the initial power-law index. If the energy spectrum is hard ($\gamma = 3$), then heating starts at the bottom end of the system due to high-energy electrons. In the case of a soft ($\gamma = 7$) impulse, the number of high-energy particles is too low to produce any noticeable heating of the deep layers. On the other hand, the higher layers are heated more effectively due to the higher number of low-energy electrons in the softer beam.

Since a converging magnetic field returns many electrons back to their source, the heating due to collisions and electric field can be reduced in a converging magnetic field by 70% in comparison with a constant one. However, a model based on indirect measurements of the magnetic field in the solar atmosphere shows a reduction of the collisional heating by only 20% in comparison with the constant magnetic field profile.

5
Proton Beam Kinetics

5.1
Proton Beam Distribution Function

We investigate the kinetics of a proton beam with its time-dependent distribution function in the phase space of depth x, velocity v, and pitch angle θ. Similarly to electrons, the proton distribution function is normalized on the beam particle density:

$$N_\text{b} = K_\text{p} \int\limits_v \int\limits_\theta f(t; z, v, \theta) d\theta \, dv \, .$$

The proton differential density (density of particles at a given depth with a given velocity per velocity unit) is defined in the same way as for electrons.

5.1.1
Effect of Coulomb Collisions on Proton Precipitation

A fast proton with energy E moving through thermal ambient plasma with a density n loses energy according to the following well-known expression (see, e.g., Emslie, 1978):

$$\frac{dE}{n dl} = -\frac{\pi e^4 Z}{E} \left(\frac{m_\text{p}}{m_\text{e}} \right) [X\Lambda + (1 - X)\Lambda'] \, , \tag{5.1}$$

where parameters Λ, Λ', and Λ'' are the Coulomb logarithms described in Chapter 3.

At the same time, collisional scattering causes the average deflection of a proton velocity vector as follows:

$$\frac{d\mu}{n dl} = -\frac{\pi e^4 Z}{2E^2} \mu \left[\left(\frac{m_\text{p}}{m_\text{e}} - 1 \right) X\Lambda + (1 - X) \left(\frac{1}{2}\Lambda'' - \Lambda' \right) \right] . \tag{5.2}$$

Equation (3.41) can be rewritten as follows:

$$\frac{dE}{d\xi} = -\frac{a_\text{p}}{E} \, , \tag{5.3}$$

Electron and Proton Kinetics and Dynamics in Flaring Atmospheres, First Edition. Valentina Zharkova.

where ξ is the column density

$$\xi = \int_0^z n(z)dz$$

and a_p is a collisional parameter for protons defined as follows:

$$a_\text{p} = \pi e^4 Z \left(\frac{m_\text{p}}{m_\text{e}}\right) \left[X\Lambda + (1-X)\Lambda'\right] .$$

Let us analyze the behavior of a precipitating proton beam governed solely by Coulomb collisions, as was done for electrons. The expression for proton collisional energy losses is similar to that for electrons, the only difference being that the collisional parameter a_p is by a factor of m_p/m_e higher than that for electrons a_e (Section 3.2.2.2). Therefore, the analysis given in Chapter 3 for electrons is applicable here as well.

As with electrons, proton energy decreases with depth as

$$E(\xi) = \sqrt{E_0^2 - 2a_\text{p}(\xi - \xi_0)} , \tag{5.4}$$

its velocity decreases as

$$v(\xi) = \left[v_0^4 - \frac{8a_\text{p}(\xi - \xi_0)}{m_\text{p}^2}\right]^{1/4} ,$$

and the proton thermalization depth can be found in the same way as for electrons:

$$\xi_\text{max} = \frac{E_0^2}{2a_\text{p}} = \frac{m_\text{p}^2 v_0^4}{8a_\text{p}} . \tag{5.5}$$

Let us analyze Eq. (5.5) for the case where protons are injected with the same energy as electrons and for the case where protons are injected at the same velocity as electrons.

If the initial energies of injected protons and electrons are equal ($E_\text{p0} = E_\text{e0}$), then their thermalization depths (ξ_maxP for protons and ξ_maxE for electrons) are related as follows:

$$\xi_\text{maxP} = \frac{E_\text{p0}^2}{2a_\text{p}} = \frac{m_\text{e}}{m_\text{p}}\frac{E_\text{p0}^2}{2a_\text{e}} = \frac{m_\text{e}}{m_\text{p}}\frac{E_\text{e0}^2}{2a_\text{e}} = \frac{m_\text{e}}{m_\text{p}}\xi_\text{maxE} . \tag{5.6}$$

At the same time, if protons and electrons are injected at the same velocities, then their thermalization depths are related as follows:

$$\xi_\text{maxP} = \frac{E_\text{p0}^2}{2a_\text{p}} = \frac{m_\text{e}}{m_\text{p}}\frac{E_\text{p0}^2}{2a_\text{e}} = \frac{m_\text{e}}{m_\text{p}}\frac{m_\text{e}^2}{m_\text{p}^2}\frac{E_\text{e0}^2}{2a_\text{e}} = \frac{m_\text{p}}{m_\text{e}}\xi_\text{maxE} . \tag{5.7}$$

Therefore, from the analysis above one can make the following two conclusions. First, protons injected with *the same energy* as electrons are thermalized at a column density of ξ_{maxP}, which is by a factor of $m_{\mathrm{p}}/m_{\mathrm{e}}$ *smaller* than the thermalization depth for electrons. Second, protons injected at *the same velocity* as electrons are thermalized at a column density of ξ_{maxP}, which is by a factor of $m_{\mathrm{p}}/m_{\mathrm{e}}$ *greater* than the thermalization depth for electrons.

The velocity distribution of protons at different depths in a pure collisional approach can be obtained using the pure collisional solution for electrons as $dN/dv = dN/dE \; dE/dv$ and substituting a_{p} for a_{e}. The resulting differential density of a proton beam can be written as follows:

$$N(v,\xi) = K\frac{m_{\mathrm{p}}^{1.5}}{\sqrt{2}}v^2\left(\frac{m_{\mathrm{p}}^2 v^4}{4} + 2a_{\mathrm{p}}\xi\right)^{-\frac{\gamma+1}{2}} \times \Theta\left\{v - \left[v_{\mathrm{low}}^4 - \frac{8a_{\mathrm{p}}(\xi-\xi_0)}{m_{\mathrm{p}}^2}\right]^{1/4}\right\} \times \Theta\left\{\left[v_{\mathrm{upp}}^4 - \frac{8a_{\mathrm{p}}(\xi-\xi_0)}{m_{\mathrm{p}}^2}\right]^{1/4} - v\right\} . \tag{5.8}$$

Differential densities of power-law proton beams with spectral indices γ of 3, 5, 7 are plotted in Figure 5.1.

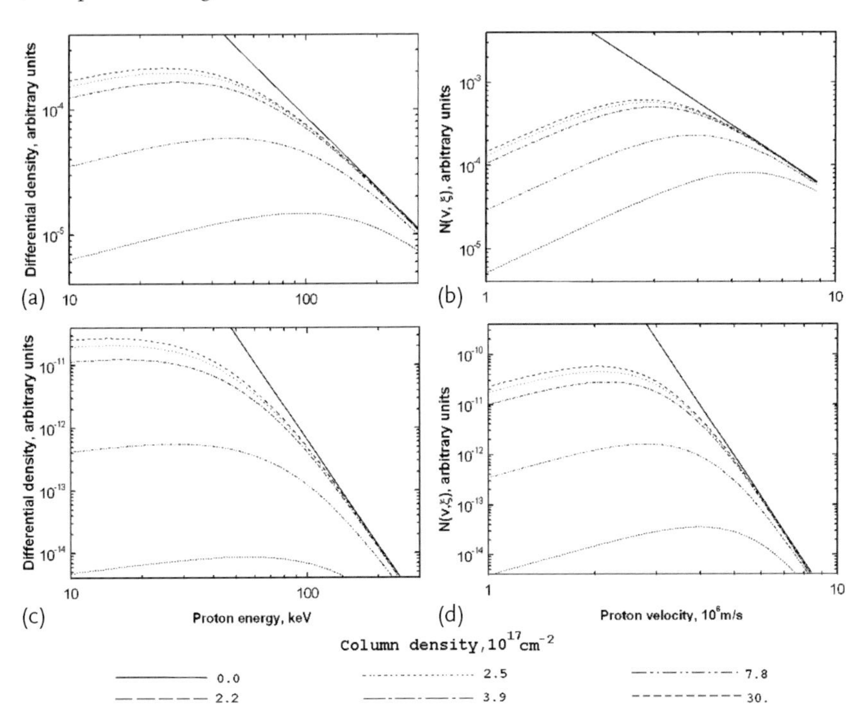

Figure 5.1 Differential densities of proton beams vs. energy (a,c) and velocity (b,d) found from the solution of the continuity equation. The initial beam spectral indices are $\gamma = 2$ (a,b) and $\gamma = 6$ (c,d). The lower-velocity cutoff in the initial spectra is $10^8\ \mathrm{cm\,s^{-1}}$ (corresponding to an energy of ~ 50 keV).

5.1.2
Effect of a Self-Induced Electric Field on Proton Precipitation

Similarly to electrons, proton precipitation into atmospheric plasma will lead to the appearance of an electric field if the proton beam is not compensated by an electron beam with the same particle flux. However, this self-induced electric field should not have a very significant effect on proton motion, for two reasons.

First, average proton energies are higher than those of electrons. This means that for the same energy flux the proton particle flux and, consequently, the corresponding electric current and electric field will be lower than those for electrons.

Second, protons are much more strongly affected by Coulomb collisions in comparison with electrons of the same energy. Taking into account that an electric field decelerates protons and electrons to the same extent, one can conclude that for protons the effect of an electric field is weaker than for electrons in comparison with the effect of collisions.

That is why we neglect the effect of electric fields on proton precipitation in this thesis, in order to focus on the effect of collisions and wave–particle interactions. However, in the case of partially neutralized beam precipitation, various electrostatic effects (e.g., ambipolar diffusion) can play a significant role and, thus, have to be taken into account. On the other hand, although an induced electric field does not significantly affect proton kinetics directly, it can be significant for other processes that, in turn, affect the proton kinetics. For instance, a self-induced electric field can support proton-beam-driven MHD wave instability, reducing the instability threshold and, consequently, increasing the effect of MHD waves on proton kinetics.

5.1.3
Effect of Magnetic Field Convergence on Proton Precipitation

The adiabatic deflection of charged particles in a magnetic field does not depend on their charge sign and mass. Therefore, in a converging magnetic field protons behave in the same way as electrons, and the analysis of electron beams presented above in Chapter 4 is applicable to proton beam precipitation.

5.1.4
Effect of Wave–Proton Interaction

It was shown that protons precipitating in the solar atmosphere can lose a noticeable part of their energy interacting with plasma waves, in particular with kinetic Alfvén waves (KAWs) (Tamres *et al.*, 1989; Voitenko, 1996). Although there are many other types of plasma waves that can be excited (e.g., magnetosonic waves, ion–cyclotron waves, etc.), KAWs have been shown to be most significant for the kinetics of proton beams because of their comparably low instability threshold (Voitenko, 1998).

Voitenko (1998) have shown how 0.1–100 MeV protons precipitating in a reconnected flaring loop lose their energy owing to their resonant interaction with KAWs, which in turn heat the ambient plasma by scattering on the background electrons owing to Cherenkov resonance. Let us consider KAW–proton interactions in an approach similar to that of Voitenko (1996, 1998).

Generally speaking, the resonance of a particle with an electromagnetic wave occurs if the wave stands in the particle's inertial frame, that is, when the particle's velocity is equal to the wave's phase velocity (see, e.g., Benz, 2002; Somov, 2000):

$$v_z = \frac{\omega}{k_z} ,$$

where ω and k_z are the wave frequency and wave number, respectively. If so, then the particle velocity on average remains constant. However, if the particle velocities are slightly lower than ω/k_z, then they are accelerated by the wave's electric field while particles with velocities slightly higher than ω/k_z are decelerated.

Therefore, if the number of particles with velocities slightly higher than ω/k_z is greater than the number of particles with velocities slightly lower than ω/k_z, then the number of decelerated particles is greater than the number of accelerated ones and vice versa. In other words, if the velocity spectrum of particles has a positive inclination close to the wave phase velocity (i.e., $\partial f/\partial v_z \;|_{v_z=\omega/k_z} > 0$), then the particles are being, on average, decelerated, losing their energy to the wave, while if $\partial f/\partial v_z \;|_{v_z=\omega/k_z} < 0$, then the particles are, on average, being accelerated, taking energy from the wave.

This simple physical picture explains how the energy of a proton beam can be converted into energy plasma waves (and vice versa). Let us consider the interaction of a proton beam with KAW turbulence.

The effect of KAWs on a proton population can be described by the so-called quasilinear diffusion coefficient: D^{QL}. If $f(x, v_z, \mu)$ is the proton distribution function, then its variation owing to the KAW–proton interaction can be expressed as follows:

$$\left(\frac{\partial f}{\partial t}\right)_{\mathrm{KAW}} = \frac{\partial}{\partial v_z} D^{\mathrm{QL}} \frac{\partial f}{\partial v_z} , \tag{5.9}$$

while the quasilinear diffusion coefficient can be written as follows:

$$D_{\mathrm{QL}}(v_z) = \frac{\pi e^2}{m^2} \int\limits_{k_z} \delta(\omega_k - k_z v_z) \mathcal{I}_0(\varrho_p) e^{-\varrho_p^2} w_{zk} dk_z . \tag{5.10}$$

Here w_{zk} is the energy spectrum of KAWs (the energy density per wave number unit), $\varrho_{\mathrm{p}} = k_\perp m_{\mathrm{p}}/(eB)v_\perp$, where $v_\perp$ is the perpendicular proton velocity $v_\perp = \sqrt{1-\mu^2}/\mu v_z$. I_0 is the zero-order modified Bessel function.

Equation (5.9) can be rewritten as follows:

$$\left(\frac{\partial f}{\partial t}\right)_{\mathrm{KAW}} = \frac{\partial D^{\mathrm{QL}}}{\partial v_z} \left(\frac{\partial f}{\partial v_z}\right) + D^{\mathrm{QL}} \left(\frac{\partial^2 f}{\partial v_z^2}\right) . \tag{5.11}$$

The first term on the right-hand side of Eq. (5.11) corresponds to the particle energy variation, while the second term describes diffusion in the proton velocity spectrum. It can be seen from Eq. (5.9) that the quasilinear diffusion coefficient is always positive. Therefore, the sign of the first derivative in the proton velocity spectrum $\partial f/\partial v_z$ determines the sign of the first term on the right-hand side of Eq. (5.11). Therefore, if the sign of $\partial f/\partial v_z$ is positive, then protons are being decelerated and vice versa. At the same time the diffusion determined by the second term on the right-hand side of Eq. (5.10) spreads the particles over the velocity spectrum.

The analysis above allows us to make one important qualitative conclusion: a proton beam injected with a positive inclination into its velocity spectrum is unstable with respect to the excitation of KAWs. If such a beam is injected, then its spectrum will be affected by KAWs, leading to the reduction of the positive slope in its spectrum.

Now let us analyze the two simple cases that are similar to those in the paper by Voitenko (1998).

First, let us consider the injection of a monoenergetic beam with an initial particle velocity of v_0, which is higher than the local Alfvén velocity V_A. It is obvious that its spectrum with depth will become plateaulike (or Π-like), with the upper velocity cutoff corresponding to the initial velocity v_0 and the lower cutoff corresponding to the local Alfvén velocity during the relaxation length l_{rx}, which is estimated to be about 10^5–10^6 cm.

Second, this Π-like spectrum can evolve further if the Alfvén velocity decreases with depth. That is, at every depth the proton velocity spectrum's lower cutoff relaxes to the local Alfvén velocity. This can be expressed as follows:

$$f(z, v_z) = \frac{2 N_{pb0} v_0}{v_0^2 - V_A^2} \theta(v_z - V_A)\theta(v_0 - v_z) ,$$

where N_{pb0} is the initial proton density.

5.1.5
Collisions versus Kinetic Alfvén Waves: the Effect on Proton Precipitation

From the previous section one can see that the KAW–proton interaction has a characteristic length of $l_{rx} \preceq 10^7$ cm. At the same time, it can be seen from Section 5.1.1 that protons with an energy of $\sim$ 0.1 MeV are thermalized at a column density of $\sim 10^{18}$ cm^{-2}, which corresponds to a length of $\sim 10^8$ cm for a plasma density of 10^{10} cm^{-3}. Hence the quasilinear relaxation of a proton beam occurs much faster than collisional energy losses.

Now let us compare the energy efficiencies of these two effects simply by comparing the energy fluxes carried by the two initially monoenergetic beams at a depth of $l_{rx} = 10^7$ cm, which corresponds to a column density of 4×10^{17} cm^{-2} and is solely governed by either collisions or KAWs. The results are shown in Table 5.1, and the variations of energy fluxes are shown in Figure 5.2. It can be seen that protons with an initial energy of 0.1 MeV (the corresponding velocity

Table 5.1 Ratio of energy flux carried by initially monoenergetic beams at a depth of l_{rx} to initial energy flux. The first column shows those for the case of a beam affected solely by Coulomb collisions; the second column shows those for the case of protons affected solely by KAWs.

	F/F_0	
E, MeV	Collisions	KAWs
0.1	0.77	0.95
0.2	0.95	0.73
1	0.99	0.55
2	~ 1	0.52

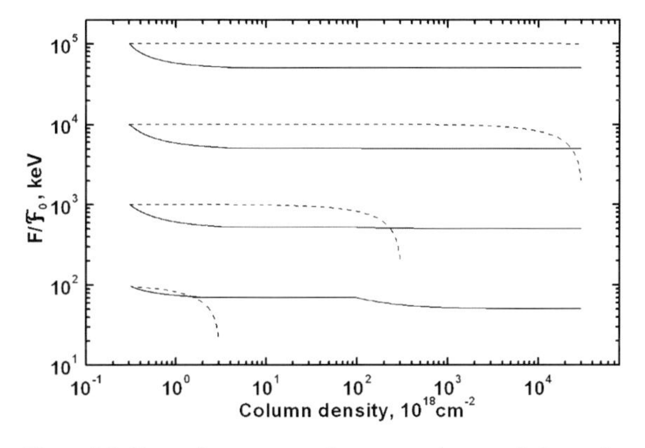

Figure 5.2 Proton beam energy fluxes vs. column densities calculated for initially monoenergetic proton beams with initial energies of 0.1, 1, 10, and 100 MeV. The solid lines show variations of energy fluxes for beams affected solely by KAWs; the dashed lines show variations of energy fluxes owing to pure Coulomb collisions.

is $4.4 \times 10^8\,\mathrm{cm\,s^{-1}}$) lose their energy mostly owing to collisions and are thermalized at a depth of $\sim 3 \times 10^{18}\,\mathrm{cm^{-2}}$. Protons injected with an energy of 1 MeV (or a velocity of $v_0 = 1.4 \times 10^9\,\mathrm{cm\,s^{-1}}$) can be noticeably affected by KAWs but also lose most of their energy owing to collisions and are thermalized at a depth of $\sim 3 \times 10^{20}\,\mathrm{cm^{-2}}$. Finally, protons injected with initial energies of 10 and 100 MeV ($v_0 = 4.4 \times 10^9$ and $1.4 \times 10^{10}\,\mathrm{cm\,s^{-1}}$, respectively) are practically unaffected by collisions during their propagation through the corona, transition region (TR), and the chromosphere but lose about half their energy owing to the interaction with KAWs in the upper corona.

However, this does not mean immediately that for energy losses of protons with energies $\succeq$ 10 MeV KAWs are more effective than collisions: the ratio of the energy efficiencies of these two mechanisms depend on the proton velocity spectrum as well as on conditions in the ambient plasma. For instance, if a beam is injected with a soft power-law spectrum instead of a monoenergetic one, then KAWs are virtually not excited and the energy losses in collisions with ambient plasma particles play a key role even for high-energy protons.

On the other hand, one must take into account that collisional energy losses depend on energy. Hence, similarly to electrons, Coulomb collisions make the proton spectrum softer with depth and can lead to the appearance of a positive slope in their velocity spectra. Therefore, the simultaneous effect of both collisions and wave–particle interactions is more complex than just a superposition of the two effects. This will be considered in the next section using a numerical approach.

5.1.6 Fokker–Planck Equation for Proton Beams

5.1.6.1 General Form of Equation

Now let us formulate the kinetic problem for a proton beam in the form of the Fokker–Planck equation along with the initial and boundary conditions.

The Fokker–Planck equation has the following form:

$$\frac{\partial f}{\partial t} + v_z \frac{\partial f}{\partial z} + \left(\frac{\partial f}{\partial t}\right)_{\text{coll}} + \left(\frac{\partial f}{\partial t}\right)_{\text{KAW}} + \left(\frac{\partial f}{\partial t}\right)_{\text{magn}} = 0 \,. \tag{5.12}$$

5.1.6.2 Collisions, KAWs and Magnetic Field

For the sake of simplicity, in the present thesis we neglect collisional pitch-angle scattering of protons considering only the variations of a longitudinal component of the velocity v_z. Similarly to electrons, we assume that the thermal energies of ambient plasma particles are much smaller than the beam proton energies. Therefore, there is no energy diffusion in a proton beam. Collisional energy losses can be accounted for as follows:

$$\left(\frac{\partial f}{\partial t}\right)_{\text{coll}} = \left.\frac{d v_z}{d t}\right|_{\text{coll}} \frac{\partial f}{\partial v_z} = \left.\frac{d E}{d t}\right|_{\text{coll}} \frac{d v_z}{d E} \frac{\partial f}{\partial v_z} \,.$$

Substituting from Eq. (3.39) $dE/dt = -2a_\text{p} n/(mv)$ into the equation above gives the following:

$$\left(\frac{\partial f}{\partial t}\right)_{\text{coll}} = -\frac{2a_\text{p} n}{m^2 v_z^2} \,. \tag{5.13}$$

Variations of proton pitch angle in a converging magnetic field can be taken into account as follows:

$$\left(\frac{\partial f}{\partial t}\right)_{\text{magn}} = v_z \frac{\sin^2\theta}{2} \frac{\partial \ln B(x)}{\partial x} \frac{\partial f}{\partial \cos\theta} \,. \tag{5.14}$$

Finally, the effect of KAWs can be taken into account as per Eq. (5.12).

Therefore, the Fokker–Planck (or Landau–Vlasov) equation for proton beam Eq. (5.13) affected by collisions with ambient plasma particles, KAWs, and a converging magnetic field is written as follows:

$$\frac{\partial f}{\partial t} + v_z \frac{\partial f}{\partial z} - \frac{2a_{\mathrm{p}}n}{m^2 v_z^2}\frac{\partial f}{\partial v_z} + \frac{2a_{\mathrm{p}}n \cos^4\theta}{m^2 v_z^3 \sin\theta}\frac{\partial f}{\partial\theta} + \frac{2a_{\mathrm{p}}n}{m^2 v_z^3}\cos^3\theta \frac{\partial^2 f}{\partial\theta^2}$$
$$+ \frac{\partial D^{\mathrm{QL}}}{\partial v_z}\left(\frac{\partial f}{\partial v_z}\right) + D^{\mathrm{QL}}\left(\frac{\partial^2 f}{\partial v_z^2}\right) + v_z \frac{\sin^2\theta}{2}\frac{\partial \ln B(x)}{\partial x}\frac{\partial f}{\partial\cos\theta}$$
$$= 0 .$$

In order to make the problem self-consistent, the initial spectrum of protons at the injection point $f(z = 0, v_z, \theta)$ is required as well as the ambient plasma parameters ($n(z)$, $T(z)$, and $B_z(z)$). Finally, in order to deduce the quasilinear diffusion coefficient, one must determine the "distribution function" for KAWs $w_k(z)$.

The latter requires an equation for KAW dynamics, which has the following form:

$$\frac{dw_k}{dt} = \gamma_k w_k , \tag{5.15}$$

where γ_k is the growth rate for KAWs with wave number k. The growth rate is determined, in turn, by the following equation:

$$\gamma_k = \sqrt{\frac{\pi}{8}}\omega_k \frac{\varrho^2}{\mathcal{K}^2}\frac{T_{\mathrm{e}}}{T_{\mathrm{i}}} g_b \frac{N_{\mathrm{b}}}{n_0} , \tag{5.16}$$

where

$$g_b = \Lambda'' \frac{T_{\mathrm{e}}}{T_{\mathrm{b}}}\left(\frac{v_{b,\max} - v_k}{v_{\mathrm{i}}}\right) e^{-0.5\frac{v_{b,\max}-v_k}{v_{\mathrm{i}}}} ,$$

with T_{b} indicating beam temperature ($T_{\mathrm{b}} = E_{\mathrm{b}}/k_{\mathrm{B}}$, where E_{b} is the beam average energy), $v_{\mathrm{b,max}}$ the beam maximum velocity, and v_{i} the thermal ion velocity.

We will consider a simplified problem looking for a stationary solution and neglecting magnetic field convergence. For this purpose the kinetic equation can be written as follows:

$$v_z \frac{\partial f}{\partial z} - \frac{2a_{\mathrm{p}}n}{m^2 v_z^2}\frac{\partial f}{\partial v_z} + \frac{\partial D^{\mathrm{QL}}}{\partial v_z}\left(\frac{\partial f}{\partial v_z}\right) + D^{\mathrm{QL}}\left(\frac{\partial^2 f}{\partial v_z^2}\right) = 0 . \tag{5.17}$$

5.2 Precipitation of Proton Beam: Numerical Simulations

5.2.1 Numerical Calculation of Proton Beam Distribution Function

As was shown in Sections 5.1.1–5.1.4, analysis of Coulomb collisions is a relatively simple problem both for analytical and numerical approaches since we neglect any response of the ambient plasma to beam precipitation. At the same time, analysis of proton–KAW interaction is more complex because it is necessary to consider simultaneously both KAW and proton dynamics.

Let us consider a steady state regime of proton beam precipitation. It was shown in Section 5.1.4 that proton–KAW interaction causes redistribution in the proton velocity spectra, leading to a flattening of parts with a positive slope. At the same time it was shown in Section 5.1.5 that the characteristic time scale of KAW–proton interactions is much shorter than the characteristic collisional time. This allows us to consider the simultaneous effect of collisions and KAWs by direct calculation of collisional energy losses while using a semianalytical scheme to accounting for KAW effects.

Now let us consider a discrete form of the proton distribution function defined as $f_{ij} = f(z_i, v_{zj})$ with f_{0j} known. In order to calculate the magnitudes of the distribution function at all meshes, $i > 0$, we will use the discrete form of Eq. (5.17) as follows:

$$f_{i+1j} = f_{ij} + \delta_{1ij} + \delta_{2ij} , \tag{5.18}$$

where δ_{1ij} is the variation of the distribution function with depth owing to Coulomb collisions and δ_{2ij} is the variation of the distribution function owing to proton–KAW interaction.

The first δ_{1ij} is calculated in a similar way to what was done for the calculation of the electron beam distribution function (Chapter 3). The second δ_{2ij} is calculated using the scheme below. Calculation of each difference between i and $i+1$ meshes are made in two steps. In the first step we calculate only δ_{1ij}, which accounts for Coulomb collisions, and derive the velocity spectrum f'_{i+1j}, which is not affected by KAWs. If there is not a positive slope in the "intermediate" spectrum f'_{i+1j}, then $f_{i+1j} = f'_{i+1j}$; otherwise the distribution for the $i+1$ step can be derived keeping in mind that the effect of KAWs leads to the disappearance of the positive slope. Mathematically, it can be expressed as follows:

$$\begin{aligned} f_{i+1j} &= f'_{i+1j} , \quad \text{if} \quad v_j < V_A , \\ f_{i+1j} &= f'_{i+1j} , \quad \text{if} \quad j > j_{cr} , \\ f_{i+1j} &= f'_{i+1jcr} , \quad \text{if} \quad V_A < v_j < v_{jcr} . \end{aligned} \tag{5.19}$$

The "breaking" velocity v_{jcr} (or the corresponding mesh number j_{cr}) can be found from the following simple equality:

$$\sum_{j=j(V_A)}^{j_{cr}} f_{i+1j} j = \sum_{j=j(V_A)}^{j_{cr}} f'_{i+1j} j ,$$

where $j(V_A)$ is the number of the velocity mesh corresponding to the local Alfvén velocity.

5.2.2
Accepted Parameters

In Chapter 2 we showed that a fast magnetic reconnection in a current sheet can result in the generation of two types of high-energy protons: those accelerated by a super-Dreicer electric field and ejected from an RCS with a power-law energy spectrum and those accelerated on slow MHD shocks and ejected with quasithermal energy distributions with a maximum at a higher-than-thermal energy. Let us now consider the precipitation either of proton beams with power-law energy spectra or of quasithermal protons replaced for the sake of simplicity by monoenergetic ones at the maximum energy.

The initial velocity spectra for proton beams injected with power-law energy spectra are defined as follows:

$$f_0(V, \theta) = K_\mathrm{p} V^{-2\gamma+1} \Theta(V - V_\mathrm{low}) \Theta(V_\mathrm{upp} - V) , \quad (5.20)$$

where V_low and V_upp are the lower and upper energy limits in the initial velocity spectra, respectively, $\Delta\mu$ is the half width of the initial pitch-angle distribution, K_p is a scaling factor (defined in a similar way as for electrons). The lower and upper velocity limits are set to be 3×10^8 cm s^{-1} (corresponding to the energy $\sim$ 40 keV) and 5×10^9 cm s^{-1} (corresponding to $\sim$ 10 MeV), respectively. The initial spectral indices γ are set at 2, 4, or 6. The initial energy fluxes carried by beam particles are set at 10^{10}, 10^{12}, or 10^{14} erg cm^{-2} s^{-1}.

As far as precipitation of the monoenergetic beam is concerned, there are two parameters that can vary: the initial particle energy and the initial energy flux.

The initial proton energy is set at 100 keV (corresponding to a velocity of $\sim$ 4.4×10^8 cm s^{-1}), 1 MeV ($\sim 1.4 \times 10^9$ cm s^{-1}), or 10 MeV ($\sim 4.4 \times 10^9$ cm s^{-1}). At the same time, the initial beam energy flux is set at either 10^{10} or 10^{12} erg cm^{-2} s^{-1}.

5.2.3
Proton Beam Distribution Functions

The distribution functions (in the form of differential density per v_z unit) for beams with initially power-law spectra are shown in Figures 5.3–5.5 for different initial spectral indices. The distribution functions for an initially monoenergetic beam are shown in Figure 5.6. Let us compare these distribution functions with those calculated analytically in the pure collision approach.

It can be seen that in the corona the simulated distributions are very similar to those calculated analytically. The slight difference at lower velocities appears to be owing to the "numerical diffusion" and can be neglected. The simulated spectra gradually become harder; and the lower the velocity, the faster this spectral hardening happens.

However, starting from some depth the distributions calculated numerically taking KAWs into account become different from those calculated for pure collisions. There are no positive slopes at velocities higher than V_A in the proton spectra calculated numerically taking into account the effect of KAWs. Instead, there are flat

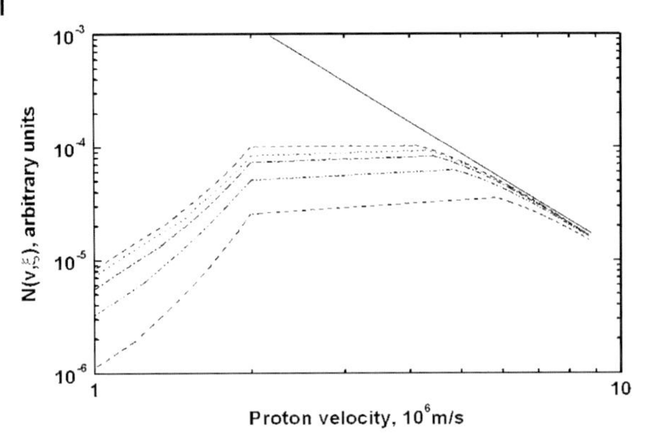

Figure 5.3 Proton distribution functions $N(\nu, \xi)$ calculated taking into account Coulomb collisions and interaction with kinetic Alfvén waves. The initial proton energy spectrum (solid line) is power law (i.e., $N(E, 0) \sim E^{-\gamma}$ or $N(\nu, 0) \sim \nu^{-2\gamma+1}$) with $\gamma = 2$. The dashed line is for a depth of 2.2×10^{18} cm^{-2}, the dotted line is for 2.5×10^{18} cm^{-2}, the dotted–dashed line is for 3.9×10^{18} cm^{-2}, the dotted–dotted–dashed line is for 7.8×10^{18} cm^{-2}, and the long-dash line is for 2.9×10^{19} cm^{-2}.

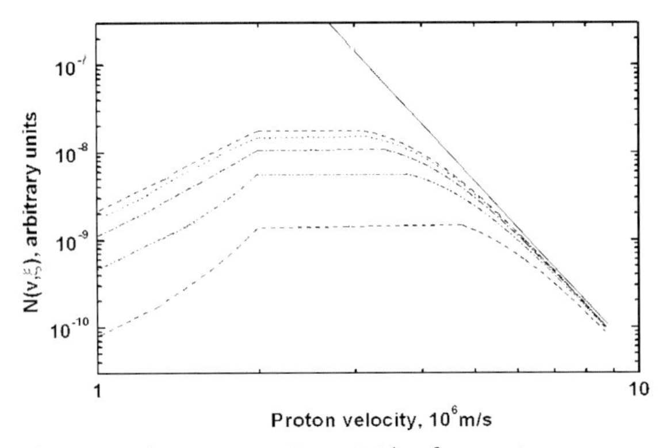

Figure 5.4 The same as in Figure 5.3 but for $\gamma = 4$.

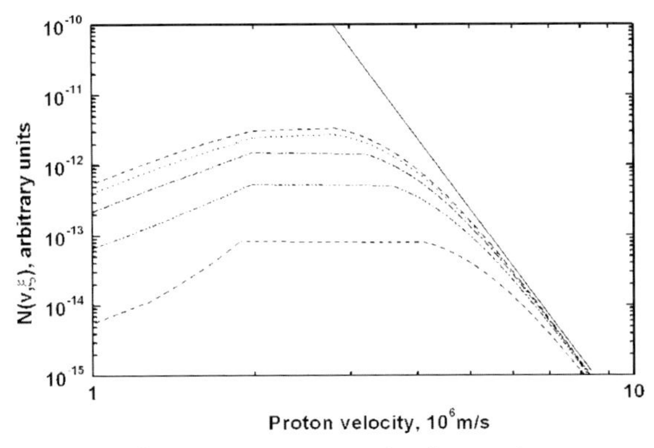

Figure 5.5 The same as in Figure 5.3 but for $\gamma = 6$.

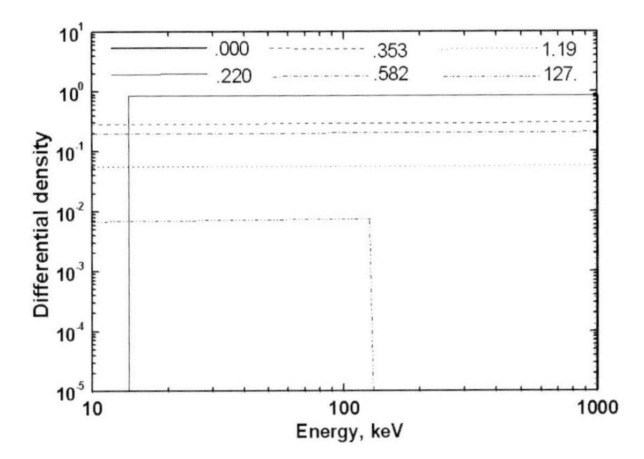

Figure 5.6 Distribution functions $N(v, \xi)$ calculated for an initially monoenergetic beam taking into account Coulomb collisions and interaction with kinetic Alfvén waves. The initial energy of protons is 1 MeV ($V_0 = 1.4 \times 10^9$ cm s^{-1}). The column densities are shown in units of 10^{18} cm^{-2}.

regions in the proton spectra between the local Alfvén speed $V_A \approx 3 \times 10^8$ cm s^{-1} and some upper velocity $\sim (5–7) \times 10^8$ cm s^{-1}, which increases with depth.

Apparently this flattening is the manifestation of the effect of KAWs described in Section 5.1.4. It was shown above that Coulomb collisions lead to the appearance of a positive slope at velocities lower than v_{extr} starting at a depth of ξ_{extr}. However, the effect of KAWs on proton beam precipitation leads to the appearance of a flat distribution instead of a positive slope.

These depths and velocities can be deduced from Eq. (5.8) by assuming $\partial N/\partial v = 0$, and this provides us with the following expressions:

$$\xi_{\text{extr}} = \frac{m_p^2 v_{\text{low}}^4}{16 a_p}, \tag{5.21}$$

$$v_{\text{extr}} = \left(8 \frac{a_p}{m_p^2} \xi\right). \tag{5.22}$$

To investigate the behavior of the "slow" populations of beam protons (accelerated as separatrix jets), we also calculated distribution functions for the initially monoenergetic beam (Figure 5.6). In fact, owing to the assumptions accepted in the considered model, the spectra of initially monoenergetic beams become Π-like immediately after injection. Therefore, they can be considered as beams with an initial spectral index equal to 0. It can be seen that these distributions remain flat at all depths where a beam exists with the upper velocity cutoff decreasing owing to Coulomb collisions as per Eqs. (5.6)–(5.7).

5.3 General Discussion of Proton and Electron Precipitation

5.3.1 Beam Spectra at Precipitation

Let us discuss the energy spectra of protons and electrons precipitating in a flaring atmosphere.

It was shown in Chapters 3 and 4 that, on average, during electron precipitation into a flaring atmosphere the beam energy spectra remain nearly power law at higher energies while at lower energies there is noticeable spectral flattening owing to Coulomb collisions. The effect of a self-induced electric field causes further energy losses and pitch-angle scattering of the precipitating electrons with lower energies that leads to further flattening of their energy spectra at lower energies.

Proton beam spectra at shallower depths are rather similar to those for electron beams. They are nearly power laws with the low-energy part being only slightly harder than the high-energy part, which is well explained by their much smaller energy losses in Coulomb collisions. Then, from a depth of about 3×10^{18}–$4 \times 10^{18}\,\text{cm}^{-2}$ the proton velocity spectra consist of two parts: a flat low-velocity part and a power-law high-velocity part.

5.3.2 Energy and Momentum Transfer

Now let us investigate the variations of beam particle density, energy, and momentum fluxes with depth. These functions can be derived from the distribution functions calculated from the continuity equation using Eqs. (3.44) and (3.47) that are valid for both protons and electrons.

Electron beam energy fluxes are shown in Figure 5.7 while those for proton beams are shown in Figure 5.8. It can be seen that electron beam energy flux variations with depth are at first determined by the initial energy flux and then by the initial spectral index. If the electron beam is weak ($F_0 = 10^8\,\text{erg}\,\text{cm}^{-2}\,\text{s}^{-1}$), then its energy deposition strongly depends on the initial spectral index (see Chapter 4 above). In order to show this, let us compare depths where the beams carry 1% of their initial energy flux. For the harder beam this depth is $\xi \approx 2 \times 10^{23}\,\text{cm}^{-2}$, which corresponds to the lower chromosphere and upper photosphere. At the same time, softer beams with initial indices of $\gamma = 5$ and 7 deposit 99% of their energy much higher, above column densities of $\sim 10^{21}$ and $\sim 3 \times 10^{20}\,\text{cm}^{-2}$, respectively, which corresponds to the lower and upper chromosphere.

Moderate beams with an initial flux of $F_0 = 10^{10}\,\text{erg}\,\text{cm}^{-2}\,\text{s}^{-1}$ deposit their energy faster owing to the return current effect. As a result, the depths where they deposit only 90% of their initial energy are reduced to $\xi \approx 5 \times 10^{21}\,\text{cm}^{-2}$ for harder beams $\gamma = 3$ and to $\sim 10^{20}\,\text{cm}^{-2}$ for softer beams ($\gamma = 5$ and 7). In the case of strong beams, this depth is further reduced to $\xi \approx 10^{20}\,\text{cm}^{-2}$ for harder beams

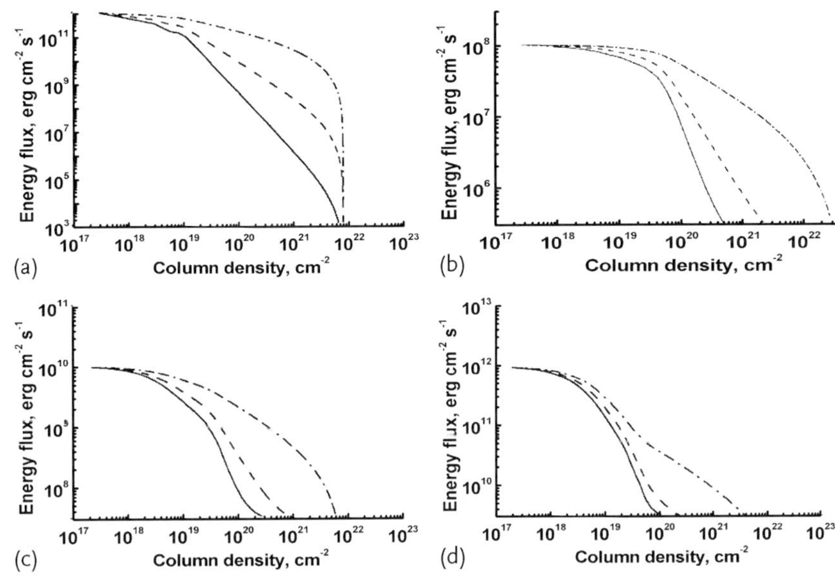

Figure 5.7 Electron beam energy flux variations with depth calculated without electric field for an initial energy flux of 10^{10} erg cm^{-2} s^{-1} (a) and with an induced electric field for initial energy fluxes of 10^8 erg cm^{-2} s^{-1} (b), 10^{10} erg cm^{-2} s^{-1} (c), and 10^{12} erg cm^{-2} s^{-1} (d) and for initial spectral indices of 7 (solid curves), 5 (dashed curves), and 3 (dotted–dashed curves).

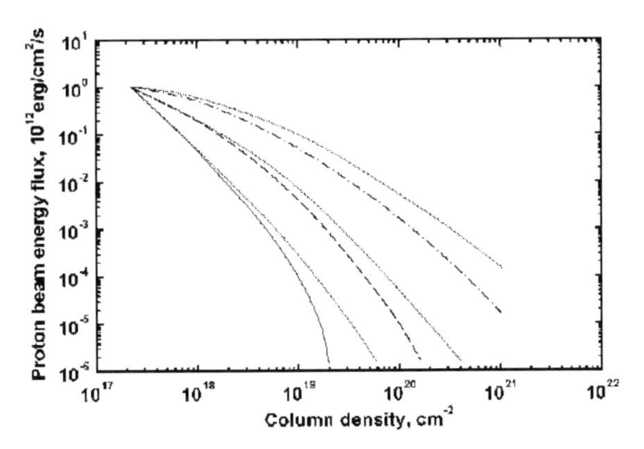

Figure 5.8 Proton beam energy flux variations caused both by Coulomb collisions and interaction with KAWs. The solid line is for $\gamma = 6$, dashed line for $\gamma = 4$, and dotted–dashed line for $\gamma = 2$. The dotted lines show energy flux variations for pure Coulomb energy losses. In all three cases the initial beam energy flux is 10^{12} erg cm^{-2} s^{-1}.

and to $\sim 7 \times 10^{19}$ cm^{-2} for softer beams. These column densities correspond to the transition region and upper chromosphere.

Therefore, electron beams with initial parameters close to those estimated from observations (Chapter 1) lose the bulk of their energy in the corona and the transition region (weaker and softer beams). Increases of the initial spectral index and inclusion of ohmic losses lead to faster energy deposition by softer beams and to the loss of a bulk of energy in the corona owing to strong ohmic dissipation.

It was shown in Section 5.1.1 that protons of the same initial energy as electrons are thermalized at a column density that is by a factor of m_p/m_e smaller than the electron thermalization depth. At the same time, protons injected with the same initial velocity as electrons penetrate to a column density that is by a factor of m_p/m_e greater than the electron penetration depth. Therefore, in order to compare protons with electrons as energy transfer agents, one must, above all, determine the initial energies/velocities of the compared particles.

The numerical simulations and analytical estimations of energy spectra of accelerated particles in a reconnecting current sheet made in Chapter 2 show that the protons can gain energy at about one or two orders of magnitude higher than electrons accelerated in the same current sheet. Most observations give a similar ratio of proton and electron energies in solar flares: about 10–100 keV ($\sim 10^{10}$ cm s^{-1}) for electrons and about 1–10 MeV ($\sim 3 \times 10^9$ cm s^{-1}) for protons (Chapter 1). Therefore, one can conclude that the ratio of the average proton penetration depth to those for electrons is about ~ 10 (Eq. (5.5)), that is, protons can penetrate to much greater depths than electrons. Furthermore, since the proton pitch-angle scattering owing to Coulomb collisions is much weaker than that for electrons, this ratio can be even higher.

In the present simulations the return current effect on proton beam precipitation has been neglected while energy losses in particle–wave interactions with kinetic Alfvén waves have been considered. As a result of the former, the depth variations of energy fluxes of proton beams are rather similar for proton beams with different initial energy fluxes. However, as a result of the latter, variations of proton beam energy fluxes with depth strongly depend on the initial spectral indices. In fact, harder proton beams ($\gamma = 2$) lose a bulk (99%) of their energy above column densities of $\sim 6 \times 10^{20}$ cm^{-2}. At the same time, softer beams with $\gamma = 4$ and 6 lose the same part of their energy faster, before a depth of $\sim 2 \times 10^{20}$ cm^{-2}. In other words, Coulomb energy losses are stronger for softer proton beams while proton energy losses in KAW excitation are stronger for harder beams.

It was pointed out in Section 1.4 of the introduction that delivery of momentum to the photosphere by fast particle beams is of particular interest for local helioseismology. To determine the momenta transferred by proton or electron beams, let us consider the following two issues.

First, the energy needed for electrons to reach the photosphere ($\xi \approx 3 \times 10^{22}$–$10^{23}$ cm^{-2}) is at least 400 keV (Eq. (3.42) and Table 3.1). Protons need at least an energy of 20 MeV to reach the same depth (Eq. (5.5)). Let us estimate the fractions of protons and electrons that reach the photosphere using the beam parameters calculated in Chapter 2 ($\gamma_P = 1.5$, $E_{\text{low}\,P} \approx 10$ keV, $\gamma_E = 2$, $E_{\text{low}\,E} \approx 1$ keV). Since the initial energy spectra are power law, one can write the following expression, which can be easily deduced from the definition of the distribution function under the assumption $E_{\text{upp}} = \infty$:

$$\frac{N(E > E_1)}{N(E > E_2)} = \left(\frac{E_1}{E_2}\right)^{-\gamma+1} . \tag{5.23}$$

Substituting the beam parameters into the equation above one can find that the fraction of electrons that reach the photosphere is about 1/400, while that of protons is about 1/40. Furthermore, by taking into account that the initial particle density in a proton beam is about eight times higher than that in electron beams, one can conclude that the number of protons bombarding the photosphere is about 100 times greater than the number of electrons reaching the photosphere. Therefore, even if the average energies $\langle E \rangle$ of protons and electrons at the photospheric level are the same, the momentum flux $M \sim N\langle E \rangle$ carried by a proton beam is 100 times greater than that carried by an electron beam.

Second, it was shown that strong electron beams (with initial energy fluxes of $F_0 \succeq 10^{10}\,\text{erg}\,\text{cm}^{-2}\,\text{s}^{-1}$) are significantly affected by an self-induced electric field, which is stronger for electron beams with a higher initial energy flux (Chapter 3). This effect can significantly reduce the electron beam penetration depth and returns the bulk of the electrons back to the injection point in the corona, as we showed in Chapter 4 in our discussion of particle numbers. At the same time, proton beams are not expected to be much affected by this effect as we discussed above, and hence their penetration depth does not depend much on their initial energy flux. Therefore, the momentum needed to account for a local helioseismic response, observed in some flares, is more likely delivered by proton beams or by mixed proton and electron beams rather than by pure electron beams.

Obviously, the ideal candidate for energy and momentum transfer is a neutralized beam, which does not lose its energy owing to return current ohmic dissipation. However, from the present acceleration models (Chapter 2) it appears that acceleration of a neutralized beam by a super-Dreicer electric field is very unlikely. Thus the only neutral beams that can appear in flaring atmospheres should come from separatrix jets ejected through the sides of current sheets by a magnetic diffusion process. Furthermore, even if such a beam is generated, its neutrality should be destroyed because electron pitch-angle scattering owing to collisions is stronger than that for protons. Therefore, electrons with the same thermalization depth as protons lose their directivity much faster and, thus, destroy the initial neutrality of the beam.

6
Hydrodynamic Response to Particle Injection

Variations in the density are considered from the hydrodynamic responses described below.

6.1
Hydrodynamic Equations

Let us now consider a hydrodynamic response of 1-D solar atmosphere to the injection of electrons or protons taking into account the continuity, momentum, and energy equations for ambient electrons and protons/ions.

The physical conditions in a flaring atmosphere can be described by a plasma density n, electron T_e and ion T_i temperatures, and vertical velocity v. All these parameters vary with vertical coordinate z or column density ξ, and time t. The ambient plasma response to the injection of high-energy particles is described by the following hydrodynamic equations (see, e.g., Fisher *et al.*, 1985a,b,c; Somov *et al.*, 1981):

a) Continuity equation:

$$\frac{\partial n}{\partial t} + n^2 \frac{\partial v}{\partial \xi} = 0 \; ; \tag{6.1}$$

b) Momentum equation:

$$\frac{\partial v}{\partial t} + \frac{1}{\mu} \frac{\partial}{\partial \xi} [n k_\mathrm{B} (T_\mathrm{i} + x T_\mathrm{e})] = \frac{4}{3} \frac{1}{\mu} \frac{\partial}{\partial \xi} \left(\eta_i n \frac{\partial v}{\partial \xi} \right) + g_\odot \; ; \tag{6.2}$$

c) Energy equation for ions:

$$\frac{n k_\mathrm{B}}{\gamma - 1} \frac{\partial T_\mathrm{i}}{\partial t} - k_\mathrm{B} T_\mathrm{i} \frac{\partial n}{\partial t} = \frac{4}{3} \eta_\mathrm{i} n^2 \left(\frac{\partial v}{\partial \xi} \right)^2 + Q(n, T_\mathrm{e}, T_\mathrm{i}) \; ; \tag{6.3}$$

Electron and Proton Kinetics and Dynamics in Flaring Atmospheres, First Edition. Valentina Zharkova.

d) Energy equation for electrons:

$$\frac{n k_{\mathrm{B}}}{\gamma - 1} \frac{\partial (x T_{\mathrm{e}})}{\partial t} - x k_{\mathrm{B}} T_{\mathrm{e}} \frac{\partial n}{\partial t} + n \chi \frac{\partial x}{\partial t}$$
$$= n \frac{\partial}{\partial \xi} \left(\kappa n \frac{\partial T_{\mathrm{e}}}{\partial \xi} \right) + P(n, \xi) - L(n, T_{\mathrm{e}}), - Q(n, T_{\mathrm{e}}, T_{\mathrm{i}}) . \qquad (6.4)$$

Here T_{i} and T_{e} are the ion and electron temperatures, respectively, n is the ambient plasma density, x is the ionization degree of the ambient plasma, γ is the adiabatic constant, k_{B} is the Boltzmann constant, κ is the thermal conductivity, η_{i} is the ion viscosity, χ is the full ionization energy of a hydrogen atom, $\mu = 1.44 m_{\mathrm{H}}$, and $g_{\odot}$ is the acceleration due to gravity of the Sun. $P(n, \xi)$ indicate the volume heating rates provided by electrons owing to collisions, by electrons owing to ohmic losses, by protons owing to collisions and by KAWs, L_{rad} is the volume radiative energy loss rate, and, finally, $Q(n, T_{\mathrm{e}}, T_{\mathrm{i}})$ is the rate of energy exchange between ions and electrons.

The solution is sought in a limited region of the solar atmosphere $\xi_{\min} \leq \xi \leq \xi_{\max}$ with the minimum boundary located at $\xi_{\min} = 2 \times 10^{17}\ \mathrm{cm}^{-2}$ and the maximum boundary deep in the photosphere at $\xi_{\max} = 2 \times 10^{24}\ \mathrm{cm}^{-2}$. The initial atmosphere is assumed to be in hydrostatic equilibrium, that is, $v(0, \xi) = 0$, and isothermal equilibrium, that is, $T_{\mathrm{e}}(0, \xi) = T_{\mathrm{i}}(0, \xi) = T_0$, where $T_0 = 6700$ K. The ionization degree x is defined by a modified Saha formula (Somov *et al.*, 1981). We also take into account the initial momenta delivered by the particles at injection (Brown and Craig, 1984) by assuming $v(0, \xi_{\min}) = P_{\mathrm{B}}/m$, where P_{B} is the sum of the momenta of all injected particles.

6.1.1 Additional Equations

The initial distribution of a plasma density is defined as follows:

$$n(0, \xi) = n_{\min} + h_0^{-1}(\xi - \xi_{\min}) , \qquad (6.5)$$

where $n_{\min} = 10^{10}\ \mathrm{cm}^3$ and h_0 is the height scale:

$$h_0 = \frac{k_{\mathrm{B}}[1 + x(T_0)] T_0}{\mu g_{\odot}} . \qquad (6.6)$$

In order to make this system self-consistent, one should write the additional expressions for the particle collision frequencies, heat conduction, and radiative losses.

The frequency of proton–electron collisions (in CGSE units) is

$$\nu_{\mathrm{pe}} = 4.8 \times 10^{-8} n \Lambda T_{\mathrm{p}}^{-3/2} \sqrt{\frac{m_{\mathrm{p}}}{m_{\mathrm{e}}}} , \qquad (6.7)$$

and the heating conduction coefficient is defined as follows:

$$\kappa_{II} = 3.2 p_e \frac{\tau_e}{m_e} , \tag{6.8}$$

where

$$\tau_e = \frac{3}{4} \frac{\sqrt{me}}{\sqrt{2\pi} n \Lambda e^4} (k_B T_e)^{3/2} . \tag{6.9}$$

Obviously, to make the problem self-consistent one needs an equation of state as follows:

$$p_e = n k_B T_e . \tag{6.10}$$

The radiative loss rate $L(n, T_e)$ is described by the analytical expression

$$L_{n,T_e} = n^2 x L(T_e) + n L_H(n, T) , \ (\mathrm{erg\,cm^{-3}\,s^{-1}}) \tag{6.11}$$

where the radiative loss function $L(T_e)$ is taken for the coronal abundances of elements in optically thin plasma (Summers and McWhirter, 1979) and $L_H(n, T_e)$ are the radiative losses in all hydrogen lines calculated for an optically thick atmosphere (Kobylinskii and Zharkova, 1996).

6.1.2 Boundary Conditions

The boundary conditions are defined as follows.

1. We assumed that there was no initial heat flux at the top boundary, that is,

$$\frac{\partial T_e(0, \xi_{min})}{\partial \xi} = \frac{\partial T_i(0, \xi_{min})}{\partial \xi} = 0 ;$$

2. The upper boundary is a free surface in the presence of the coronal pressure, that is,

$$\frac{\partial v(t, \xi_{min})}{\partial \xi} = \frac{4}{3} \frac{1}{n \eta_i} \{ p[(t, \xi_{min}) - p_{cor}[z(t, \xi_{min})] \} , \tag{6.12}$$

where $p(t, \xi_{min}) = n k_B (T_i + x T_e)$ and $p_{cor}[z(0, \xi_{min})] = n_{min} k_B [1 + x(T_0)] T_0$, where the ionization degree x is defined by a modified Saha formula (Somov *et al.*, 1981).

This approach allows one to simulate not only a strong increase of the temperature in the corona caused by chromospheric evaporation as reported by many other hydrodynamic simulations (Fisher *et al.*, 1985b,c; Zharkova and Zharkov, 2007) and observed in blue shifts of soft X-ray and UV lines (see, e.g., Antonucci *et al.*, 1982). This simulation also reveals the formation of low-temperature condensation in the lower chromosphere traveling as a shock toward the photosphere and observed with red shifts in H_α and other chromospheric lines (see, e.g., Canfield *et al.*, 1984; Ichimoto and Kurokawa, 1984). This shock is exactly the region in a flaring atmosphere from where H_α emission originates as it is often revealed by red shifts observed in the H_α emission in flares.

6.2 Hydrodynamic Responses to Heating by Electron Beams

6.2.1 The Heating Functions by High Energy Particles

The energy deposition functions, or heating rates, for these particles are calculated using distribution functions found from the full kinetic approach by solving the Fokker–Planck equation for electrons losing their energy in collisions and ohmic heating (Chapter 4) and protons in the generation of kinetic Alfvén waves (KAWs) and their dissipation via Cherenkov's resonance (Gordovskyy *et al.*, 2005)

The volume heating rates by all kinds of high-energy particles simultaneously presented in a flaring atmosphere are calculated from particle distribution functions as a vertical gradient of the beam energy flux:

$$S(\xi) = -n(\xi)\frac{d(F_e(\xi) + F_p(\xi) + F_j(\xi))}{d\xi}, \tag{6.13}$$

where $\xi = \int_{z_{min}}^{z_{max}} n(z)dz$ is the column density, that is, the number of the ambient particles in the area of 1 cm^2 on a line of sight from the height z_{min} to z_{max}, $n(z) = n(\xi)$ is a total density of the ambient plasma at a given height, and $F_e(\xi)$, $F_p(\xi)$, and $F_j(\xi)$ are the energy fluxes carried by fast electrons, "fast" protons, and "slow" protons (of separatrix jets), respectively. We do not include energy losses by slow electrons since these are negligible compared to other particles (Gordovskyy *et al.*, 2005). The heating rate per particle of the ambient plasma $P(\xi)$ is related to the volume heating rate $S(\xi)$ as $P(\xi) = S(\xi)/n(\xi)$.

6.2.2 Simulated Heating Functions

Let us compare the heating rates of the three kinds of particles considered in Section 6.2.1. The time scale within which each kind of particles can reach the photosphere is about 1 s for electron beams, 2–5 s for proton beams, and 10–20 s for protons of separatrix jets for a standard loop length of 10^9 cm (Gordovskyy *et al.*, 2005; Zharkova and Zharkov, 2007). Therefore, the propagation time for each kind of particle is short enough to contribute to the ambient plasma heating and to form a shock, or a lower-temperature condensation, appearing as a result of the hydrodynamic response to these particle injections.

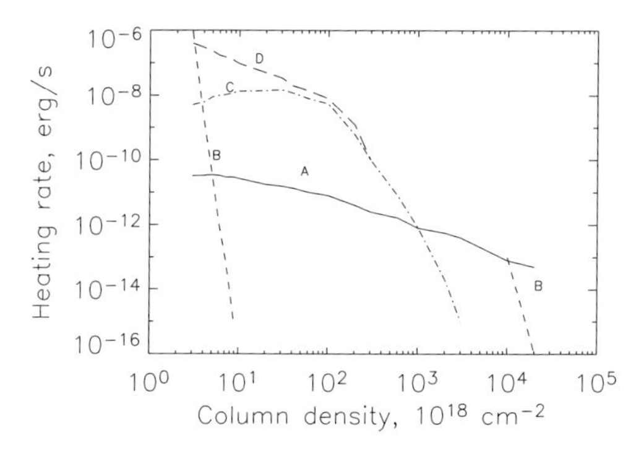

Figure 6.1 Heating rates for different populations of accelerated particles: curve A for a weak hard electron beam, curve B marks heating by "slow" jet protons, curve C corresponds to a strong and soft electron beam, and curve D to a "fast" proton beam.

The heating rates simulated from the full kinetic approach are presented in Figure 6.1: curve A – for an electron beam with an initial energy flux of $F_0 = 1.4 \times 10^8$ erg cm^{-2} s^{-1}, a spectral index of $\gamma = 2$, and a lower energy cutoff of 16 keV; curve B – for separatrix jet protons with initial energies of $E_0 = 1$ MeV and an initial energy flux of 4×10^{11} erg cm^{-2} s^{-1}; curve C – for an electron beam with an initial energy flux of $F_0 = 10^{10}$ erg cm^{-2} s^{-1}, spectral index of $\gamma = 5$, and lower energy cutoff 16 keV; curve D – for a proton beam with an initial energy flux of $F_0 = 4 \times 10^{10}$ erg cm^{-2} s^{-1}, spectral index of $\gamma = 1.5$, and lower energy cutoff of 40 keV; and curve B for separatrix jet protons with initial energies of $E_0 = 1$ MeV and an initial energy flux of 4×10^{11} erg cm^{-2} s^{-1}.

It can be seen that the heating by electron or proton beams with power-law energy distributions is strongly dependent on the initial beam parameters: softer and weaker beams deposit their energy mainly in the corona and upper chromosphere while harder and more powerful beams deposit more energy deeper in the lower chromosphere (compare curves A and C for hard and soft electron beams and curve D for a soft proton beam in Figure 6.1) (Gordovskyy *et al.*, 2005).

Electron beams are considered to deposit their energy more evenly in depth compared to the proton ones (compare curves A and C with D). Proton beams deposit the bulk of their energy via generation of KAWs with their subsequent dissipation in the Cherenkov resonance at depths of the flaring atmosphere, where their velocities are higher than the local Alfvén ones (Gordovskyy *et al.*, 2005). Note that the heating by KAWs induced by protons has two regions where this condition stands: in the upper corona because of their initial exponential distributions (first curve B) and at the lower chromosphere due to a reduction of the local Alfvén speed and of the proton exponential distributions (second, spikelike curve B).

The heating of the upper atmosphere by proton beams can be even more noticeable after Coulomb collisional losses are taken into account, which will make the proton distributions in the chromosphere even more exponential. While strongly affecting heating by proton or electron beams of the corona and upper chromosphere (before the column depths of 10^{20} cm^{-2}), the collisions do not change significantly the heating of the lower chromosphere where the Cherenkov resonance is dominant (Gordovskyy *et al.*, 2005). Another heating mechanism considered for beam electrons is ohmic dissipation in a self-induced electric field that contributes significantly to the heating of the coronal levels but again does not affect the lower atmosphere heating (Zharkova and Gordovskyy, 2005a).

6.2.3
Hydrodynamics Caused by Electron Beams

The results of hydrodynamic heating by electron beams with different initial energy fluxes (10^{10}–10^{12} erg cm^{-2} s^{-1} and spectral indices of 3 and 5 are plotted in Figures 6.2 and 6.3. At the initial moments after electron beam injection into a cool atmosphere at an initial temperature of 6700 K, the electron temperature starts to grow very quickly in the upper chromospheric levels. At 1 s the electron temperature profile resembles slightly the electron heating function in the chromosphere. However, very quickly this similarity disappears because of the nonlinear process

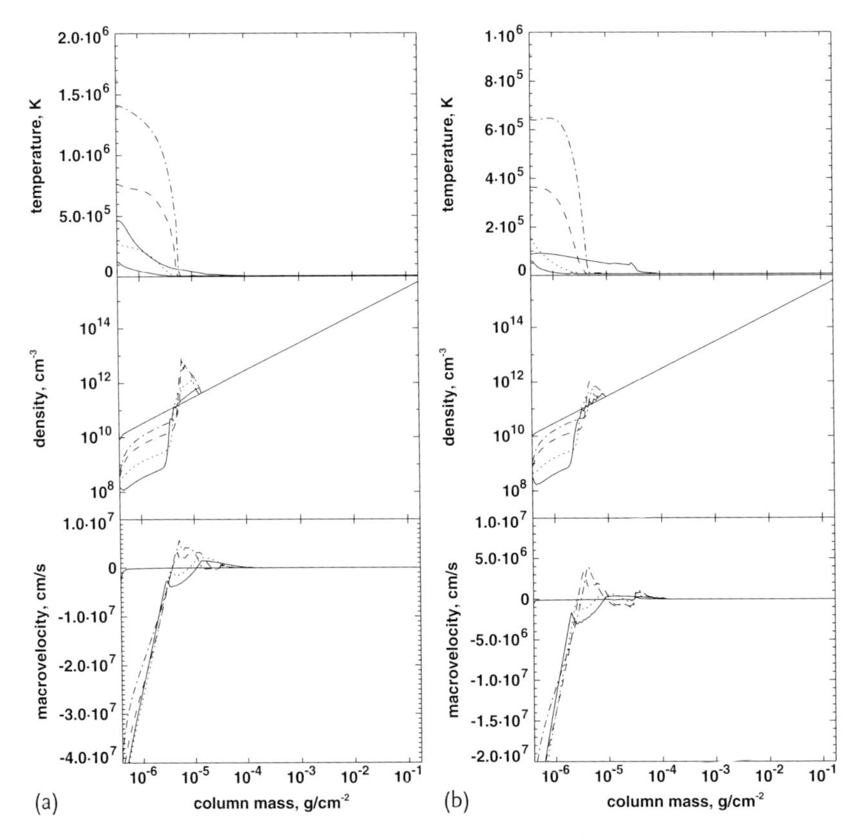

Figure 6.2 Hydrodynamic response of ambient plasma to injection of electron beam with an initial energy flux of $F = 10^{10}$ erg cm^{-2} s^{-1} and spectral index of $\gamma = 3$ (a) and 5 (b).

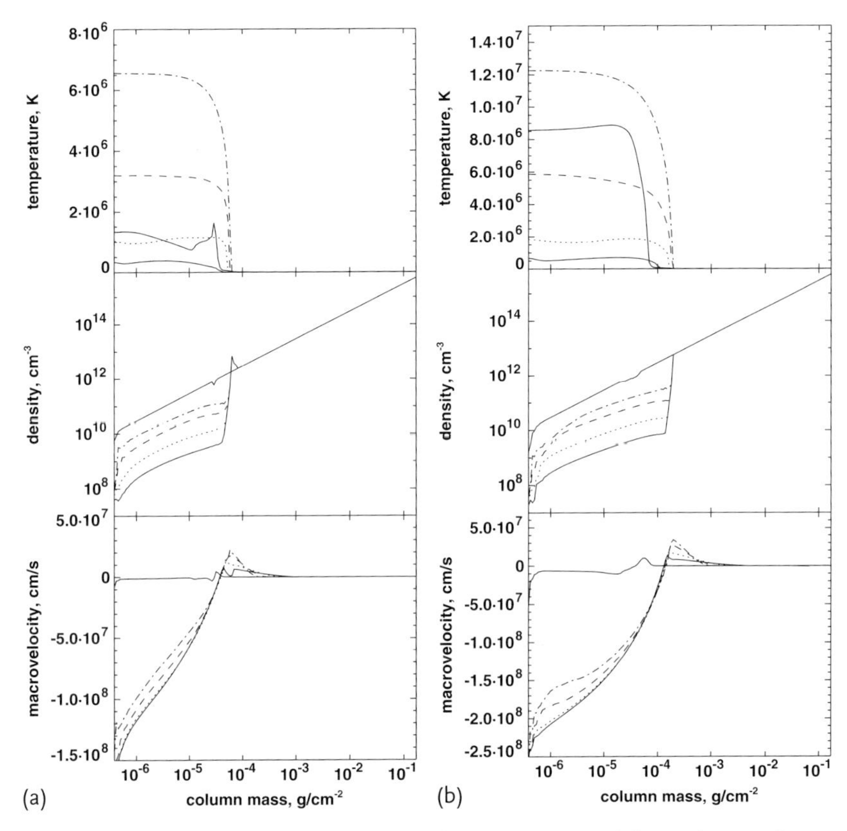

Figure 6.3 Hydrodynamic response of ambient plasma to injection of electron beam with an initial energy flux of $F = 10^{12}$ erg cm^{-2} s^{-1} and $\gamma = 5$.

of radiative cooling and thermoconductivity. In a range of temperatures less than 10^5 K the radiative cooling balances the heating by beam electrons. Let us look separately at the temperature, density, and macrovelocity variations after heating by beam electrons.

6.2.3.1 Temperature Variations

Following electron beam onset, it heats the ambient plasma so that the ambient temperature increase in the corona of both components (electrons and ions) approaches a few tens of MK, with the ions' temperature being twice as small as the electrons' (cf. Figures 6.2–6.4) (Somov *et al.*, 1981), while the macrovelocity of the explosive evaporation into the corona reaches 1200 km s^{-1}.

In the region of the heating function maximum the ambient plasma temperature is not high and is smaller than the critical temperature of 2.5×10^4 K, corresponding to maximum losses in the radiative cooling. Then in the coronal region with higher temperatures ($> 10^6$), electron thermoconductivity results in a fast increase

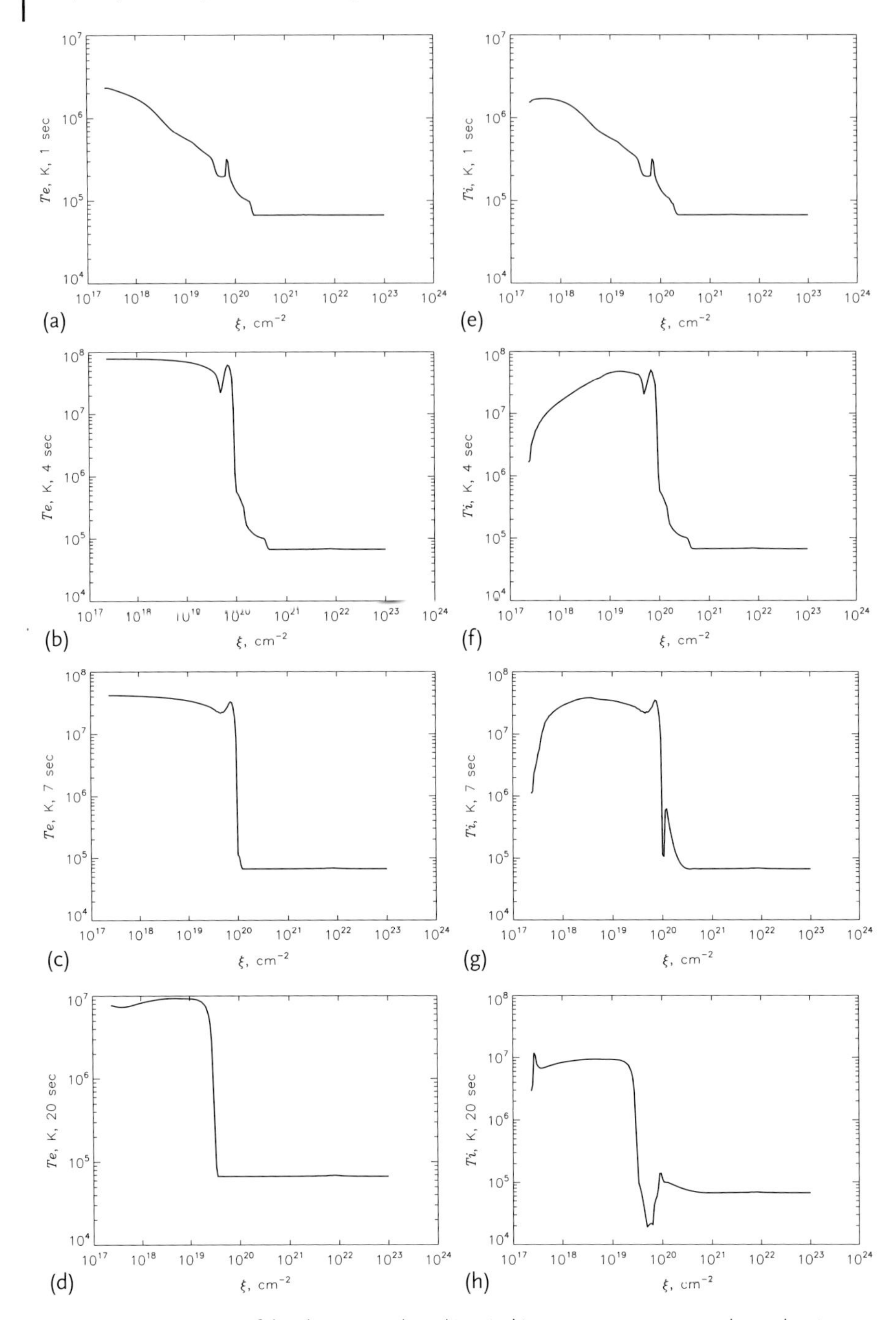

Figure 6.4 Comparison of the electron (a–d) and ion (e–h) temperatures versus column density in a hydrodynamic atmosphere heated by an electron beam with the following parameters: $F_0 = 2 \times 10^{12}$ erg cm^{-2} s^{-1} and $\gamma = 5$.

of the temperature and in the formation of a narrow transition region at depths of $3.3 \times 10^{19}\,\mathrm{cm}^{-2}$, with a sharp drop in electron (and ion) temperatures.

After the heating by beam electrons stops, the temperature in the hot coronal region starts decreasing very quickly because of the transfer of thermal energy via conductivity to the transition region where this energy is emitted in various radiations. However, after the ambient electron temperature decreases further, the conductivity drops very dramatically (by $\approx T_e^{2.5}$) and the further temperature decrease occurs because of radiative losses. At the same time, the ambient plasma density decreases because this region expands, which slows down the radiative cooling. Because of a decrease in the gas pressure in the high-temperature part, at 15 s there is a noticeable shift of the flare transition region upward until this region completely disappears when the temperature approaches 2.5×10^5 K.

The ion temperature (Figure 6.4) follows the electron temperature only in the lower-temperature region, while in the higher-temperature coronal region the ion temperature is substantially delayed compared to the electron one. Also, if the heating by beams is substantial, then the ion temperature becomes several-fold higher than the electron, or even by an order of magnitude. This points to a need to consider two-temperature plasma treatment instead of a single temperature one, as is often accepted in many hydrodynamic codes. However, at the cooling stage the ion and electron temperatures quickly achieve a balance.

The difference between the ion and electron temperatures gained during heating by beam electrons can be observed in the temporal profiles of soft X-ray emission where the plasma cooling rate because of a reduction in the pure electron temperature cannot explain the extended emission observed. This extended soft X-ray emission can only be reconciled with the assumption that the extension is caused by ambient electron cooling in the process of leveling the electron and ion temperatures, for example, when electrons gain extra energy from ions.

6.2.3.2 **Density Variations**

As pointed out above, very fast density variations occur during precipitation of beam electrons into a flaring atmosphere. At 1 s after beam onset there is a substantial decrease of the density in the upper corona. At the same time the density increases slightly at lower atmospheric levels with a small amount of condensation forming below the transition region ($\xi \geq 1.9 \times 10^{19}\,\mathrm{cm}^{-2}$), which moves downward to the photosphere. Because of thermal instability (the increase in density causes an increase in radiative losses and a quick decrease in the temperature, leading to further plasma compression), cool condensation forms with densities approaching $10^{14}\,\mathrm{cm}^{-3}$. The increase in gas pressure in the high-temperature region causes this condensation to move downward at a velocity of about $90\,\mathrm{km\,s}^{-1}$ at 2.5 s for an electron beam with an initial energy flux of $10^{10}\,\mathrm{erg\,cm}^{-2}\,\mathrm{s}^{-1}$ and spectral index of $\gamma = 3$, which rises to $110\,\mathrm{km\,s}^{-1}$ for an electron beam with $10^{11}\,\mathrm{erg\,cm}^{-2}\,\mathrm{s}^{-1}$ and the same spectral index.

After this time the temperature increase in the corona decreases sharply and a transition region is established about a column density of $\xi \geq 2 \times 10^{19}\,\mathrm{cm}^{-2}$. The cold dense condensation keeps moving downward at decreasing speed and reaches a height of about 350 or 80 km above the photosphere 60 s after beam onset for beams with energy fluxes of 10^{10} and 10^{11} erg cm^{-2} s^{-1}, respectively.

A fast expansion of the high-temperature region into the corona leads to a substantial decrease in the ambient density by approximately two orders of magnitude compared to the preheated density. The expanding plasma velocity can reach up to 1100 km s^{-1} 15 s after beam onset, reaching heights of about 10^5 km while the temperature drops to 10^5 K. At this stage the main energy losses are in UV radiation, while at later stages, when the temperature decreases further, the losses shift to optical emission (Somov *et al.*, 1981). The authors argue that at maximum heating by electron beams the main part of their energy is converted into radiation of low-density condensation and not into soft X-rays or UV emission.

The increase of the spectral index shifts the main energy deposition into upper atmospheric levels, leading to a stronger temperature increase and denser low-temperature condensation for harder beams ($\gamma = 3$) compared to softer ones (cf. Figure 6.2a and 6.2b). Although the macrovelocities of the chromospheric plasma evaporate into the corona, the velocities of the low-temperature condensation are twice as high for harder beams.

The increase of the initial energy flux of the beam with the same spectral index of $\gamma = 3$ (Figure 6.3) reveals a stronger heating of both coronal and chromospheric levels for beams with a larger energy flux. Also, the lower-temperature shock is formed at deeper atmospheric levels if it is caused by more powerful electron beams (cf. Figure 6.3a and 6.3b).

6.2.3.3 **Macrovelocity Variations**

Above the critical depth of 1.9×10^{19} cm^{-2}, which is the maximum of the electron heating function for pure collisional heating (Chapter 3), electron heating is compensated by radiative cooling. Here the temperature increase ceases while forming the first minimum coinciding with the first maximum in the ambient plasma density, which occasionally occurs at 2.5–5 s after beam onset. This condensation lasts for another couple of seconds, being heated by thermal fronts for both sides, although in some cases (stronger softer beams) this condensation can last longer and propagate deeper into the chromosphere.

Another story is about the second maximum in the condensation profile appearing at 2.5 s, as shown in Figures 6.2 and 6.3 (bottom plots). This condensation forms as a result of thermal instability, which starts at a lower temperature and leads to higher ambient plasma density. Simulations show that during the nonlinear stage this condensation develops in a nonlinear regime close to $p = \text{const}$. At the same time, the mass of this condensation increases during its motion as a shock into the deep photosphere at a macrovelocity of about 90 km s^{-1} (see the velocity plots in Figures 6.2 and 6.3), although the shock front is smoothed by the artificial viscosity. This condensation moves toward the column depths of the quiet photosphere at a velocity that decreases with propagation depth to a few kilometers per second as measured from the Doppler velocity in the NI 6768 Å line.

6.2.4
Hydrodynamics Formed by Mixed Electron and Proton Beams

In this section we simulate the heating and hydrodynamic response in a flaring atmosphere extended towards the photosphere and beneath caused by the beam with parameters derived for the powerful X-class flare of 14 December 2006 (Chapter 12) in an attempt to see if lower-temperature shocks could reach the photosphere and how deep they could push chromospheric and photospheric plasma into the solar interior.

As can be noticed from Figures 6.5 and 6.6, the more powerful the beams become, the more significant is the difference in their hydrodynamic responses, in general, and in the generation of lower-temperature shocks, in particular. The electron temperature, ambient density, and macrovelocity variations caused by a pure electron beam with parameters derived for the flare of 14 December 2006 (Matthews *et al.*, 2011) (flux 2×10^{12}, spectral index about 5) (Figure 6.5) reveal within the first 4 s a fast increase in electron temperatures up to 10^7 K and a decrease in the ambient density by two orders of magnitude.

Note that heating by a more powerful beam, presented in Figure 6.5 in comparison with Figures 6.2 and 6.3, leads to rather high velocities (up to 10^8 cm s^{-1}) of chromospheric evaporation simulated for the upper chromosphere (column depth up to 5×10^{19} cm^{-2}). While below this depth a lower-temperature shock forms

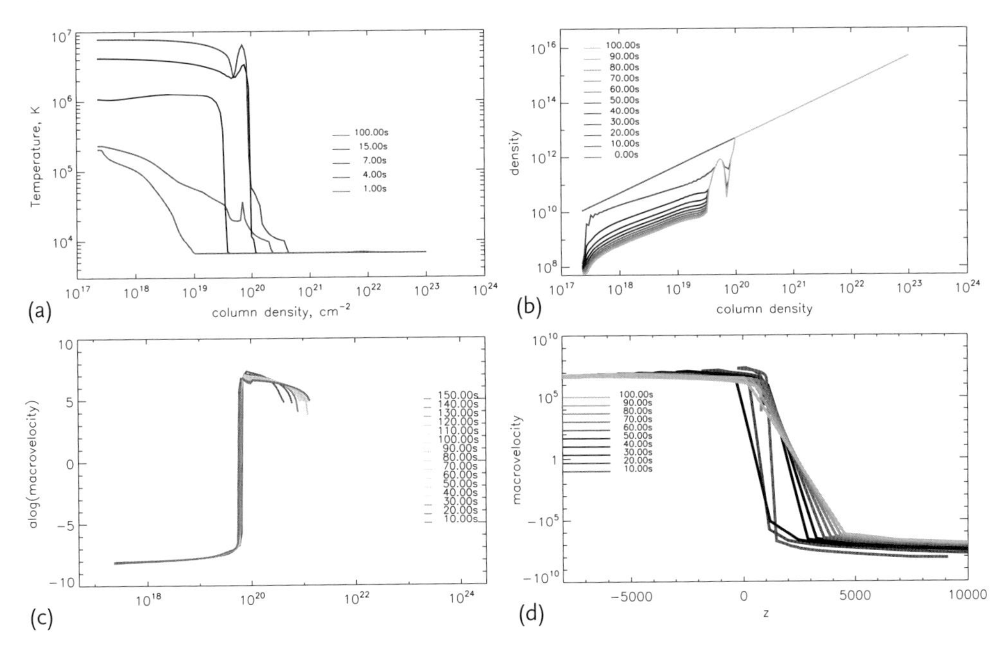

Figure 6.5 Snapshots of hydrodynamic models of electron beam injection with $F_0 = 2 \times 10^{12}$ erg cm^{-2} s^{-1}; electron temperature (a), ambient density (b), and macrovelocity profiles (c,d) vs. linear height above the photosphere. For a color version of this figure, please see the color plates at the beginning of the book.

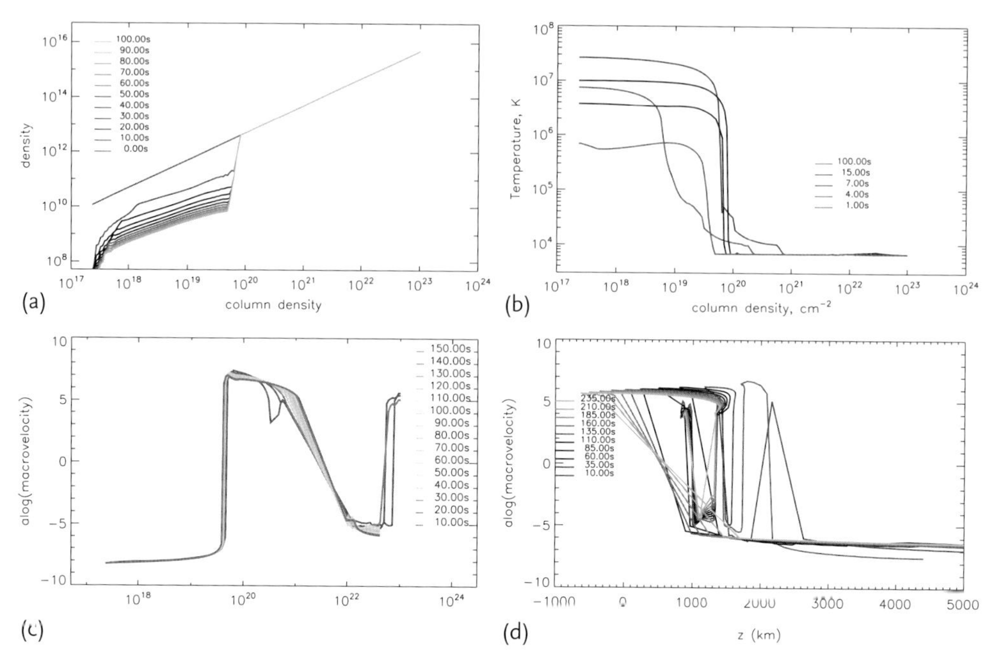

Figure 6.6 Snapshots of hydrodynamic models of mixed beam injection (70% protons and 30% electrons in energy flux) with $F_0 = 2 \times 10^{13}\,\mathrm{erg\,cm^{-2}\,s^{-1}}$; electron temperature (a), ambient density (b), and macrovelocity profiles (c,d) vs. linear height above the photosphere. For a color version of this figure, please see the color plates at the beginning of the book.

moving at velocities of up to $10^7\,\mathrm{cm\,s^{-1}}$ and terminating at velocities of up to $10^5\,\mathrm{cm\,s^{-1}}$ at a depth of just below $10^{21}\,\mathrm{cm^{-2}}$, or a linear depth of about that of the photosphere. As indicated earlier (Siversky and Zharkova, 2009b; Zharkova and Zharkov, 2007), this is obviously related to the fact that beam electrons deposit at the upper chromosphere the maximum of their energy in collisions and ohmic losses.

By increasing the initial energy flux of the injected particles, for example by mixing an electron beam with a proton one assuming there are 70% protons and 30% electrons (in total particle energy flux) (Figure 6.6), the energy deposition maximum shifts from the upper to the lower chromosphere and the photosphere. This slightly increases the electron temperature to 2×10^8 K in the ambient plasma evaporating into the corona and sweeps the coronal plasma faster, reducing the ambient density by two orders of magnitude. Simultaneously, this mixed beam reduces the macrovelocties of chromospheric evaporation into the corona from 10^8 in the first 10 s to a factor of 10^5 after 130 s. At the same time, the lower-temperature shock is now split into two shocks: one moves from the lower corona toward a chromospheric column depth of $10^{19}\,\mathrm{cm^{-2}}$, terminating at depths of $10^{20}\,\mathrm{cm^{-2}}$, and the second one appears at a column depth of $4 \times 10^{20}\,\mathrm{cm^{-2}}$ and terminates at a depth of 0.5×10^{21} cm^{-2}. The maximum velocity approaches 10^6 cm s^{-1} in the first shock and 10^5 in the second one.

The further increase of the initial energy flux and momentum by consideration of beams of protons (90% of flux) and electrons (10% of flux), while keeping the velocities of chromospheric evaporation at the same level (up to 10^8 cm s^{-1}) as that of electrons, can make the first lower-temperature shock much wider, spreading from a column depth of 8×10^{19}–10^{22} cm^{-2} with the second shock starting at a depth of 5^{22} cm^{-2} and terminating at a velocity of a few units of 10^5 cm s^{-1} at a column depth of $\times 10^{23}$ cm^{-2}, which corresponds to about 3×10^8 cm linear depth beneath the photosphere in the solar interior.

The motion of these single or double shocks is likely to play a key role in producing seismic waves observed as sunquakes occurring in the solar interior and their occasional links with Moreton waves occurring in the chromosphere, discussed in Chapter 12.

6.2.5
Momenta Delivered by Beams and Hydrodynamic Shocks

6.2.5.1 Momentum Delivered by a Proton or Electron Beam

The momentum $P_{\mathrm{e,p}}$ delivered in pure collisions by an electron or proton beam with a spectral index of $\gamma_{\mathrm{e,p}}$ and a lower energy cutoff of $E_{\mathrm{low}_{\mathrm{e,p}}}$ can be evaluated as (Zharkova and Gordovskyy, 2005c)

$$P_{\mathrm{e,p}} = \sqrt{2m_{\mathrm{e,p}}}K \int_{E_{\mathrm{low_{e,p}}}}^{\infty} E^{-\gamma_{\mathrm{e,p}}+0.5} dE \,, \tag{6.14}$$

where $m_{\mathrm{e,p}}$ is the electron or proton mass, E_1 is the lowest energy in the relevant particle spectrum (theoretical lower energy cutoff), and K is the normalization constant, which can be found from the total number of measured particles $N_{\mathrm{e,p}}$:

$$N_{\mathrm{e,p}} = K \int_{E_{\mathrm{low_{e,p}}}}^{\infty} E^{-\gamma_{\mathrm{e,p}}} dE = \frac{K}{\gamma} E_{\mathrm{low_{e,p}}}^{-\gamma+1} \,, \tag{6.15}$$

where $E_{\mathrm{low}_{\mathrm{e,p}}}$ is the measured lowest energy of electrons or protons. Hence, without taking into account pitch-angle scattering, the momentum delivered by a proton beam can be evaluated by substituting the constant K found from N_{p} into the equation for $P_{\mathrm{e,p}}$ and performing the integration. This will result in

$$P_{\mathrm{e,p}} \simeq \sqrt{2m_{\mathrm{p}}}N_{\mathrm{e,p}} \frac{\gamma_{\mathrm{e,p}}}{(\gamma_{\mathrm{e,p}}+0.5)} \frac{E_1^{-(\gamma_{\mathrm{e,p}}-0.5)}}{E_{\mathrm{low_{e,p}}}^{-(\gamma_{\mathrm{e,p}}-1)}} \,. \tag{6.16}$$

If we assume that $E_1 = E_{\mathrm{low}_{\mathrm{e,p}}}$, then the momentum delivered by electrons/protons without pitch-angle scattering can be evaluated as

$$P_{\mathrm{e,p}} \simeq \sqrt{2m_{\mathrm{e,p}}}N_{\mathrm{e,p}} \frac{\gamma_{\mathrm{e,p}}}{\gamma_{\mathrm{e,p}}+0.5} E_{\mathrm{low_{e,p}}}^{-0.5} \,. \tag{6.17}$$

It should be emphasized that this is the upper limit of the momentum carried downward to the photosphere by electrons or protons since it is calculated without taking into account pitch-angle scattering and wave dissipation for protons or ohmic dissipation for electrons, which can reduce its magnitude by a few factors (Gordovskyy *et al.*, 2005). However, it allows for the comparison of the momenta delivered by high-energy particles with those measured from the downward motions in dopplergrams and from time-distance (TD) diagrams.

6.2.5.2 Momentum Delivered by Hydrodynamic Shock

Let us also evaluate the momentum delivered by a hydrodynamic shock using the following simple formula:

$$P_{\text{hd}} = \sum_{t} m v(t) \,, \tag{6.18}$$

where the summation is done over time from 0 to τ, where τ is the duration of the impact causing the seismic waves, m is the mass of the plasma delivering the momentum related to the flare area A where the momentum is deposited, V is the starting velocity of the plasma at the moment of impact, and t is the duration of the impact.

For the known plasma mass density $\rho = m_{\text{H}} n$ where n is the particle density per volume defined from hydrodynamic solutions, this equation can be rewritten as follows:

$$P_{\text{hd}} = \sum_{t} m v(t) \approx \rho A v^2 \tau \,, \tag{6.19}$$

where ρ is the average density of the plasma delivering the momentum, A is the flaring area where the momentum is deposited, v is the averaged velocity of the plasma propagation at the moment of impact, and τ is the duration time of the impact. These momenta will be used in Chapter 12 for the evaluation of seismic responses associated with solar flares.

6.2.6
Comparison of Ambient Heating by Electrons and Protons for 28 October 2003 Flare

The variations in temperature, density, and macrovelocity simulated for hydrodynamic responses are plotted for an electron beam for the flare of 28 October 2003 with parameters derived from HXRs (Figure 6.7a–c) or for a mixed proton/jet beam with parameters derived from γ-ray emission (Figure 6.7d–f).

In order to evaluate the transfer into ambient plasma of the momenta and energy carried by different kinds of particles after their injection from the top into a flaring atmosphere, let us compare their heating functions and resulting hydrodynamic responses following Zharkova and Zharkov (2007).

We assume that protons or electrons are accelerated in a reconnecting current sheet (RCS) with a strong longitudinal magnetic field, so that they can then be ejected as power-law beams, being completely or partially separated into opposite

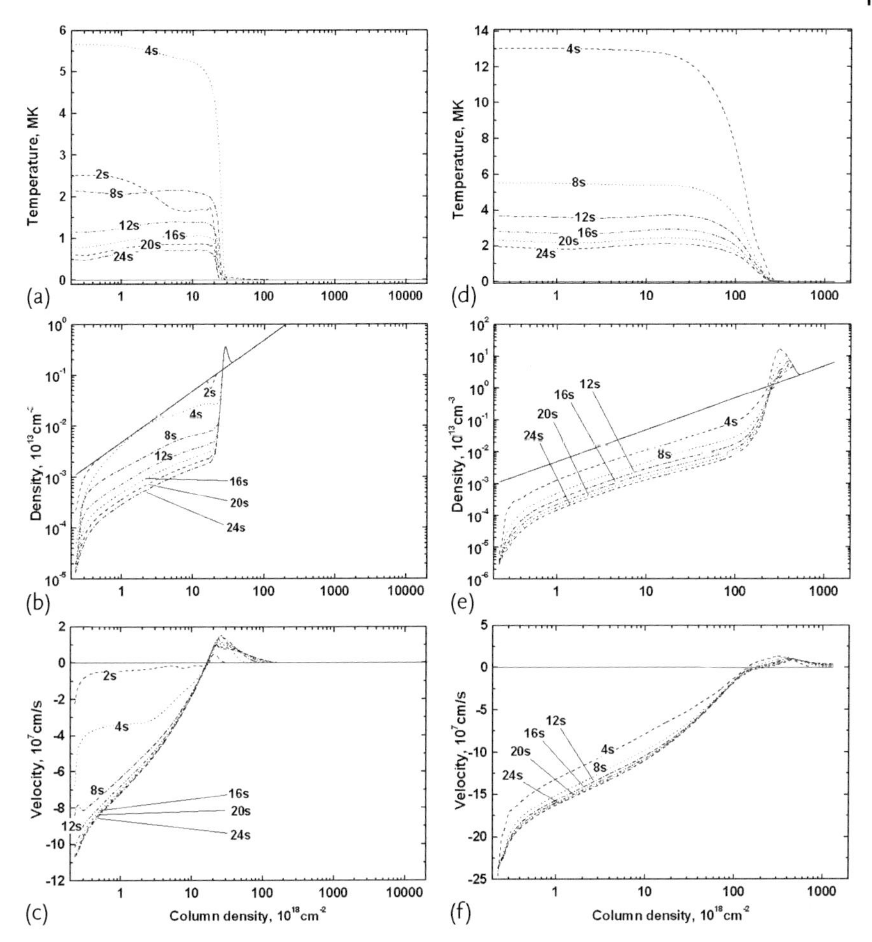

Figure 6.7 Hydrodynamic responses of a flaring atmosphere caused by pure electrons (a–c) or by mixed proton and electron beams (d–f) with parameters of protons deduced from γ-rays and electrons from X-rays from RHESSI and KORONAS: (a,d) ambient electron temperatures; (b,e) densities; (c,f) macrovelocities. The numbers on the graphs show times in seconds after beam injection.

footpoints of the same loop (Zharkova and Gordovskyy, 2004, 2005b). In addition to electron or proton beams, let us consider particles accelerated along the separatrices of the RCS with Maxwellian (thermal) energy distributions shifted to higher energies (Gordovskyy *et al.*, 2005). As a result we consider four kinds of high-energy particles whose energy losses are converted into ambient plasma heating: fast electron beams and fast proton beams with power-law energy distributions and slow electrons and protons of separatrix jets with thermal energy distributions.

To maximize a deposition at lower atmospheric levels, let us compare hydrodynamic responses caused by a hard electron beam with $\gamma = 3.5$ and an initial energy flux of 10^{11} erg cm^{-2} s^{-1} (which is higher and harder than measured) with those by proton beams/jets with spectral indices of about 3 and initial energy fluxes of 10^{12} erg cm^{-2} s^{-1}, as deduced from the HXRs and γ-rays in Section 7.3.3.1. We include in the heating functions collisional losses by both electron and proton beams, ohmic losses only for an electron beam, and Cherenkov resonance for thermal-like protons. Two heating functions are considered: by a pure electron beam and by a proton beam combined with quasithermal jet protons. The variations in temperature, density, and macrovelocity simulated for hydrodynamic responses are plotted for an electron beam for the 28 October 2003 flare with parameters derived from HXRs (Figure 6.7a–c) or for a mixed proton/jet beam with parameters derived from γ-ray emission (Figure 6.7d–f).

The presented hydrodynamic results caused by beam electrons agree with those by Fisher *et al.* (1985b,c); Somov *et al.* (1981). For an intense hard beam as deduced from the HXR data from KORONAS (Kuznetsov *et al.*, 2006) there is a noticeable decrease (rarefying) in the total plasma density and a strong temperature increase in the corona accompanied by explosive evaporation of the chromospheric plasma into the corona (upward macro motion) occurring in response to the beam injection.

Starting from a column depth of 2×10^{19} cm^{-2} and a collisional stopping depth for lower-energy electrons of 12 keV accepted in our simulations, a low-temperature condensation is formed, moving as a shock at velocities of about 100–200 km s^{-1}, which are higher than the local sound velocity (Fisher *et al.*, 1985b,c; Somov *et al.*, 1981). However, such a shock, even one produced by a powerful electron beam with a spectral index of about 3 (Figure 6.7e,f), appears rather high in the upper chromosphere between column depths of 2×10^{19} and 2×10^{20} cm^{-2}, as can be seen in Figure 6.7d–f.

These two motions of ambient plasma (upward and downward) were reported in all previous hydrodynamic simulations (Somov *et al.*, 1981; Nagai and Emslie, 1984; Fisher *et al.*, 1985a,b,c) and have been widely investigated from blue- and red-shifted spectral measurements in UV and H_α emission, respectively. For some events or beam parameters, these can be nearly equal (see, e.g., Zarro *et al.*, 1988), while for many other beams, if the electron ohmic losses are taken into account (Zharkova and Gordovskyy, 2006), only the blue shifts are observable without the red ones.

As can be seen from Figure 6.7, the hydrodynamic responses to the injection of pure electron or mixed proton/jet beams are substantially different. An electron beam injected for 10 s produces a smaller (by a factor of 2–2.5) temperature increase in the corona (top plot of Figure 6.7a), a smaller (by an order of magnitude) density depression of the coronal plasma into the chromosphere (middle plot of Figure 6.7a), and smaller (by a factor of 2–2.5) evaporation velocities (bottom of Figure 6.7a) compared to those by mixed proton/jet beams (right-hand plots in Figures 6.7 and 6.8).

In addition, the mixed proton/jet beam forms a lower-temperature shock, which is spread much deeper into the lower chromosphere between column depths of 2×10^{20} and 8×10^{21} cm^{-2} (see Figure 6.8 for a closer view of the shock in the first 100 s). The velocities of the shock induced by the mixed proton beam are also

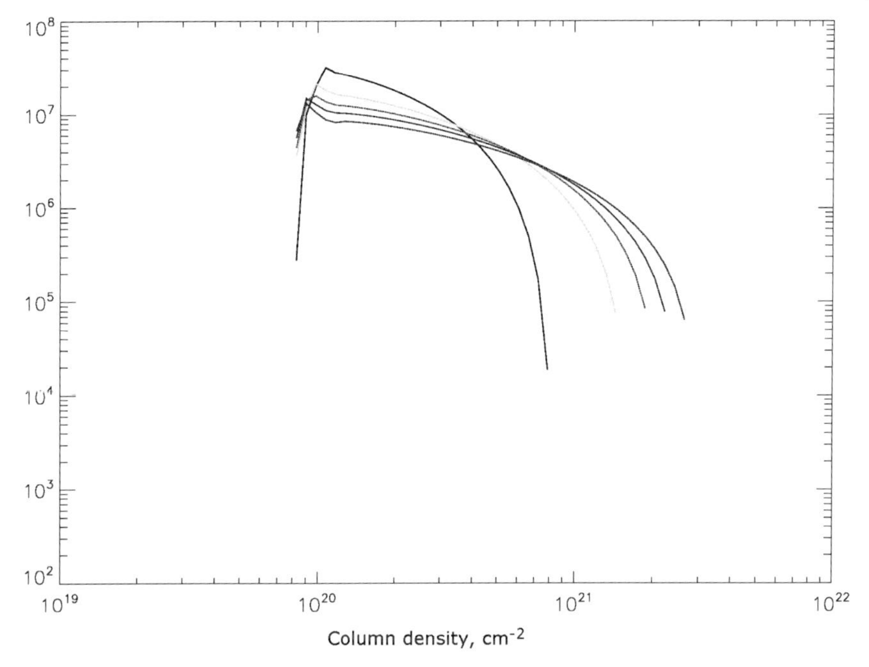

Figure 6.8 Close-up of Figure 6.7 of temporal variations in macrovelocity in centimeters per second in a lower-temperature condensation, or a hydrodynamic shock (*y*-axis) vs. column depth in cm^{-2} (*x*-axis), appearing in response to injection of mixed proton/electron beams (Figure 6.7d–f), after 10 s (black line), 30 s (red line), 50 s (burgundy line), 70 s (navy line), and 100 s (blue line). For a color version of this figure, please see the color plates at the beginning of the book.

higher (by a factor of 2–2.5) than those induced by a pure electron beam. These macrovelocities induced by the proton beam decrease in regions with a column density of 5×10^{21} cm^{-2} to a few kilometers per second compared with those of less than 1 km s^{-1} for one induced by electrons. The momentum induced by this shock is transferred to the photosphere within a much shorter time scale – under 1 min – because it is formed at much deeper and denser atmospheric levels. This is also confirmed by the temporal profile of the simulated macrovelocity variations at the lower edge of the shock (the far right points in the distributions in Figure 6.8) induced by proton beams, which shows an increase, to a few kilometers per second, of the edge macrovelocities, within a minute, which resembles the observed Doppler velocity variations measured for source S1 in the 28 October 2003 flare (Zharkova and Zharkov, 2007).

A proton-induced shock deposits its momentum from depths of (5–8) $\times 10^{21}$ cm^{-2} to the very dense plasma beneath that is delivered at a velocity of about 2 km s^{-1} through about 120 km of the solar atmosphere before approaching the column depth of 4×10^{23} cm^{-2} for the Ni line-formation region. By contrast, an electron-induced shock is required to travel from a column depth of (2–5) $\times 10^{20}$ cm^{-2}, or about 350 km, at the same or twice the lower velocity. This will result not only in the twice smaller momentum delivered by this shock compared to the proton-induced one but also in the delay of 3–6 min before this shock from the chromosphere can reach the region with Ni-line formation measuring the Doppler velocities.

6.3
Case Study of a Hydrodynamics of the 25 July 2004 Flare

6.3.1
Observations

The M2.2-class flare of 25 July 2004 occurred in NOAA Active Region 10652, started at 13:37 UT, ended at 13:55 UT, and had its maximum at 13:49 UT. Note that information about the optical importance of the flare is absent (no patrol observations are available).

Since TRACE instruments have a higher spatial resolution, the TRACE 195 Å images are used for the precise identification of flaring-loop elongations and the locations where the loops are likely to be embedded in the photosphere. This provides further information about loop connectivity in the corona. The temporal evolution of the flaring loops is shown in Figure 6.9d–f. The loops were brighter at 13:40:48 UT (image TRACE 195 Å) at the maximum of HXR emission, which was followed by a rather complex loop structure that appeared at 13:41:38 UT, when HXR emission started to decrease.

A superposition of the HXR sources (dashed and solid contours), H_α emission kernels with TRACE images in 195 and 1600 Å, are presented in Figure 6.9. We used the HXR source images for 12–25 and 25–50 keV bands, which were reconstructed with the PIXON algorithm (Metcalf *et al.*, 1996; Puetter and Pina, 1994) due to the clearer shape of the sources. Correct positions of the sources were confirmed by a comparison of the images reconstructed by both CLEAN and PIXON algorithms.

6.3.1.1 Magnetic Field Variations

For an analysis of a magnetic structure of flaring region SOHO/MDI magnetograms obtained with 1 min cadence are used. Figure 6.10a shows a longitudinal magnetogram at 13:40:02 UT overplotted with the locations of H_α emission kernels. The H_α sources appeared at the opposite sides of the magnetic neutral line (MNL), which in turn coincides with the filament location (Figure 6.10b).

From the 1600 Å images by TRACE (Figure 6.9) and the magnetic field locations (Figure 6.10) one can see that the flare under investigation was, in fact, a two-ribbon flare. One of the ribbons is associated with the HXR sources but the other is not. The ribbons are located on opposite sides of the inversion line, or MNL, that coincides with the filament (Figure 6.10b). This points to a set of loops embedded in the

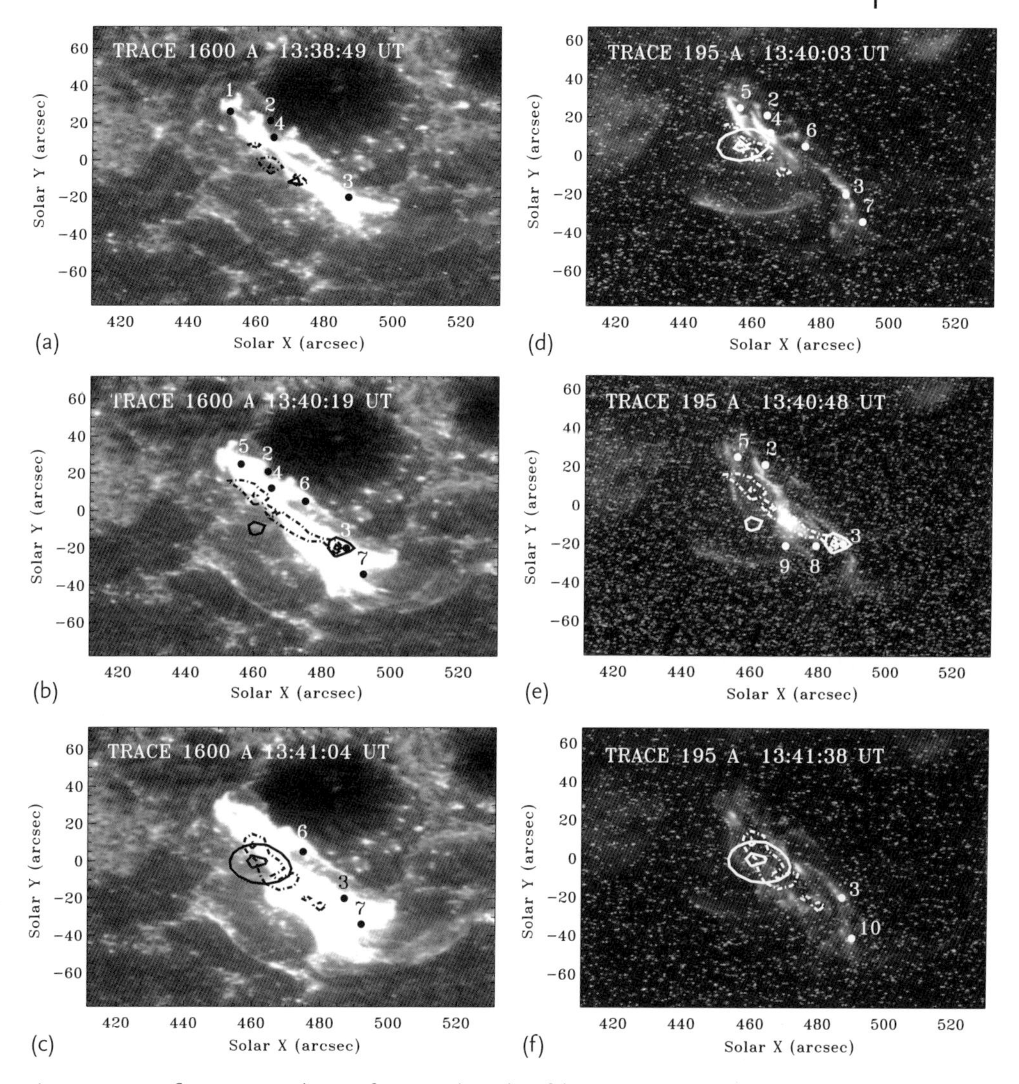

Figure 6.9 Main flare event: evolution of H_α emission kernels and HXR sources vs. the TRACE image of the same region in 195 and 1600 Å. H_α emission kernels are marked as black points whose number corresponds to the order of their onset. The HXR sources are plotted as the contours at levels of 50 and 90% from their maxima. The dashed and solid contours represent, respectively, 12–25 and 25–50 keV bands.

photosphere from opposite sides of the MNL with a strong shift of the embedded locations with opposite polarities strongly elongated with respect to the MNL.

In a few footpoints the HXR sources appear simultaneously (within an observational cadence of 5 s) with the H_α kernels, while in other footpoints H_α emission is not associated spatially with HXR sources and appears 10–20 s later than any HXR sources in the other locations.

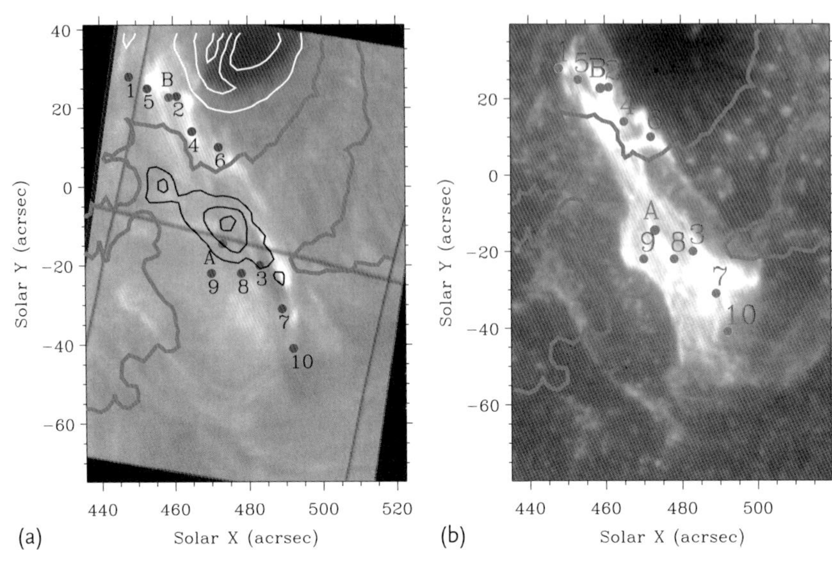

Figure 6.10 (a) The 1600 Å image at 13:40:02 UT with locations of H_α emission (kernels) marked by white dots. "A" and "B" show the locations of preflare kernels. (b) The same H_α kernels (black dots) overlaid on the H_α-line image. The black and white contours are negative and positive magnetic fields, respectively. A magnetic inversion line or magnetic neutral line (MNL) is marked by the gray line in both panels.

6.3.1.2 **Morphology of Flaring Kernels in the Main Event**

The spatial locations of the HXR sources and H_α sources with respect to the MNL and their temporal variations often indicate the sites of energy deposition associated with a magnetic reconnection and transport of this energy into deeper atmospheric levels by high-energy particles (Priest *et al.*, 2001). The magnetic reconnection events are likely to occur in succession in each loop along the ribbon (Priest *et al.*, 2001).

Particles (electrons and protons) accelerated during this reconnection precipitate downward into the loop legs, losing their energy in collisions with ambient particles (Brown, 1971) and ohmic losses in the electric field induced by the particles themselves (Zharkova and Gordovskyy, 2005a, 2006). This process produces HXR emission from the corona and chromosphere and H_α emission in the chromosphere while the thermoconduction transfers (via a hydrodynamic response) heating by a beam to the whole atmosphere. Moreover, in the footpoints with HXR emission there are signs of a transient increase, or reversible changes, of its magnetic field (Zharkova *et al.*, 2005a) that are likely to be caused by the precipitating electrons (van den Oord, 1990).

The magnetic field changes recorded for this flare along with the variations of the measured HXR photon flux are presented in Figure 6.11 with a close-up plot of the spectral index variations at lower energies (< 40 keV) and the electron flux corrected for ohmic energy losses. This shows that the line-of-sight (LOS) magnetic field changes were definitely associated with the flare phenomena, revealing

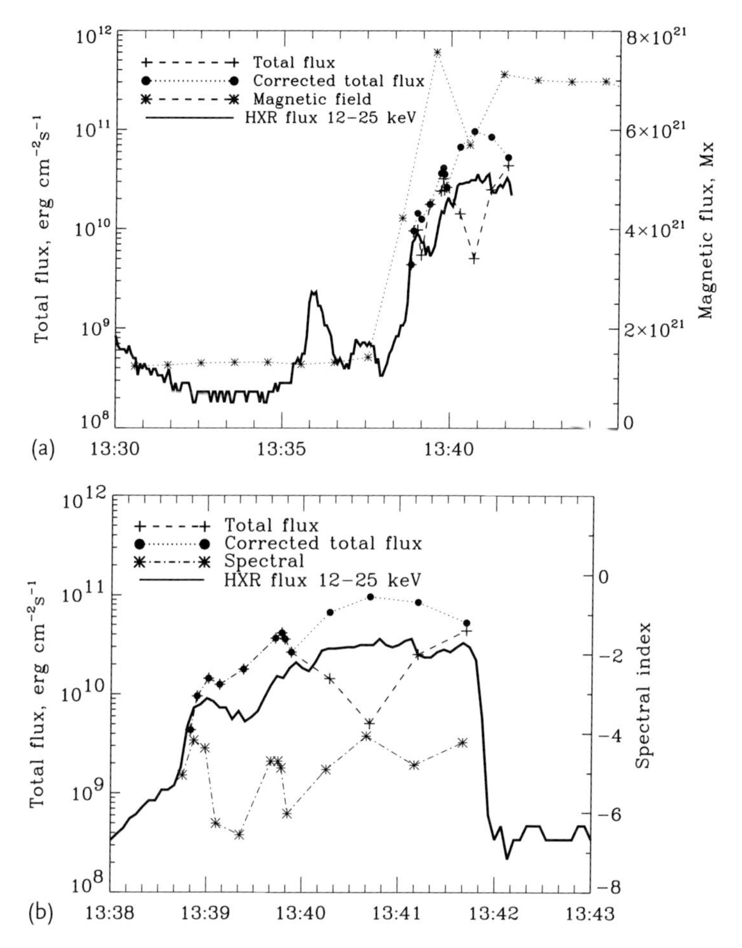

Figure 6.11 (a) Initial electron energy fluxes (dashed line with pluses), magnetic flux variations (dotted–dashed line with asterisk), energy fluxes of beam electrons corrected for the induced electric field losses (dotted line with black disks) in the main flare event are plotted vs. light curve in an energy band of 12–25 keV (solid curve). (b) The close-in plots of the same initial electron energy fluxes and electron flux corrected for the induced electric field losses based on variations of spectral indices of HXR photon spectra at lower energies, below 40 keV (the dotted–dashed line with asterisk) in the main flare event. The solid line represents a light curve in an energy band of 12–25 keV.

the magnetic flux increase in close temporal correlation with the variations of the HXR photon flux. However, the cadence of magnetic field measurements was 1 m, meaning these variations can only show us the magnetic field variations on a scale larger than 1 min. Thus, the LOS magnetic flux shows steady growth during the flare bursts while only for the third burst at about 13:40:48 UT is there a noticeable magnetic field reduction (up to 26 G/pixel) resembling the reversible "transient" changes of magnetic field reported by Zharkova *et al.* (2005a). A discussion about a

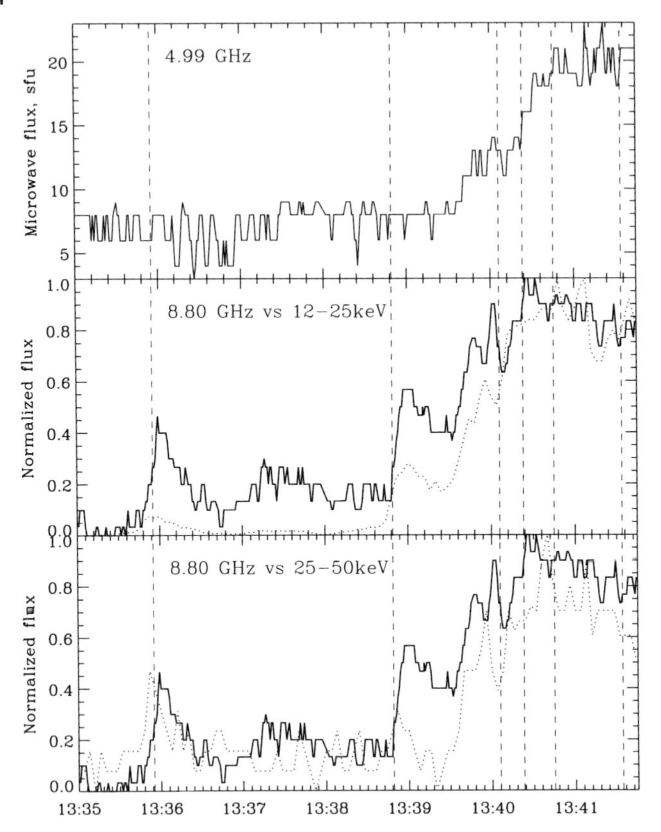

Figure 6.12 Temporal evolution of microwave flux by Radio Solar Telescope Network (RSTN) and HXR flux (RHESSI). Microwave flux is shown by the solid line, HXR flux by the dotted line. The dashed lines correspond to an event of fast rising intensity on H_α images during the 25 July 2004 flare.

relationship between the magnitude of this transient magnetic field and the electric field induced by precipitating electrons is presented in Chapter 4.

The light curves of HXR emission of this flare observed in different bands by the RHESSI (counts corrected for pile-up and albedo effects obtained with a 4 s time resolution) and the microwave flux by RSTN (obtained with a 1 s temporal resolution) are presented in Figure 6.12 and clearly show a burstlike structure. There are two events observed for the light curves: the preflare one occurring around 13:35:00 UT and the main event occurring after 13:37:00 UT. The spikes of HXR and radio emission were also accompanied by small-scale chromospheric events observed in H_α light curves and images discussed in Zharkova *et al.* (2011b).

6.3.1.3 **Correction of HXR Photon Energy Flux for Ohmic Losses**

The parameters of an electron beam energy distribution can be obtained from the RHESSI photon energy spectrum plotted in Figure 6.13 using the imaging spec-

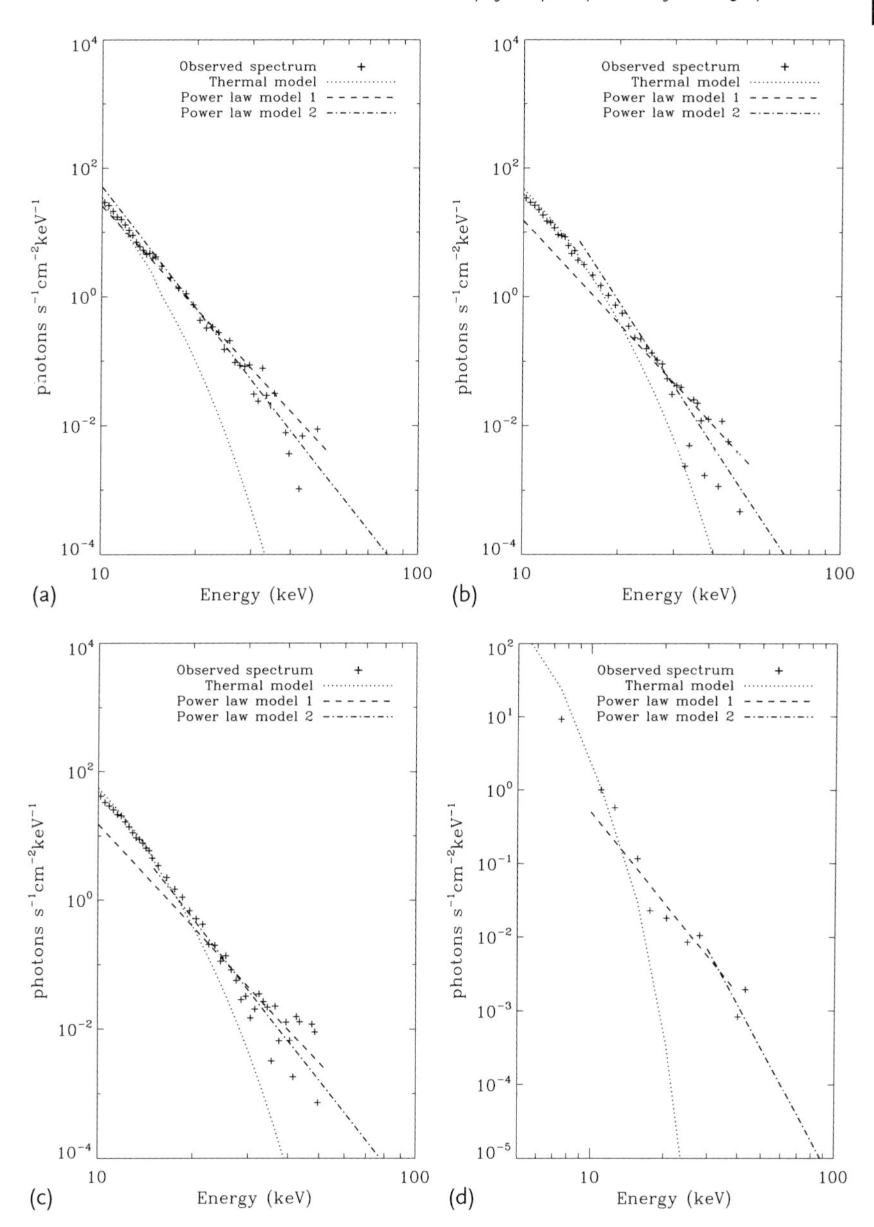

Figure 6.13 The main flare event: HXR photon energy spectra obtained for the first and third HXR bursts in energy bands from 12–70 keV. The integration times cover the first burst at 13:38:52–13:38:56 UT (a), 13:38:56–13:39:00 UT (b), 13:39:00–13:39:04 UT (c), and the third burst at 13:40:38 UT with an integration time of 4 s (d). The fitting of the thermal part into the deduced energy spectra at energies below 30 keV carried by taking into account the albedo effect is shown in each plot by a dotted line.

troscopy method (Aschwanden *et al.*, 2004), which works as follows. The image for each energy bin of the spectrum is reconstructed using the PIXON algorithm known to have a high photometric accuracy (Metcalf *et al.*, 1996; Puetter and Pina, 1994). Then we count the photons in each image constructed for different energy bins and plot their numbers as a function of energy of the relevant bin, which presents the observed HXR spectrum. Normally, this spectrum is power law with a spectral index (γ) and lower energy cutoff of 20 keV. Then the initial energy flux of an electron beam is deduced from the area under the HXR photon curve from the energy 20 keV to the upper energy observed. And the spectral index of the beam electron is smaller by one than that of the HXR photon energy spectrum (Brown, 1971).

However, more powerful flares often reveal an elbow-type power-law energy spectrum model with two spectral indices: a lower one at energies below 50 keV and a higher one at energies above 50–70 keV. These double power-law photon spectra are shown to be produced by powerful electron beams if their energy losses during precipitation include not only Coulomb collisions but also ohmic losses in the electric field carried by beam electrons (Zharkova and Gordovskyy, 2006; Zharkova *et al.*, 2010). In this case the electric field of the beam leads to photon spectrum flattening at lower energies and much lower energy flux deduced from the elbow-type photon spectrum than those from the single power-law one.

A partial ionization degree can contribute to HXR spectrum flattening at lower energies (Kontar *et al.*, 2002; Su *et al.*, 2009); however, this effect is already considered in our Fokker–Planck approach applied for the calculation of electron beam and HXR parameters (Zharkova *et al.*, 2010). This is treated via the coefficient for frequency of collisions, which varies with depth as per the simulations of hydrodynamic atmospheres (Zharkova and Zharkov, 2007), which in turn requires to calculate the variations of ambient plasma ionization with depth. In fact, in our model there are about 70 points in depth of the chromosphere and photosphere with a partial ionization degree, which can cause a small HXR spectrum flattening that exceeds those considered by either Kontar *et al.* (2002) or Su *et al.* (2009). Furthermore, in our Fokker–Planck approach (Zharkova *et al.*, 2010), unlike previous authors, we consider relativistic HXR cross-sections that have been proven to provide the most correct HXR photon spectra (Kontar *et al.*, 2006).

Therefore, the spectral flattening in HXR photon spectra found from the proposed Fokker–Planck approach is produced by the joint effects of ohmic losses and partial ionization, which are already accounted for in the simulated differences in spectral indices (Zharkova and Gordovskyy, 2006). Then, in order to deduce an electron spectrum from an HXR photon one, one needs to keep in mind that beam electrons emitting these photons have a single power law with the index equal to the spectral index at upper energies as per Fokker–Planck simulations (Zharkova and Gordovskyy, 2006). Then, the correct initial energy flux of beam electrons is deduced by extending a higher-energy elbow of the photon spectra back to lower energies and by measuring the area under the new, extended, photon energy spectrum.

This correction increases the total energy flux carried by the electron beam as plotted in Figure 6.11 by the dashed line with dark circles. It can be noted that the total flux corrected for the ohmic losses anti-correlates with the spectral index at lower energy. This is well expected because it corresponds to the photon spectrum flattening caused by the self-induced electric field discussed in Section 7.3.2 (see Figure 7.4). The hydrodynamic model calculated using this correction will be used in Chapter 10 for simulation of H_α-line emission produced in this flare.

Since the difference between the lower and upper spectral indices is dependent on the initial energy flux of beam electrons (Zharkova and Gordovskyy, 2006), this photon spectrum flattening can explain the soft–hard–soft variations of the photon spectra observed in some flares by an increase in the initial flux of electron beam from low to high energy and back to low again, for example by a peak in the HXR photon light curve. This method is used to deduce the spectral index and initial energy flux of beam electrons from their HXR photon spectrum.

6.3.2
Hydrodynamics of Ambient Plasma

Let us consider the hydrodynamics of a flaring atmosphere during the first burst of the 25 July 2004 flare, which occurred at about 11:39:00 UT. We assume that the flaring atmosphere was heated by an electron beam with the parameters derived from HXR observations and corrected for ohmic loss effects (Zharkova *et al.*, 2011b) as described above. This particular time was selected because H_α kernels were also observed in the flaring atmospheres in a close temporal proximity with HXR emission, suggesting that nonthermal processes in these events should dominate the thermal ones.

If we use the initial energy fluxes derived from HXR energy spectra (Figure 6.11), then the hydrodynamics of a flaring atmosphere could be very similar to those simulated for a preflare event with the ambient temperature being twice higher. While the fitting of the thermal spectrum into the HXR energy spectrum derived in Figure 6.13 reveals the mean electron kinetic temperature to be about 24, 30, and 29 MK for consecutive intervals of 4 s starting from 13:38:52 UT. This can only be achieved by increasing the initial energy flux of beam electrons with the effect of a self-induced electric field and increasing its spectral index to one (6.2) derived at higher energies. Only in such a way is it possible to fit the observed thermal temperatures since heating by a beam with the same energy flux but harder spectral indices (4 or 5) produces a smaller increase of the temperature in the corona than those measured in Figure 6.13a,b.

After a beam onset, beam electrons heat the ambient plasma so that the ambient temperature increase in the corona of both components (electrons and ions) approaches a few tens of MK with the ions' temperature being half as much as the electrons', while the macrovelocity of the explosive evaporation into the corona reaches 1200 km s^{-1}, although the thermal temperatures derived from these spectra are still higher than those simulated in hydrodynamic models for the relevant times after beam injection (Figure 6.14). This discrepancy was removed after the

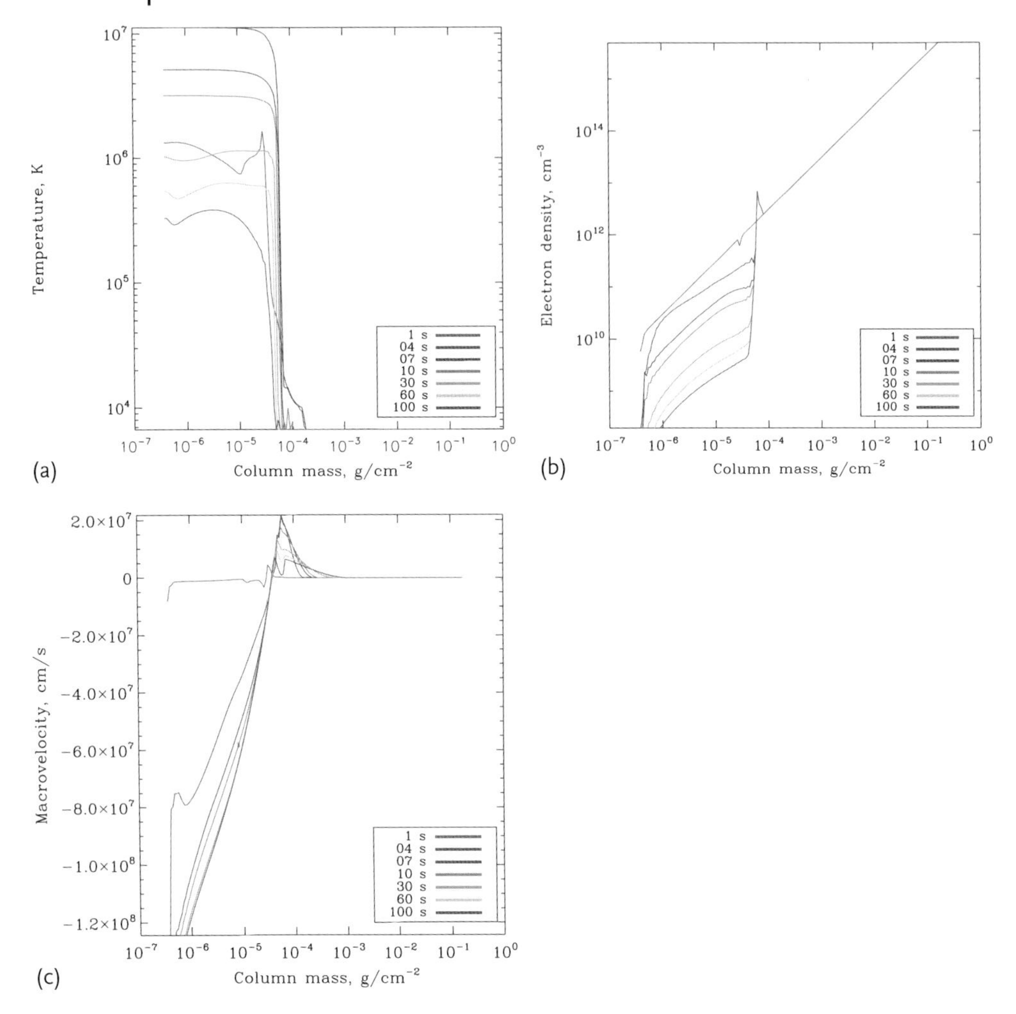

Figure 6.14 Temperature (a), density (b), and macrovelocity (c) variations in the main flare event calculated from a hydrodynamic response to the injection of an electron beam with parameters derived from HXR emission. For a color version of this figure, please see the color plates at the beginning of the book.

albedo effect, calculated for isotropic beams (in pitch angles), was excluded since in this flare the electron beams seemed to be very anisotropic. As a result, the electron temperatures are reduced by about a factor of two for each moment, which allowed them to fit very closely the temperature profiles in the corona simulated for 5–15 s after beam onset.

After the beam is stopped, the ambient plasma continues with chromospheric evaporation and an increase of the coronal density back to the preflare magnitudes while becoming cooled by radiation for another hour or so. Also, in the first 10–15 s, a lower-temperature condensation forms in the chromosphere, moving with supersonic speed toward the photosphere, which produces low-temperature optical emission in the first burst of the main event. This condensation is formed in the chromosphere, while in the preflare event it appeared just below the transition region. It has a much higher temperature, up to 10^4 K, and a density of up to 10^{14} cm^{-2} compared with the preflare event (7×10^4 and 3×10^{13} cm^{-2}, respectively), and its macrovelocity approaches 200 km s^{-1}. The increased temperature in this condensation defines the time scale of an increase in the contribution to generation of H_α emission by impacts with thermal electrons that competes with the contribution to this emission and total hydrogen ionisation caused by non-thermal electrons.

The resulting H_α emission simulated for this hydrodynamic model is compared with observations for this flare in Chapter 10.

6.4 Conclusions

In this chapter we investigated hydrodynamic responses of flaring atmospheres to the injection of different agents – electron beams, proton beams, proton jets, and their mixture. We established that some of them are able to deliver the required energy and momentum to the flaring atmosphere: power-law electron beams, power-law proton beams, and quasithermal protons with energies below 200 MeV occurring through acceleration by a super-Dreicer electric field in reconnecting current sheets on top of a flaring atmosphere (Siversky and Zharkova, 2009a; Zharkova and Agapitov, 2009; Zharkova and Gordovskyy, 2004). Electron beams are considered to deposit their energy in collisions and ohmic dissipation, proton beams – in collisions and in the generation of kinetic Alfvén waves with their subsequent dissipation in a Cherenkov resonance (Gordovskyy *et al.*, 2005).

The hydrodynamic responses to a precipitating power-law electron beam or to a proton beam mixed with the protons of separatrix jets were investigated. The hydrodynamic response to the injection of pure electron, pure proton, or mixed proton/electron beams were found to be substantially different. An electron beam injected for 10 s produced a smaller (by a factor of 2–2.5) temperature increase in the corona (Figure 6.7a, top), a smaller (by an order of magnitude) density depression of the coronal plasma into the chromosphere (Figure 6.7a, middle), and a smaller (by a factor of 2–2.5) evaporation velocities (Figure 6.7a, bottom) compared to those induced by mixed proton/jet beams (Figures 6.7b and 6.8).

In addition, a mixed electron/proton or proton/jet beam forms a low-temperature shock, which is spread much deeper into the lower chromosphere between column depths of 2×10^{20} and 8×10^{21} cm^{-2} (see Figure 6.8 for a closer view of the shock in the first 100 s). The velocities of the shock induced by a mixed proton beam are also higher (by a factor of 2–2.5) than those induced by a pure electron beam. These macrovelocities induced by the proton beam decrease in the region of a column density of 5×10^{21} cm^{-2} to a few kilometers per second compared with those less than 1 km s^{-1} for one induced by electrons. Also, the temporal profile of macrovelocity variations at the lower edge of a shock (far right points in distributions in Figure 6.8) induced by proton beams shows an increase to a few kilometers per second of the edge macrovelocities within 1 min that resembles those measured for source S1 in the 28 October 2003 flare (Zharkova and Zharkov, 2007).

The energetic protons (power laws combined with Maxwellian ones from the separatrix jets) were shown to deliver high momenta, allowing them to form hydrodynamic shocks much deeper in a flaring atmosphere compared to a pure electron beam. This allows proton-formed shocks to travel shorter distances to the photosphere and to the interior over much shorter times, which can result in observable seismic waves occurring nearly simultaneously with high-energy emissions, as observed in many seismic sources associated with flares.

The momenta and start times delivered to the photosphere by different kinds of high-energy particles with parameters deduced from HXR and γ-ray emissions, as well as by hydrodynamic shocks caused by these particles, will be compared in Chapter 12 with those measured from TD diagrams in the sources for some flares.

7
Hard X-Ray Bremsstrahlung Emission and Polarization

7.1
Introduction

In recent years the theory describing the generation of bremsstrahlung HXR emission has progressed significantly in many directions by improving the mechanisms for emitting this radiation in pure Coulomb collisions (Brown, 1971; Brown *et al.*, 2006; Kontar and Brown, 2006; Syrovatskii and Shmeleva, 1972a), in collisions and converging magnetic fields (Leach and Petrosian, 1981; McClements, 1992a), or in combined collisional and ohmic energy losses in electric fields induced by precipitating electrons (Zharkova and Gordovskyy, 2005a, 2006; Zharkova *et al.*, 1995, 2010). Another advancement occurred from a consideration of relativistic bremsstrahlung cross-sections (Kontar *et al.*, 2006) and taking into account various aspects of photospheric albedo effects while deriving mean electron spectra from observed bremsstrahlung photon spectra (Kontar *et al.*, 2006).

As detected from RHESSI observations (Holman *et al.*, 2003; Lin *et al.*, 2003), HXR photon energy spectra observed from powerful flares often have double power laws revealing noticeable flattening toward lower photon energies that confirm earlier findings by SMM observations (Benz, 1977; Brown and Loran, 1985). This flattening leads to the soft–hard–soft temporal profile of photon spectra indices measured below energies of 35 keV (Grigis and Benz, 2004). This flattening was shown to be proportional to the initial energy flux of beam electrons and their spectral indices (Zharkova and Gordovskyy, 2006), which confirmed observations by Grigis and Benz (2004). The lower energy spectral flattening was first interpreted by an increase of the lower cutoff energy (Sui *et al.*, 2005), which was later overturned by Sui *et al.* (2007a) in favor of ohmic losses caused by self-induced electric fields of precipitating electrons (Zharkova and Gordovskyy, 2006).

In addition, pitch-angle distributions of the populations of electrons responsible for HXR emission are expected to be highly variable and dependent on the location of flares on the solar disk. This pitch-angle anisotropy of precipitating particles can be detected by polarimetric observations in flares located near the solar center to the limb. Such observations are rather scattered since they were carried out over many years in different energy bands: for example, at 15 keV in early rocket observations

Electron and Proton Kinetics and Dynamics in Flaring Atmospheres, First Edition. Valentina Zharkova.

by Tindo *et al.* (1970, 1972a,b), at 16–21 keV later by Tramiel *et al.* (1984), at 100–350 keV in recent RHESSI observations (Suarez-Garcia *et al.*, 2006), and at 200–400 keV in ACT observations by Boggs *et al.* (2006). The polarimetric observations indicated a high variability of HXR polarization degrees in different events: from a small percentage (Tramiel *et al.*, 1984) up to 40% (Suarez-Garcia *et al.*, 2006; Tindo *et al.*, 1976), or even above 50% as observed for the flare of 23 July 2002 (McConnell *et al.*, 2003) or 75% as observed by KORONAS payload (Zhitnik *et al.*, 2006). The bulk of observations also reveal a noticeable increase of the HXR polarization and directivity in flares located closer to the limb, although some deviations from this trend have recently been observed by RHESSI (Suarez-Garcia *et al.*, 2006). A similar increase toward the limb was reported in HXR directivity at 20 keV obtained from observations of a large number of flares by Kašparová *et al.* (2007).

In this chapter we apply the Fokker–Planck solutions for electron-beam precipitation into a flaring atmosphere to the calculation of HXR emission and polarization for impulsive and steady injection regimes by taking into account various energy loss and angle scattering mechanisms discussed in Chapter 4.

7.2
Stokes Parameters for HXR Emission

As was shown in Chapter 4, scattering of beam electrons on ambient particles is often strongly anisotropic, especially in the presence of external magnetic and self-induced electric fields. In this case the resulting HXR bremsstrahlung emission deviates from isotropic ones, becoming anisotropic and requiring one to simulate Stokes parameters defining HXR intensity I and linear polarization Q, while the circular polarization of HXR emission by beam electrons is shown to be negligible (Zharkova *et al.*, 1995).

The Stokes parameters of X-ray emission from a homogeneous source in the solar atmosphere (observed from Earth) can be defined by the following equation (Elwert and Haug, 1970; Brown, 1972; Leach and Petrosian, 1983:

$$\begin{bmatrix} I \\ Q \end{bmatrix}(h\nu, \theta) = \frac{n_{\mathrm{i}} V}{2\pi \mathcal{R}^2} \int_{h\nu}^{\infty} v(E) d E \int_{-1}^{1} f(E, \mu) d\mu \times \int_{0}^{2\pi} \begin{bmatrix} \sigma_I \\ \sigma_Q \end{bmatrix}(h\nu, \theta, E, \mu, \varphi) d\varphi \,, \tag{7.1}$$

where I and Q are the Stokes parameters of bremsstrahlung radiation, $h\nu$ and θ are the energy and the propagation angle (with respect to the magnetic field, i.e., pitch angle) of X-ray photons, $n_{\mathrm{i}} = n$ is the concentration of background ions, V is the source volume, $\mathcal{R}$ is the astronomical unit, $\sigma_{I,Q}$ are parameters describing the bremsstrahlung probability (Sections 7.2.2 and 7.2.3), and φ is the electron azimuthal angle. The factor 2π is related to the normalization condition since the distribution function of electrons is assumed to be azimuthally symmetrical. In this case, another Stokes parameter (U) equals zero (Bai and Ramaty, 1978; Leach and Petrosian, 1983).

One can assume that the X-ray source is represented by a cylindrical magnetic tube in which the magnetic field is aligned along the tube axis. All the parameters of the plasma, magnetic field, and accelerated particles depend only on coordinate z along the tube. In this case, the total (integrated over all layers of the source) emission is determined by the equation

$$\begin{bmatrix} I \\ Q \end{bmatrix}(h\nu, \theta) = \frac{S}{2\pi R^2} \int_0^{\xi_{\max}} d\xi \int_{h\nu}^{\infty} v(E) dE \int_{-1}^{1} f(\xi, E, \mu) d\mu \times \int_0^{2\pi} \begin{bmatrix} \sigma_I \\ \sigma_Q \end{bmatrix}(h\nu, \theta, E, \mu, \varphi) d\varphi , \tag{7.2}$$

where S is the cross-section area of a flux tube. In a converging magnetic field, the cross-section area varies (due to magnetic flux conservation) as $S(\xi) \sim B^{-1}(\xi)$, and the concentration of the accelerated particles also varies as $n_e(\xi) \sim B(\xi)$. Thus these variations can compensate each other and are not considered here.

Using the Stokes parameters, the degree of linear polarization can be calculated as

$$\eta = \frac{Q}{I} \tag{7.3}$$

and the angule distribution of HXR emission can be characterized by the directivity parameter (Leach and Petrosian, 1983)

$$D(\theta) = \frac{I(\theta)}{\langle I \rangle} , \tag{7.4}$$

where $\langle I \rangle$ is the emission intensity averaged over all propagation angles. The 3-D integrals over the components of the electron velocity in Eq. (7.2) are calculated using the Monte Carlo method (Zharkova *et al.*, 2010).

7.2.1 Geometry of Observations

Electrons precipitating in a flaring loop located somewhere on the Sun produce H-ray bremsstrahlung emission in some direction defined by the position of the momentum vector by electrons $\boldsymbol{P}$ and the momentum vector of emitted photons $\boldsymbol{K}$ seen under the angle ψ from the Earth. Since the loop is magnetic, the directions of both vectors are measured with respect to a magnetic field $\boldsymbol{H}$ of the loop. The two most popular geometries of observations are presented in Figure 7.1.

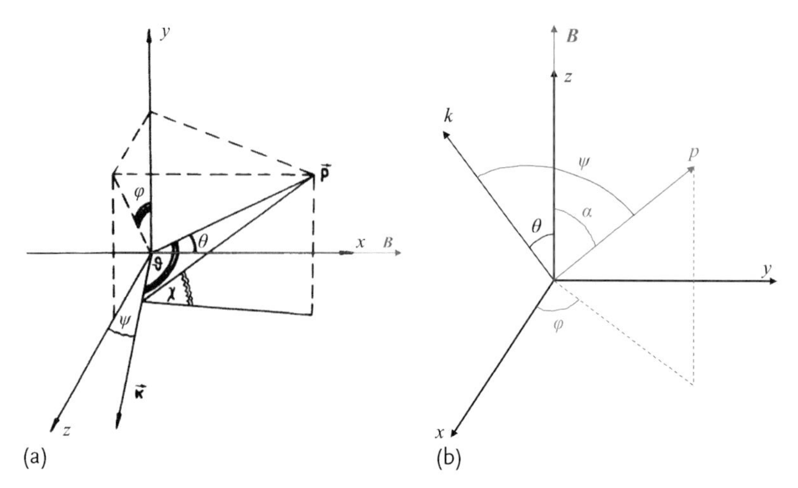

Figure 7.1 Geometry of observations of HXR emission from a flare under viewing angle ψ with respect to electron and photon momenta and magnetic field direction: (a) geometry considered by Nocera *et al.* (1985) and Zharkova *et al.* (1995); (b) geometry accepted by Syniavskii and Zharkova (1994) and Zharkova *et al.* (2010). See text for more details. For a color version of this figure, please see the color plates at the beginning of the book.

In the geometry used by Nocera *et al.* (1985) the magnetic field $\boldsymbol{H}$ is directed along the x-axis, the photon vector $\boldsymbol{K}$ is located in the x–z plane, that is perpendicular to magnetic field direction, while in the geometry by Elwert and Haug (1970) the observer looks at the loop from the direction of magnetic field (z-direction). The viewing angle ψ is the angle between the $\boldsymbol{K}$ vector and the z-axis, θ is the pitch angle of electron vector $\boldsymbol{P}$, and ϕ is the azimuthal angle of electrons.

Then the unit vectors are defined as follows:

$$\hat{H} = \frac{\boldsymbol{H}}{|\boldsymbol{H}|} = (1, 0, 0) \; ; \tag{7.5}$$

$$\hat{K} = \frac{\boldsymbol{K}}{|\boldsymbol{K}|} = (\sin\psi, 0, \cos\psi) \; ; \tag{7.6}$$

$$\hat{P} = \frac{\boldsymbol{P}}{|\boldsymbol{P}|} = (\cos\theta, \sin\theta\cos\phi, \sin\theta\sin\phi) \; , \tag{7.7}$$

where the vector components in Cartesian coordinates (x, y, z) are presented in brackets.

Then the angle ϑ between vectors $\hat{\boldsymbol{P}}$ and $\hat{\boldsymbol{K}}$ is

$$\cos\vartheta = \hat{\boldsymbol{P}} \cdot \hat{\boldsymbol{K}} = \cos\theta\sin\psi + \sin\theta\cos\psi\sin\phi \; . \tag{7.8}$$

The cross-sections of bremsstrahlung emission polarized in directions perpendicular and parallel to the $(\boldsymbol{H}, \boldsymbol{K})$ plane can be defined as

$$\sigma_{\perp}^{\mathrm{H}} = \sigma_0 C \left\{ A + B \frac{[\hat{P} \cdot (\hat{H} \times \hat{K})]^2}{(\hat{H} \times \hat{K})^2} \right\} ,$$

$$\sigma_{\parallel}^{\mathrm{H}} = \sigma_0 C \left\{ A + B \frac{[(\hat{P} \times \hat{K}) \cdot (\hat{H} \times \hat{K})]^2}{(\hat{H} \times \hat{K})^2} \right\} .$$

Using definitions (7.5)–(7.7) one can derive the Cartesian components of the products above as follows:

$$\hat{H} \times \hat{K} = (0, -\cos \psi, 0) , \quad (\hat{H} \times \hat{K})^2 = \cos^2 \psi ; \tag{7.9}$$

$$\hat{P} \cdot (\hat{H} \times \hat{K}) = -\sin \theta \cos \phi \cos \psi , \quad [\hat{P} \cdot (\hat{H} \times \hat{K})]^2 = \sin^2 \theta \cos^2 \phi \cos^2 \psi , \tag{7.10}$$

so that the cross-section in the perpendicular direction $\sigma_{\perp}^{\mathrm{H}}$ is converted to

$$\sigma_{\perp}^{\mathrm{H}} = \sigma_0 C [A + B (\sin \theta \cos \phi)^2] . \tag{7.11}$$

Similarly

$$\hat{P} \times \hat{K} = (\sin \theta \cos \phi \cos \psi, \sin \theta \sin \phi \sin \psi - \cos \theta \cos \psi , \\ - \sin \theta \cos \phi \sin \psi) , \tag{7.12}$$

$$(\hat{P} \times \hat{K}) \cdot (\hat{H} \times \hat{K}) = -(\sin \theta \sin \phi \sin \psi - \cos \theta \cos \psi \cos \psi) , \tag{7.13}$$

so that the cross-section in the parallel direction $\sigma_{\parallel}^{\mathrm{H}}$ is converted to

$$\sigma_{\parallel}^{\mathrm{H}} = \sigma_0 C [A + B (\sin \theta \sin \phi \sin \psi - \cos \theta \cos \psi)^2] . \tag{7.14}$$

The elementary cross-section σ_0 is defined as follows:

$$\sigma_0 = \frac{1}{137} \frac{1}{2\pi} \frac{mc^2}{E_{\mathrm{low}}^2} r_0^2 ,$$

with r_0 being the classical electron radius.

7.2.2 Nonrelativistic HXR Cross-Sections

First let us consider a very simplified case where cross-sections are not dependent on pitch angles, as is done in many papers (see, e.g., Kontar *et al.*, 2005). As usual, we assume a stationary injection of beam electrons into a flaring atmosphere and use the solutions of a stationary Fokker–Planck equation for precipitation of power-law beam electrons with a lower energy cutoff of 12 keV and upper energy cutoff of 386 keV (Zharkova and Gordovskyy, 2005a) as considered in the first part of Chapter 4.

During precipitation beam electrons are scattered on ambient particles and produce HXR bremsstrahlung emission, which is defined by electron distribution functions and their angle-independent cross-sections as follows:

$$I(\varepsilon) = 2\pi A_x K \int_{\xi_{\min}}^{\xi_{\max}} \int_{\varepsilon}^{\infty} \int_{-1}^{1} f(\xi, \eta, \mu) \eta \sigma^{\mathrm{H}}(\eta, \varepsilon) d\mu \, d\eta \, d\xi \,, \tag{7.15}$$

where x is the ionization degree and constant A_x is defined as follows:

$$A_x = \frac{S}{4\pi R^2} \frac{2E_0}{m_e} \,,$$

where S is the flare area and R is the distance to the observer. HXR bremsstrahlung differential cross-sections σ^{H} are defined by those parallel or perpendicular to the direction of the magnetic field as follows:

$$\sigma^{\mathrm{H}}(\varepsilon, \psi) = \frac{d^2\sigma_{\|}}{d\Omega \, d(h\nu)} + \frac{d^2\sigma_{\perp}}{d\Omega \, d(h\nu)} \,, \tag{7.16}$$

where HXR bremsstrahlung differential cross-sections σ^{H} are defined by those parallel or perpendicular to the direction of the magnetic field described below. For isotropic HXR bremsstrahlung emission the differential cross-sections of electron scattering in directions parallel and perpendicular to the magnetic field, $\sigma^{\mathrm{H}}_{\|}$ and $\sigma^{\mathrm{H}}_{\perp}$, averaged over pitch angles, can be calculated as follows (Elwert and Haug, 1970):

$$\begin{aligned} \frac{d^2\sigma_{\|}}{d\Omega \, d(h\nu)} &= C[A + B(1 - \mu^2)]\sigma_0 \,, \\ \frac{d^2\sigma_{\perp}}{d\Omega \, d(h\nu)} &= C A \sigma_0 \,, \end{aligned} \tag{7.17}$$

where μ is the cosine of a viewing angle accepted as a pitch angle θ. Parameters A, B, and C are defined by the following equations:

$$\begin{aligned} A &= \frac{\eta - \varepsilon/2}{\eta(\eta - \varepsilon)} \ln \frac{\sqrt{\eta} + \sqrt{\eta - \varepsilon}}{\sqrt{\eta} - \sqrt{\eta - \varepsilon}} - 1 \,, \\ B &= \frac{\frac{3}{2}\varepsilon - \eta}{\eta(\eta - \varepsilon)} \ln \frac{\sqrt{\eta} + \sqrt{\eta - \varepsilon}}{\sqrt{\eta} - \sqrt{\eta - \varepsilon}} + 3 \,, \\ C &= \frac{1}{\eta\varepsilon} \frac{1 - \exp\left(-\frac{2\pi\alpha c\sqrt{m_e}}{\sqrt{2\eta}\, E_{\mathrm{low}}}\right)}{1 - \exp\left(-\frac{2\pi\alpha c\sqrt{m_e}}{\sqrt{2(\eta - \varepsilon)}\, E_{\mathrm{low}}}\right)} \,, \end{aligned}$$

where ε is the dimensionless photon energy $h\nu/E_{\mathrm{low}}$.

It must be noted that the polar diagrams for such nonrelativistic cross-sections have the strongest HXR emission in the direction perpendicular to electron motion (Zharkova *et al.*, 1996), although for lower-energy electrons HXR emission becomes increasingly directed at $\pm 45°$ toward the electron motion vector.

7.2.3
Relativistic Angle-Dependent Cross-Sections

The further improvement in the calculation of HXR emission can be achieved by considering the angle dependent cross-section for relativistic electrons. In this case we extend the upper cutoff energy of beam electrons up to 1.2 MeV, leaving the lower cutoff energy the same (12 keV). Then the parameters σ_I and σ_Q in Eqs. (7.1)–(7.2) can be expressed using the following bremsstrahlung cross-sections (Bai and Ramaty, 1978; Leach and Petrosian, 1983):

$$\sigma_I(h\nu, \theta, E, \mu, \varphi) = \frac{d^2\sigma_\perp}{d(h\nu)d\Omega} + \frac{d^2\sigma_\parallel}{d(h\nu)d\Omega} , \tag{7.18}$$

$$\sigma_Q(h\nu, \theta, E, \mu, \varphi) = \left[\frac{d^2\sigma_\perp}{d(h\nu)d\Omega} - \frac{d^2\sigma_\parallel}{d(h\nu)d\Omega}\right] \cos 2\chi , \tag{7.19}$$

where $d^2\sigma_\parallel/[d(h\nu)d\Omega]$ and $d^2\sigma_\perp/[d(h\nu)d\Omega]$ are respectively the bremsstrahlung cross-sections for photons with polarization parallel and perpendicular to the emission plane, that is, to the plane given by vector $\boldsymbol{P}$ of the electron impulse and wave vector $\boldsymbol{K}$ of the X-ray emission. The Stokes parameter Q is usually calculated with respect to the normal plane, that is, to the plane given by the magnetic field and the wave vector of X-rays. The parameter χ is the angle between the normal and emission planes:

$$\cos 2\chi = 1 - \frac{2\sin^2\alpha \sin^2\varphi}{\sin^2\psi} ,$$

where ψ is the angle between the vector of electron impulse and the wave vector of X-rays; this angle is determined by the equation

$$\cos\psi = \cos\alpha\cos\theta + \sin\alpha\cos\varphi\sin\theta . \tag{7.20}$$

The expressions for bremsstrahlung cross-sections for relativistic electron velocities are given in the articles by Gluckstern and Hull (1953a) and Bai and Ramaty

(1978):

$$\begin{aligned}
\frac{d^2\sigma_{\parallel}}{d(h\nu)d\Omega} = & \frac{Z^2}{8\pi}\frac{r_0^2}{137}\frac{p_1}{p_0}\frac{Q_E}{h\nu}\left\{\frac{8\sin^2\psi\,(2\Gamma_0^2+1)}{p_0^2\Delta^4}\right. \\
& - \frac{(5\Gamma_0^2+2\Gamma_0\Gamma_1+5)}{p_0^2\Delta^2} - \frac{(p_0^2-k^2)}{T^2\Delta^2} + \frac{2(\Gamma_0+\Gamma_1)}{p_0^2\Delta} \\
& + \frac{L}{p_0 p_1}\left[\frac{4\Gamma_0\sin^2\psi\,(3k-p_0^2\Gamma_1)}{p_0^2\Delta^4}\right. \\
& + \frac{2\Gamma_0^2(\Gamma_0^2+\Gamma_1^2)-(9\Gamma_0^2-4\Gamma_0\Gamma_1+\Gamma_1^2)+2}{p_0^2\Delta^2} \\
& \left. + \frac{k\,(\Gamma_0^2+\Gamma_0\Gamma_1)}{p_0^2\Delta}\right] \\
& + \frac{\varepsilon^T}{p_1 T}\left[\frac{4}{\Delta^2}-\frac{7k}{\Delta}-\frac{k\,(p_0^2-k^2)}{T^2\Delta}-4\right]-\frac{4\varepsilon}{p_1\Delta} \\
& + \frac{1}{p_0^2\sin^2\psi}\left[\frac{2L}{p_0p_1}\left(2\Gamma_0^2-\Gamma_0\Gamma_1-1-\frac{k}{\Delta}\right)\right. \\
& \left.\left. - \frac{4\varepsilon^T(\Delta-\Gamma_1)^2}{p_1 T} - \frac{2\varepsilon(\Delta-\Gamma_1)}{p_1}\right]\right\}, \qquad (7.21)
\end{aligned}$$

$$\begin{aligned}
\frac{d^2\sigma_{\perp}}{d(h\nu)d\Omega} = & \frac{Z^2}{8\pi}\frac{r_0^2}{137}\frac{p_1}{p_0}\frac{Q_E}{h\nu}\left\{\frac{-(5\Gamma_0^2+2\Gamma_0\Gamma_1+1)}{p_0^2\Delta^2} - \frac{(p_0^2-k^2)}{T^2\Delta^2}\right. \\
& - \frac{2k}{p_0^2\Delta} + \frac{L}{p_0p_1}\left[\frac{2\Gamma_0^2(\Gamma_0^2+\Gamma_1^2)-(5\Gamma_0^2-2\Gamma_0\Gamma_1+\Gamma_1^2)}{p_0^2\Delta^2}\right. \\
& \left. + \frac{k\,(\Gamma_0^2+\Gamma_0\Gamma_1-2)}{p_0^2\Delta}\right] + \frac{\varepsilon^T}{p_1T}\left[\frac{k}{\Delta}-\frac{k\,(p_0^2-k^2)}{T^2\Delta}+4\right] \\
& - \frac{1}{p_0^2\sin^2\psi}\left[\frac{2L}{p_0p_1}\left(2\Gamma_0^2-\Gamma_0\Gamma_1-1-\frac{k}{\Delta}\right)\right. \\
& \left.\left. - \frac{4\varepsilon^T(\Delta-\Gamma_1)^2}{p_1T} - \frac{2\varepsilon(\Delta-\Gamma_1)}{p_1}\right]\right\}. \qquad (7.22)
\end{aligned}$$

In the above formulas, Z is the ion charge, r_0 is the classical electron radius, and k is the normalized photon impulse:

$$k = \frac{h\nu}{m_e c^2},$$

Γ_0 and Γ_1 are respectively the normalized electron energies before and after scattering:

$$\Gamma_0 = \frac{E}{m_e c^2} + 1, \quad \Gamma_1 = \Gamma_0 - k,$$

p_0 and p_1 are respectively the normalized electron impulses before and after scattering:

$$p_0 = \sqrt{\Gamma_0^2 - 1}\,, \quad p_1 = \sqrt{\Gamma_1^2 - 1}\,,$$

and the angle ψ is determined by Eq. (7.20). The remaining parameters in Eqs. (7.21)–(7.22) are defined as

$$T = |\boldsymbol{p}_0 - \boldsymbol{k}| = \sqrt{p_0^2 + k^2 - 2p_0 k \cos\psi}\,,$$

$$L = \ln\frac{\Gamma_0\Gamma_1 - 1 + p_0 p_1}{\Gamma_0\Gamma_1 - 1 - p_0 p_1}\,,$$

$$\varepsilon = \ln\frac{\Gamma_1 + p_1}{\Gamma_1 - p_1}\,, \quad \varepsilon_T = \ln\frac{T + p_1}{T - p_1}\,,$$

$$\Delta = \Gamma_0 - p_0\cos\psi\,.$$

The factor Q_E is the Coulomb correction (Koch and Motz, 1959):

$$Q_E = \frac{\beta_0}{\beta_1}\frac{1 - e^{-(2\pi Z/137)/\beta_0}}{1 - e^{-(2\pi Z/137)/\beta_1}}\,,$$

where

$$\beta_0 = \frac{p_0}{\Gamma_0}\,, \quad \beta_1 = \frac{p_1}{\Gamma_1}\,.$$

7.3 Simulation Results

7.3.1 Time-Dependent Hard X-Ray Photon Spectra for a Short Impulse

Observations at the highest temporal resolution of HXR emission show that short millisecond spikes of HXR intensity increases are likely to be produced by an injection of very short impulses of electron beams (Charikov *et al.*, 1996, 2004; Kiplinger *et al.*, 1983). We simulated with the Fokker–Planck approach the precipitation of beam electrons with an energy range between 12 keV and 1.2 MeV injected within a very short time scale into a flaring atmosphere and then calculated their bremsstrahlung HXR emission at various depths and integrated them in depth for various times shown in Figure 7.2.

The time scale of an HXR impulse, δt_{HXR}, is accepted to be about 20 ms, which is determined by the relaxation time of the atmosphere $t_{\mathrm{r}} \approx 70$ ms established earlier in Chapter 4. The bremsstrahlung cross-sections are taken in a relativistic form (see Gluckstern and Hull, 1953a). The time profile of HXR intensity shown in Figure 7.2a is produced by an impulse with length $\delta t_{\mathrm{e}} = 1.7$ ms. Simulations

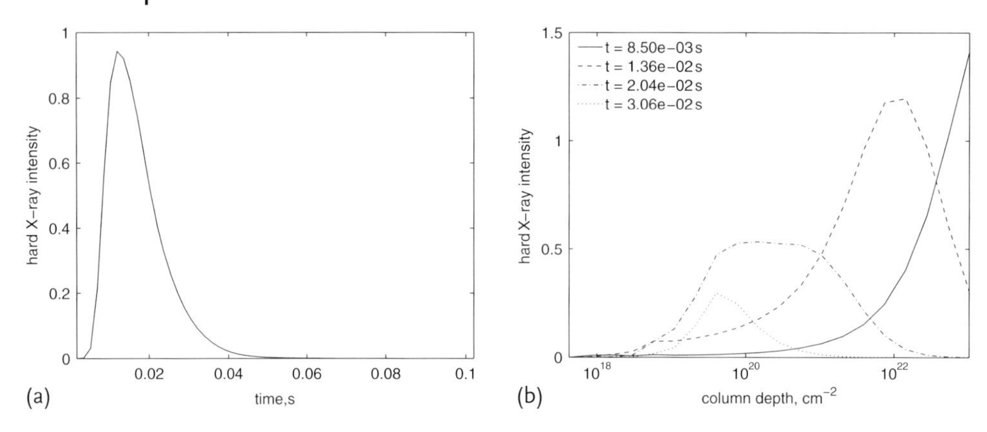

Figure 7.2 Intensity of HXR emission burst (in arbitrary units): (a) temporal profile integrated over precipitation depths and (b) temporal evolution of HXR impulse at various precipitation depths. Beam parameters: initial energy flux 10^{11} erg cm^{-2} s^{-1} and spectral index 3.

show that as long as $\delta t_e \ll \delta t_{\text{HXR}}$, the HXR time scale depends only on the atmosphere parameters and not on the length of the initial electron impulse (Siversky and Zharkova, 2009b).

The evolution of the spatial profiles of HXR intensity (Figure 7.2b) resembles the evolution of the heating function by electrons precipitating in collisions with ambient particles into a flux tube without a strong magnetic field convergence, as presented in Chapter 4. The HXR spike emission first appears in the photosphere from energy deposition by beam electrons with the highest energy then moving quickly to the upper atmospheric levels where electrons with the lowest energy are fully thermalized.

The emission appears first at the bottom when high-energy electrons reach this depth and gradually move upward, increasing in intensity when lower-energy electrons arrive at the upper chromosphere and start to thermalize. After reaching a depth of $\sim 2 \times 10^{19}$ cm^{-2} (the stopping depths of electrons with a lower cutoff energy), the intensity of the emission decreases and, finally, the emission vanishes.

The temporal profile reflects this spatial evolution in an asymmetric bell-type shape, slightly extended after the maximum owing to collisional depletion caused by lower-energy electrons. Ohmic losses are found not to have sufficient time to be established, as discussed in Chapter 4. Therefore, short impulses of HXR emission can be good probes for tracing scenarios of electron precipitation into flaring atmospheres in collisional energy losses.

7.3.2
HXR Emission with Nonrelativistic Cross-Sections for Steady Injection

Very often HXR emission observed at the temporal resolution of HXR observations is about 1–2 s (Lin *et al.*, 2003) and can last for a long time, up to 1 h, indicating that beams are being steadily injected with much longer time scales. Therefore, more

Table 7.1 Photon spectral indices δ_{low} and δ_{high} vs. electron-beam spectral indices γ.

Flux, erg cm^{-2} s^{-1}	10^8		10^{10}		10^{12}	
γ			$\delta_{low}/\delta_{high}$			
3	2.3	2.6	1.7	2.8	1.2	3.0
5	4.1	4.4	2.9	4.8	1.9	5.1
7	6.3	6.5	4.4	6.9	3.2	7.0

it is appropriate to consider HXR emission produced by beam electrons with relativistic cross-sections in the downward and upward directions since in Chapter 4 we showed that a self-induced electric field forms these two streams of electrons that exist as long as the injection continues.

Now let us consider steady electron-beam precipitation affected by both Coulomb collisions and self-induced electric fields. The distribution functions for this case are calculated in the first part of Chapter 4. Using these functions and Eqs. (7.2) and (7.17) we calculated the bremsstrahlung photon spectra for different initial energy fluxes and different indices of the initial beam energy spectra (Figure 7.3) where the indices of lower- and higher-energy parts were found to be different. These indices of the low- and high-energy parts of the resulting HXR spectra are presented in Table 7.1. Note that there is an HXR spectrum rollover at higher photon energies related to the accepted electron's upper cutoff energy of 384 keV, which can be remedied by expanding the upper cutoff energy to 1.2 meV, as considered later in this chapter.

The spectra corresponding to beams injected with low initial energy fluxes (10^8 erg cm^{-2} s^{-1}) reveal nearly single power-law shapes with the indices being close to those following from the pure collisional precipitation model, that is, $\delta \approx \delta_{coll} = \gamma - 1$. However, photon spectra corresponding to stronger beams ($F_0 = 10^{10}$ or 10^{12} erg cm^{-2} s^{-1}) reveal double power-law (elbow-type) spectra with spectral indices for lower energies, λ_{low}, being rather different from those for higher ones, λ_{high}. It can also be seen that the high-energy parts of these spectra ($E > 50$–70 keV) reveal their indices as being close to those following from the pure electric field model of electron precipitation, that is, $\delta_{high} \approx \delta_{elec} = \gamma$ (Chapter 3), while a lower-energy part of the spectra is much harder. The photon indices in the low-energy part (δ_{low}) are dependent on the electron beam spectral indices γ and the beam initial flux, that is, the bigger the initial flux, the lower the photon index δ_{low} (Figure 7.4 and Table 7.1). Therefore, one can conclude that increases of the initial energy flux of an injected beam leads to substantial hardening of the lower-energy part of the photon spectra, while the higher-energy part becomes slightly softer (in comparison with spectra corresponding to the pure collisional model).

Although spectral indices vary smoothly with energy, it is possible to determine the energy that separates the softer high-energy part and the harder low-energy part, or the "break energy". This energy also appears to be noticeably dependent

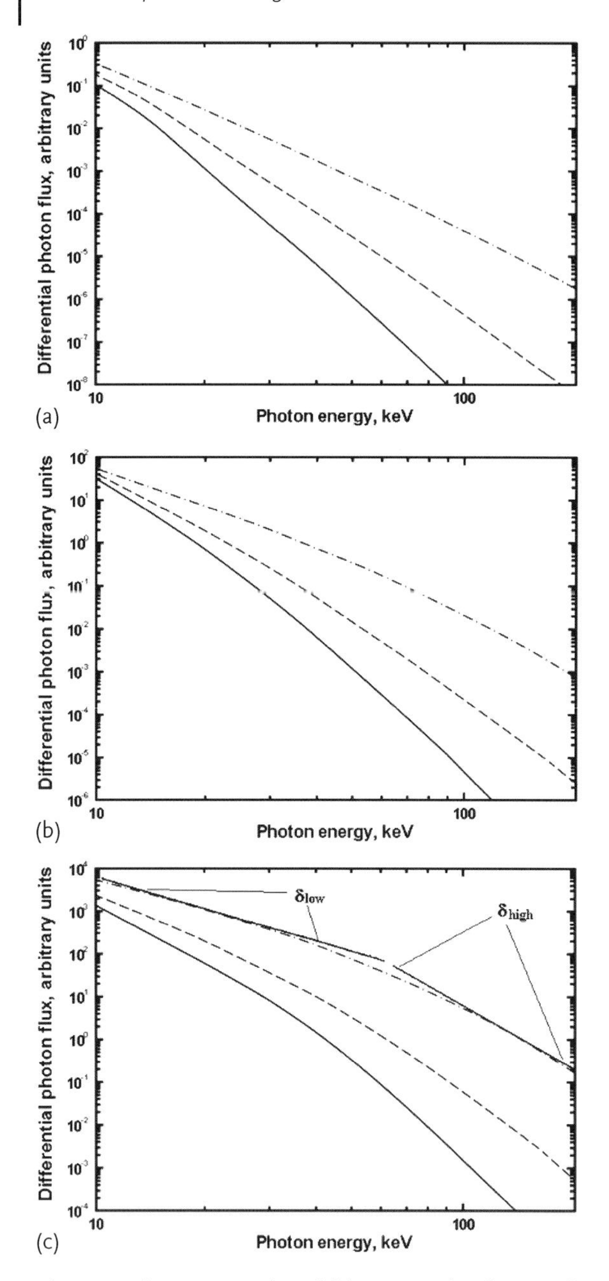

Figure 7.3 Photon spectra from full kinetic steady solutions of beams with initial energy fluxes of 10^8 erg cm^{-2} s^{-1} (a) and 10^{12} erg cm^{-2} s^{-1} (b).

on the beam initial energy flux as well as slightly on the beam initial spectral index. Thus, for initial fluxes of 10^{10} erg cm^{-2} s^{-1} this energy is about 25–30 keV for $\gamma = 3$ and increases to 30–40 keV for $\gamma = 7$, while for an initial flux of

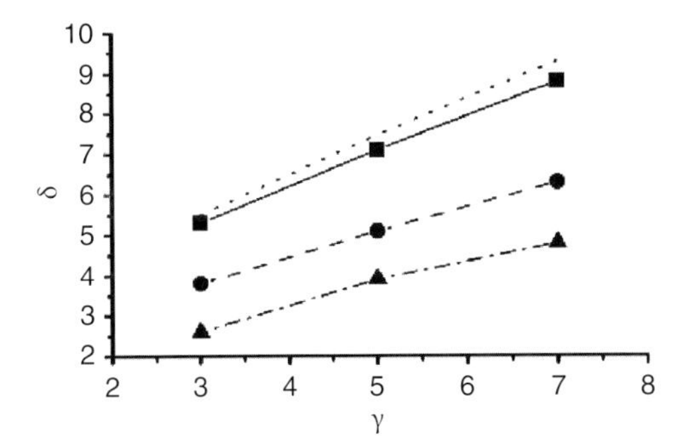

Figure 7.4 Photon spectral indices δ_{high} (solid line) and δ_{low} (dashed and dotted–dashed lines) vs. electron indices γ for initial energy fluxes of 10^8 (squares), 10^{10} (circles), and 10^{12} erg cm^{-2} s^{-1} (crosses).

10^{12} erg cm^{-2} s^{-1} this energy reaches 55–60 and 60–70 keV for $\gamma = 3$ and $\gamma = 7$, respectively. An analysis of the mean electron flux spectra simulated in Section 3.2 shows that these break energies coincide with the energies of the "bump" in the distribution functions appearing as a result of electrons decelerated by a self-induced electric field and returning to the source. Apparently, the deficit of electrons with energies lower than the energy of this bump leads to a significant hardening of the lower-energy photon spectra.

Therefore, one can conclude that both the hardening of a lower-energy part of the spectra and the softening of a higher-energy part are related to the effect of a self-induced electric field (Chapters 3 and 4). This naturally explains why a low-energy part of photon spectra becomes harder with an increase of the beam initial energy flux and the beam initial spectral index: they both lead to an increase of the initial magnitude of the electric current carried by precipitating electrons.

As a result, the photon spectra calculated using the numerical simulations for strong beams ($F_0 \succeq 10^{10}$ erg cm^{-2} s^{-1}) presented in Figure 7.5 are in good agreement with the bremsstrahlung photon spectra observed by RHESSI in solar flares (see Figure 7.6, courtesy of Sui *et al.*, 2007a).

Therefore, we have grounds for suggesting that the broken power-law HXR spectra (shown in Figure 7.6) observed in solar flares (Section 1.3) are produced by the precipitation of strong beams in a flaring atmosphere. Furthermore, this can explain the variations of HXR spectral indices during the flare. Many observers show that, typically, the low-energy part of bremsstrahlung spectra becomes harder in the rising phase of a flare and then becomes softer with flare decay (see, e.g., Fletcher and Hudson, 2002; Grigis and Benz, 2005). Naturally, this can be explained by the increase of the beam energy flux during the rising phase followed by its decrease during the decay.

Thus we have analyzed spectral observations of the 23 July 2002 flares done with RHESSI. The spectra for this flare were produced by Holman *et al.* (2003) and clearly reveal a low-energy flattening that appears during the rising phase of the flares (Figure 7.7).

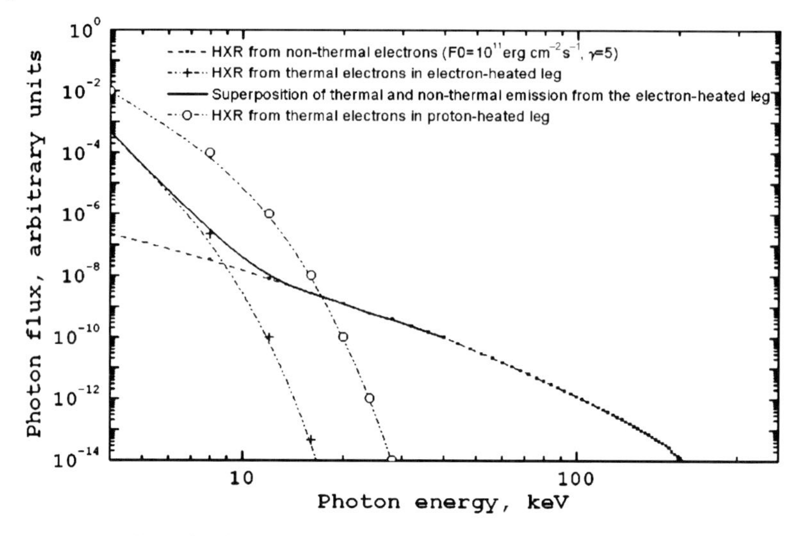

Figure 7.5 Thermal and nonthermal HXR spectra for separate precipitations of proton and electron beams with the parameters used in experiments 3 and 4 (Chapter 4).

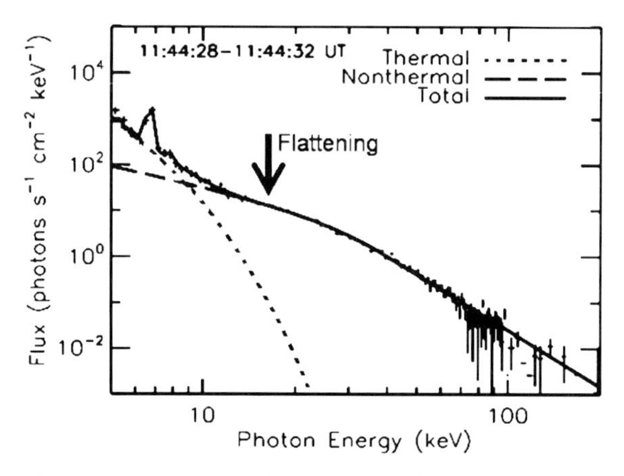

Figure 7.6 RHESSI HXR photon spectrum for 2 June 2002 flare. Thermal and nonthermal components are separated under the assumption that the thermal component corresponds to a temperature of 17 MK. Courtesy of Sui *et al.* (2007a).

The high-energy spectral indices δ_{high} were found to be about 6.2 for both times (00:23:44 and 00:25:32 UT), corresponding to an initial beam spectral index γ of about 6.8. The observed low-energy indices δ_{low} were 5.2 at the first time and 4.8 at the second time. According to the results obtained above, these photon spectral indices can be produced by a single electron beam with an initial energy flux of $\sim 1 \times 10^9$ at 00:23:44 UT to $\sim 4 \times 10^9\,\text{erg}\,\text{cm}^{-2}\,\text{s}^{-1}$ at 00:25:32 UT. The break energy in both observed spectra was about 30–35 keV, which is also in good agreement

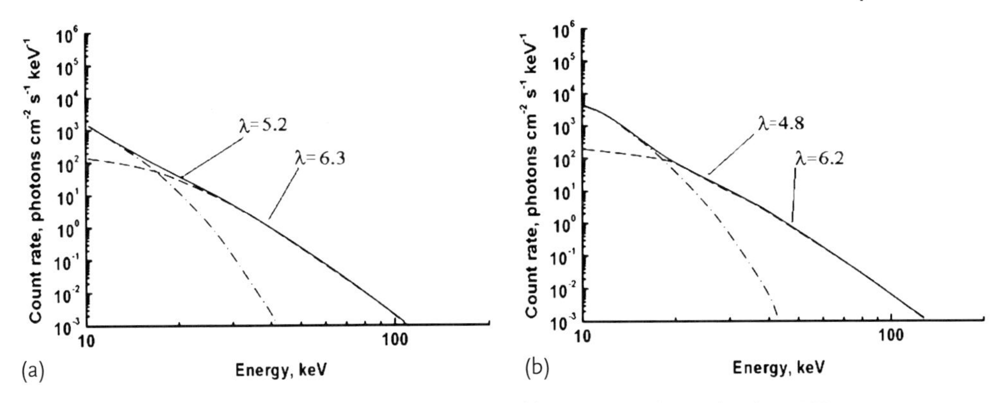

Figure 7.7 Low-energy flattening in nonthermal bremsstrahlung spectra observed with RHESSI in flares of 23 July 2002: (a) 00:23:44 UT; (b) 00:25:32 UT. Courtesy of Holman *et al.* (2003).

with those in the simulated spectra for a single precipitating beam with the given parameters.

7.3.3
HXR Emission with Relativistic Cross-Sections for Steady Injection

Observations in some flares of HXR emission of up to a few tens of MeV suggest that the HXR is produced by bremsstrahlung emission of beam electrons scattering on the ambient particles. This indicates that beam electrons can have energies well above 1 MeV, which requires one to consider relativistic cross-sections for HXR emissions. Also, this approach will allow us to correct the HXR photon spectrum rollover at the higher photon energies found above for nonrelativistic cross-sections and to calculate the directivity and polarization of HXR emission seen from different viewing angles, or different locations of a flare on the solar disk.

In this approach we will use the time-dependent solutions of the Fokker–Planck equation after they have reached the setady state, as discussed in Chapter 4. In order to distinguish the effects of various energy loss mechanisms, let us simulate HXR emission produced by the same beam electrons for specific models of energy losses: collisions (C), collisions and electric field (CE), collisions and converging magnetic field (CB), and collisions, electric field, and converging magnetic field (CEB).

7.3.3.1 Depth Variations of HXR Bremsstrahlung Emission

Let us first compare HXR bremsstrahlung intensity and polarization produced at various atmospheric depths by beam electrons plotted in Figure 7.8 for power-law spectral indices of $\gamma = 3$ (red lines) and $\gamma = 7$ (blue lines). Figure 7.8a–f represents the X-ray photon spectra for the different simulation models and propagation directions, for three distances from the electron injection point. Note that the curves for different electron power-law indices are drawn using a different scaling factor because their intensities differ significantly. Figure 7.8g–l shows the polar-

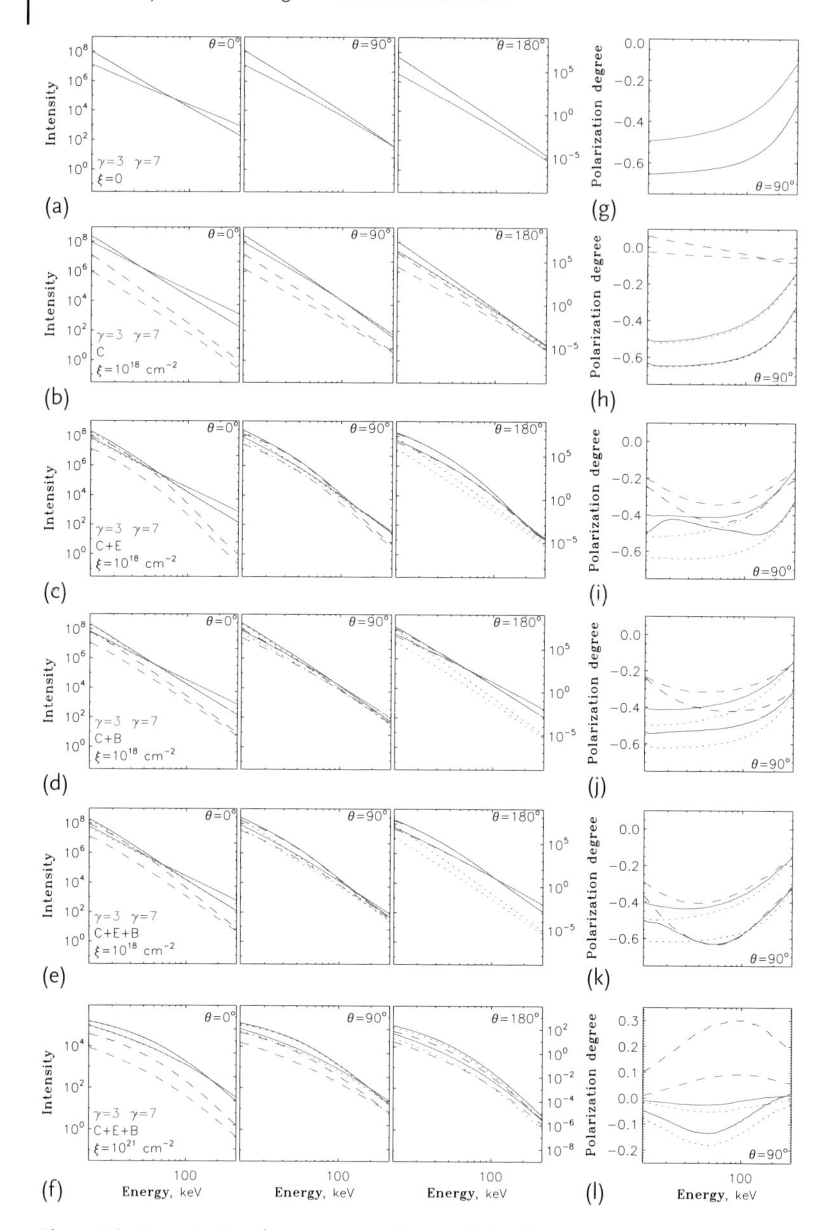

Figure 7.8 Intensity (in relative units) and polarization of HXR emission for different energies. The initial power-law indices of the accelerated particles are $\gamma = 3$ (red lines) and $\gamma = 7$ (blue lines). The intensity plots for $\gamma = 3$ and $\gamma = 7$ are drawn by using a different scaling; the intensity values are shown in the left (for $\gamma = 3$) and right (for $\gamma = 7$) margins, respectively. Dotted line: emission of downward propagating particles (with $\mu > 0$); dashed line: emission of upward propagating particles (with $\mu < 0$); solid line: total emission (downward + upward). Panels (a)–(f) correspond to the different depths (ξ) indicated in the legends. The factors taken into account are collisions (C), self-induced electric field (E), and convergence of the magnetic field (B). For a color version of this figure, please see the color plates at the beginning of the book.

ization of HXR emission calculated for a pitch angle of $\theta = 90°$, where it is the highest.

The spectrum of emission from an initial electron beam (at the injection point) with a power-law distribution at $\gamma = 3$ (Figure 7.8a,l) reveals that HXR emission at injection is emitted preferably in the direction of electron propagation ($\theta \simeq 0°$). However, with scattering processes of electrons increasing at every depth, electrons become scattered at different angles and can emit much more isotropically.

The intensity plots in Figure 7.8b–e illustrates the effects on emission of different factors of this scattering (collisions, self-induced electric field, converging magnetic field) at various distances from the injection point in the upper corona.

Figure 7.8b illustrates the effect on HXR emission of collisions with pitch-angle scattering by downward (dotted line) and upward (dashed line) moving particles. The effect of a self-induced electric field on the appearance of returning electrons (moving with $\mu < 0$) is demonstrated in Figure 7.8c. It is evident that the inclusion of an electric field considerably changes the resulting HXR emission spectra, with the effect being strongest at energies of < 60 keV but still noticeable at higher energies (up to 300 keV), even if a beam's upper energy cutoff is 1.2 MeV. The upward emission ($\theta = 180°$) is clearly defined by returning electrons, and the emission intensity (especially at energies of about 30–100 keV) is much higher than for cases of collisional losses with anisotropic scattering but without a return current.

As a result, at a depth of 10^{18} cm^{-2} the HXR emission spectrum becomes essentially different from a power-law one: the spectrum apparently flattens at lower energies while keeping the power-law shape at higher ones (Zharkova and Gordovskyy, 2006; Zharkova *et al.*, 2010). This leads to the case where the coronal emission from precipitating electrons becomes an order of magnitude higher than those in the chromosphere (compare Figure 7.8b,h and 7.8f,l) while the shape of the energy spectrum is not changed much with depth. The emission from the returning electrons (dashed line) in the chromosphere is much smaller than for the precipitating ones. Emission from returning electrons becomes closer to those from precipitating ones with an increase in the viewing angle.

HXR spectral flattening at low energies is also visible at a viewing angle of $\theta = 90°$, although it is much weaker. For the downward emission ($\theta = 0°$), the contribution of the downward moving electrons always far exceeds that of the upward moving electrons. Nevertheless, even in this case the returning electrons affect the HXR spectrum shape at energies below 50 keV.

The effect of a converging magnetic field is seen at its best in Figure 7.8d,j. The convergence effect is strongest at $\theta = 180°$, where the emission at higher energies is formed completely by the reflected electrons, and at lower energies (< 30 keV) the contribution of the reflected particles dominates. The emission intensity is about two times higher than in the case of pure collisions (Figure 7.8b,h), and the spectrum is harder. For transversal propagation ($\theta = 90°$), the contributions of the downward and reflected electrons are comparable, and the spectra of emission produced by these electron populations are nearly the same. For downward emission propagation ($\theta = 0°$), the contribution of the reflected electrons to X-ray emission is negligible.

The intensities and polarization calculated, including all the energy loss effects, for example collisions, electric field, and magnetic field convergence, are plotted in Figure 7.8e,k. One can see that the two factors that lead to the formation of the upward electron flux are now combined: returning electrons increase the upward emission intensity at lower energies (< 100 keV), while those reflected by the converging magnetic field increase the population of electrons in a high-energy tail. In fact, the converging magnetic field removes the spectral flattening at the lower energies, as was seen in Figure 7.8c,i (for a beam with $\gamma = 3$). The emission spectra at all viewing directions become similar to that in Figure 7.8d.

The emission spectra for deeper layers (in the chromosphere) are plotted in Figure 7.8f,l. Our calculations show that the spectrum shape is almost independent of the simulation model. This means that the effects of a converging magnetic field or a self-induced electric field become less significant compared to collisions, which dominate as the mechanism responsible for the HXR emission at this level. The effect of a converging magnetic field for softer beams with $\gamma = 7$ is similar to that for harder electron beams with $\gamma = 3$ (Figure 7.8d,j), both qualitatively and quantitatively. For the directions $\theta = 0°$ and $\theta = 90°$, the contribution of the emission produced by downward electrons dominates. For the direction $\theta = 180°$, the contributions of the emission produced either by downward or upward (scattered) electrons are almost equal.

The relative contributions of returning electrons to HXR emission are greater for softer beams (blue lines in Figure 7.8). In the model, taking into account the effects of collisions and a self-induced electric field (Figure 7.8c,i), the relative contribution of the returning electrons is more significant for the beam with $\gamma = 7$ than for the beam with $\gamma = 3$. This is seen especially well for the viewing angles $\theta = 90°$ (where the emission at 20–100 keV is produced mainly by the return current electrons) and $\theta = 0°$ (where contributions of the upward and downward electrons are nearly the same at 20–50 keV); at $\theta = 180°$, the emission from the return current electrons strongly dominates both for the hard and soft beams and the spectral flattening at low energies is observed.

The plots of HXR emission caused by electron beam scattering in the model with collisions + return current + magnetic field convergence (Figure 7.8e,k, blue lines) are similar to those calculated for the model with collisions + return current (Figure 7.8c,i). This points out that for softer beam ($\gamma = 7$), the effect of a self-induced electric field much exceeds all the other factors. In particular, the converging magnetic field does not remove the spectral flattening effect entirely (although it reduces the energy range where the flattening takes place). The HXR polarization will be significantly affected as well, as discussed in Section 7.3.4.

In deeper layers of a flaring atmosphere (Figure 7.8f,l), the emission parameters again become almost insensitive to the simulation model, because the parameters of the precipitating particles are determined mainly by collisions only. However, it should be noted that for the beam with $\gamma = 7$, only a small fraction of particles can reach the chromosphere. This results in much lower intensities of the HXR emission produced at larger precipitation depths by softer electrons than by harder ones (compare the magnitudes of HXR intensities in Figure 7.8a–f and 7.8g–l).

7.3.3.2 Integrated HXR Bremsstrahlung Emission

The HXR bremsstrahlung intensities calculated by integration in depths and pitch angles for relativistic electron cross-sections and for different positions of flaring atmospheres on the solar disk are plotted in Figure 7.9.

The total emission and polarization from the considered coronal magnetic tube with emission parameters integrated over all layers using Eq. (7.2) are shown in Figures 7.9 and 7.10. Similarly to Figure 7.8, the emission parameters for electron beams with different power-law indices in Figure 7.9 are shown in different colors (red for $\gamma = 3$ and blue for 7). Note that in Figure 7.9, the intensity plots for $\gamma = 3$ and $\gamma = 7$ are drawn using different scaling (shown on the left axis for $\gamma = 3$ and on the right axis for 7).

One can notice from Figure 7.9 that for a harder beam there is a strong dependence on a magnetic field convergence (compare models with a constant magnetic field – C, C+E – with those that take into account a magnetic field convergence – C+B, C+E+B). In a converging magnetic field, the emission intensity is lower than in a homogeneous magnetic field because magnetic mirroring reduces the number of electrons reaching the deepest layers of the solar atmosphere. For softer beams, such a dependence is less significant as the number of electrons reaching the deepest layers is small in any case due to collisional damping. A small spectral flattening at lower energies can be noticed, although it is less pronounced than for emission from individual layers in the corona (compare with Figure 7.8); the "break" energy – for the C+E+B model and for the upward emission – is found to be about 100 keV for beams with $\gamma = 3$ and about 50 keV for $\gamma = 7$. This confirms the analytical predictions for the simplified precipitation model produced by Zharkova and Gordovskyy (2006).

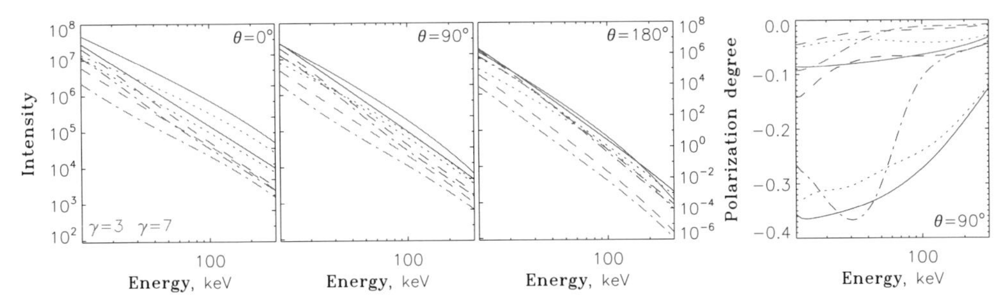

Figure 7.9 Integrated by column depth intensity (in relative units) and polarization of HXR emission for different energies. The initial power-law indices of the beam electrons are $\gamma = 3$ (red lines) and $\gamma = 7$ (blue lines). The intensity plots for $\gamma = 3$ and $\gamma = 7$ are drawn using different scaling: the intensity magnitudes are shown in the left (for $\gamma = 3$) and right (for $\gamma = 7$) margins, respectively. Different lines correspond to different simulation models: solid line – pure collisions (C), dotted line – collisions and return current (C+E), dashed line – collisions and converging magnetic field (C+B), dash-dotted line – all factors (C+E+B). For a color version of this figure, please see the color plates at the beginning of the book.

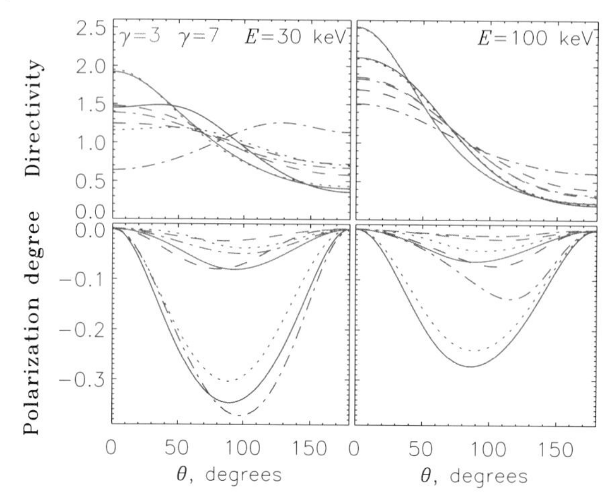

Figure 7.10 Directivity and polarization of HXR emission for different observation directions. The emission parameters are integrated over all layers of the coronal magnetic tube. The initial power-law indices of the accelerated particles are $\gamma = 3$ (red lines) and $\gamma = 7$ (blue lines). Different lines correspond to different simulation models (Figure 7.9). For a color version of this figure, please see the color plates at the beginning of the book.

7.3.4
HXR Bremsstrahlung Directivity and Polarization for a Steady Beam Injection

7.3.4.1 Depth Variations of HXR Bremsstrahlung Polarization

With respect to the depth variation of the electron distributions discussed in Chapter 4, the polarization degree viewed in the perpendicular direction (Figure 7.8g–l) reveals also a strong dependence on precipitation depth and beam parameters.

At the injection point (Figure 7.8a), the polarization degree steadily decreases (by an absolute value) with energy for both hard and soft beams. At an intermediate precipitation depth, for a harder beam ($\gamma = 3$, red lines) in a model with pure collisions (Figure 7.8b), the polarization degree has a maximum near the lowest photon energy (at about 15 keV) and decreases with energy increases.

For the same beam parameters ($\gamma = 3$) and at the same depth, the model with a magnetic field convergence (Figure 7.8d) shows an increase of the polarization degree for higher energies (the polarization degree decreases with energy slower than without the converging magnetic field); this is due to the reflected electrons. In contrast, the return current (Figure 7.8c) decreases the polarization degree at lower energies, so the hard electron beam produces emission with nearly constant ($\eta \simeq -0.4$) polarization in the energy range 12–100 keV. The combination of these two factors (Figure 7.8e) results in a weakly expressed polarization maximum that takes place at an energy of about 30 keV. At deeper layers (Figure 7.8f), the polarization degree decreases in all the models considered since the electron distribution becomes more isotropic, in general, owing to collisions and pitch-angle diffusion.

Polarization of emissions from beams with $\gamma = 7$ (blue lines) is higher than those from beams with $\gamma = 3$. Partially, this is caused by the different numbers of high-energy particles and by the properties of bremsstrahlung cross-sections. For initially injected beams (Figure 7.8a), the difference in polarizations is caused only by the two mentioned factors. Another reason is the increased degree of scattering of beam electrons for $\gamma = 7$, for example, a larger number of lower-energy electrons moves not downward but in the perpendicular direction where the polarization is presented. This, in turn, happens owing to a joint effect of collisions and ohmic losses (Zharkova and Gordovskyy, 2005a, 2006).

A converging magnetic field (Figure 7.8d) affects emissions from beams with $\gamma = 3$ and $\gamma = 7$ in the same way. The effect of a return current (Figure 7.8c) for soft beams is much more significant than for harder ones: the polarization degree is reduced significantly at the lower energies, so we can see the polarization maximum at about 150 keV. The combined effect of collisions, converging magnetic field, and return current (Figure 7.8e) produces the well-defined polarization maximum at 50–60 keV; the polarization degree at that point exceeds (by an absolute value) 60%. A similar polarization maximum, but shifted to lower energies (about 50 keV) and with lower magnitude, can be seen even at deeper layers (Figure 7.8f).

In general, for other viewing angles the depth variations of HXR polarization produced by a hard beam are rather complex (Figure 7.11). Note that the precipitation depth of emitting electrons is the key factor in defining a magnitude of polarization produced. The highest polarization degree (−46% at 30 keV and −37% at 100 keV, for beams with $\gamma = 3$) for an initially injected collimated beam is observed in the transversal ($\theta = 90°$) propagation direction. As the electron beam propagates downward, the polarization dependence on the propagation direction becomes asymmetric, although in general the polarization maximum does not deviate much from $\theta \simeq 90°$. The polarization values depend on the simulation model. For models with collisions and a return current (Figure 7.11a), at 30 keV the polarization degree steadily decreases with depth shifting the maximum to lower viewing angles. At 100 keV, the polarization degree first increases slightly and then decreases again.

A converging magnetic field increases the rate of electron isotropization, thus leading to a decrease in HXR polarization with depth (Figure 7.11b,c) happening much faster than in a homogeneous magnetic field. At the deepest layers, we can see small positive polarization values at propagation angles $> 90°$. This is the effect of reflected particles with large pitch angles.

Softer electron beams (Figure 7.11d–f) demonstrate a similar dependence of the HXR polarization on the depth and the simulation model. The polarization degree is higher now (up to −62% at 30 keV and −58% at 100 keV). As with harder beams, the evolution of the polarization degree with depth is governed mainly by the convergence of the magnetic field, but the effect of the return current is stronger now (compare Figure 7.11e,f).

It can be seen the HXR polarization calculated even with all energy losses above can still reach 60% at the right viewing angle (90°) (Figures 7.8 and 7.11). This helps one to understand that the position of a flaring atmosphere on the solar

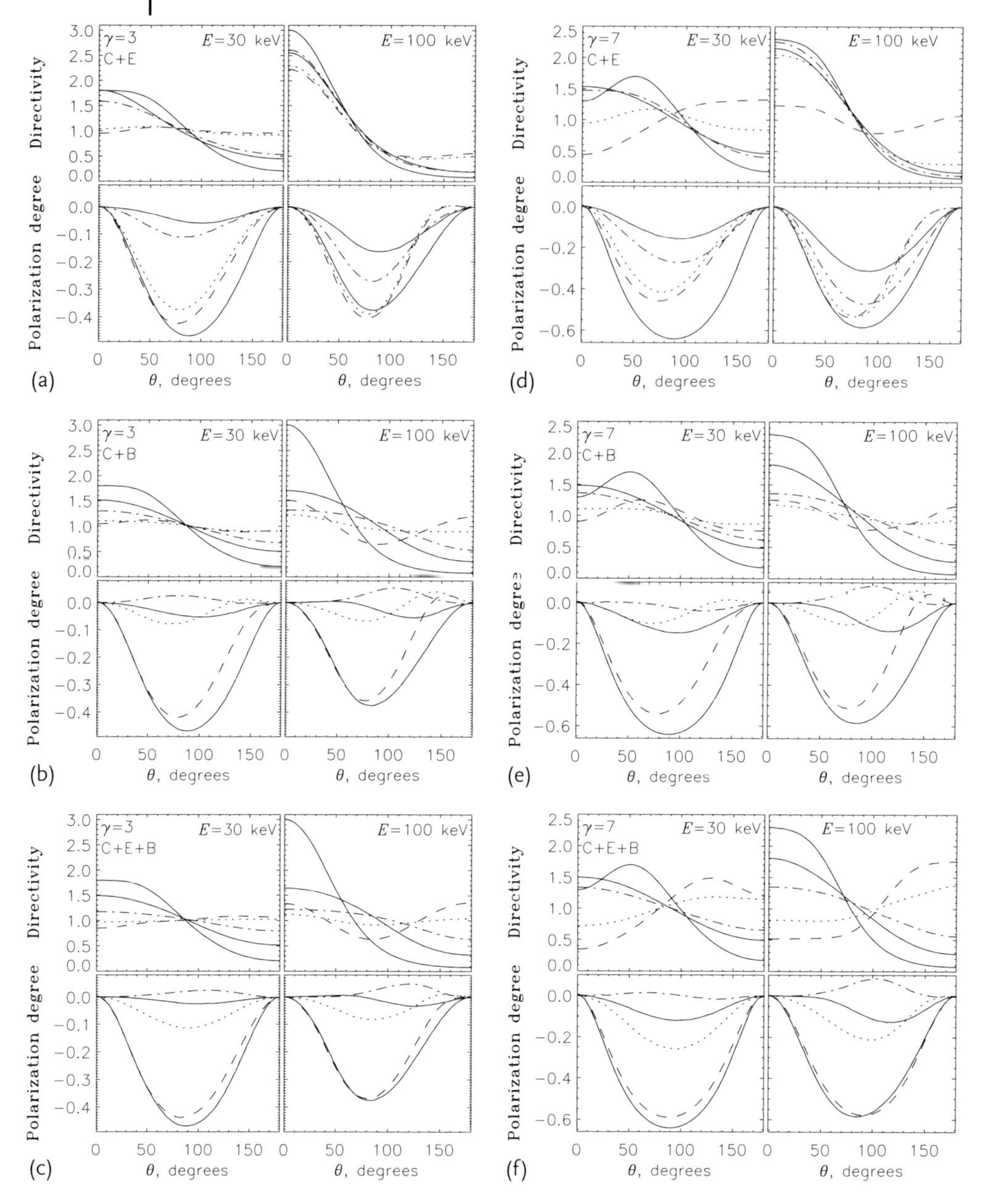

Figure 7.11 Directivity and polarization of HXR emission for different propagation directions. The electron distribution functions are obtained by taking into account different factors (Figure 4.15), the initial power-law indices of beam electrons are $\gamma = 3$ (a–c) and $\gamma = 7$ (d–f). Solid line: $\xi = 0$, dashed line: $xi = 10^{18}\ \mathrm{cm}^{-2}$, dotted line: $\xi = 10^{19}\ \mathrm{cm}^{-2}$, dashed–dotted line: $\xi = 10^{20}\ \mathrm{cm}^{-2}$, dashed–double dotted line: $\xi = 10^{21}\ \mathrm{cm}^{-2}$.

surface and a viewing angle from which it is observed can be the factors defining the total observed outcome. For example, this can explain the very high polarization degrees (up to 50%) reported for the 23 July 2002 flare (McConnell *et al.*, 2003), which was located very close to the limb and so seen from a viewing angle of close to 90° and at a small integration depth due to being seen from a side.

7.3.4.2 Depth Variation of HXR Bremsstrahlung Directivity

The variations with depth of the directivity of X-ray emission can be explored using details from Figure 7.11. First, we consider the case of $\gamma = 3$ (Figure 7.11a–c) where the directivity is highest. If one includes only the effects of collisions and a self-induced electric field (Figure 7.11a), then the changes of the X-ray directivity reflect the evolution of electron isotropization via collisions and return current. At the injection point ($\xi = 0$), the emission is highly directed and is radiated mainly downward. As long as the stream of returning electrons develops at deeper column depths ($\xi = 10^{18}$ and $10^{19}\,\text{cm}^{-2}$), the intensity of emission into the upward hemisphere (toward an observer) increases due to the increased number of particles moving upward (Chapter 4).

Thus, at lower energies (30–40 keV), the emission becomes almost isotropic. At deeper layers of the solar atmosphere ($\xi = 10^{20}$ and $10^{21}\,\text{cm}^{-2}$), the HXR photon flux decreases due to a decreased total number of electrons, which lose their energy in collisions. This results, in turn, in the density and the energies of the returning electrons decreasing considerably, and the X-ray emission (at 30 keV) becomes again directed mainly downward. At higher energies (> 100 keV), this decrease is much less noticeable because only particles with energies much higher than 100 keV can reach this energy at a given depth. These higher-energy electrons do not lose their energy fast enough because the ohmic losses become negligible owing to reduced electron numbers while collisional losses are reduced because of the ambient plasma's becoming partially ionized, which can precipitate to deeper atmospheric levels.

Figure 7.11b shows the HXR directivity and polarization for the effects of collisions and a converging magnetic field. One can see that at a lower energy (30 keV), the directivity pattern steadily becomes more isotropic when the electron beam propagates toward deeper atmospheric layers. At a higher energy (100 keV), the process of isotropization is slower. At a level of $\xi = 10^{18}\,\text{cm}^{-2}$, we can see the emission of two beams – direct and mirrored ones.

In Figure 7.11c, collisions, return current, and magnetic field convergence are taken into account. In this case the relative contribution of upward moving particles to emission increases compared with two previous cases. Now the upward emission can even dominate (at $\xi = 10^{18}$ and $10^{19}\,\text{cm}^{-2}$ at an energy of 30 keV, and at $\xi = 10^{18}\,\text{cm}^{-2}$ for an energy of 100 keV).

It was noted already in Sections 7.3.3.1 and 7.3.4.1 that the effects of a self-induced electric field become much stronger for softer electron beams (with $\gamma = 7$). This conclusion is also confirmed by the directivity plots (Figure 7.11d–f). In Figure 7.11d (collisions+return current), one can see that the X-rays are emitted mainly upward at $\xi = 10^{18}\,\mathrm{cm}^{-2}$ and 30 keV; the upward emission at the same depth at 100 keV is comparable to the downward one. The converging magnetic field (Figure 7.11e) has almost the same effect as for the harder beam. In Figure 7.11f (collisions+return current+magnetic field convergence), the upward emission strongly dominates at the $\xi = 10^{18}$ and $10^{19}\,\mathrm{cm}^{-2}$ levels for both considered energies. These effects need to be incorporated into the integrated polarization in order to be compared with those measured by RHESSI instruments.

7.3.4.3 Integrated HXR Bremsstrahlung Directivity and Polarization

The integrated HXR directivity and polarization are found to be even more sensitive to beam parameters and viewing angles (Figure 7.10). Although one can see that for harder beams, the electric field almost does not change the integrated directivity (in comparison with the purely collisional model) but reduces the integrated polarization at both energies considered. The converging magnetic field reduces the directivity and polarization significantly, and the results for the C+E+B model become similar to those for C+B.

For softer beams the effect of an electric field is much stronger. We can see that the combination of a converging magnetic field and return current changes the directivity at 30 keV, so the integrated emission is directed mainly upward. The integrated polarization in such a case can become higher than for the purely collisional model. At 100 keV, the downward emission always dominates, but the upward emission in C+E+B model is considerably stronger than in the other three models. It can be seen from Figures 7.9 and 7.10 that the maximal polarization degree (about −36%) is achieved for a beam with $\gamma = 7$, in a model with collisions, return current, and converging magnetic field, at an energy of about 30 keV and in the transversal propagation direction.

Such results are due to the fact that for a hard electron beam ($\gamma = 3$) of medium intensity ($F = 10^{10}\,\mathrm{erg\,cm^{-2}\,s^{-1}}$), the parameters of HXR emission are defined by the electrons, which do not induce a strong electric field and lose their energy mainly by collisions, that is, in other words, harder beams' X-ray emission is generated mainly in the deepest layers of the solar atmosphere (transition region and chromosphere). For electron beams with $\gamma = 7$, only a small fraction of particles can reach the chromosphere. As a result, the X-ray emission from such beams is generated mainly in the corona, where the effect of the self-induced electric field is very significant.

The current study has also established that the measurements of HXR integrated polarization combined with those of photon flux can derive much more precise electron beam parameters and give much deeper insight into the mechanisms of particle transport into deeper atmospheric levels than pure emission measurements.

7.4
Comparison with Observations

Let us test the simulation results by the available observations of HXR bremsstrahlung emission, polarization, and directivity.

7.4.1
HXR Bremsstrahlung Photon Spectra

Now let us compare the deduced photon energy spectra and figure out how they fit the theoretical calculations (Figure 7.12). It can be seen that the observed energy spectra for the flares of 20 and 23 July 2002 have well-defined double power-law energy spectra with the lower-energy part being flattened to much lower spectral

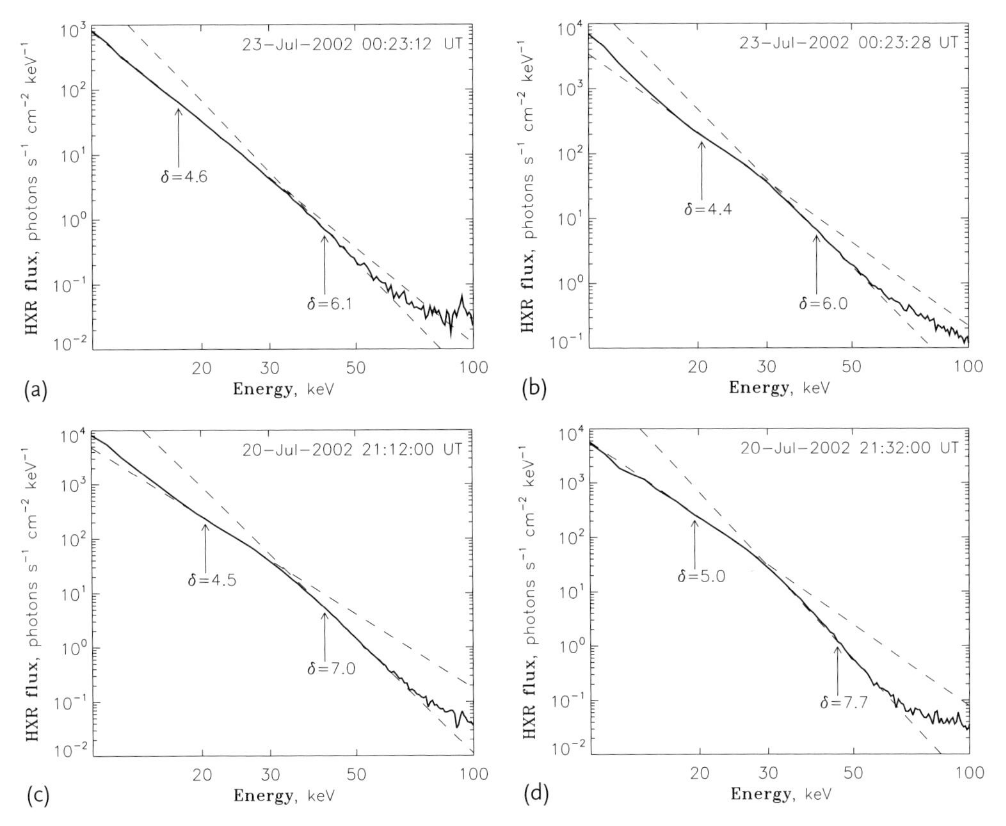

Figure 7.12 HXR double power-law spectra observed by RHESSI for flares of 23 July 2002 (a,b) and 20 July 2002 (c,d) for times plotted in top right corners of each plot. Panels (a) and (b) correspond, respectively, to electron energy fluxes of 2×10^{10} and 5×10^{10} erg cm^{-2} s^{-1} for an electron spectral index of $\gamma = 7$. Panels (c) and (d) correspond, respectively, to energy fluxes of 1×10^{11} and 5×10^{10} erg cm^{-2} s^{-1}. The straight lines show the fits by single power-law spectra for lower-energy and higher-energy parts.

indices (e.g., 4.6 and 4.4 for the 23 July flare (Sui *et al.*, 2007a) and 4.5 and 5.0 for the 20 July flare) compared to the higher-energy parts (6.1–6.2 for the 23 July flare (Holman *et al.*, 2003) and 7–7.7 for the 20 July flare). The observed HXR photon spectra can be well fitted by the simulated ones produced by the distribution functions of precipitating and returning electrons for the relativistic cross-sections and taking into account the self-induced electric field, which causes photon spectrum flattening at lower energies (Zharkova and Gordovskyy, 2006), as discussed below.

The theoretical spectra of HXR bremsstrahlung photon spectra have been shown to split into two parts: lower- and higher-energy ones, with spectral indices being much lower at the lower energy parts than in the higher ones (see Figure 11 in Zharkova and Gordovskyy, 2006). The break energy corresponds to the electric field induced by lower-energy electrons that lose their energy in collisions and ohmic losses. It can be seen that the difference in the spectral indices δ_l of photon spectra at a lower-energy part is greater for beam electrons with a higher initial spectral index γ and initial energy flux F_0. The photon spectral indices in the higher-energy part δ_u are lower than for the initial spectral indices of beam electrons γ for lower initial energy fluxes F_0 of beam electrons or equal to γ for beams with a higher initial energy flux of 10^{12} erg cm^{-2} s^{-1}.

This means that for the flare of 23 July 2002, where the lower-energy index changes from 4.6 to 4.4 in Figure 7.12a,b, respectively (Holman *et al.*, 2003), one observes a growth of the initial energy flux of beam electrons in the two consecutive times by a factor of 2 from 2×10^{10} to 4×10^{10} erg cm^{-2} s^{-1}, which is confirmed by the light curves for this flare (Holman *et al.*, 2003). This increase in the electron energy flux for the two observed time instances results in a stronger flattening of the photon spectrum for the second instance compared with the first one.

A spectral difference of the lower and upper spectral indices δ_l and δ_u in the observed photon spectra for these instances (fit by the simulated ones plotted by solid dashed lines in Figure 7.12) increased from 1.5 (6.1–4.6) to 1.6 (6.0–4.4) (Zharkova *et al.*, 2010). For the 20 July flare the spectral index difference in the photon spectra produced by beam electrons, plotted in Figure 7.12c,d, increased from 4.5 to 5.0, confirming that the energy flux of beam electrons that were produced increased in time from 1×10^{11} to 5×10^{11} erg cm^{-2} s^{-1} as per Figure 11 and Table 2 in the paper by Zharkova and Gordovskyy (2006). Thus we calculate that the observed spectral flattening at lower energies and the soft–hard–soft pattern in the lower energy spectral indices of photon spectra can be naturally explained by the low–high–low variations of the initial energy flux of beam electrons not only in this but in other flares as well (Zharkova *et al.*, 2010, 2011b).

It is interesting to note that the above-mentioned spectral flattening of the X-ray spectra at lower energies caused by returning beam electrons is qualitatively similar to the expected influence of the so-called albedo effect (e.g., Bai and Ramaty, 1978; Kontar *et al.*, 2006). A part of the photons emitted downward should be reflected from the photosphere up to the observer, thus distorting the primary emission spectra. The albedo contribution should be highest for a photon energy of about 30–50 keV (owing to the energy dependence of the Compton scattering probability) and for the events occurring near the solar disk center (for geometrical reasons).

The same can be said about the contribution of returning electrons. Indeed, in both cases we are dealing with the reflection process, but either the photons or the precipitating electrons (and photons emitted by them) change their propagation direction.

7.4.2
HXR Bremsstrahlung Directivity and Polarization

Another property of HXR bremsstrahlung emission that can help one better understand the mechanisms of electron precipitation into a flaring atmosphere is linear polarization calculated from Eq. (7.3) and directivity obtained from Eq. (7.4). In order to simulate a better fit to observations, we have considered an electron beam with all energy losses (C+B+E, for example, collisions, convergence, and ohmic losses) and with wider and narrower pitch-angle dispersion $\Delta\mu = 0.2; 0.02$ in the initial distribution defined by Eq. (4.28). The polarization calculated for HXR bremsstrahlung photon energies of 20 and 200 keV for different models of electron-beam precipitation and flare locations on the solar disk is plotted in Figure 7.13 and the directivity in Figure 7.14. The simulations are tested by numerous observations of HXR polarization and directivity plotted in each figure with their error bars and are discussed below.

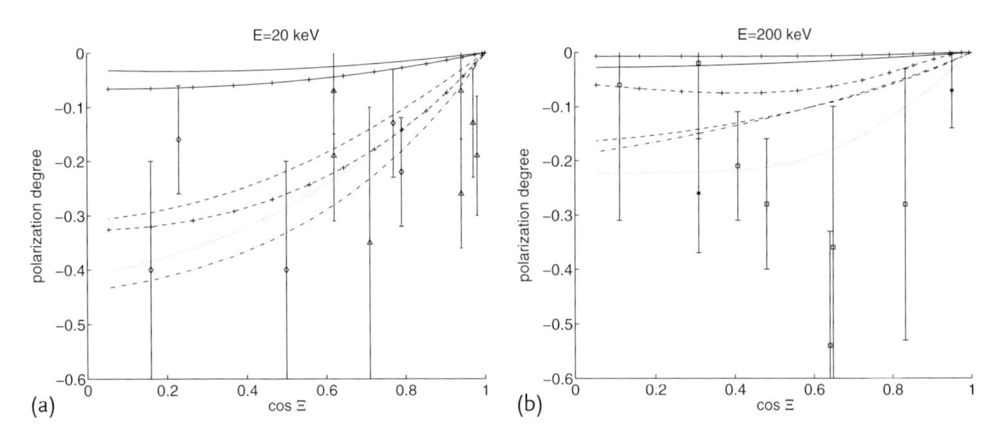

Figure 7.13 Comparison of simulations of HXR bremsstrahlung polarization for 20 keV (a) and 200 keV (b) produced for different position angles on the solar disk ($\cos \Xi = 1$ in the disk center and 0 on the limb) by a wider electron beam with $\Delta\mu = 0.2$ and energy flux of 10^{10} erg cm^{-2} s^{-1} ($\gamma = 3$): C+E model: solid line, C+E+B model: solid line with crosses; $\gamma = 7$: C+E model: dashed line, C+E+B model: dashed line with crosses) and by a more collimated ($\Delta\mu = 0.02$) electron beam (C+E model) with $\gamma = 7$ and initial energy fluxes of 10^{10} (dotted–dashed lines) and 10^{12} erg cm^{-2} s^{-1} (dotted lines). The observation results are plotted as follows: diamonds correspond to observations of Tindo *et al.* (1970, 1972a,b) at 15 keV, triangles to those of Tramiel *et al.* (1984) at 16–21 keV, squares to those of Suarez-Garcia *et al.* (2006) at 100–350 keV, and asterisks to those of Boggs *et al.* (2006) at 200–400 keV.

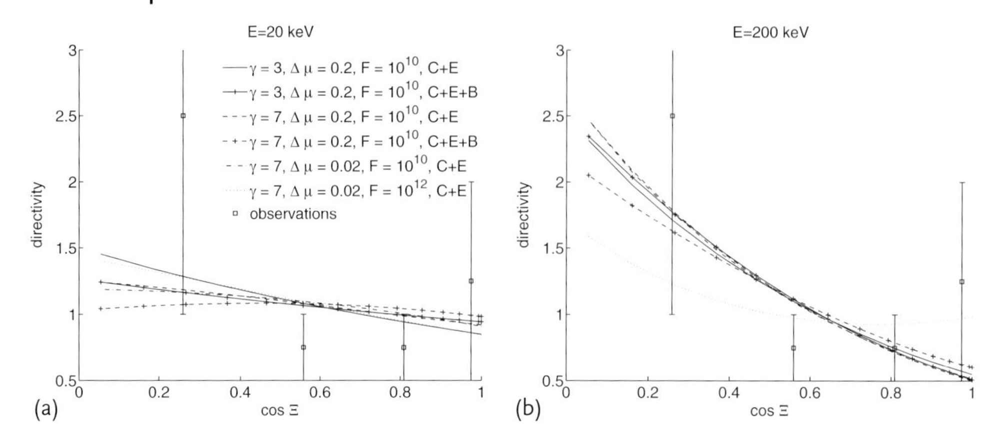

Figure 7.14 Comparison of directivity estimations inferred for energies 20–200 keV from a statistical survey of different flare positions at the solar disk (cos $\Xi = 1$ in the disk center and 0 on the limb) at $\sim$ 20—100 keV by Kašparová *et al.* (2007) with the calculated directivity ($I_\theta / I_{\text{mean}}$) of HXR bremsstrahlung emission produced by model electron beams with the same parameters as in Figure 7.13.

Note that polarization is higher by up to a factor of 2 for lower-energy photons (20 keV) than from higher-energy ones (200 keV), which can be caused by a smaller number of high-energy photons compared to lower-energy ones because of the power-law dependence on energy. For both photon energies (20 and 200 keV) the simulated HXR polarization smoothly increases toward the solar limb. For beam electrons with wider pitch-angle dispersion $\Delta\mu = 0.2$ this increase is bigger for a larger spectral index of 7 compared to 3 and for C+E+B models taking into account magnetic field convergence in addition to collisions plus electric field (C+E). The same is valid for polarization in a high-energy band of 200 keV produced by beams with a higher energy flux of 10^{12} erg cm^{-2} s^{-1} (Figure 7.13b). While for polarization at a lower energy of 20 keV, the larger the initial energy flux of beam electrons, the lower becomes the polarization (see Figure 7.13a). The latter is a clear electric field effect that causes more beam electrons to return back to the corona and bigger the spectral flattening for $\gamma = 7$ as discussed above. For narrower electron beams HXR polarization increases by an additional 30–40% for flares located closer to the limb.

For a comparison of HXR polarization with available observations we considered it to be a good fit of our theoretical curves to the observed polarization if the observational error bars appeared between the simulated polarization curves. The comparison is complicated by the fact that many authors who measure polarization do not always present the spectral indices of their photon spectra but only their polarization degree. Thus we have not yet had a chance to compare all the simulations one to one with the observations for the same flare, but only in bulk for many flares. This is why we present here the distribution of polarization degree depending on the position of a flare on the solar disk for a set of electron-beam indices $\gamma = 3$ and 7 and energy fluxes and plot the observed polarization on the same graph. It is clearly seen that in most cases the measured polarization degrees are located well between the theoretical curves, which confirms rather well the validity of the applied models of electron precipitation.

Although some discrepancies occur between the simulations and observations located at the disk center, which can be caused by loop tilts, resulting in the beam electrons' viewing angles being greater than their position angle on the disk. Nevertheless, the fit of simulations to the observations for both 20 and 200 keV is significantly improved by consideration of a narrower electron beam leaving only one measurement at $\cos \Xi \approx 0.55$ not covered by the simulated curves. The only measurement that is not fit by the simulations is that for the viewing (heliocentic) angle of about 50°, which can be caused by an even narrower beam, say with $\Delta\mu = 0.008–0.01$, or by the albedo effect, which, according to Bai and Ramaty (1978), can also increase the visible polarization.

The fit of our theoretical predictions of HXR bremsstrahlung directivity to those measured by Kašparová *et al.* (2007) is plotted in Figure 7.14. The simulations are carried out for both wide and narrow electron beams with $\Delta\mu = 0.2$ and 0.02 for the same beam parameters as discussed for polarization (Figure 7.13). Note that the directivity of lower-energy photons of 20 keV is lower and less dispersed over the locations on the solar disk than those of higher-energy ones of 200 keV. The directivity is highest, approaching 2.5 at the solar limb while smoothly decreasing to 0.5 at the solar disk center. The directivity of harder beams ($\gamma = 3$) and for models with pure collisions or collisions and electric field is higher than that of softer ones ($\gamma = 7$), as well as for models considering also a magnetic field convergence. The directivity of higher-energy photons produced by a narrower beam with a higher initial energy flux becomes much lower than for a wider beam.

The fit of the observed directivity to the theoretical curve of directivity produced for C+E models by a narrower electron beam with a higher energy flux of 10^{12} erg cm^{-2} s^{-1} is rather satisfactory for higher-energy emission (Figure 7.14b). It must also be emphasized again that in the current paper we have not taken into account any albedo effects and only considered the nonisotropic pitch-angle scattering of beam electrons during their precipitation in the presence of a converging magnetic field and self-induced electric field. The effect of particle kinetics is reflected in their emission and, in particular, polarization and directivity.

However, when interpreting solar X-ray observations, albedo and electric field effects can substantially change the HXR directivity and polarization since they act on HXR emission in a similar way. This explains why Kašparová *et al.* (2007) have concluded that the observed deviations of the RHESSI flare X-ray spectra from the power-law ones can be well explained by the albedo effect, similar to our current and previous conclusions about the electric field effect (Zharkova and Gordovskyy, 2006). At the same time Sui *et al.* (2007a) have found that the albedo correction is usually not sufficient to explain observations, and the remaining difference is most probably due to the return current. A better differentiation of the albedo and return current effects can be achieved, on the one hand, by using the spatially resolved observations of very compact X-ray sources (the albedo source should be larger than the primary X-ray source). On the other hand, simultaneous simulations of albedo and return current contributions are urgently required.

7.4.3 Relationships between Electron and HXR Photon Spectra and Electron Numbers

7.4.3.1 Particle Number

A stationary injection of electrons beams with the same spectral index γ leads to the formation of returning electrons that complete an electric circuit in a flaring loop that exists as long as injection continues. This circuit allows the vast majority of beam electrons to replenish within a time scale required for subrelativistic electrons at a velocity of $0.3c$, where c is the speed of light, to reach their stopping depth at a distance of 10^7-10^9 cm from the top of the loop where injection occurs, which gives about 0.3–0.003 s for travel in one direction.

This means that within 1 s the same electron can travel up to a few hundred times to the loop bottom and back to its top while producing HXR emission. This significantly reduces the requirements for the total number of beam electrons needed to produce the observed HXR emission in flares from $10^{36}-10^{37}$ particles per second to a very acceptable $10^{34}-10^{35}$, which constitutes only about 1–10% of the particle number at the coronal density.

7.4.3.2 HXR Spectral Evolution at Lower Energies

Also note that simulated HXR photon spectra often have double power-law (elbow-type) distributions, similar to the photon spectra deduced from RHESSI observations (Holman *et al.*, 2003) with spectra flattening toward lower energies (Zharkova and Gordovskyy, 2006).

The spectral indices at lower energies measured at 20 keV, δ_{low}, and those at higher energies measured at energies above 80 keV, δ_{high}, were found to be dependent on the beam initial flux and the electron beam spectral indices γ (Figures 7.4 and 7.9). δ_{high} slightly decreases with the initial energy flux and the electron spectral index γ (compare curves with dark circles, triangles, and dark squares). In contrast, the spectral indices δ_{low} reveal a noticeable decrease compared to the electron indices γ with an increase of the electron flux and, especially, with an increase of γ as seen in Figure 7.4.

This is more apparent in the differences between the higher and lower photon indices, $\delta_{\text{high}} - \delta_{\text{low}}$ plotted vs. the spectral indices γ and initial energy fluxes F_0 in Figure 7.4 (Zharkova and Gordovskyy, 2006). The differences grow significantly with increases of the initial energy flux F_0, that is, with an increase of the absolute magnitudes of the self-induced electric field. They also grow with increases of the electron spectral index γ (Figure 7.4). The increase is more sensitive to beam intensity, or to the magnitude of the induced electric field, than to the beam spectral index (compare the curves in Figure 7.4), showing some saturation for stronger and softer beams.

These dependences of HXR lower- and higher-energy spectral indices on the energy fluxes of beam electrons caused by energy losses in their electric field can explain the so-called soft–hard–soft patterns for fluxes above 35 keV reported by many observers (e.g., see Fletcher and Hudson, 2002; Grigis and Benz, 2004).

At flare onset, the initial fluxes are smaller, as are the self-induced electric fields induced by injected electrons. As a result, HXR photon spectra have indices close to δ_{high} without any flattening (soft spectra). At the flare maximum, energy fluxes of electron beams are much higher, and as result they induce much higher electric fields of their own. This leads to a strong flattening of their photon spectra at lower energies with an index of δ_{low} (hard spectra) because of the joint effect of collisions and electric field, as discussed above (Zharkova and Gordovskyy, 2006). After the flare maxima, the beam fluxes decrease again, which will result in smaller induced electric fields, which in turn will lead to the photon spectra returning to those with original spectral indices of δ_{high} (soft spectra as at the flare onset). Thus, the triangle temporal profile of the energy flux of injected beam electrons naturally leads to soft–hard–soft patterns in the spectral indices of observed HXR photon spectra at lower energies (below 40–60 keV).

Thus an electron beam with a single power-law energy spectrum is shown to produce double power-law HXR photon spectra if the beam's energy flux is high enough. In practical terms, while interpreting HXR observations in the case of double power-law HXR photon spectra, a correction of their energy flux is required to account for this spectral flattening at lower energies. This means that, in order to calculate the energy flux of beam electrons producing these spectra, one needs to extend to lower energies a photon spectrum at higher energies and derive the total energy flux carried by these beam electrons (Zharkova *et al.*, 2011b). This correction will allow one to avoid an incorrect interpretation of the beam parameters derived from observed HXR photon spectra.

8
Microwave Emission and Polarization

8.1
General Comments

HXRs and radio emission are considered to be the main diagnostic tools for high-energy particles accelerated in the corona. These types of emission are produced by different mechanisms and usually in different parts of a flaring region. Thus, X-rays and radio emission carry different (sometimes complementary) information about the parameters of the accelerated particles.

The simultaneous observations of HXR and microwave (MW) emissions in footpoints of solar flares are often closely correlated in time, pointing to their common origin (Aschwanden, 2005; Bastian *et al.*, 1998). There is a high likelihood of HXR and MW radiation being produced by the same population of nonthermal electrons (Kundu *et al.*, 2001a,b, 2004; Vilmer *et al.*, 2002; Willson and Holman, 2003). Assuming that both kinds of emission are caused by the same population of electrons, the mechanisms of transport affecting HXR and MW emission are substantially different: MW radiation is related to gyrosynchrotron emission of high-energy electrons with energies from a few tens of kiloelectronvolts (Kundu *et al.*, 2001a) up to several megaelectronvolts (Bastian, 1999; Kundu *et al.*, 2004), while HXR radiation is often produced by electrons with much lower energies, from 10 to 300 keV (see, e.g., Holman *et al.*, 2003; Lin *et al.*, 2003).

However, diagnostics of the processes and conditions in solar active regions using MW observations meets serious difficulties caused by the fact that its characteristics depend on a number of parameters, such as plasma density, magnetic field strength, energy and density of accelerated particles, as well as the shape of electron distribution functions at different depths. At present, most studies use simplified expressions for gyrosynchrotron emission parameters (e.g., Dulk and Marsh, 1982). Those expressions were obtained for the case where the particles have isotropic (or, sometimes, weakly anisotropic) distribution on the pitch angle and a simple (e.g., power-law) distribution on energy.

In spite of a large number of the separate simulations for interpretation of HXR or MW emission (see for example, reviews by Holman *et al.*, 2011; Krucker *et al.*, 2008a,b; White *et al.*, 2011, and references therein), there are no simultaneous simulations accounting for these two kinds of emission with the same distribution

Electron and Proton Kinetics and Dynamics in Flaring Atmospheres, First Edition. Valentina Zharkova.

of electrons. Even knowing that HXR and MW emission are formed by different mechanisms and can be produced by the same population of high energy electrons, it is a rather challenging task to describe their simultaneous variations in depth and time, which result in different locations, intensities and areas of HXR and MW emission. This imposes some strict limitations onto the proposed models of electron precipitation requiring to correctly account for electron energy losses such as: Coulomb collisions, ohmic losses and anisotropic scattering, as well as for the changes of the electron pitch angles caused by a magnetic field converging with depth.

Although many existing models of particle acceleration and precipitation considering not only collisional but other energy loss mechanisms unavoidably report anisotropic particle distributions that vary significantly with precipitation depth (e.g., see distribution functions in Zharkova *et al.*, 2010, hereafter Paper I). Fleishman and Melnikov (2003a,b) have shown that the pitch-angle anisotropy of accelerated particles, even if it is unable to support a coherent wave amplification, can change significantly the emission intensity (up to orders of magnitude) and polarization (the sign of circular polarization). Thus, approximated electron distributions cannot be used in calculations of gyrosynchrotron emission. Instead, one must firstly consider exact formulas for MW emission that are rather cumbersome and computationally expensive, and, secondly, utilize more realistic distribution functions of accelerated electrons that can account simultaneously for both HXR and MW emissions in solar flares.

In Chapters 4 and 5 we already determined electron distribution functions at different levels of a coronal magnetic loop based on hydrodynamic models of ambient plasma heated by precipitating beam electrons. The electron distribution functions were obtained numerically by solving the time-dependent Fokker–Planck kinetic equation for times of steady injection for collisional and ohmic energy losses in a converging magnetic loop.

In this chapter we continue to investigate electron-beam precipitation by extending the simulations to gyrosynchrotron emission in a MW range. Note that the emission in metric and decimetric ranges is also of great interest for the diagnostics of accelerated particles, acceleration mechanisms, and structure of active regions. However, at these frequencies, the emission is caused mainly by different kinds of coherent plasma mechanism (e.g., Dulk, 1985, and references therein), which involves nonlinear processes and is not yet sufficiently developed (Aschwanden, 2005). At the same time, MW emission is known to be mainly caused by incoherent radiation of accelerated electrons gyrating in the magnetic field; the theory of such a process is well developed and will be used in this chapter.

8.2 Evaluation of Models for Electron Precipitation

Thus we consider a quasistationary injection when distributions of electron beams become well established across all precipitation depths, for a stream of returning electrons where the depth distributions of both beams do not change over time. Such a state is usually achieved in about 70–200 ms after injection onset (Siversky and Zharkova, 2009b). We consider models with pure collisions (C), collisions and electric field (CE), collisions and magnetic field convergence (CB), and with all the above losses (CEB) in order to investigate the relative contribution of the different processes to the formation of electron distributions and their resulting MW emission.

In our simulations we used the following parameters: electron energy range $12\,\text{keV} < E < 1200\,\text{keV}$, initial pitch-angle half-width of an injected electron beam $\Delta\mu = 0.2$, and the magnetic field characteristic depth was located at $\xi_c = 10^{20}\,\text{cm}^{-2}$ (Siversky and Zharkova, 2009b). The initial energy flux of energetic electrons F was taken to be 10^{10} and $10^{12}\,\text{erg}\,\text{cm}^{-2}\,\text{s}^{-1}$, the initial power-law index γ was taken to be 3 and 7 (the corresponding beam densities n_b at the upper boundary are given in Table 8.1). While the electron densities for $F = 10^{12}\,\text{erg}\,\text{cm}^{-2}\,\text{s}^{-1}$ seem very high (especially for the softer beams), we should note that this is caused mainly by the relatively small low-energy cutoff of the power-law distribution (12 keV) used in this work, so the total electron density is mainly the density of low-energy electrons, which do not make a significant contribution to gyrosynchrotron radiation. Moreover, the numbers stated in Table 8.1 refer to the injection point, while the particle number decreases with distance traveled (see, e.g., Zharkova and Gordovskyy, 2005a; Zharkova and Kobylinskii, 1993). The magnetic convergence factor B_c/B_0 (the ratio of magnetic fields at a characteristic depth and the loop top) varied from 1 (no convergence) to 10.

Examples of electron energy spectra at a column depth of $5 \times 10^{19}\,\text{cm}^{-2}$ occurring between the injection point and the transition region are shown in Figure 8.1. Figure 8.1a,b demonstrate the effect of an electric field induced by beam electrons, which results in the formation of returning electrons directed upward, which reduce the return current formed by the ambient plasma electrons. For higher-energy fluxes F of injected beams, this effect becomes stronger. The energy of the returning particles can reach $\sim 100\,\text{keV}$ for $F = 10^{10}\,\text{erg}\,\text{cm}^{-2}\,\text{s}^{-1}$ and $\gtrsim 500\,\text{keV}$ for $F = 10^{12}\,\text{erg}\,\text{cm}^{-2}\,\text{s}^{-1}$. For $F = 10^{12}\,\text{erg}\,\text{cm}^{-2}\,\text{s}^{-1}$, in a certain energy range, the

Table 8.1 Densities of precipitating electrons (cm^{-3}) at injection point for different energy fluxes and power-law indices of electron beams.

F, erg cm^{-2} s^{-1}	$\gamma = 3$	$\gamma = 5$	$\gamma = 7$
10^{10}	2.2×10^7	5.0×10^7	6.0×10^7
10^{12}	2.2×10^9	5.0×10^9	6.0×10^9

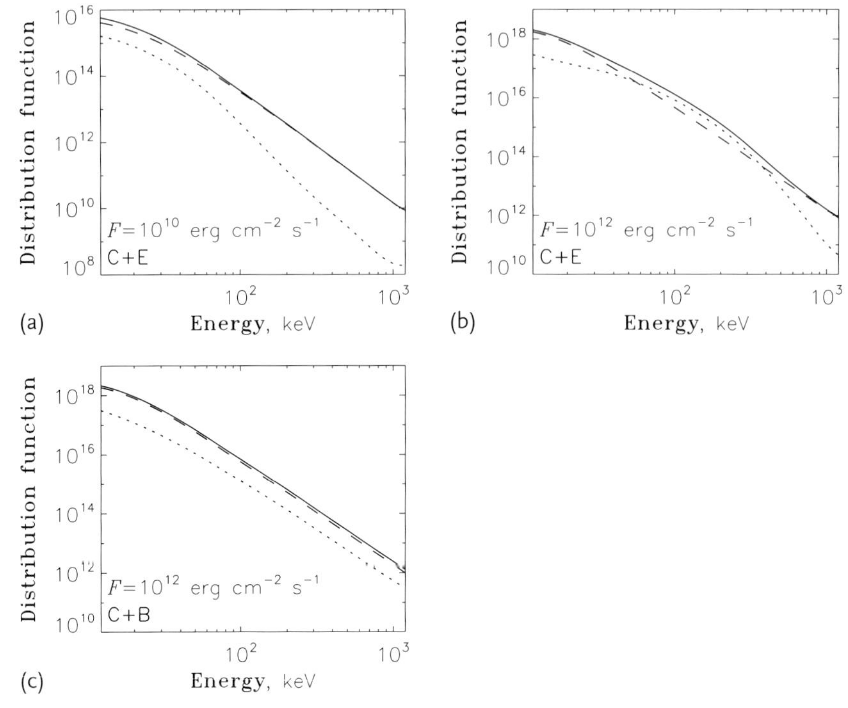

Figure 8.1 Electron distribution functions (integrated over pitch angles) obtained by solving the Fokker–Planck equation at a column depth of $\xi = 5 \times 10^{19}\,\mathrm{cm}^{-2}$. Dashed line: electrons moving downward (with $\mu > 0$); dotted line: electrons moving upward (with $\mu < 0$); solid line: resulting distribution function (downward + upward). The initial power-law index of beam electrons is $\gamma = 3$, the initial energy flux F is equal to 10^{10} or $10^{12}\,\mathrm{erg\,cm^{-2}\,s^{-1}}$. The factors taken into account are collisions (C), self-induced electric field (E), and convergence of magnetic field (B).

number of electrons moving upward can even exceed the number of those electrons moving downward.

The effect of a converging magnetic field for electron beams with an energy flux of $F = 10^{12}\,\mathrm{erg\,cm^{-2}\,s^{-1}}$ is demonstrated in Figure 8.1c; for beams with a lower energy flux, the effect is very similar. One can see that the number of returning electrons is now smaller than in models with a self-induced electric field. However, unlike a self-induced electric field, a converging magnetic field affects particles at all energies. Thus, it forms the upward-directed flow of high-energy particles whose pitch angles exceed the loss-cone values, and this effect is shown to be of primary importance for MW emission. The combined effect of a self-induced electric field and converging magnetic field results in the formation of an upward-directed electron flow (beam) with the whole spectrum of energies from low to high.

8.3 Gyrosynchrotron Plasma Emissivity and Absorption Coefficient

The exact equations for gyrosynchrotron emissivity j_σ and absorption coefficient $\varkappa_\sigma$ have the following form (Eidman, 1958, 1959; Melrose, 1968; Ramaty, 1969):

$$j_\sigma = \frac{2\pi e^2}{c}\frac{N_\sigma \nu^2}{1+T_\sigma^2}\sum_{s=-\infty}^{\infty}\int\left[\frac{T_\sigma(\cos\theta - N_\sigma\beta\mu) + L_\sigma \sin\theta}{N_\sigma\beta\sin\theta\sqrt{1-\mu^2}}J_s(\lambda) + J_s'(\lambda)\right]^2$$
$$\times\, \beta^2(1-\mu^2) f(\boldsymbol{p})\delta\left[\nu(1 - N_\sigma\beta\mu\cos\theta) - \frac{s\nu_B}{\Gamma}\right] d^3\boldsymbol{p}\ ; \qquad (8.1)$$

$$\varkappa_\sigma = -\frac{2\pi e^2}{N_\sigma(1+T_\sigma^2)}\sum_{s=-\infty}^{\infty}\int\left[\frac{T_\sigma(\cos\theta - N_\sigma\beta\mu) + L_\sigma \sin\theta}{N_\sigma\beta\sin\theta\sqrt{1-\mu^2}}J_s(\lambda) + J_s'(\lambda)\right]^2$$
$$\times\, \beta(1-\mu^2)\left[\frac{\partial f(\boldsymbol{p})}{\partial p} + \frac{N_\sigma\beta\cos\theta - \mu}{p}\frac{\partial f(\boldsymbol{p})}{\partial\mu}\right]$$
$$\times\, \delta\left[\nu(1 - N_\sigma\beta\mu\cos\theta) - \frac{s\nu_B}{\Gamma}\right] d^3\boldsymbol{p}\ , \qquad (8.2)$$

where ν is the emission frequency, ν_B is the electron cyclotron frequency, N_σ, T_σ, and L_σ are the refraction index and the components of the polarization vector, respectively, θ is the angle between the wave vector and the magnetic field, p and β are the electron momentum and dimensionless speed, respectively, $\mu = \cos\alpha$, α is the electron pitch angle, $\Gamma = (1-\beta^2)^{-1/2}$ is the relativistic factor, $J_s(\lambda)$ and $J_s'(\lambda)$ are the Bessel function and its derivative over the argument λ,

$$\lambda = \frac{\nu}{\nu_B}\Gamma N_\sigma\beta\sin\theta\sqrt{1-\mu^2}\ . \qquad (8.3)$$

The electron distribution function $f(\boldsymbol{p})$ satisfies the normalization condition

$$\int f(\boldsymbol{p}) d^3\boldsymbol{p} = n_e\ , \qquad (8.4)$$

where n_e is the number density of energetic electrons.

The refraction index of electromagnetic waves in plasma satisfies the dispersion equation

$$N_\sigma^2 = 1 - \frac{2V(1-V)}{2(1-V) - U\sin^2\theta + \sigma\sqrt{\mathcal{D}}}\ , \qquad (8.5)$$

where

$$\mathcal{D} = U^2\sin^4\theta + 4U(1-V)^2\cos^2\theta\ , \qquad (8.6)$$

$$U = \frac{\nu_B^2}{\nu^2}\ , \quad V = \frac{\nu_p^2}{\nu^2}\ , \qquad (8.7)$$

and ν_p is the electron plasma (Langmuir) frequency. For an ordinary wave (O mode), $\sigma = +1$; for an extraordinary wave (X mode), $\sigma = -1$. The parameters

T_σ and L_σ equal

$$T_\sigma = \frac{2\sqrt{U}(1-V)\cos\theta}{U\sin^2\theta - \sigma\sqrt{\mathcal{D}}}\,, \tag{8.8}$$

$$L_\sigma = \frac{V\sqrt{U}\sin\theta + T_\sigma UV\sin\theta\cos\theta}{1-U-V+UV\cos^2\theta}\,. \tag{8.9}$$

Equations (8.1) and (8.2) contain 3-D integrals over $d^3\boldsymbol{p}$. Using the properties of the δ-function, we can reduce those integrals to 1-D integrals over the parallel (to the magnetic field) component of the momentum vector p_z:

$$\begin{aligned}\left\{\begin{matrix} j_\sigma \\ \varkappa_\sigma \end{matrix}\right\} &= \frac{4\pi^2 e^2}{c}\frac{N_\sigma \nu}{1+T_\sigma^2}\left\{\begin{matrix} 1 \\ -\frac{c}{N_\sigma^2\nu^2} \end{matrix}\right\} \\ &\times \sum_{s=1}^{\infty}\int\limits_{p_{z\,\min}}^{p_{z\,\max}}\left[\frac{T_\sigma(\cos\theta - N_\sigma\beta\mu) + L_\sigma\sin\theta}{N_\sigma\beta\sin\theta\sqrt{1-\mu^2}}J_s(\lambda) + J_s'(\lambda)\right]^2 \\ &\times \left\{\begin{matrix} f(\boldsymbol{p}) \\ \frac{1}{\beta}\left[\frac{\partial f(\boldsymbol{p})}{\partial p} + \frac{N_\sigma\beta\cos\theta - \mu}{p}\frac{\partial f(\boldsymbol{p})}{\partial\mu}\right]\end{matrix}\right\} p^2(1-\mu^2)\,dp_z\Bigg|_{p=p(p_z)}\,,\end{aligned} \tag{8.10}$$

where the momentum value $p(p_z)$ at every point of the resonance curve is found from the resonance condition:

$$p(p_z) = m_e c\sqrt{\left(\frac{N_\sigma p_z\cos\theta}{m_e c} + \frac{s\nu_B}{\nu}\right)^2 - 1}\,, \tag{8.11}$$

$\mu(p_z) = p_z/p(p_z)$, and the integration limits $p_{z\,\min}$ and $p_{z\,\max}$ correspond to the boundaries of the interval where such a solution exists:

$$\begin{aligned} p_{z\,\min,\max} &= m_e c\frac{(s\nu_B/\nu)N_\sigma\cos\theta \mp \sqrt{N_\sigma^2\cos^2\theta + (s\nu_B/\nu)^2 - 1}}{1 - N_\sigma^2\cos^2\theta}\,, \\ N_\sigma^2\cos^2\theta &+ \left(\frac{s\nu_B}{\nu}\right)^2 - 1 > 0\,. \end{aligned} \tag{8.12}$$

When obtaining Eq. (8.10), it is taken into account that for electromagnetic waves (with $N_\sigma < 1$) the resonance condition can be satisfied only at $s \geq 1$. The gyrosynchrotron emissivity and absorption coefficient (8.10) are calculated by numerical integration and consequent summation of the series over s. In this work, the integrals are evaluated using the Romberg method (Press *et al.*, 1997). The series summation is stopped when a necessary accuracy (10^{-5}) is achieved. For the Bessel functions, we use the approximate formulas proposed by Wild and Hill (1971); the analysis made by Fleishman and Kuznetsov (2010) has shown that these formulas provide very high accuracy while reducing the computation time considerably, in comparison with the exact Bessel functions.

The kinetic model (numerical solution of the Fokker–Planck equation) provides the electron distribution function $f(\boldsymbol{p})$ as an array of values on a grid in (E, μ)

space derived from the Landau–Fokker–Planck approach described in Chapter 4. Since the integration nodes in the numerical integration method, as a rule, do not coincide with the gridpoints, the resonance values of the distribution function and its derivatives were calculated using bilinear interpolation and the necessary transformations from (E, μ) space to (p, μ) space were applied.

8.4 Gyrosynchrotron Emission from a Homogeneous Source

8.4.1 Depth Variations of MW Emission

Let us calculate the emission parameters under the assumption of a homogeneous source. In this case, the equation of radiation transfer for MW emission has a simple solution and the emission intensity (observed at the Earth) is equal to

$$I_\sigma = \frac{S}{\mathcal{R}^2} \frac{j_\sigma}{\varkappa_\sigma} (1 - e^{-\varkappa_\sigma L}) \tag{8.13}$$

for each magnetoionic mode σ. In the above expression, S is the visible source area, L is the source depth along the line of sight, and $\mathcal{R}$ is the astronomical unit. The formulas for the gyrosynchrotron plasma emissivity j_σ and absorption coefficient $\varkappa_\sigma$ are given in Section 8.4. The polarization degree is defined as

$$\eta = \frac{I_X - I_O}{I_X + I_O}, \tag{8.14}$$

where I_O and I_X are the intensities of the ordinary and extraordinary modes, respectively.

For estimations we use the following parameters: visible source area $S = 1.8 \times 10^{18}\,\text{cm}^2$, source depth $L = 6 \times 10^8\,\text{cm}$, ambient (thermal) plasma density $n_0 = 2 \times 10^9\,\text{cm}^{-3}$, and magnetic field strength at the considered depth $B = 370\,\text{G}$. Note that in models with a converging magnetic field, the field strength must vary with depth, which is another factor (in addition to the variations of the distribution function) affecting the emission parameters. However, in this section we assume that all the source parameters, except the electron distribution functions, remain the same in all calculations.

The accepted magnitude of the ambient plasma density is relatively low and uncommon for the active regions of the solar corona. Moreover, this plasma density can be comparable to or even less than the particle density of the energetic electron component for very soft beams (Table 8.1). However, our aim here is to rule out the Razin suppression and investigate the "pure" gyrosynchrotron radiation in a wide spectral range, including relatively low frequencies. The effect of higher plasma density is discussed in Section 8.4.1.4, where it is shown that all the conclusions made for the above-mentioned ambient density can be applied to a wider range of coronal plasma densities.

The emission is characterized by its frequency ν and the angle θ between the wave vector and the magnetic field. We use a coordinate system where the angle $\theta = 0$ corresponds to the direction of the injected electron beam, that is, toward the solar surface. Therefore, the emission observed from Earth (neglecting reflection and scattering) must have $\theta > 90°$. We consider the cases of $\theta = 100°$ (almost transversal propagation that should be typical for limb flares) and $\theta = 140°$ (which, if we take into account the variety of shapes and tilts of the coronal magnetic loops, can take place in flares located in various parts of the solar disk).

8.4.1.1 Effect of a Magnetic Field Convergence

The spectra and polarization of emissions produced by different distributions of energetic particles (corresponding to different heights) for an initial energy flux of $F = 10^{10}\,\mathrm{erg\,cm^{-2}\,s^{-1}}$ and initial power-law index of $\gamma = 3$ are shown in Figure 8.2a for the case where only collisions are taken into account while either the self-induced electric field or a convergence of magnetic field are neglected. One can see that both the emission spectra and polarization are almost the same for all depths. For the viewing angle $\theta = 100°$, polarization is found to be very low. For $\theta = 140°$, in the optically thin part of the spectrum (where the emission intensity decreases with frequency), the X mode dominates and polarization is positive.

By taking into account the magnetic field convergence in addition to collisions, the frequency distributions of MW intensity and polarization are changed dramatically (Figure 8.2b). For $\theta = 100°$, the MW emission intensity (especially in the optically thin frequency range) increases with depth, and its spectral peak shifts toward higher frequencies. As a result, at the maximal depth considered in the model, the emission intensity exceeds by more than an order of magnitude the intensity for the pure collisional model. This is caused by the increase with depth of a number of particles having pitch angles of around $\alpha \simeq 90°$. Near the injection point, polarization in the optically thin frequency range corresponds to the O mode, which is typical for a beamlike distribution (Fleishman and Kuznetsov, 2010; Fleishman and Melnikov, 2003a). In fact, we have here a two-stream distribution where both beams, a downward moving precipitating one and an upward moving returning one, are highly collimated. As the depth increases, the polarization is gradually dominated by the X mode, which is typical for loss-cone distributions.

A similar trend in polarization (contribution of X mode increasing with depth) can be seen for the viewing angle $\theta = 140°$. However, in this case the emission intensity decreases with depth, due to a decrease in the number of particles having large pitch angles ($\alpha \gtrsim 140°$), while near the footpoint the returning particles are mainly concentrated around 90°. Nevertheless, even at the deepest layer considered, the emission intensity is noticeably higher than for the pure collisional model, meaning that there are more energetic electrons with appropriate pitch angles in the emitting region of the corona.

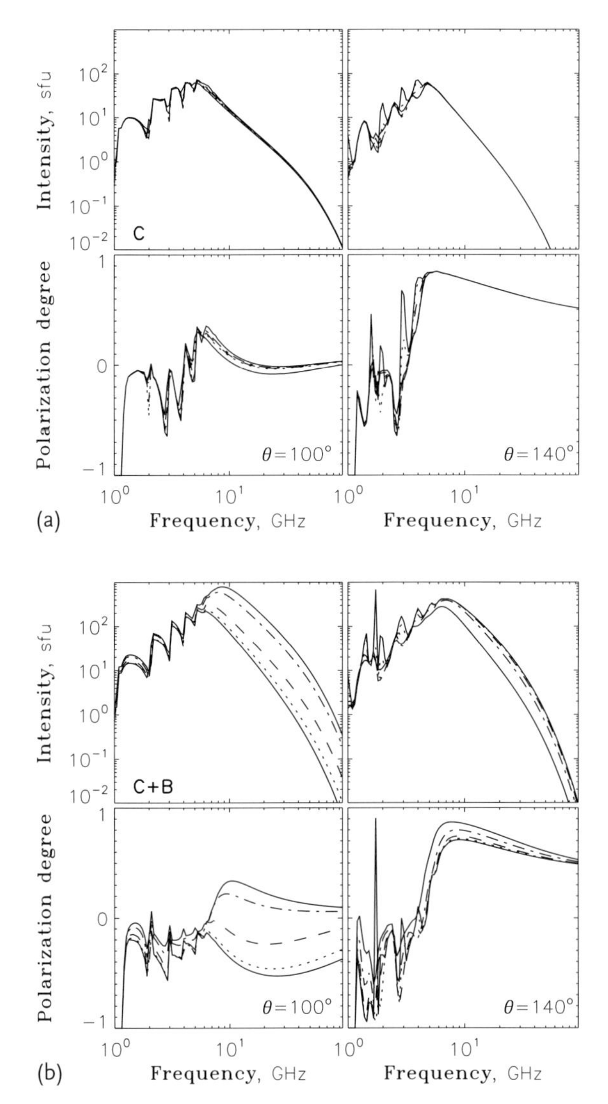

Figure 8.2 Intensity and polarization of gyrosynchrotron emission from a homogeneous source (for different viewing angles). The used electron distribution functions are obtained by solving the Fokker–Planck equation with different factors taken into account (Figure 8.1). The initial power-law index $\gamma = 3$, the initial energy flux $F = 10^{10}\,\mathrm{erg\,cm^{-2}\,s^{-1}}$, and other simulation parameters are given in the text. The different lines correspond to the electron distributions at different column depths from the injection point: solid line: $\xi = 2.4 \times 10^{17}\,\mathrm{cm^{-2}}$, dotted line: $\xi = 1.8 \times 10^{18}\,\mathrm{cm^{-2}}$, dashed line: $\xi = 5.5 \times 10^{18}\,\mathrm{cm^{-2}}$, dashed–dotted line: $\xi = 1.6 \times 10^{19}\,\mathrm{cm^{-2}}$, and dashed–triple dotted line: $\xi = 5.0 \times 10^{19}\,\mathrm{cm^{-2}}$.

8.4.1.2 **Effects of a Self-Induced Electric Field**

As we described in Paper I, the electric field induced by a precipitating beam is linearly dependent on the initial particle flux (and, therefore, their energy flux) at the top boundary and integrally on the electron distributions in energy and pitch angles at various depths, which obviously will obviously reflect various energy loss mechanisms considered in the current study.

From the electron distributions plotted in Figure 8.1 it can be noticed that, for weaker electron beams with an energy flux of $F = 10^{10}\,\mathrm{erg\,cm^{-2}\,s^{-1}}$, the effect of a self-induced electric field is relatively weak. As a result, for weaker beams, inclusion of the self-induced electric field (in addition to collisions and the magnetic field convergence) does not affect much the emission spectra and polarization; they remain nearly the same as for pure collisions and collisions with a converging magnetic field, as shown in Figure 8.2a and b, respectively.

However, for a more intense hard electron beam, with an initial energy flux of $F = 10^{12}\,\mathrm{erg\,cm^{-2}\,s^{-1}}$ and power-law index of $\gamma = 3$, the emission spectra and polarization become significantly different (Figure 8.3). At first, one can note that for the purely collisional model (Figure 8.3a) and for the model with collisions and a converging magnetic field (Figure 8.3b), with the increase of the beam's initial energy flux the MW emission intensities become higher by two orders of magnitude than for beams with an energy flux of $F = 10^{10}\,\mathrm{erg\,cm^{-2}\,s^{-1}}$, and the spectral peaks are shifted toward higher frequencies. Depth variations of the emission parameters are qualitatively the same as for weaker beams (Figure 8.2a,b).

When collisions and a self-induced electric field are taken into account (Figure 8.3c), emission patterns change. We established earlier that the effect of a self-induced electric field becomes stronger with an increase of the energy flux F (Figure 8.1) and spectral index (Siversky and Zharkova, 2009b; Zharkova and Gordovskyy, 2006). This effect is also observed in MW emission and polarization, plotted in Figure 8.3c. The self-induced electric field results in the immediate (right near the injection point) formation of a return current from the beam electrons scattered at pitch angles larger than 90°. This current is maintained for the entire time of the beam's steady precipitation, that is, the particle distribution remains nearly unchanged with depth having two beams (precipitating and returning) traveling in opposite directions. These two beams create a close electric circuit linking the injection site at the top with the whole loop, where beam electrons precipitate.

Interesting effects come from exploring this MW emission from different viewing angles. For example, for a viewing angle of $\theta = 100°$ (which is close to the transversal direction), the maximal intensity of MW emission is twice as high as that for the purely collisional model. Furthermore, the spectral peaks are continually shifted further toward higher frequencies with depth, and the polarization in the optically thin part of the spectrum reveals the domination of the O mode (beamlike distribution). For a viewing angle of $\theta = 140°$ (looking at the loop from the side), the effect of the self-induced electric field becomes much stronger because the electrons of the return current are contributing more to the emission. It can be noted that the emission intensity is increased by a factor of about 5 while the polarization

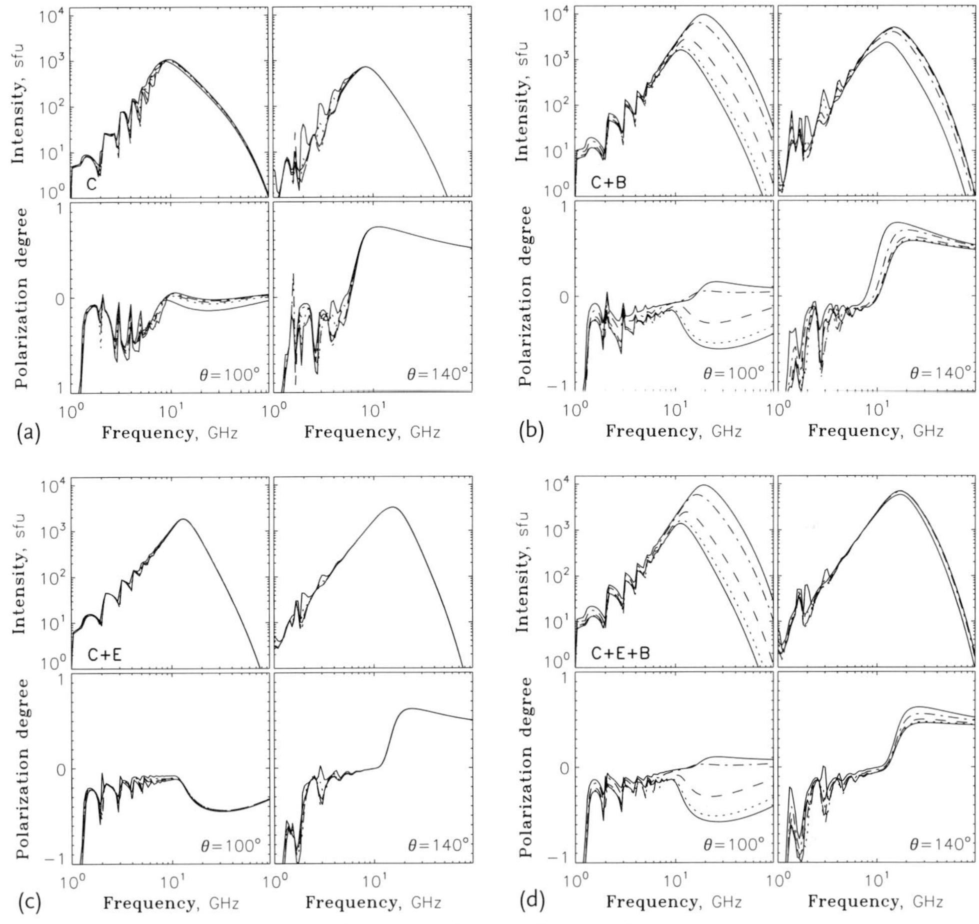

Figure 8.3 The same as in Figure 8.2, for $\gamma = 3$ and $F = 10^{12}$ erg cm^{-2}.

degree has decreased, in comparison with the purely collisional model. This is because the returning electrons were shown in Paper I to have pitch angles between 90 and 180°, with a maximum close to the latter (Zharkova and Gordovskyy, 2006), meaning that more emission is radiated in oblique directions than in transversal ones.

The electric field affects MW emission and polarization calculated for models that include all three effects (collisions, self-induced electric field, and converging magnetic field) in a similar way, as shown in Figure 8.3d. For a viewing angle of $\theta = 100°$ (a limb flare), the effect of a self-induced electric field is relatively weak (in comparison with the effect of a converging magnetic field), so that both the spectra and polarization look similar to those in Figure 8.3b, left image. However, for oblique viewing at $\theta = 140°$, the combined effect of the self-induced electric field and converging magnetic field results in a noticeable increase of the emission intensity (cf. right images in Figure 8.3b–d). Also note that the intensity of emission at $\theta = 140°$ almost does not decrease with depth now (contrary to what is shown in Figures 8.2b and 8.3b). This is because the self-induced electric field drags particles with pitch angles $\alpha \simeq 90°$ toward larger pitch angles. The importance of this effect is shown in Section 8.4.2.

8.4.1.3 **Effect of a Power-Law Index of Beam Electrons**

Since the effects of an electric field are more pronounced in HXR emission for softer beams (Siversky and Zharkova, 2009b; Zharkova and Gordovskyy, 2006; Zharkova *et al.*, 2010), let us compare the effect of a spectral index change of an intense electron beam on MW emission and polarization. The simulation results for a softer electron beam with an initial power-law index of $\gamma = 7$ (energy flux of $F = 10^{12}\,\mathrm{erg\,cm^{-2}\,s^{-1}}$) are plotted in Figure 8.4, which we compare with those for a hard beam plotted in Figure 8.3. For MW emission, the effect of an electric field is more difficult to investigate because the increase of a power-law index results in a sharp decrease of the emission intensity and a shift of a spectral peak toward lower frequencies (where the spectrum is dominated by the harmonic structure).

Nevertheless, by comparing Figure 8.4a (purely collisional model) and b (with collisions and self-induced electric field) one can notice that at greater depths, the account for a self-induced electric field results in a clearly visible increase of the emission intensity: by a factor of 2–3 for $\theta = 100°$ and by a factor of $\gtrsim 5$ for $\theta = 140°$. That is, the relative effect of a self-induced electric field is nearly the same as for harder beams. A noticeable feature of soft beams is that for $\gamma = 7$, the return current sets up slower than for $\gamma = 3$. In addition, the effect of collisions is more important now. The effect of a converging magnetic field for beams with $\gamma = 7$ is qualitatively similar to that for beams with $\gamma = 3$, with a slight increase of the number of electrons mirrored by a magnetic field back to the top for softer

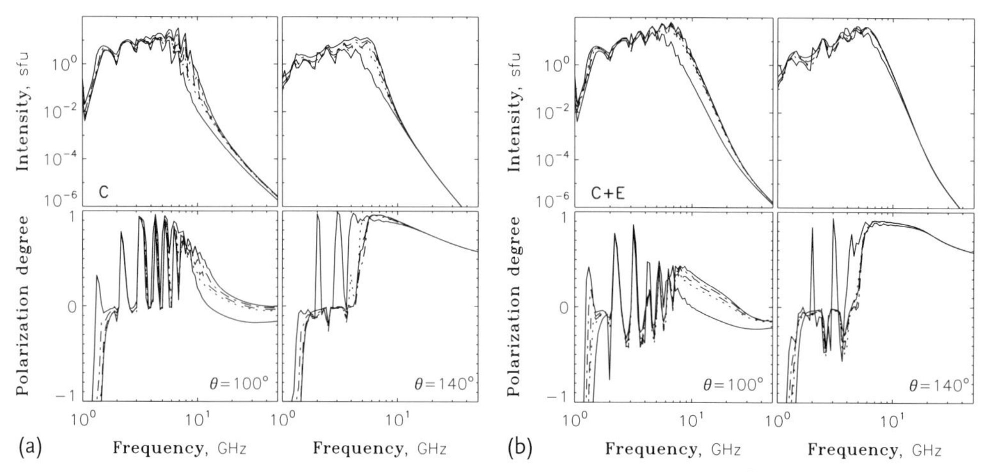

Figure 8.4 The same as in Figures 8.2 and 8.3, for $\gamma = 7$ and $F = 10^{12}\,\mathrm{erg\,cm^{-2}}$.

beams, related to their faster collisional scattering to larger pitch angles above the loss cone.

8.4.1.4 Effect of Ambient Plasma Density

The effect of background (thermal) plasma on gyrosynchrotron radiation has been investigated in many papers (e.g., Fleishman and Melnikov, 2003b; Klein, 1987; Melnikov *et al.*, 2008). First of all, the emission cannot propagate if its frequency is below the cutoff frequency (which is equal to the electron plasma frequency for the O mode and typically slightly higher for the X mode). In addition, the intensity of gyrosynchrotron radiation decreases considerably (Razin, 1960a,b) if its frequency is below the Razin frequency $\nu_R = 2\nu_p/(3\nu_B)$, where ν_p and ν_B are the electron plasma and cyclotron frequencies. In previous sections, we neglected the Razin suppression (by using a relatively low plasma density) in order to investigate in detail the effects of various variations of the distributions of energetic electrons. The effects of depth variations of the ambient density and temperature will be investigated in Section 8.5.2.

Figure 8.5 shows the emission spectra and polarization for five different values of the thermal plasma density (the magnetic field and source geometry are the same as in Figures 8.2–8.4). The initial parameters of the electron beam are $F = 10^{12}\,\mathrm{erg\,cm^{-2}\,s^{-1}}$ and $\gamma = 3$, and the plots in Figure 8.5 correspond to the distribution function at a depth equivalent to $\xi = 5.0 \times 10^{19}\,\mathrm{cm^{-2}}$. Two simulation models are considered (both including the collisions and converging magnetic field, but with or without self-induced electric field). One can see that the increasing plasma density makes the low-frequency (optically thick) slope of the spectrum

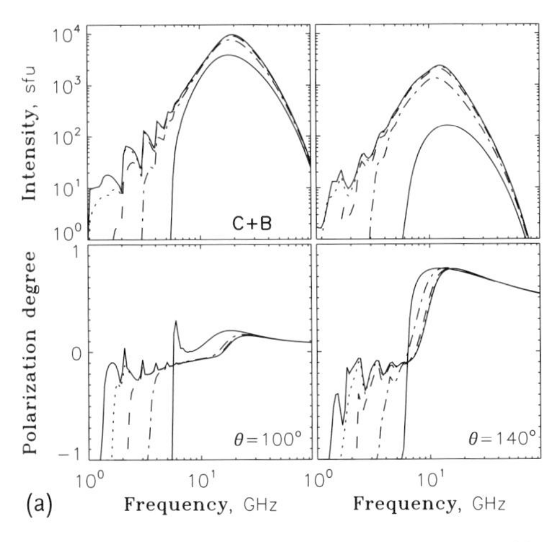

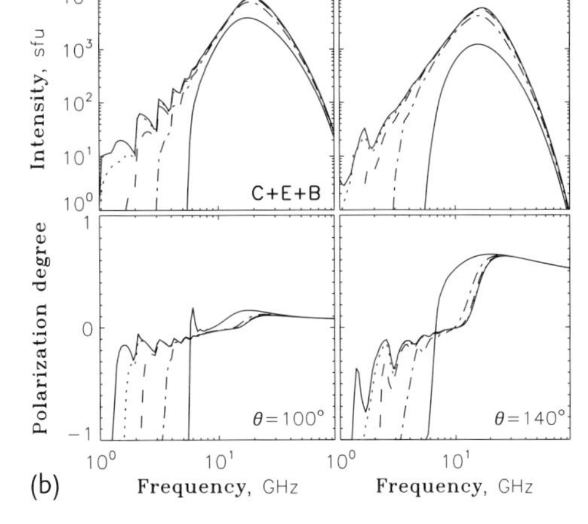

Figure 8.5 Spectra and polarization of gyrosynchrotron emission from a homogeneous source (for two viewing angles: from above at 100° and from the side at 140°). The different lines in each plot correspond to different magnitudes of the thermal plasma density n_0: solid line: $3 \times 10^9\,\mathrm{cm^{-3}}$, dotted line: $10^{10}\,\mathrm{cm^{-3}}$, dashed line: $3 \times 10^{10}\,\mathrm{cm^{-3}}$, dashed–dotted line: $10^{11}\,\mathrm{cm^{-3}}$, dashed–triple dotted line: $3 \times 10^{11}\,\mathrm{cm^{-3}}$. The other simulation parameters and the distribution of energetic electrons are described in the text.

more steep, and the low-frequency cutoff of the spectrum increases. On the other hand, the high-frequency (optically thin) part of the spectrum remains, in general, unaffected. The thermal plasma effects reach the optically thin range only for the highest considered plasma density ($3 \times 10^{11}\,cm^{-3}$) when we have $\nu_R = 15.6\,GHz$, which is close to the spectral peak of the "pure" gyrosynchrotron emission. The emission spectra for other electron distributions (like those shown in Figures 8.2–8.4) will be affected by the increasing plasma density in a similar way.

Figure 8.5a,b also allows us to analyze the effect of a self-induced electric field. The results are similar to those for low-density plasma (Figure 8.3c,d). We can see that for the viewing angle $\theta = 100°$ (nearly across the magnetic field), the self-induced electric field has almost no effect for all thermal plasma densities. On the other hand, for the oblique viewing angle $\theta = 140°$, inclusion of the self-induced electric field in the model results in a noticeable increase of the emission intensity. It is interesting to note that this effect becomes even stronger for the higher plasma densities: the self-induced electric field increases the maximal emission intensity by an order of magnitude for $n_0 = 3 \times 10^{11}\,cm^{-3}$ and by a factor of about 3 for $n_0 = 10^{11}\,cm^{-3}$, in contrast to a factor of about 2 for $n_0 = 3 \times 10^{9}\,cm^{-3}$. Thus we can conclude that, while variations in thermal plasma density can significantly affect the shape of gyrosynchrotron spectra at lower frequencies, the general effects of varying beam electron distributions (due to collisions, converging magnetic field, and self-induced electric field) on those spectra remain rather similar to those for ambient plasmas with different densities.

8.4.2
Gyrosynchrotron Emission from a Whole Coronal Magnetic Tube

Let us investigate now MW emission from the whole coronal magnetic tube (spatially integrated in depth), as observed by instruments without imaging capabilities.

As one can deduce from the previous section, variations of the electron distributions with precipitation depth affects noticeably the parameters of MW emission generated at different layers of a coronal magnetic tube. In addition, inhomogeneity of the magnetic field itself results in significant variations of the emission parameters with height, showing that with an increase of the magnetic field magnitude, the emission intensity increases and the spectral peak shifts toward higher frequencies. As a result, regions with the strongest magnetic field (near the footpoints) make a dominant contribution to the MW emission of an active region, especially at high frequencies. The question is how a self-induced electric field will affect this MW emission and polarization.

We assume that all the parameters of the emission source depend only on the z-coordinate, a linear distance along the magnetic tube related to the column density as described in Section 8.4. We also assume that the tube is relatively thin and the angle θ between the line of sight and the tube axis (magnetic field) does not differ much from 90°, so we can neglect variations of the source parameters along a single ray trajectory and consider the emission source as a superposition of separate quasihomogeneous sources with the visible areas $dS = D(z)dz$, where $D(z)$

is the tube diameter at height z. In this case, the total emission intensity can be estimated as

$$I_\sigma = \frac{1}{\mathcal{R}^2} \int_{z_1}^{z_2} \frac{j_\sigma(z)}{\varkappa_\sigma(z)} \left[1 - e^{-\varkappa_\sigma(z) L(z)}\right] D(z) dz \,. \tag{8.15}$$

Due to the requirement of the magnetic flux conservation, the tube diameter varies with depth as $D(z) = D(z_0)\sqrt{B(z_0)/B(z)}$ and the source depth at a given height can be estimated as $L(z) = D(z)/\sin\theta$. In our calculations, we used the following source parameters: tube diameter at the footpoint $D_0 = 5000$ km, magnetic field at the footpoint $B_0 = 780$ G, distance between the electron injection point and the transition region $z_c = 10\,000$ km, which corresponds to the thermal plasma density $n_0 = \xi_c/z_c = 10^{11}\ \mathrm{cm}^{-3}$ (note that $n_b \ll n_0$ now).

In this section we aim to investigate how the evolution of energetic particle distribution during propagation affects MW emission. This influence is expected to be qualitatively similar for all feasible source geometries. On this basis, possible variations in the parameters of the plasma, magnetic field, and energetic particles across the magnetic tube in the above model are neglected. Also the loop curvature is neglected as well, that is, the viewing angle θ is assumed to be constant along the tube.

The spatially integrated emission spectra are shown in Figure 8.6 for the different simulation models (initial energy fluxes of energetic electrons of $F = 10^{12}\ \mathrm{erg\,cm^{-2}\,s^{-1}}$, initial power-law indices of $\gamma = 3$ and 7). For models C and C+E (pure collisions and collisions + self-induced electric field), the variation in the magnetic field with depth is taken into account when calculating the emissivity

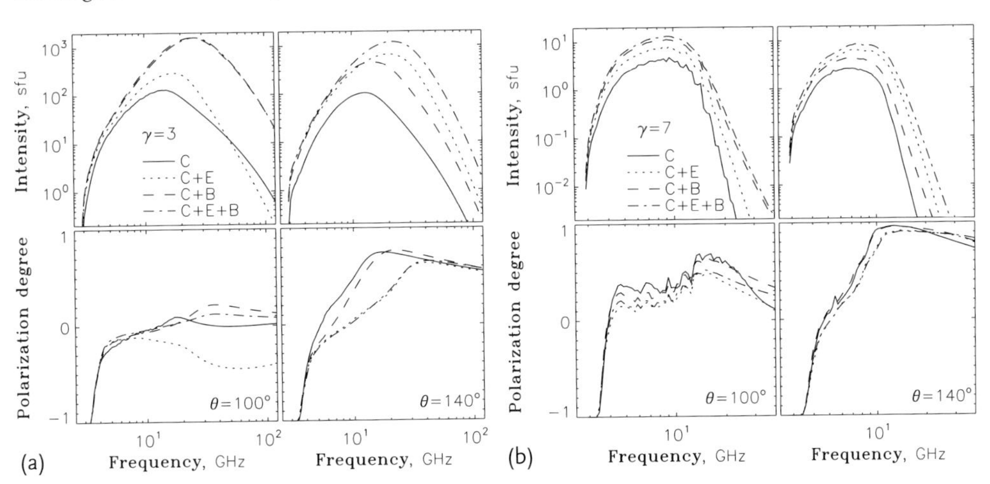

Figure 8.6 Intensity and polarization of gyrosynchrotron emission from a whole coronal magnetic tube (from different viewing angles). The used electron distribution functions are obtained by solving the Fokker–Planck equation; the adopted parameters and source geometry are given in the text. The different lines correspond to different simulation models (Figure 8.1).

and absorption coefficient while the effect of a converging magnetic field on the distribution function is neglected.

Figure 8.6a shows the calculation results for a hard electron beam ($\gamma = 3$). Firstly, we can notice that for the viewing angle $\theta = 100°$, the main factor affecting the emission is the magnetic field convergence. Models with a converging magnetic field (C+B and C+E+B) provide much higher emission intensity than the model without this factor (C and C+E). The effect of the self-induced electric field is relatively weak and is visible only if we "turn off" the magnetic field convergence: the model with the return current (C+E) provides a higher peak intensity but a steeper intensity decrease with frequency than the purely collisional model (C); the difference in polarizations is noticeable as well. In the two models with magnetic field convergence but either with (C+E+B) or without (C+B) a self-induced electric field, the MW intensity and polarization are almost the same.

On the other hand, for $\theta = 140°$, the effect of a self-induced electric field exceeds that of a converging magnetic field: the model with collisions and self-induced electric field (C+E) provides a higher emission intensity than the model with collisions and magnetic field convergence (C+B). And the highest MW intensity is reached when taking into account the combined effect of a self-induced electric field and a converging magnetic field in addition to collisions (C+E+B). Also, a self-induced electric field has a dominating effect on emission polarization: we can see that the polarization plots are grouped in pairs according to whether the self-induced electric field is considered (C+E, C+E+B) or not (C, C+B).

The combined effect of a self-induced electric field and a converging magnetic field turns the electrons moving upward to such pitch angles that their distribution maximum takes place at about $\mu = -0.7$–0.8 (Zharkova and Gordovskyy, 2006), so that the direction where the particles emit the MW (and HXR) radiation corresponds to the intermediate viewing directions ($\theta \simeq 120$–$150°$). As a result, the account for the self-induced electric field (in the C+E+B model) results in an increase of the maximal intensity by a factor of about 3 (in comparison with the C+B model) and in a noticeable shift of the spectral peak from 15 to 23 GHz.

Figure 8.6b shows the calculation results for a softer electron beam ($\gamma = 7$). The emission intensity is now far less than for $\gamma = 3$. The effects of the converging magnetic field and self-induced electric field are qualitatively similar to those for harder beams. However, the relative differences between different models are now somewhat smaller – for example, for the viewing angle $\theta = 140°$, the maximal intensities for the C+B and C+E+B models differ by a factor of about 2. The polarization plots reveal very small difference between the different simulation models. All the above features are caused by the fact that the evolution of soft electron beams during propagation is dominated by collisions, that is, only a small fraction of the energetic particles can reach the loop footpoints (where the effects of particle anisotropy on emission are strongest). In addition, the stronger (in comparison to other factors) collisions produce more isotropic electron distributions than for harder beams. As a result, the relative importance of both the converging magnetic field and self-induced electric field (in comparison to collisions) decreases with an increase in the particle power-law index. Nevertheless, for beams with

$\gamma \simeq 7$, the effect of these factors on MW emission intensity is noticeable and must be considered when interpreting observations.

For relatively weak flares and limb events, these factors are proved to be negligible. However, for more powerful flares (caused by electrons with higher electron fluxes and higher[1] spectral indices) located not very far from the disk center, the self-induced electric field is shown to produce a substantial effect on emission intensity, spectrum shape, and polarization, which can provide a much closer fit to the MW observations of solar flares, like those presented in Section 8.5.

8.5
Comparison with Observations

8.5.1
Flare of 23 July 2002

We consider the flare of 23 July 2002. This X4.8 class flare was observed by many ground-based and space-borne instruments (see e.g., Lin *et al.*, 2003, and other articles published in the same journal issue). The Nobeyama Radio Polarimeter (Nakajima *et al.*, 1985) provided the spectra of the solar MW emission. Figure 8.7 shows a sample MW spectrum in the rise phase of the flare. The hard X-ray observations (Holman *et al.*, 2003) indicate that the energetic electrons at that time had a double power-law spectrum with spectral indices of $\gamma_L \simeq 4.5$–5.5 (below the break energy) and $\gamma_H \simeq 6.5$–7.5 (above the break energy) and an energy flux $F \gtrsim 10^{12}\,\mathrm{erg\,cm^{-2}\,s^{-1}}$. The magnetograms reveal the presence of two spots (loop footpoints) with magnetic fields of about -650 and $+500$ G (Zharkova *et al.*, 2005a) by a distance of about 10 000 km. The height of the magnetic loop can be estimated as 10 000–20 000 km, and the loop diameter near the footpoints was on the order of several thousand kilometers. The rise phase was chosen for analysis because at that time, as we expect, the emission was produced mainly by the precipitating and reflected energetic electrons, while at the latter phases of the flare the accumulated (near the top of the magnetic loop) electrons gave a significant contribution.

The loop occurred at the heliographical coordinates 13° S 72° E, so for the radial (perpendicular to the solar surface) magnetic field we obtain the angle between the magnetic field and line-of-sight of about 110°. However, the observations of γ-ray emission (Hurford *et al.*, 2003a) and HXR polarization (Emslie *et al.*, 2008) suggest that there was a tilt of about 30–40°, so the actual viewing angle with respect to the magnetic field was about 140–150°. We consider here both cases of the loop orientation (with and without the tilt).

We use the source model similar to the one described in the previous section (a loop curvature is neglected). The emission is assumed to be a sum of the emissions

1) For the same energy flux, softer electron beams produce stronger X-ray emission since they release the bulk of their energy at higher atmospheric layers, while the energy of harder beams goes mainly to plasma heating in the lower chromosphere (Zharkova and Gordovskyy, 2006), thus producing weaker X-ray flares.

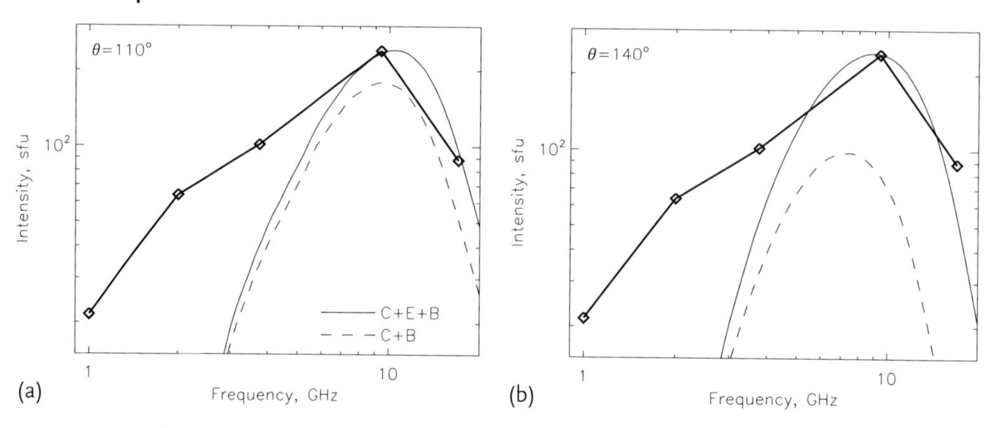

Figure 8.7 NoRP microwave emission spectrum for the 23 July 2002 flare (at 00:26:00 UT), together with the theoretical predictions.

from the two magnetic tubes with the footpoint magnetic fields of $B_1 = -650$ G and $B_2 = +500$ G, respectively. The footpoint tube diameters satisfy the relation $D_1/D_2 = \sqrt{|B_2/B_1|}$. The distance from the electron injection point to the transition region is taken to be $z_c = 20\,000$ km and the plasma density equals $n_0 = 5 \times 10^{10}\,\text{cm}^{-3}$ for both tubes. The energy flux of the accelerated electrons is $F = 10^{12}\,\text{erg}\,\text{cm}^{-2}\,\text{s}^{-1}$ and the power-law index $\gamma = 5$.

The theoretical curves at Figure 8.7a are calculated for the viewing angle $\theta = 110°$ (i.e., without a tilt) and tube diameters at the footpoints $D_1 = 6000$ km and $D_2 = 6850$ km. In Figure 8.7b, we assume a tilt, so the viewing angle is $\theta = 140°$ and the tube diameters are taken to be $D_1 = 8000$ km and $D_2 = 9120$ km. We can see that the model we used allows us to reproduce the maximal emission intensity and the spectral peak frequency. With decreasing frequency, the theoretical model predicts a faster decrease in emission intensity than is really observed. This is because our model does not consider variations in the plasma density across the magnetic tube: as was shown above, the emission frequency is limited from below by the local plasma frequency (which was equal to 2 GHz in our simulations); the Razin effect also suppresses the radiation at low frequencies. In real magnetic loops, a decrease in the plasma density with distance from the loop axis (across the loop) would allow the MW emission to be generated at lower frequencies, so the spectrum would look like a superposition of different spectra (corresponding to the different plasma densities) in Figure 8.5. The simulation made by Melnikov *et al.* (2008) confirms that the account for the plasma inhomogeneity allows us to reproduce the observed MW spectra with a flat low-frequency slope.

The effect of the self-induced electric field is clearly seen for both viewing directions as plotted in Figure 8.7a,b. As we showed in Section 8.4.1, the self-induced electric field (combined with the magnetic field convergence) increases the number of electrons moving upward with the intermediate pitch angles (between $\alpha = 90°$ and $\alpha = 180°$). In turn, this results in an increase in the MW emission intensity at in the viewing directions $\theta \sim \alpha$. For viewing angles close to $\theta = 90°$, this effect is relatively weak (Figure 8.7a). However, for larger viewing angles (e.g., the case shown in Figure 8.7b), a neglect of the self-induced electric field decreases the calculated emission intensity by more than a factor of 2 (up to 10). Thus, the diagnostics of energetic particles based on models without a self-induced electric field would overestimate by this factor the source size (or the abundance of energetic electrons).

The simulations carried out for both viewing angles can reasonably account for the MW observations of this flare. On the one hand, our simulations do not require a loop tilt in the 23 July 2002 event, on the other hand they do not exclude it either. However, for MW emission with a viewing angle of $\theta = 110°$ a smaller source size is required, and models with a converging magnetic field with and without an electric one give similar results. At the same time, the model simulations of MW emission for $\theta = 140°$ without an electric field are much lower than those observed; the observations can be fit pretty closely only if the electric field is taken into account. Thus, the interpretation of the source properties based on MW observations only does not allow us to select the preferred viewing angle. This can be related mainly to the uncertainties of observations (such as magnetic field measurements, loop size estimations, etc.) with some contribution related to the relative simplicity of the model.

8.5.2 Flare of 10 March 2001

The 1B/M6.7 flare of 10 March 2001 associated with a coronal mass ejection (CME) occurred in active region 9368 (N27W42) and was previously studied in detail in a number of papers (Altyntsev *et al.*, 2008; Chandra *et al.*, 2006; Ding, 2003; Liu *et al.*, 2001; Uddin *et al.*, 2004).

8.5.2.1 Light Curves of MW and HXR Emissions

As shown by Altyntsev *et al.* (2008), there were broadband short pulses present in the whole frequency range 1–80 GHz of NoRP (see time profiles in Figure 8.8a). The MW time profiles in total intensity at different frequencies were very similar. Three main peaks are detectable in the MW burst with a total duration of about 40 s (Figure 8.8a). The first MW peak at about 04:03:30.90 (*all times hereafter are UT*) was the lowest one; the second (central) MW peak at 04:03:40 with a total duration of about 5 s was the highest one. The third MW peak had a double structure with two subpeaks occurring at 04:03:44.90 and 04:03:46.80, respectively.

The time profiles of the polarized MW emission (Stokes V component) around the central peak are shown in Figure 8.8b. They coincide with the light curves shown by Altyntsev *et al.* (2008). The MW emissions of all three peaks had a positive (right-handed) circular polarization (RCP) at low frequencies (1–3.75 GHz), and their polarization at higher frequencies (9.4–35 GHz) was negative (left-handed, LCP), as Figure 8.8b shows.

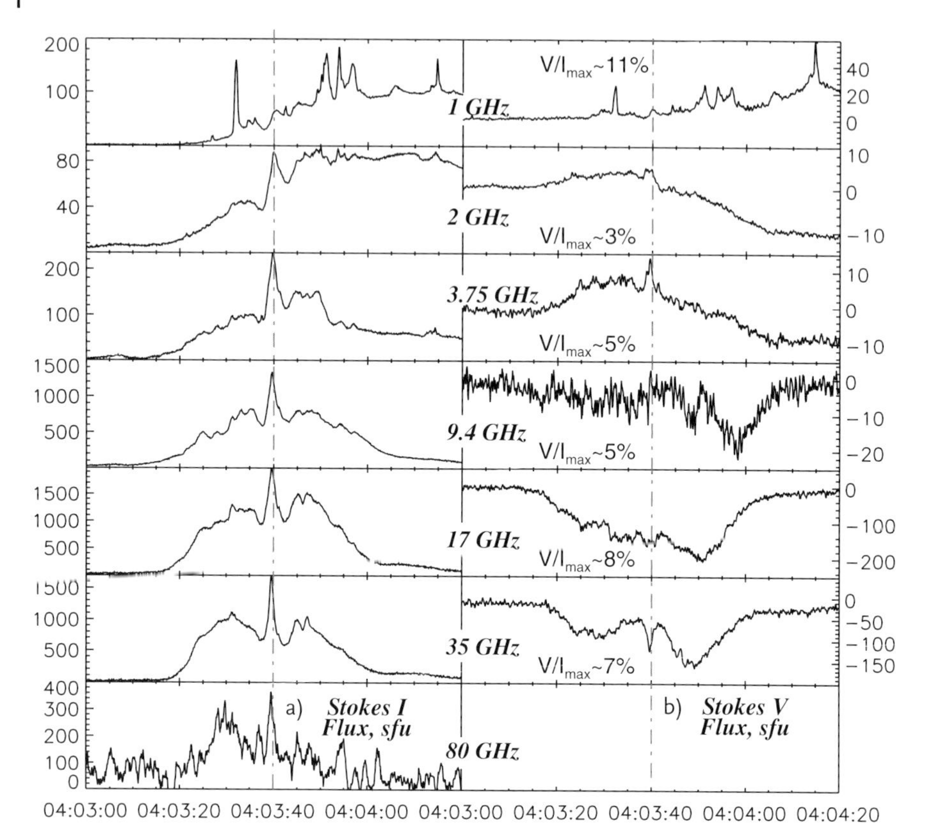

Figure 8.8 The 2001 March 10 event. Time profiles of the average brightness temperature at 1–80 GHz (NoRP) in total intensity (a) and polarization (b). (V/I_{max}) is the maximum degree of polarization at the central peak (04:03:40). The numbers at the y-axes are in the solar flux unit (sfu).

In order to understand the dynamics of high-energy particles in this flare, let us also use the whole energy range of the HXR emission recorded by Yohkoh/HXS (Hard X-Ray Spectrometer) and estimate parameters of the electron beam, thus extending studies carried out by Altyntsev *et al.* (2008) and Chandra *et al.* (2006) that were confined to the MW emission.

This flare was observed by Yohkoh/HXT (Hard X-Ray Telescope) in four energy channels: L, M1, M2, and H. Thus the available energy bands of HXR channels allowed us to derive photon spectra up to 600 keV. The HXS light curve presenting an integrated flux within a range of 20–657 keV is plotted along with HXT and NoRP data as shown in Figure 8.9. The HXS light curve is plotted as a histogram to reveal its temporal correspondence with the HXR and NoRP data. Sometimes the HXS time bin could contain emissions integrated over two peaks seen by HXT and at 17 GHz (for example, this happened at about 04:03:47). Fortunately, the temporal variations of HXR and NoRP emissions during the main peak resemble the HXS light curve.

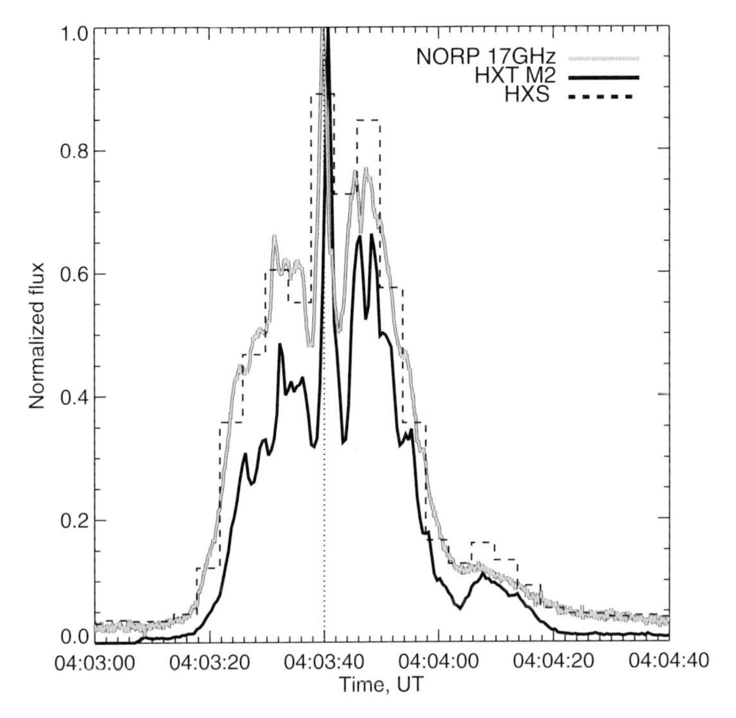

Figure 8.9 Light curves of the HXR burst registered in the HXT/M2 channel (33–53 keV) and in the HXS record integrated over the whole band (20–657 keV, histogram) along with the microwave burst registered by NoRP at 17 GHz. The vertical dotted line marks the main peak under consideration at 04:03:40 UT.

The time profiles of the HXR and 17 GHz emissions have a close temporal correlation (Figure 8.9). Nevertheless, there are some delays of HXR emission relative to MWs. They amount to a fraction of a second at high frequencies and reach 1–2 s at low frequencies (Altyntsev *et al.*, 2008).

8.5.2.2 MW and HXR Images

The flare was triggered by an interaction between two loops, one of which was a small, newly emerging loop, the other a large overlying one (Chandra *et al.*, 2006). The two loops formed a "three-legged" configuration in which the magnetic field had a "bipolar + remote unipolar" structure. The footpoints of the small loop (main flare source) can be seen in HXR/H and MW images (Figure 8.10b) whose position is denoted by the white dotted frame in Figure 8.10a. The magnetogram taken at 04:48:30 was compensated for the differential rotation to 04:03:40 (the most prominent peak time). The length of the loop visible in an EIT (SOHO Extreme Ultraviolet Imaging Telescope) 195 Å image is 10 000 km according to Altyntsev *et al.* (2008). During the flare, the NoRH beam size was about $20.6'' \times 17.3''$.

The bulk of the MW emission at 17 GHz during the main peak was generated from a small loop. At the onset of the burst an LCP source located in the N-polarity magnetic field (Figure 8.10b) was observed. An RCP source appeared north of the

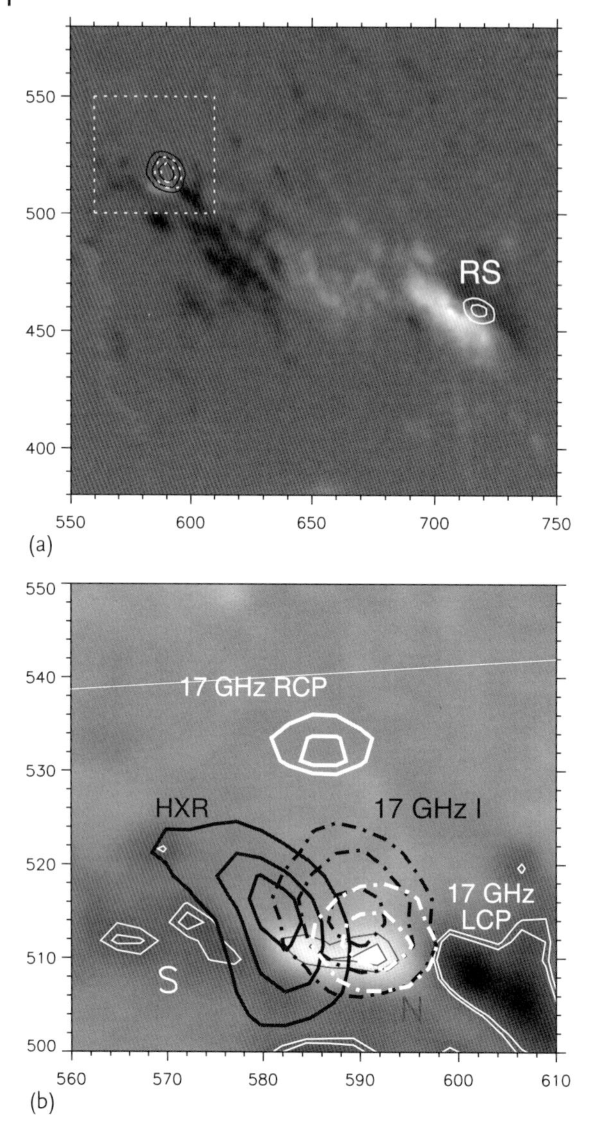

Figure 8.10 (a) Contours of 17 GHz (NoRH, 04:04) flare sources (solid black contours: Stokes I; solid white contours: Stokes V, RCP; dotted white contours: Stokes V, LCP) superimposed on an MDI magnetogram (04:48:01.61, light areas represent N polarity, dark areas S polarity). The axes show hereafter arc seconds from the solar disk center. (b) The enlarged flare site in (a) is depicted by the broken frame. Contours of 17 GHz (NoRH, 04:03:40). Thick black contours correspond to the HXR source (Yohkoh/HXT/H, 04:03:40.25), thin solid gray contours mark N polarity [levels 600, 800 G], and thin white contours mark S polarity [levels −600, −500 G].

LCP source at 04:03:27 in the S-polarity magnetic field, meaning that they both were polarized in the sense of the O mode.

The magnitude of the field in the spot labeled "N" (solid thin gray contours in Figure 8.10b) reached 800 G and for the S-polarity field labeled "S" it reached 600 G. The large loop had footpoints in the main and remote (RS) sources (Figure 8.10a). RS appeared at 04:03:51 and had a right circular polarization at 17 GHz. The right polarization in the main source disappeared after 04:03:52.

At the onset of the flare (04:03:12) a single compact source was seen only in all energy channels of Yohkoh/HXT. HXR images obtained with a 4 s cadence in the L channel showed that the source expanded and got elongated by 04:04:05. In the H channel, the compact source was transformed into a looplike one by 04:03:25, and by 04:04:05 it became compact again. The distance between the centroids of the radio and HXR sources was about 10″.

The sizes of the sources were obtained in all four Yohkoh/HXT energy channels by fitting them with ellipses at half magnitude for the positions as if these sources would be located at the solar disk center to exclude projection effects. The estimated sizes of the HXR sources were $16'' \times 12.6''$, $18'' \times 12.2''$, $15.4'' \times 12''$, and $12'' \times 16.8''$ for the L, M1, M2, and H channels, respectively.

8.5.2.3 **Deduced Parameters**

Our estimations of the new HXR emission gave a total energy flux of about 1.5×10^{11} erg cm^{-2} s^{-1} for energies from 10 to 100 keV and $\gamma = 2.4$, which was very close to that estimated by Altyntsev *et al.* (2008). We extended the energy range to 200 keV and calculated the spectral index $\gamma = 3.12$, and the break energy was about 80 keV. Considering that the true spectrum of accelerated particles is determined by particles in an energy range of 100–200 keV, we prolonged the spectrum to the low-energy range and calculated a total energy flux of $F = 10^{12}$ erg cm^{-2} s^{-1}.

This difference in spectral indices and energy fluxes for lower- and higher-energy bands resembles very closely the HXR photon spectra produced in models of electron beam precipitation with a self-induced electric field, which causes their flattening toward lower energies (Zharkova and Gordovskyy, 2006). This indicates a need to consider a model of electron-beam precipitation taking into account ohmic losses in addition to collisions and to adopt for the beam parameters those derived for the higher-energy range.

We needed to use the corrected magnitudes of magnetic field in the footpoints, which increased with the new calibration to 600–800 G, compared with magnitudes of up to 200 G accepted in a previous study by Altyntsev *et al.* (2008) (the importance of correcting the magnetic field strength to reconcile MW and HXR spectra was recently demonstrated by Kundu *et al.* (2009)). Also, the calculations of HXR and MW emissions in the current paper were carried out for a visible HXR source area of about 150 arcsec^2 (Section 8.5.2.2). Given the fact of broadband MW emission in a wide frequency range (1–80 GHz) and the simultaneous presence of HXR emission from 10 to 200 keV and assuming they both were produced by the same populations of electrons, we needed to accept an energy range of beam electrons of 12 keV to 1.2 MeV.

Another improvement over the previous simulation model was related to a consideration of a highly inhomogeneous flaring atmosphere where density and temperature variations are derived as a result of the hydrodynamic response to heating caused by the injection of beam electrons (Zharkova and Zharkov, 2007). This model led to significant depth variations not only of the physical conditions in the ambient plasma but also of the beam electrons themselves, which was contrary to homogeneous models of both ambient and beam electrons considered by Altyntsev *et al.* (2008).

8.5.3 Simulated HXR and MW Emission

Both HXR and MW emissions are assumed to be produced by the same population of power-law beam electrons with a spectral index of 3 and energies ranging from 12 keV to 1.2 MeV as derived in Section 8.5.2.3, which are steadily injected into a flaring atmosphere with a pitch-angle dispersion of 0.2. The electron distribution functions for different models of energy losses (C, CE, CB, and CEB) are discussed in detail in Siversky and Zharkova (2009b), Zharkova *et al.* (2010), and Kuznetsov and Zharkova (2010). These define electron-beam dynamics and pitch-angle distributions at every precipitation depth. The main issue in electron distributions is the formation of returning electrons either by magnetic field convergence or by a self-induced electric field of precipitating beam electrons. These two beams, precipitating and returning ones, produce either HXR emission in scattering on ambient plasma particles or braking in electric field and MW gyrosynchrotron emissions from their gyration in strong converging magnetic fields.

8.5.3.1 HXR Emission and Directivity

For HXR emission we calculate photon emission spectra integrated over all atmospheric depths, polarizations, and directivities normalized on the average intensity over all angles for relativistic cross-sections with pitch-angle and viewing-angle dependence as described below.

$$\begin{bmatrix} I \\ Q \end{bmatrix}(h\nu, \theta) = \frac{A}{2\pi \mathcal{R}^2} \int_0^{\xi_{\max}} d\xi \int_{h\nu}^{\infty} v(E) dE \int_{-1}^{1} f(\xi, E, \mu) d\mu$$
$$\times \int_0^{2\pi} \begin{bmatrix} \sigma_I \\ \sigma_Q \end{bmatrix} (h\nu, \theta, E, \mu, \varphi) d\varphi \,, \qquad (8.16)$$

where A is the cross-section area of a magnetic tube, R is the astronomical unit, σ_I, σ_V, I, and Q are the Stokes parameters related to the intensity and linear polarization of the emission.

Then a degree of linear polarization can be defined as follows:

$$\eta = \frac{Q}{I} \,, \qquad (8.17)$$

and the distribution of any emission in the angles θ can be described by a directivity:

$$D(\theta) = \frac{I(\theta)}{\langle I \rangle}, \tag{8.18}$$

where $\langle I \rangle$ is the emission intensity averaged over all angles. The 3-D integrals over the electron velocity in Eq. (8.16) are calculated using a Monte Carlo method.

Simulations of HXR photon spectra for different precipitations models (C, CB, CE, and CEB) including those emitted downward (to the photosphere) and upward (to the observer) and for different magnitudes of viewing angles are plotted in Figure 8.11 from the top to the bottom, respectively. On top of the simulated energy spectra we overplotted by a solid line the energy spectrum deduced from the observations (Section 8.5.2.2).

The plots of HXR emission simulated for different directions of electron propagation (downward with pitch-angle cosines $\mu > 0$ and upward with $\mu < 0$) reveals that in the corona the majority of electrons move in the downward direction while only half as many move upward. In the chromosphere, it is the opposite: most electrons move upward and only a smaller fraction of them keep moving downward. Thus, for this flare both effects – albedo and ohmic losses – have to be very significant at upper (coronal) precipitation depths.

The spectra calculated for the CB model (collisions and converging magnetic field) show the highest intensity compared to CE and CEB models (Zharkova *et al.*, 2010). This is because the electric field induced by precipitating electron beams prevents the particles from reaching deeper layers where the bulk of electrons (with lower cutoff energy) can emit HXR photons, thereby reducing the intensity. The CE model provides a smaller intensity than a CB model, while the intensities for the CEB model are much smaller than those calculated for the models CB or CE. This happens because the combined effect of the two factors (convergence and electric fields) significantly reduces the number of particles that are able to precipitate into deeper atmospheric layers where the bulk of electrons can emit HXR photons. Zharkova *et al.* (2010) have shown that for $F = 10^{10}\,\mathrm{erg\,cm^{-2}\,s^{-1}}$, the results were the opposite: the CE model provides a higher HXR intensity than the CB model. This confirms the conclusion by Zharkova *et al.* (2010) that for $F = 10^{12}\,\mathrm{erg\,cm^{-2}\,s^{-1}}$ the effect of a self-induced electric field is much stronger than for a weaker beam.

It can also be noted that the observed HXR photon spectrum has the best fit by HXR emission emitted upward and downward (the full albedo effect) under a viewing angle of 180° by an electron beam with given parameters for the CE (collisions and electric field) and CEB (collisions, electric field, and converging magnetic field with convergence equal to 3) precipitation models. The HXR spectrum observed in this flare reveals a noticeable flattening toward lower energies that is better reproduced by a CE model, indicating a significant effect of a self-induced electric field in energy losses by beam electrons. The observed and simulated energy spectra reveal the best fit for a viewing angle of 180° (the direction toward the observer looking from the top). Given the location (N27W42) of the flare, one can assume that the flaring loop emitting HXR photons is tilted toward the observer by 10–40°.

At the same time the directivity of HXR emissions (see Figure 8.12) is found to range between 2 and 3 for electrons propagating downward and about 0.5 for electrons moving upward. This points to a clear anisotropy of the electron beam producing this HXR emission with dominance of the downward moving electrons. In order to see this emission from the top where the observer is placed in our models, HXR photons must be reflected by the photosphere, e.g., the photospheric albedo effect should play a significant role (Kontar *et al.*, 2006). In the present study

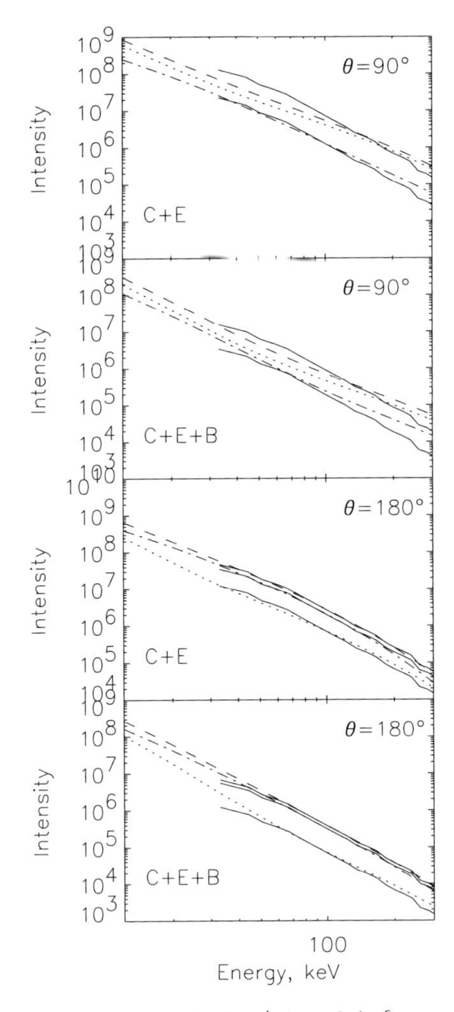

Figure 8.11 Intensity (in relative units) of HXR emission spectra for different energies, models, and propagation directions. Dashed–dotted line: upward propagating particles, dotted line: downward propagating particles, dashed line: total emission (downward + upward), solid line: observational data.

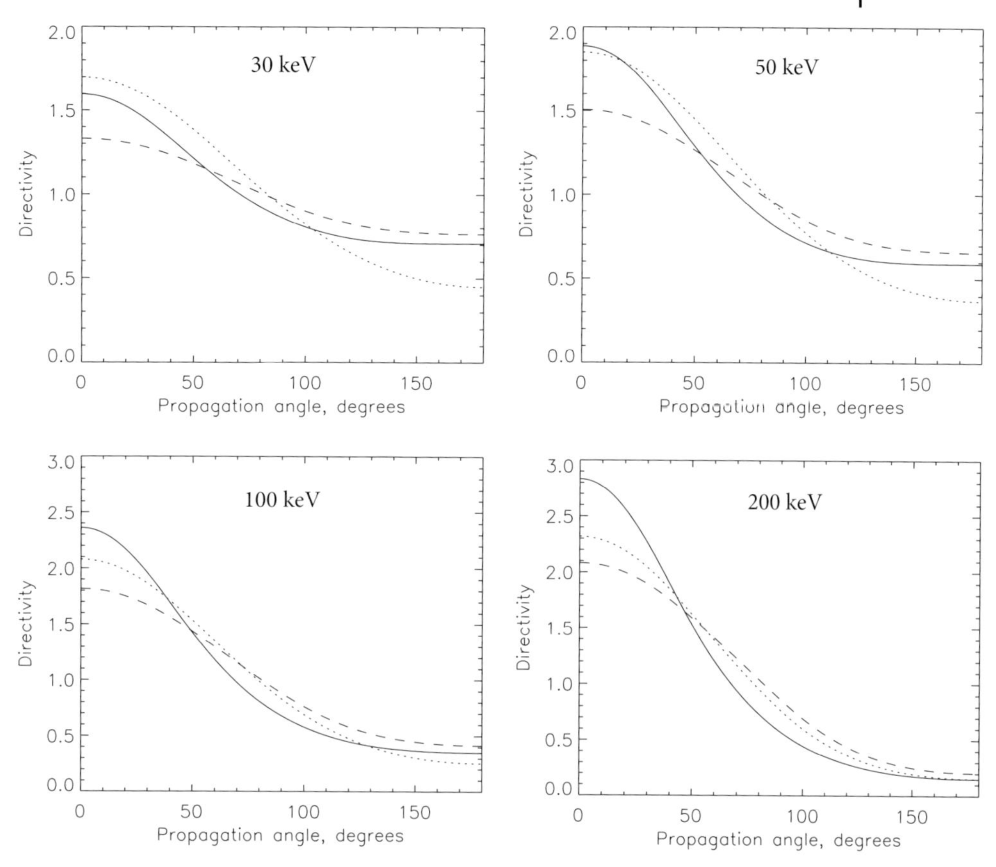

Figure 8.12 Directivity of HXR emission calculated for electron energies of 30, 50, 100, and 200 keV for different models of electron energy losses and propagation directions. Solid line: CE model, dotted line: CB, dashed line: CEB model. Propagation direction indicates pitch angle.

we have assumed a 100% albedo coefficient for downward moving emissions, while in reality this can have a more complicated dependence on the energy of electrons (Kontar *et al.*, 2006) and their pitch-angle distribution. However, this effect requires one to consider a strong electron anisotropy, described above, contrary to isotropic electron distributions, considered by Kontar *et al.* (2006) discussed in a forthcoming paper.

8.5.3.2 MW Emission

We now calculate the MW emission parameters. The model of the emission source is similar to that used by Kuznetsov and Zharkova (2010): we assume that the parameters of the coronal magnetic tube (such as the plasma density, magnetic field, and the parameters of the accelerated electrons) depend only on the z-coordinate, a linear distance along the tube. Also, we assume that the tube is relatively thin and the propagation angle θ does not differ much from $90°$, so that the source can be considered a nearly homogeneous one along the line of sight (variations of the source parameters along a single ray trajectory are negligible). In this case, the emission from a whole magnetic tube equals the sum of emission from separate quasihomogeneous sources (corresponding to different heights), that is (Kuznetsov and Zharkova, 2010):

$$I_\sigma = \frac{D}{\mathcal{R}^2} \int_0^{z_{max}} \frac{j_\sigma(z)}{\varkappa_\sigma(z)} \left[1 - e^{-\varkappa_\sigma(z)L}\right] dz \,. \tag{8.19}$$

Here I_σ is the emission intensity (observed from Earth) of the magnetoionic mode σ, D and L are the visible diameter of the magnetic tube and the source depth along the line of sight, respectively, and $\mathcal{R}$ is the astronomical unit. The formulas for the gyrosynchrotron plasma emissivity j_σ and absorption coefficient $\varkappa_\sigma$ are given by Kuznetsov and Zharkova (2010). The polarization degree is defined as

$$\eta = \frac{I_X - I_O}{I_X + I_O} \,, \tag{8.20}$$

where I_O and I_X are the intensities of the ordinary and extraordinary modes, respectively. The directivity of MW emission is calculated similarly to that of HXR emission, Eq. (8.18).

However, unlike the previous paper (Kuznetsov and Zharkova, 2010), in the present study we use the temperature and density of a flaring model derived from hydrodynamic simulations (Zharkova and Zharkov, 2007) with the distance z from the injection point varying from 0 to $z_{max} \simeq 10\,000$ km, the thermal plasma density varying from 2×10^9 to 2×10^{13} cm^{-3}, and a magnetic field strength of $B = 780$ G at a characteristic column depth of $\xi_c = 10^{20}$ cm^{-2} (or $z = z_{max}$). In models with a converging magnetic field, the field strength varies with depth from the top of the emission source to this characteristic depth (Siversky and Zharkova, 2009b), which is another factor (in addition to the variations of a distribution function) affecting the MW emission parameters; we used the models with the convergence factors $B_{footpoint}/B_{top} = 2$ and 3. In the CE model, variation of the magnetic field with height (with the convergence factor 3) was considered when calculating the plasma emissivity and absorption coefficients, but the effect of a converging magnetic field on electron distribution was neglected. Also, we assumed that the loop width was $D = 10\,000$ km and the source depth along the line of sight was $L = 10\,000$ km; these parameters agree with the imaging observations and provide the best agreement of the calculated MW spectra with the observed ones (see below).

MW emission was simulated for the following models described in Section 8.2 (CB, CE, and CEB) and for different viewing angle values (Figure 8.13). Note that for the viewing angle $\theta = 110°$, the main factor affecting MW emission is magnetic field convergence. MW intensities have a rather flat maximum at about 20 GHz approaching the magnitude above 10^3 sfu for models with a magnetic field convergence (CB and CEB models) and reducing the maximum below 10^3 sfu to 10 GHz without it (CE model). In models including a magnetic field convergence without

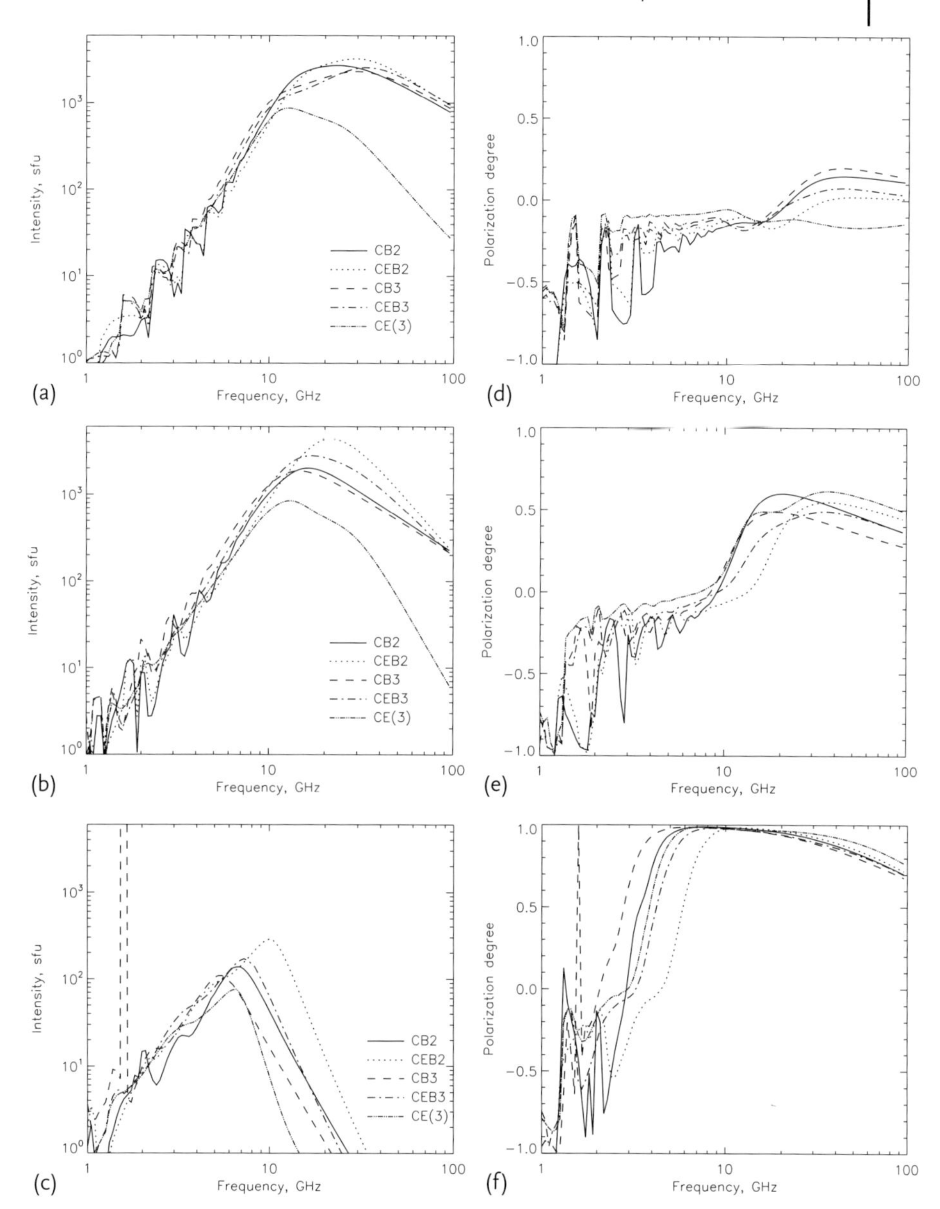

Figure 8.13 Intensity (a–c) and degree of polarization (d–f) of MW emission spectra calculated for the different precipitation models and propagation directions: for viewing angles of 110° (a,d); 140° (b,e) and 170° (c,f). Numbers in the model abbreviations define the factor of magnetic field convergence, for example CB2 or CEB3, while the number in parentheses in CE(3) denotes simulations for DF calculated for the CE model with magnetic field added in the MW intensity calculations.

(CB) or with (CEB) an electric field, the plots of MW intensity and polarization are rather close with the intensities decreasing for a convergence factor of 3 compared to 2. The models with a converging magnetic field (CB and CEB) provide much higher emission intensity than the model without this factor (CE).

The effect of a self-induced electric field is relatively weak and is visible only if the magnetic field convergence is equal to zero, for example, the model with a return current (CE) provides a lower peak intensity and a steeper intensity decrease toward higher frequencies than the collisional model with magnetic convergence (CB). The MW intensities at higher energies are power laws with spectral indices of about 2.4 for CEB models and higher than 4 for the CE model. The MW frequency distributions toward lower frequencies reveals also power laws with a lower spectral index of 2, demonstrating a strong harmonic structure from frequencies of 8 GHz toward zero.

The polarization produced by beams for this viewing angle of 110° in CE models always has a negative sign and ranges in a small percentage interval, while the polarization produced by CB or CEB models with a convergence factor of 2 or 3 is positive (X mode dominates) for higher frequencies approaching a small percentage and crossing zero between 10 and 11 GHz, after which it becomes negative (O mode dominates). The harmonic structure in MW polarization is more pronounced than that in intensity, and it increases with the growth of the magnetic field convergence.

For a viewing angle of $\theta = 140°$, the effect of a self-induced electric field becomes much more significant than the effects of a converging magnetic field: models with collisions and a self-induced electric field (CEB) provide a higher MW intensity than models with collisions and a magnetic field convergence (CB). The maximum MW intensity increases for models with an electric field and a magnetic field convergence (CEB) compared with those at 110° and frequency of maximum shifts from 11 to 12 GHz for 140°. The power-law spectrum at high frequencies has spectral indices ranging from 2.4 to 2.8 for CEB3 and CEB2 models, respectively. While the lower-energy part of MW intensity resembles the power-law distributions found for 110° with wavelike oscillations with slightly larger amplitudes.

A self-induced electric field has a significant effect on MW polarization, which now changes the sign for all models, and the frequency of this change shifts to magnitudes below 10 GHz. The MW polarization degree for CB models is higher than those with an electric field (CEB) approaching 50% at 11 GHz for the CB2 model with a convergence factor of 2 and reducing to 40% for CB3 with a convergence factor of 3. Polarization for CEB models becomes slightly lower than that for CB models with a rather flat distribution toward higher frequencies and strong wavelike oscillation toward lower frequencies with zero points between 9 and 10 GHz.

For a viewing angle of 170° the effect of a self-induced electric field is still noticeable for all the models, although the model with collisions, a converging magnetic field, and a self-induced electric field (CEB) provides a slightly higher MW intensity than the model with collisions and electric field (CE) only. However, the maximum MW intensity is significantly (by an order of magnitude) reduced for all the models with maxima ranging from 7 to 10 GHz for CE and CEB models, respectively. The power-law spectra at high frequencies become much softer having spectral indices of 4–5, while the lower-energy part of a MW intensity resembles the power-law distributions found for 110° with a harmonic structure less pronounced than for other viewing angles.

Polarization also changes significantly compared to the other viewing angles approaching about 100% from frequencies 3–5 GHz for CB3 and CB2 models, shifting to 4–6 for the CEB3 and CEB2 models, respectively. The polarization sign change is shifted to frequencies of 2 GHz for CB3 and 4 GHz for CEB2. A strong harmonic structure with very large amplitudes is still present at lower frequencies.

These distributions of MW intensity and polarization can be understood after considering the combined effect of a self-induced electric field and converging magnetic field, which turns upward moving electrons to such pitch angles that their distribution maximum occurs at about $\mu = -0.8$ (Zharkova and Gordovskyy, 2006). Hence the direction in which the particles emit the MW (and HXR) radiation corresponds to the intermediate viewing directions ($\theta \simeq 120-150°$). As a result, taking into account a self-induced electric field (in the CEB models) results in an increase in the maximal MW intensity by a factor of 3 (in comparison with the CB model) and in a noticeable shift of the spectral peak from 10 to 20–30 GHz.

The spectral index at high frequencies (for 140°) is −2.4 for the CEB3 model and −2.8 for the CEB2 model. This is close to the value of −3 derived from the observed MW intensity distribution in frequency. At low frequencies, the spectral index is about 3, which is higher than the observed one. This can mean that, though the source is large and inhomogeneous, in the numerical model the largest contribution of MW emission comes from a compact region near the loop footpoint.

8.5.3.3 **MW Emission Directivity**

Now let us consider the MW directivity plots in Figure 8.14 calculated for the same parameters as MW intensities in Figure 8.13.

Note that the directivity is also strongly dependent on the model applied in calculations. In all models there are two preferential directions for MW emission: downward and upward, with the maximum position varying for different models. For CE models calculated for 20 GHz (Figure 8.14a) and 30 GHz (Figure 8.14b) most of the MW emission is emitted in the direction toward the photosphere with a maximum at about 30°. This maximum is twice as high as the maximum MW emission occurring in the direction of 135° (e.g., toward the observer from the top). For models that include collisions and a magnetic field convergence the directivity of MW emission at 20 GHz (Figure 8.14a) still has maxima at the same viewing angle, but the magnitudes of the maxima are reduced in both directions, keeping the ratio between the downward emission and the upward emissions at about 2 as for the CE model. For MW emission at 30 GHz (Figure 8.14b) the ratio between the downward and upward emissions increases to 2.5 for the CE model, approaching 3 for the CB2 model. However, CB3 models show some isotopization of MW emission between 30 and 130°, where no preferential direction is found, while inclusion of the electric field in the CEB3 model restores this by a small ratio.

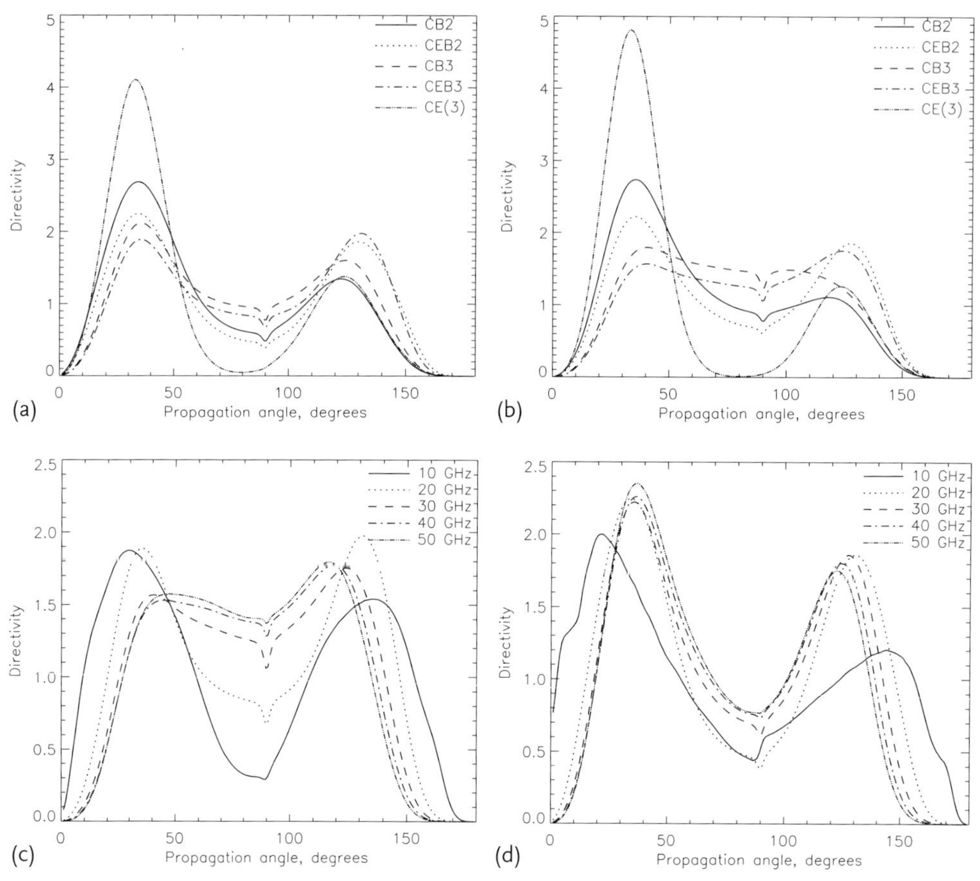

Figure 8.14 (a,b) Directivity of integrated MW emission spectra vs. different propagation directions simulated for CB, CE, and CEB models (indicated in plots) with convergence factors of 2 and 3 for 20 GHz (a) and 30 GHz (b). (c,d) MW directivity vs. propagation directions calculated for different frequencies (indicated in plots) with CEB3 model (c) and CB2 model (d).

A comparison of the CEB model simulations for different convergence factors (CEB3 in Figure 8.14c and CEB2 in Figure 8.14d) reveals that the directivity is strongest for a model with a convergence factor of 2 with the downward-to-upward emission ratio approaching 4, while for a convergence of 3 it only reaches 1.5–1.8 for the lowest frequency. There is also a visible reduction in directivity with increased frequency for any convergence, with the differences being much higher for a convergence factor of 3 (Figure 8.14c). These properties of MW directivity, in addition to intensities and polarization, help us to diagnose the electron-beam precipitation scenario in this flare.

The simulations of kinetics of a single electron beam with a wide energy range of 12 keV to 10 MeV precipitating into a flaring atmosphere being heated by this beam via hydrodynamic response are applied to simultaneously interpret the observed MW and HXR emission.

8.5.3.4 Hard X-Ray Emission

A comparison of the distributions vs. energy of simulated and observed HXR photon emissions (downward, upward, and their sum) derived from the YOHKOH instrument is presented in Figure 8.11 for the two viewing angles of 90 and 180° and two sets of models: collisions and electric field (CE) and collisions, electric field, and magnetic field convergence of 3 (CEB). It is obvious that the observed HXR photon spectrum cannot be fit by any models for the viewing angle of 80° derived from MW observation by Altyntsev *et al.* (2008). However, we can assume that HXR and MW emissions come from different parts of the loop, and this discrepancy between the derived viewing angles can be related to a loop curvature.

Although the agreement between HXR observations by the YOHKOH payload and the proposed models (shown in Figure 8.15 on a linear scale to amplify the differences) is much better for simulations carried out for a viewing angle of 180° (e.g., upward along the magnetic field) for models that include an electric field (CE and CEB). It is evident from the residuals that the fit is better for the CE model (collisions and electric field losses) combining upward and downward emissions (the full albedo effect).

This fit can indicate that, first, albedo effects are rather important in the interpretation of HXR emission produced by well-collimated beams, in addition to isotropic ones proposed earlier (Kontar *et al.*, 2006); and second, the observed HXR emission is likely to come from a part of a flaring atmosphere not affected by a magnetic field convergence, for example, from column densities larger than the characteristic column density discussed in Section 8.2.

8.5.3.5 MW Emission and Polarization

The observed distribution in frequency of MW emission and polarization at nine frequencies was fit by the MW emission simulated for different viewing angles (180, 140, and 110°) and for different models, including collisions, electric field, and magnetic field convergence, but only a viewing angle of 140°. We managed to obtain a reasonable fit to the observations, as presented in Figure 8.16.

The observed MW spectrum and polarization shown by asterisks were created using the NoRP data plotted in Figure 8.8a and the polarization variations were taken from Figure 8.8b. The model used for fitting the observations in the paper by Altyntsev *et al.* (2008) is represented in the plot by the dotted–dashed line. The errors in MW emission measurements where they were specified by observers (e.g., for 80 GHz – 40%) are shown by bars. For the remaining frequencies of MW observations the errors were estimated from a sum of the statistical and observational errors giving total errors of about 25%. Errors in polarization for most frequencies are about 10% except for the following frequencies: 1 GHz, where polarization is measured with 40% uncertainty, and 35 GHz, with 100%.

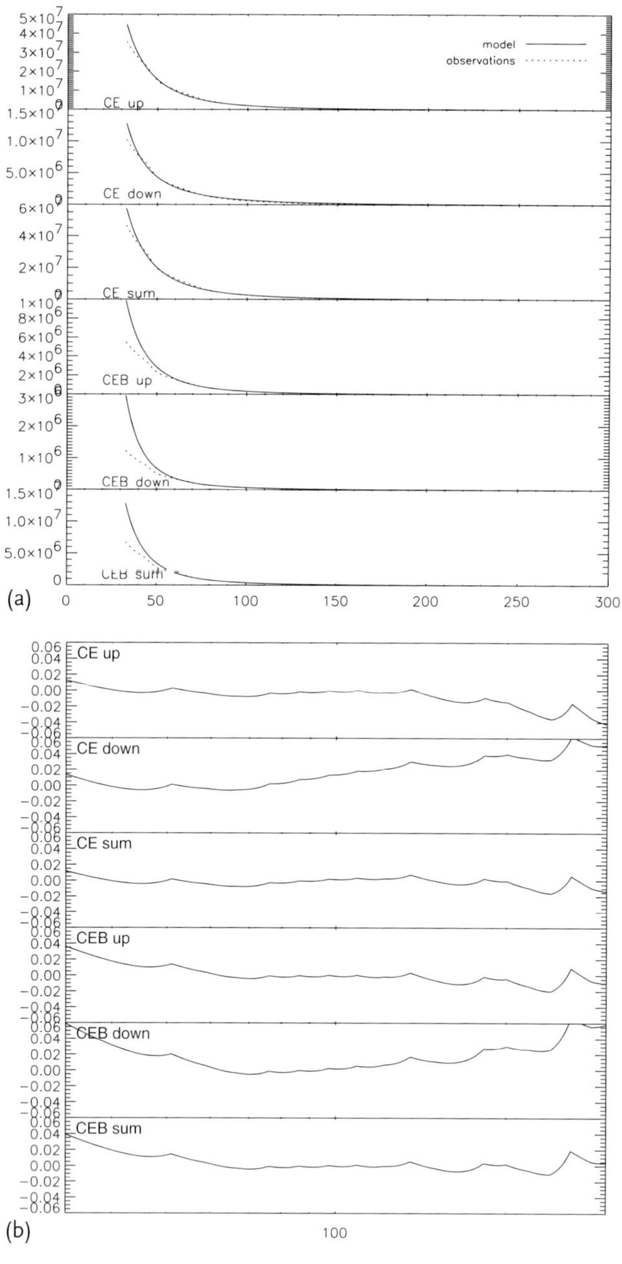

Figure 8.15 Comparison of simulated upward, downward, and total HXR intensities with the observations (a) and their residuals (b) calculated (along the magnetic field direction) for different beam precipitation models (CE and CEB) and viewing angle 180°.

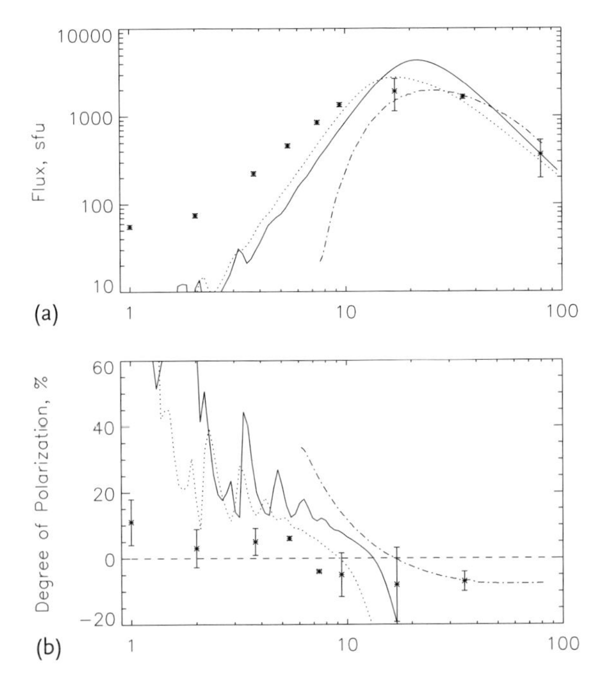

Figure 8.16 Comparison of fitting observations of MW intensity (a) and degree of polarization (b) by our CEB2 (dotted-line curves) for a magnetic field convergence of 2 and CEB3 (solid-line curves) models with a convergence of 3 vs. the model presented by Altyntsev *et al.* (2008) (dashed–dotted line). In both rows asterisks mark the observed spectra.

From a hydrodynamic model used to define the temperature and density variations with depth one can estimate the height of a semicircular loop producing this flare as approximately 10 000 km. Then the distance between the MW and HXR intensity centroids can be estimated at about 10″, which fits very well the locations of MW and HXR contours.

The MW emission and polarization simulated in Section 8.5.3.2 revealed a smooth part at lower and medium frequencies and a harmonic part at lower frequencies. In the current paper we are concerned with the smooth part of MW emission and polarization simulated for convergences of 2 and 3 (CEB2 and CEB3) and different viewing angles. For a viewing angle of 110°, which in our model is equal to 70° in the model of Altyntsev *et al.* (2008), the simulated MW intensity is not sensitive to variations of the magnetic convergence factor and is smaller than the observed one by more than 20% for both lower and higher frequencies.

However, for a viewing angle of 140° the model simulations of MW emission and polarization are able to reproduce very closely the main features of the observed MW emission. These include the magnitude of the MW emission maximum close to that observed, the frequency of the MW spectrum maximum appearing at about 17 GHz, a smooth decrease of MW emission toward lower frequencies with a spectral index of about 2 (still higher than observed and power-law MW emission with a spectral index of 2.8 toward higher frequencies). The MW polarization simulated for models CEB2 and CEB3 for the same viewing angle of 140° also fit reasonably well the observed polarization at lower frequencies, although they were twice as high as observations for higher frequencies. The CEB2 model also reproduces closely (9 GHz) the observed frequency (7 GHz) of the polarization reversal.

To quantify our fits in MW intensities, we carried out a few statistical tests for the likelihood of each curve (CEB2 and observations, CEB3 and observations) using the SPSS statistical package. The tests included Kendall tau and Spearman correlation tests (Isobe *et al.*, 1986; Kendal and Gobbons, 1990) appropriate for other than normal distributions, which we have for MW emission. We also calculated χ^2 coefficients (Chernoff and Lehmann, 1954) for the following CEB2 and CEB3 sets: the full number of nine measurements, for seven reduced measurements (excluding two points at the lowest frequencies), then for six and five (reducing the measurements subsequently by one from lower frequencies). The results of these tests are presented in Table 8.2. For comparison, we also include critical values of χ^2 for given degrees of freedom corresponding to significance levels of 90 and 99%.

It can be observed from Table 8.2 that the intensities for both models (CEB2 and CEB3) reveal a strong positive correlation with MW observations for all nine frequencies, but the intensity increases to a full dependence (correlation coefficients approach 1.0) for a reduced number of measurements of six and five for CEB3 or five for CEB2, although for nine points the correlation is slightly better for the CEB2 model. The calculated χ^2 for each data set (CEB2 and CEB3) from the full to the reduced measurement numbers plotted in Table 8.2 allows us to discriminate between the models. The CEB3 model produces a χ^2 of 3.86 for nine points (or degree of freedom, DF, of 8), which is reduced to 1.49 for five points, or DF = 4).

The χ^2 coefficients for simulated and observed polarization data obtained for eight observational points (excluding 80 GHz), or 7 degrees of freedom, produced $\chi^2 = 1.33$ for the CEB2 model and 2.13 for the CEB3 model, which reveals a similar close (above 95%) fit to observations for the model with a convergence of 2 because their χ^2 lie between the critical values for confidence levels above 0.90 and below 0.99 (Chernoff and Lehmann, 1954). This analysis of both MW intensity and polarization allows us to conclude that model simulations with any convergence

Table 8.2 Significance coefficients for Kendall tau and Spearman correlation and χ^2 statistics obtained for likehood tests between observed and simulated MW intensities for magnetic convergences of 2 (CEB2) and 3 (CEB3) vs. the tabulated critical values of χ_0^2 for given significance levels.

Points of CEB	χ_0^2 0.90	χ_0^2 0.99	Kendall CEB3	Spearman CEB3	χ^2 CEB3	Kendall CEB2	Spearman CEB2	χ^2 CEB2
9	3.49	1.65	0.944	0.944	3.86	0.889	0.950	2.56
7	2.20	0.87	0.867	0.943	2.21	0.733	0.829	0.331
6	1.61	0.55	1.000	1.000	1.87	0.800	0.900	0.128
5	1.06	0.30	1.000	1.000	1.49	1.000	1.000	0.031

fit reasonably well the MW observations up to a confidence level of 90%. However, the CEB2 model with a convergence of 2 fits observations up to 95% for all nine points and up to a confidence level of 99% with a reduced number of seven points and lower.

Therefore, it is safe to conclude that with the electron-beam parameters derived from HXR emission, the best agreement with MW observations is achieved for CEB models with a reasonably small convergence factor of 2, a magnetic field magnitude in the photosphere derived from MDI ($B = 780$ G), and a viewing angle of 140° (Figure 8.16). The results of the simulations also allow us to estimate more accurately the inclination of the loop in those parts that emit the relevant types of emission: HXR or MW.

Note that the observed MW emission from this flare is better fit by our simulated MW emission seen at a viewing angle of 140°, unlike the 80° viewing angle (equal to 100° in the current model) deduced in earlier simulations (Altyntsev *et al.*, 2008) and the 180° angle obtained for HXR emission in the current paper. As discussed above, the difference between the viewing angles derived for MW emission from our models and those by Altyntsev *et al.* (2008) can be the result of the oversimplified model used in the latter and a limited number of measurements.

However, the difference between MW and HXR viewing angles found in the current paper can be explained within our model by the assumption that the flaring loop did not stand perpendicular to the local horizontal plane in the flare location but was slightly tilted (by about 40°) toward the solar disk center. By comparing the directivity of MW and HXR emissions from Figure 8.16 one can observe that MW emission has a preferential direction of 30° downward and 130–140° upward.

The downward emission exceeds by a factor of 2–4 the upward emission. If all these downward photons are fully reflected by the photosphere (albedo effect) with the same properties (angles and energies), then they can contribute to a reasonable fit of MW emission for a viewing angle of 140° compared with 180° from HXR spectra, and this difference in viewing angles can be simply the result of directivity effects on HXR and MW intensities from beam electrons. One can also speculate that the loop curvature in the observed flare might contribute to this angle difference for viewing MW and HXR emissions to confirm which one needs to extend our model to a semicircular loop, which can be done in the future.

8.6 Conclusion

In this chapter, we investigated the energy spectra and polarization of gyrosynchrotron MW emission generated by anisotropic electron beams in flaring atmospheres with density and temperature gradients. The electron distributions were selected from the Fokker–Planck kinetic propagation/precipitation model, which takes into account energy losses and directivity changes of beam electrons in collisions with ambient particles, in converging magnetic fields, and in a self-induced electric field.

We added the combined effect of self-induced electric field and converging magnetic field in order to explain the MW emission in flares with strong anisotropy of HXR emission. The investigation revealed the following:

- Magnetic field convergence affects in a similar way electron beams with different energy fluxes or power-law indices. The MW emission intensity for viewing angles around $\theta \simeq 90°$ increases with increasing distance from the injection point of beam electrons, and for the larger viewing angles ($\theta \gtrsim 140°$) it decreases. For $\theta \simeq 90°$, the polarization sign reversal with depth is also observed, exposing the increasing production of the X mode.
- The effect of returning electrons of a beam is negligible for beams with relatively weak electron fluxes ($F \lesssim 10^{10}\,\text{erg}\,\text{cm}^{-2}\,\text{s}^{-1}$), while it becomes very important for electron beams with $F \gtrsim 10^{12}\,\text{erg}\,\text{cm}^{-2}\,\text{s}^{-1}$.
- Inclusion of a self-induced electric field effect in simulation models (for electron energy fluxes $F \simeq 10^{12}\,\text{erg}\,\text{cm}^{-2}\,\text{s}^{-1}$) results in a noticeable increase in MW emission intensity, especially at large viewing angles ($\theta \gtrsim 140°$). The MW polarization for beamlike distributions is shown to have a sign change at frequencies of 8–10 GHz, being negative for higher frequencies and positive for lower ones.
- The combined effect of a self-induced electric field and converging magnetic field reveals a noticeable (up to a factor of 10) increase in the emission intensity (for viewing angles of $\theta \simeq 140$–$150°$), in comparison with those models that consider only a collision factor, especially at deeper precipitation layers (near the loop footpoints). This is caused by the increased contribution of emission of returning electrons – not only by those reflected by the converging magnetic field but also by those turned back to the corona by the self-induced electric field.
- Thus, consideration of a self-induced electric field is especially important when interpreting the MW emission of powerful flares located close to the solar disk center. We found that including this factor in a simulation model (in addition to magnetic field convergence and collisions) results in significant changes in emission intensity, spectral shape, and polarization that need to be taken into account if the beam parameters are to be derived. This raises the issue of the importance of simultaneous observations of MW and HXR emissions in flares.

For the flares of 23 July 2002 and 10 March 2001, which showed a strong anisotropy of injected electrons, the observed MW and HXR emissions fit the simulations described above. The observed HXR energy spectrum was shown to be a double power law that fit very closely the photon spectrum simulated for those models that include a self-induced electric field. The MW emission simulated for different models was fit to the observed distribution in frequency, revealing that only models that combine collisions and electric field effects with pitch-angle anisotropy were able to reproduce closely the main features of the observed MW emission: peak frequency, a smooth decrease of MW emission toward lower frequencies, and the correct frequency of the MW polarization reversal.

- By estimating from the hydrodynamic model the height of a semicircular loop as being about 10 000 km and by taking into account the inclination of the loop in HXR emission, the distance between the MW and HXR intensity centroids was about 10″, which fits the observations very well.
- HXR emission simulated for relativistic angle-dependent cross-sections accounting for different directions of electron propagation (downward and upward) revealed that in the corona the majority of electrons move in the downward direction with half as many moving upward, while in the chromosphere most electrons move upward and only a smaller fraction keeps moving downward. Thus, for this flare the effects of electrons' magnetic mirroring and ohmic losses were significant.
- The observed HXR spectrum revealed a noticeable flattening toward lower energies, below 50 keV, indicating a significant effect of a self-induced electric field in beam electron energy losses. The observed HXR photon spectrum was best fit by the model simulated for the collisional plus ohmic loss precipitation model of electron beam with an initial energy flux of 10^{12} erg cm^{-2} s^{-1} and a spectral index of about 3. This includes emission emitted upward and downward (the full albedo effect) in the direction of 180° (toward the magnetic field direction), which for this flare location indicated a loop tilt of about 40° toward the solar disk center.
- The observed MW emission distribution in frequency revealed a better fit of the model combining collisions and electric field effects with a moderate magnetic field convergence of 2 (CEB2). This model can reproduce closely the main features in observed MW emission: maximum magnitude and frequency at about 17 GHz, a smooth decrease of MW emission toward lower frequencies with a spectral index of about 2, a spectral index of 2.8 for the higher-energy part of MW emission, and a frequency of reversal of MW polarization.
- Both models CEB2 and CEB3 showed a reasonable fit to observed polarization magnitudes, although the CEB2 model showed a fit to the frequency of MW polarization reversal that was close to the observed one.
- The best fit was achieved for MW emission from a viewing angle of 140°, in contrast to one of 80° (corresponding to 100° in our model) derived previously (Altyntsev *et al.*, 2008), and of 180° derived from the HXR emission in this study. Within our model limits, this difference, first of all, reflects an inclination of about 40° of a flaring loop toward the solar disk center that accounts for the HXR viewing angle and, second, indicates the difference in preferential directivity of HXR and MW emissions caused by electron scattering effects in the presence of self-induced electric and converging magnetic fields. In addition, a flaring loop curvature in the upper atmosphere where the MW emission is formed can also contribute to this viewing angle difference.

9
Langmuir Wave Generation by Electron Beams

9.1
Electron Beams and Their Stability

Precipitation of high-energy-beam electrons with negative power-law energy distributions by scattering in Coulomb collision with ambient particles can account for the hard X-ray (HXR) power-law photon spectra often observed in solar flares (Brown, 1971; Syrovatskii and Shmeleva, 1972b). This HXR emission is often accompanied by noticeable type III bursts of radio emission simultaneously observed in flares (Chernov, 2006; Zheleznyakov and Zaitsev, 1970, and references therein). The latter was first interpreted analytically by a quasilinear relaxation on Langmuir turbulence (Zheleznyakov and Zaitsev, 1970) or other types of turbulence (Diakonov and Somov, 1988) caused by the instability of beam electrons having energy distributions with a positive slope, for example $\partial f/\partial V > 0$, although positive slopes below a lower cutoff energy E_0 are difficult to observe in HXR or radio emission since they are obscured by emissions coming from thermal electrons.

Simulations of particle acceleration in a reconnecting current sheet for the likely coronal acceleration mechanism (Zharkova and Gordovskyy, 2005b) have shown energy/velocity distributions with a double exponent. The energy spectra of ejected particles have a positive slope at lower energies and a negative slope at higher ones with the maxima occurring at energies close to the lower cutoff energies in HXR photon spectra. The recent self-consistent particle-in-cell (PIC) simulations of the particle energy spectra formed in a diffusion reconnection region confirmed the formation of electron energy spectra with a positive slope and the occurrence of Langmuir waves at ejection from current sheets (Siversky and Zharkova, 2009a,b). Depending on reconnection scenarios, these beams can be injected either as short pulses or as a steady stream of electrons.

Hannah *et al.* (2009), considering collisional and particle–wave interactions for a short (1 s) impulse of beam electrons with a positive energy slope below the lower cutoff gained at acceleration energy, showed that these electrons could generate noticeable Langmuir turbulence that flattens the initial positive slope in electron distributions. The authors showed this effect to be important for mean electron spectra deduced from HXR emission and for the interpretation of dips appearing in energy distributions combining power-law and thermal energy spectra. However,

Electron and Proton Kinetics and Dynamics in Flaring Atmospheres, First Edition. Valentina Zharkova.

this conclusion needs to be tested for longer injection times since HXR and MW emissions can last in flares for tens of minutes.

While for a steady injection of beam electrons their precipitation into lower atmospheric levels will result in the transformation of their energy distributions into those with positive slopes by collisional energy losses in the ambient plasma (Emslie and Smith, 1984; Syrovatskii and Shmeleva, 1972b). These positive slopes can be further enhanced by particle reflection from a magnetic mirror (Leach and Petrosian, 1981) or by an electric field induced by the precipitating beam (Knight and Sturrock, 1977; McClements, 1992b; Zharkova and Gordovskyy, 2006). These different types of energy losses have been found to be important at different depths or times of beam precipitation into flaring atmospheres (Siversky and Zharkova, 2009b).

The first attempts to evaluate turbulence generated at lower atmospheric levels by electron beams with single negative power-law distributions above a lower cutoff energy were made by Emslie and Smith (1984) and Hamilton and Petrosian (1987), who described analytically the particle–wave interaction of beam electrons affected by Coulomb collisions with ambient particles in an atmosphere with static exponential density gradients. The collisions of beam electrons were found to create energy distributions with positive slopes that became unstable and generated very intense Langmuir waves (Emslie and Smith, 1984). Particle–wave interactions at any precipitation depth were shown to have a noticeable but not dominant effect on electron distributions that cannot be observed in HXR or MW emissions since the electron energy transferred to the waves is strongly absorbed by the ambient plasma, leading to its additional (less than 8%) heating (Hamilton and Petrosian, 1987; McClements, 1987). At the same time a fusion of two Langmuir waves into a transverse one at twice the plasma frequency (Emslie and Smith, 1984) can produce noticeable gyrosynchrotron emissions often observed in flares (Chernov, 2006).

More detailed investigations of the role of electron collisions on the generation of plasma waves in converging magnetic loops for different U ratios of plasma-to-gyro frequencies revealed the appearance of either ordinary (O-mode) or extraordinary (X-mode) longitudinal plasma waves near the resonance of magnetoionic modes (Hamilton and Petrosian, 1990). For $U < 1$ the resonance occurs with the O mode producing slow plasma waves, while for $U > 1$ it occurs for the X mode producing Langmuir waves. The growth rates of these waves were found to decrease strongly with ambient temperature increases or decreases in the ratio of the beam to ambient density.

McClements (1989) simulated heating of a simple hydrostatic atmosphere by Langmuir waves produced by an electron beam losing energy in collisions, ohmic losses, and particle–wave processes. He showed that, contrary to the findings of Emslie and Smith (1984), the collisional depletion of low-energy electrons combined with ohmic losses does not produce a two-beam instability at any precipitation depth unless the electron distributions have already reached a plateau in the acceleration process prior to injection.

However, a steady density gradient assumed for hydrostatic atmospheres by McClements (1989) is not a good assumption for flares, which are very dynamic events that show a sharp increase in density over the transition region (see, e.g., Somov *et al.*, 1981; Zharkova and Zharkov, 2007). Therefore, given the fact that a stationary injection of electron beams produces a substantial electric field in the corona (Zharkova and Gordovskyy, 2006), one question remains: If both collisional and ohmic losses are considered for dynamic flaring atmospheres, will they still result in the formation of electron energy spectra with positive slopes at deeper levels and in the generation of Langmuir waves?

9.2
Basic Equations

An electron beam injected into a flaring atmosphere is represented by electron distribution $f(v, x, t)$, where t is time, x is a 1-D coordinate (i.e., depth), and v is velocity along the x-coordinate. Waves are represented by their energy spectra $W(v, x, t)$, where v is the phase velocity of plasma waves and the meaning of W for waves is identical to the distribution function $f(v, x, t)$ for electrons. Using a quasilinear approach for interactions between particles and Langmuir waves (Hamilton and Petrosian, 1987; Zheleznyakov and Zaitsev, 1970), including a self-induced electric field, the simultaneous equations for f and W are solved (McClements, 1989):

$$\left(\frac{\partial}{\partial t} + v\frac{\partial}{\partial x}\right) f - \frac{e\mathcal{E}}{m_e} v^2 \frac{\partial}{\partial v}\left(\frac{f}{v^2}\right) = \frac{e^2\omega_p^2}{m_e}\frac{\partial}{\partial v}\left(\frac{\ln v/v_e}{v^2} f\right)$$
$$+ \frac{\pi\omega_p}{m_e n}\frac{\partial}{\partial v}\left(v\,W\frac{\partial f}{\partial v}\right) + \frac{4\pi e^4 n \ln\Lambda_b}{m_e^2}\frac{\partial}{\partial v}\left(\frac{f}{v^2}\right), \tag{9.1}$$

$$\left(\frac{\partial}{\partial t} + 3\frac{v_e^2}{v}\frac{\partial}{\partial x}\right) W = e^2\omega_p^2\frac{\ln v/v_e}{v} f + \frac{\pi\omega_p}{n} v^2 W \frac{\partial f}{\partial v} - \frac{\pi e^4 n \ln\Lambda_p}{m_e^2 v_e^3} W, \tag{9.2}$$

where n, v_e, and ω_p are, respectively, the density, electron thermal velocity, and plasma frequency of the background plasma and $\ln\Lambda_b$ and $\ln\Lambda_p$ are Coulomb logarithms for beam and plasma electrons, respectively. $\mathcal{E}$ is the self-induced return current electric field, which is calculated as follows (McClements, 1992b; Siversky and Zharkova, 2009b)

$$\mathcal{E}(t, x) = \frac{e}{\sigma(x)}\int_{v_{min}}^{v_{max}} dv\, v\, f(v, x, t), \tag{9.3}$$

where $\sigma(x)$ is the classic conductivity of the ambient plasma.

The distribution function f of an electron beam at the point of injection is taken in the following form obtained for electron acceleration in a model current sheet

(Zharkova and Gordovskyy, 2005b):

$$f(x = x_{\min}, v, t) = f_n \frac{E^{\kappa}}{E^{\kappa+\delta} + E_0^{\kappa+\delta}}, \tag{9.4}$$

where $E = m_e v^2/2$ is the electron energy. For energies greater than the energy of the maximum, or lower cutoff energy, E_0, the electron spectrum is negative power law with the index δ, that is, $f \propto E^{-\delta}$, while the low-energy part of the spectrum ($E < E_0$) is positive power law with the index κ.

9.2.1
Method of Solution and Model Parameters

For electron–wave interactions a quasilinear approach is utilized, as suggested by Zheleznyakov and Zaitsev (1970), so that Eqs. (9.1) and (9.2), together with Eq. (9.3), are solved numerically using the summary approximation method (Siversky and Zharkova, 2009b), taking Eq. (9.4) as the boundary condition. The electron beam is assumed to be steadily injected into deeper atmospheric levels with a single power-law distribution in energy, that is, $\partial f/\partial E \leq 0$, which means that $\kappa = 0$ in Eq. (9.4).

The injected beam parameters are as follows: high-energy index $\delta = 3$ and 7, cutoff energy $E_0 = 12$ keV, beam energy flux $F_0 = (1–100) \times 10^{10}$ erg cm^{-2} s^{-1}, and the low-energy index κ varies from 5 to 0. Density and temperature profiles of the ambient plasma are adopted from the hydrodynamic atmosphere heated by the beam (Zharkova and Zharkov, 2007) as shown in Figure 9.1 for beam parameters $F_0 = 10^{11}$ erg cm^{-2} s^{-1} and $\delta = 3$.

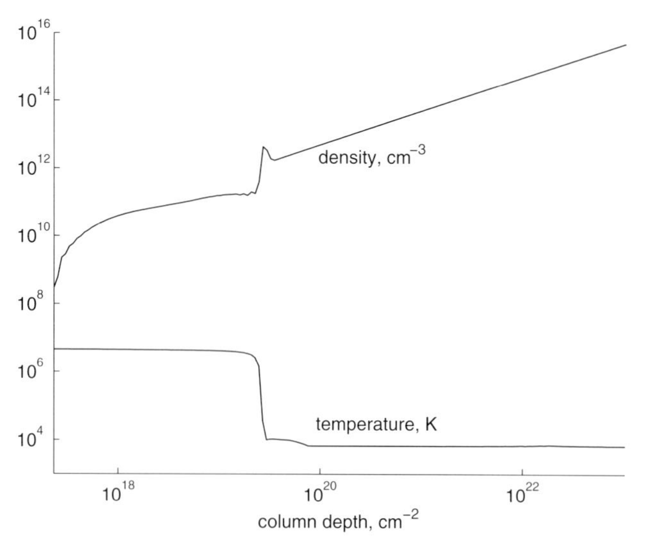

Figure 9.1 Sample of density and temperature distributions of ambient plasma calculated using a hydrodynamic model (Zharkova and Zharkov, 2007) heated by electron beam with initial energy flux $F_0 = 10^{11}$ erg cm^{-2} s^{-1} and power-law index $\delta = 3$ of the beam electron distribution.

9.3 Results and Discussion

9.3.1 Electric Field Effects on Langmuir Turbulence

9.3.1.1 Selection of Initial Beam Distribution

Let us first select an electron beam injected with initial energy distributions with a positive slope having index $\kappa = 5$ at energies lower than 10 keV (Eq. (9.4)) predicted by some acceleration models (Zharkova and Gordovskyy, 2005b) (Figure 9.2) and explore the Langmuir turbulence it produces (Figure 9.2).

It turns out that there is fast collisional depletion (flattening) at lower energies of such initial electron distributions (Figure 9.2a) caused by fast resonant interaction with Langmuir waves, which resembles the findings by Zheleznyakov and Zaitsev (1970) and Hamilton and Petrosian (1990). This depletion occurs at a short distance of 3 cm and a very short time scale of $\sim 3 \times 10^{-8}$ s which is similar to the 40 plasma periods obtained by Karlický *et al.* (2008). This means that with sufficient accuracy the initial distribution for electron beams can be selected with $\kappa = 0$, so we can vary only δ.

9.3.1.2 Electron Distribution During Precipitation

A hydrodynamic response forms steep gradients and temperature variations in the transition region in ambient plasma (Figure 9.1), which in turn forms a maximum in an electric field in this region, leading to a significant increase of energy losses by beam electrons in this electric field (Emslie, 1980; Siversky and Zharkova, 2009b; Zharkova and Gordovskyy, 2006) and a strong effect on electron-beam distributions

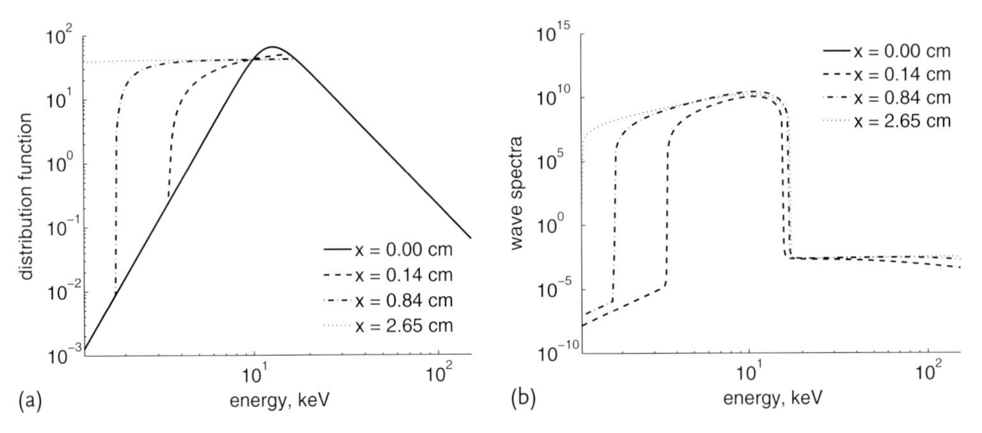

Figure 9.2 (a) Distribution function (in arbitrary units) of beam electrons with spectral index $\delta = 3$ and initial energy flux 10^{11} erg cm^{-2} s^{-1} at various depths from injection point calculated at 3×10^{-8} s after injection onset. (b) Energy spectra (in arbitrary units) of Langmuir waves generated by this electron beam at various depths from the injection point.

and their HXR emission. In particular, the electric field induced by a precipitating beam is shown to effectively decelerate precipitating electrons and to accelerate them upward, forming an electric circuit along the whole loop (Zharkova *et al.*, 2010).

It was shown earlier that in a pure collisional precipitation, described by the continuity equation (Syrovatskii and Shmeleva, 1972b), an electron beam with a negative power-law spectrum gains a positive slope at lower energies with maxima at about $\delta \times E_{\text{low}}$ (Zharkova and Gordovskyy, 2006). This is different from the results obtained by Hamilton and Petrosian (1987) for small-scale nonhomogeneous atmospheres and by McClements (1989) for a hydrostatic atmosphere, which can be explained by different physical conditions between hydrostatic and hydrodynamic atmospheres (Somov *et al.*, 1981; Zharkova and Zharkov, 2007).

The inclusion of a self-induced electric field is shown to steepen these positive slopes in electron distributions, making them appear at upper precipitation levels compared to pure collisions (Zharkova and Gordovskyy, 2006) (cf. solid and dashed lines in Figure 9.3). In this case, the beam electron density is found to be higher at upper atmospheric levels (in the corona) and lower at lower levels (in the chromosphere), in comparison with the pure collisional precipitation of a beam (Siversky and Zharkova, 2009b; Zharkova and Gordovskyy, 2006) (see Figure 9.3a,b).

The presence of such positive slopes in electron distributions leads to the generation of plasma waves in the corona and the transition region (Emslie and Smith, 1984), which are proven to be Langmuir waves (Hamilton and Petrosian, 1990). The effects of energy losses in Langmuir waves on electron distributions gained in collisional and ohmic losses for beams with various parameters are presented in Figure 9.4.

The effect of an electric field on electron energy distributions for various depths calculated for energy losses in collisions and Langmuir wave generation is demonstrated in Figure 9.4a–c. The distributions simulated for collisions plus waves but without an electric field (solid lines) are substantially different from those simu-

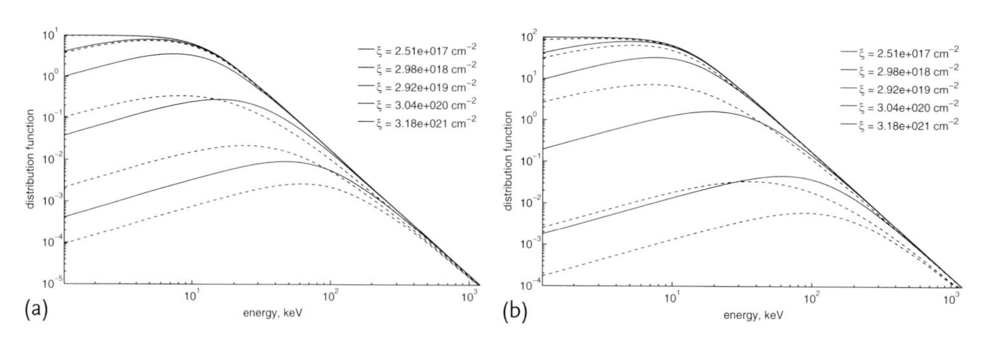

Figure 9.3 Electron differential distributions (in s cm^{-4}) for collisions without (solid) and with (dashed) electric field for beams with $\delta = 3$ and $F_0 = 10^{10}$ erg cm^{-2} s^{-1} (a) and $F_0 = 10^{11}$ erg cm^{-2} s^{-1} (b). The curves are shown from top to bottom for column depths ξ, shown in legend starting from $\xi = 2 \times 10^{17}$ cm^{-2} for the uppermost ones (both solid and dashed lines).

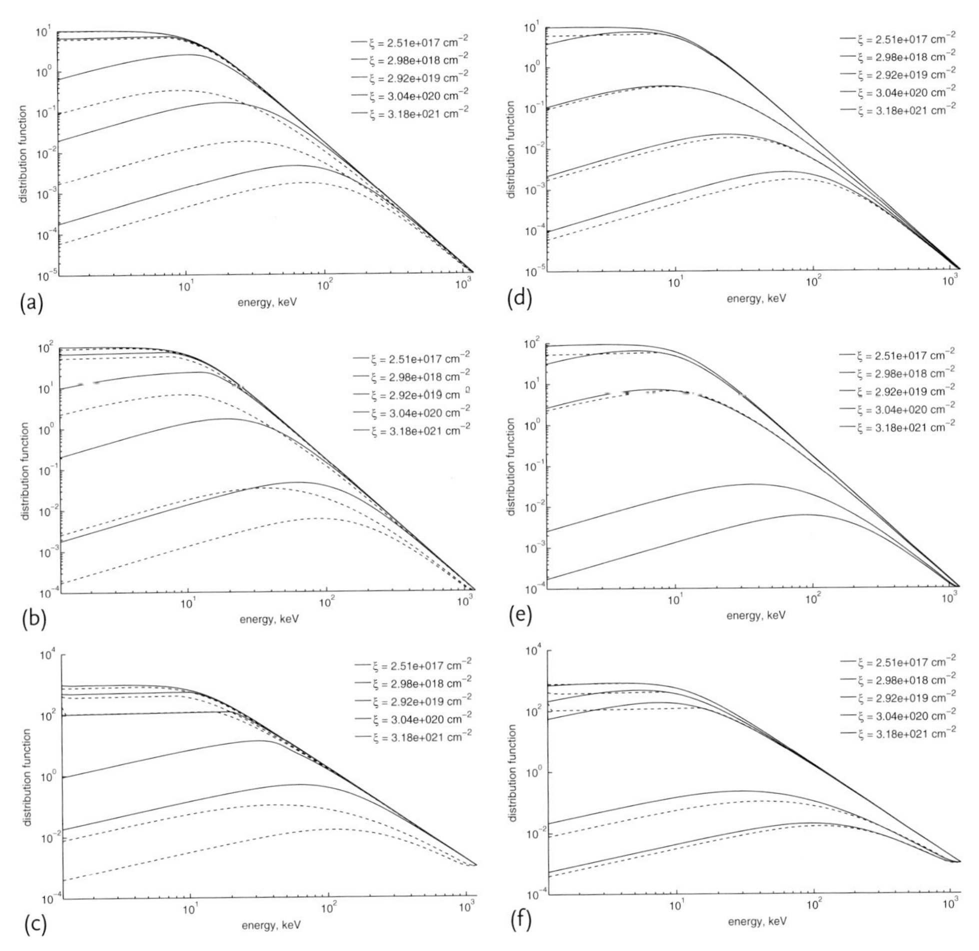

Figure 9.4 (a–c) Beam electron distributions (in $\mathrm{s\,cm^{-4}}$) for collisions and waves but without (solid lines) and with (dashed lines) an electric field. (d–f) Beam electron distributions for collisions and electric field without (solid) and with (dashed) Langmuir waves for $F_0 = 10^{10}\,\mathrm{erg\,cm^{-2}\,s^{-1}}$ (a,d), $F_0 = 10^{11}\,\mathrm{erg\,cm^{-2}\,s^{-1}}$ (b,e), and $F_0 = 10^{12}\,\mathrm{erg\,cm^{-2}\,s^{-1}}$ (c,f). The curves are shown from top to bottom for column depths ξ shown in the legend starting from $\xi = 2 \times 10^{17}\,\mathrm{cm^{-2}}$ for the uppermost ones (both solid and dashed).

lated with an electric field (dashed lines). The difference becomes much stronger for more intense beams (cf. Figure 9.4a,d and 9.4c,f). Thus, unlike hydrostatic atmospheres (McClements, 1989), the electric field in hydrodynamic atmospheres sharpens electron energy distributions with maxima caused by collisions and reduces the column depths where positive slopes can be formed.

The inclusion of Langmuir waves has a small effect on electron distributions for electron beams with low initial energy fluxes if both collisions and electric field are considered (dashed lines in Figure 9.4d–f), in contrast to those simulated in the absence of Langmuir waves (solid lines). This explains why the consideration of Langmuir turbulence without an electric field can have a small effect on HXR emission produced by moderate beams as found in the previous estimations (Hamilton and Petrosian, 1987) and simulations (McClements, 1989).

However, for more intense beams with energy fluxes of about 10^{12} erg cm^{-2} s^{-1} the inclusion of Langmuir waves causes a noticeable reduction of electron numbers at lower energies (cf. Figure 9.4d,f). This difference is less noticeable at upper atmospheric levels but becomes larger at lower ones, which can result in some reduction of the electron numbers deduced from HXR photon counts if Langmuir waves are considered.

9.3.1.3 Langmuir Waves

As found in Section 9.3.1.2, the initial single negative power-law velocity distributions of electron beams injected into an ambient plasma are transformed into distributions with a positive slope at lower energies. Electrons with such distributions excite, due to normal or anomalous Doppler resonance, low- and high-frequency resonant Langmuir oscillations with a phase velocity of v_{ph} (Tsytovich, 1970).

If Langmuir waves grow in the region where the slope is positive and there is a greater number of faster particles ($v > v_{\text{ph}}$) than slower ones, then a greater amount of energy is transferred from the fast particles to the wave, giving rise to exponential wave growth at a growth rate of γ_{w} (Tsytovich, 1970):

$$\gamma_{\text{w}} =\approx \frac{\pi}{2}\frac{n_{\text{b}}}{n}\left(\frac{v_{\text{s}}}{\Delta v_{\text{s}}}\right)^2 \omega_{\text{pe}} , \tag{9.5}$$

where n_{b} is the beam density, n is the ambient plasma density, v_{s} and Δv_{s} are, respectively, the mean vertical velocity and the mean spread of this velocity of beam electrons, and ω_{pe} is the electron plasma frequency.

The Langmuir instability grows if the growth rate of waves is greater than the collisional dumping rate γ_{col} defined as (Ginzburg and Zhelezniakov, 1958)

$$\gamma_{\text{col}} \approx \frac{5.5n}{T_{\text{e}}^{3/2}} \ln\left(10^4 \frac{T_{\text{e}}^{2/3}}{n^{1/3}}\right) \approx 80 n T_{\text{e}}^{-3/2} , \tag{9.6}$$

which is appropriate for the coronal values of plasma density n and electron temperature T_{e}. Thus, the ratio $\Gamma = \gamma_{\text{w}}/\gamma_{\text{col}}$ defines the growth rate of Langmuir waves above the collisional damping, for example when $\Gamma \gg 1$, then the waves are most effectively generated.

There are two kinds of beam pairs that can produce two-beam instability and cause Langmuir waves: (i) a direct electron beam and ambient plasma electrons and (ii) a direct beam and the beam of returning electrons mixed with thermal ones. The interaction of the first set of beams on the generation of Langmuir waves throughout precipitation depths from the corona to the chromosphere is presented in Figure 9.5a–d, and the effects of the second set, including the beam associated with a self-induced electric field, in Figure 9.5e–h.

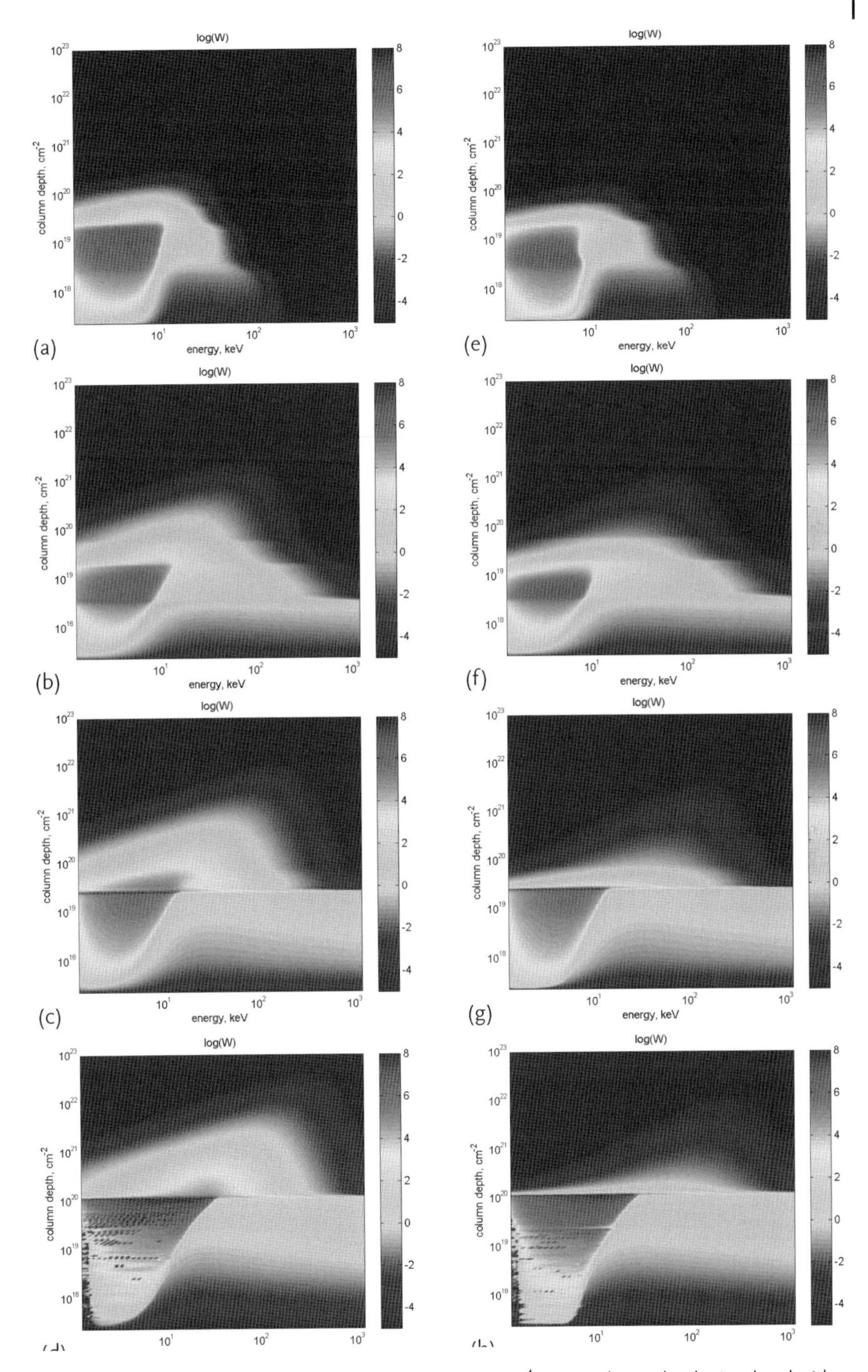

Figure 9.5 3-D density of Langmuir wave energy (erg cm^{-4} s) vs. column depth simulated without (a–d) and with (e–h) a self-induced electric field for beams with the following parameters: (a,e) $\delta = 7$ and $F_0 = 10^{10}$ erg cm^{-2} s^{-1}; (b,f) $\delta = 3$ and $F_0 = 10^{10}$ erg cm^{-2} s^{-1}; (c,g) $\delta = 3$ and $F_0 = 10^{11}$ erg cm^{-2} s^{-1}; (d,h) $\delta = 3$ and $F_0 = 10^{12}$ erg cm^{-2} s^{-1}. For a color version of this figure, please see the color plates at the beginning of the book.

At upper atmospheric levels (or at lower column densities, in the corona) the effect of Langmuir waves generated without taking into account electric field effects is clearly noticeable at all the coronal depths for energies below 10 keV for beams with an initial energy flux of $F_0 = 10^{10}$ erg cm^{-2} s^{-1} while shifting to a few tens of kiloelectronvolts for more intense beams ($F_0 = 10^{12}$ erg cm^{-2} s^{-1}). The restriction in energy is likely to occur because only electrons with energies below or about 10 keV are effectively scattered in the corona since the coronal plasma is a thin target (Brown, 1971).

There is a sharp peak of wave energy density observed for any beams just below the transition region at about 1×10^{19} cm^{-2} for beams with an initial energy flux of 10^{10} erg cm^{-2} s^{-1} or deeper at 3×10^{19} cm^{-2} and 1×10^{20} cm^{-2} for beams with an initial energy flux of 10^{11} and 10^{12} erg cm^{-2} s^{-1}, respectively. The occurrence of this peak coincides with the peaks in the density and temperature profiles of the ambient plasma heated by beams (Zharkova and Zharkov, 2007) (Figure 9.1). A sharp enhancement in the plasma density and temperature just below the transition region causes an increase of the wave growth rate represented by the second (collisional) term on the right-hand side of Eq. (9.2), which is seen as a peak in Figure 9.5.

The spread of Langmuir waves at the coronal depths is more extended for beams with energy distributions having higher spectral indices (Figure 9.5a) or higher initial energy fluxes (Figure 9.5h). For beams with a higher initial energy flux of 10^{12} erg cm^{-2} s^{-1} or higher spectral index of 7, the Langmuir wave energy density above the transition region is higher than for a less intense or harder beam.

This occurs because softer beams effectively lose much of their energy in the upper atmosphere, so their average velocities v_S, which affect the growth rate of Langmuir waves (Eq. (9.5)), are higher in the corona. Furthermore, the beam density of a softer beam is higher in the corona, while the ambient plasma density caused by a hydrodynamic response (Zharkova and Zharkov, 2007) is lower than for harder beams, which increases the wave growth rate. The density of more intense beams is, in aggregate, also higher (from the normalization condition) (Siversky and Zharkova, 2009b), while the ambient plasma density in the corona becomes much lower than for less intense beams, leading again to an increased growth rate of the Langmuir waves (according to Eq. (9.5)).

At deeper layers, in the lower corona and in the upper chromosphere, the generation of Langmuir waves extends to electrons with higher energies (Figure 9.5d,h). More intense and harder beams produce a more extended region (extended toward the lower chromosphere) where Langmuir waves are generated. Also, the wave energy density (color-coded in erg cm^{-4} s on the right-hand side of each plot) is higher for such beams compared to weaker or softer beams.

This happens because at chromospheric depths harder and more intense beams have higher densities and narrower spreads (Δv_S) (Eq. (9.5)) compared to softer and weaker beams, which again leads to a higher growth rate and denser Langmuir turbulence, although at these depths collisional damping of the Langmuir waves becomes rather strong, resulting in the waves' being effectively scattered to larger spatial regions.

However, if the electric field of the beam is considered, the regions with Langmuir waves become smaller. The wave energy peaks move downward to the lower corona in the upper atmosphere (above the depth of the density maximum) and upward to the upper chromosphere (below the depth of the density maximum). The higher a beam's initial energy flux and the lower its spectral index, the narrower the region where Langmuir waves are generated, although the density of these waves becomes higher, that is, the waves become more intense while localized in a smaller region.

We understand that if the electric field is included, the energy density of Langmuir waves becomes lower at deeper atmospheric levels because the direct beam density also becomes lower compared to the case with collisions only (Figure 9.4). This is caused by a larger number of electrons' losing their energy in the upper atmosphere and returning to the injection site at the top with much wider distribution in pitch angles (Siversky and Zharkova, 2009b; Zharkova *et al.*, 1995). This significantly reduces the beam density in the chromosphere and, thus, the Langmuir wave density and spread at these levels.

The effects of an electric field can significantly reduce the intensity of type III MW emission generated by Langmuir turbulence, compared to a collisional case, for which the simulated MW emission intensity in the collision-plus-waves model was found to exceed the observed ones by an order of magnitude (Emslie and Smith, 1984).

There is another interesting effect derived from the current simulation for very strong beams with an initial flux of 10^{12} erg cm^{-2} s^{-1} appearing for any models (with and without an electric field). This is the formation of high-hybrid Langmuir waves in the form of well-defined periodic structures (zebra-type patterns), clearly seen in Figure 9.5d,h. The patterns are much more numerous and extend to all the coronal depths for models without an electric field (Figure 9.5d), while they shift to upper coronal levels and have fewer patterns for the model with an electric field (Figure 9.5h).

For model electron beams with high energy flux and very narrow spread over the pitch angle θ ($\mu = \cos\theta$ is about unity), these periodic oscillations in Langmuir wave generation are believed to be defined by the cyclotron resonance of beam electrons with derived distributions. Electrons with such velocity distributions can also excite, due to anomalous Doppler resonance, resonant Langmuir oscillations with frequencies of $\omega_2 = \min(\omega_{pe}, \omega_{He})$, where ω_{pe} and ω_{He} are the electron plasma and gyrofrequencies (see discussion in Chernov, 2006).

Low- and high-frequency modes of Langmuir oscillations form high-amplitude periodic nonlinear waves, seen as well-defined patterns in Figure 9.5d,h. The inclusion of an electric field significantly increases the spread of electron beams (Siversky and Zharkova, 2009b; Zharkova *et al.*, 1995), which substantially reduces the growth rate for Langmuir waves in general and for high-hybrid Langmuir waves in particular, as shown in Figure 9.5d.

The corresponding spectrum of electromagnetic waves excited resonantly by a current of potential waves can also form an equidistant spectrum, or zebra structure, of electromagnetic radio radiation in plasma. This structure can account for the generation of type IV MW bursts (Chernov, 2006) often observed in flares.

9.4 Conclusions

In this chapter we have compared the effects of collisional and ohmic energy losses of beam electrons with those caused by the electron energy exchange with Langmuir waves generated by the beams during their precipitation into hydrodynamic atmospheres.

Beams with a positive slope in the initial energy distribution reveal a fast flattening of their energy spectra on a very short spatial and temporal scale, which is caused by the formation of Langmuir waves and their suppression by a strong self-induced electric field.

The precipitation of beam electrons into a flaring atmosphere with density and temperature gradients results in the transformation of electron energy/velocity distributions at some depths into distributions with maxima that have positive slopes for energies below these maxima and still negative power laws for energies above it (Emslie and Smith, 1984; Syrovatskii and Shmeleva, 1972b).

High-energy electrons with velocity distributions having positive slopes, gained during their precipitation, can also excite low- and high-frequency resonant Langmuir waves. At upper atmospheric levels, in the corona, the generation of Langmuir waves affects electrons with energies below 10 keV for weak beams and below a few tens of keV for the most intense ones. The energy density of Langmuir waves at these levels is higher for softer or more intense beams.

Also, there is a sharp peak in Langmuir wave density observed at about $3 \times 10^{19}\,\mathrm{cm}^{-2}$ for the beams with an initial energy flux of $F_0 = 10^{10}\,\mathrm{erg\,cm}^{-2}\,\mathrm{s}^{-1}$, which moves into a deeper level of $2 \times 10^{20}\,\mathrm{cm}^{-2}$ for beams with an initial energy flux of $F_0 = 10^{12}\,\mathrm{erg\,cm}^{-2}\,\mathrm{s}^{-1}$. The occurrence of this peak coincides with peaks in the density and temperature profiles of the ambient plasma, which shift much deeper into the chromosphere for more intense beams.

At deeper layers, in the lower corona and in the upper chromosphere, the generation of Langmuir waves extends to electrons with higher energies of up to 100 keV. More intense and harder beams produce more extended region toward the lower chromosphere where Langmuir waves are generated. Also, the wave energy density is higher for more intense and harder beams.

The inclusion of a self-induced electric field, in addition to collisions, in the energy losses by beam electrons was shown to decrease the depths where distributions with maxima are formed and increase the energy of these maxima. This in turn increases the number of beam electrons at upper atmospheric levels, in the corona, and decreases it in the chromosphere, in comparison with those deduced for purely collisional precipitation.

As result, our simulations show that, contrary to previous findings (McClements, 1989), if the energy losses in a self-induced electric field are included, in addition to collisions, Langmuir waves are found to appear in a much narrower space than in the collision-plus-waves models and have a narrower energy range. In the presence of an electric field, a region with Langmuir waves in the corona is also shifted downward to the lower corona, while in the chromosphere it is shifted upward to the upper chromosphere.

The higher the beam initial energy flux and the lower the spectral index, the narrower the regions where Langmuir waves are generated. The density of these waves becomes higher, that is, the waves become more intense. Therefore, the electric field of an electron beam plays a very important role in suppressing the generation of Langmuir turbulence at all atmospheric levels by reducing the spread and increasing the density of these waves at all atmospheric depths above and below the transition region.

The effects of an electric field can significantly reduce the intensity of type III MW emission generated by Langmuir turbulence, compared with the collision-plus-waves case, for which the simulated MW emission intensity in the collision-plus-waves model was found to exceed the observed ones by an order of magnitude (Emslie and Smith, 1984).

Additionally, very distinct patterns of high-hybrid Langmuir waves have been found to be generated by electrons from very strong beams in the form of well-defined periodic (zebra-type) structures. They occur in the coronal part of the atmosphere and have a limited number of patterns for both parameters: atmospheric depth and electron energy.

The inclusion of a self-induced electric field reduces the number of these patterns in energy and in depth, leading to a shift of their occurrences to higher atmospheric levels in the corona and higher energies. These structures can account for the zebra patterns observed in type IV MW bursts.

10
Nonthermal Hydrogen Emission Caused by Electron Beams

10.1
Introduction

The interpretation of optical hydrogen and white-light emission in flares is still a complicated problem that has not been fully resolved after several decades of observations. This leads to the conclusion that a white-light flare is part of a general process of activation of solar flares that embraces all levels of the flaring atmosphere through the injection of high-energy particle beams. The effects of these beams on higher atmospheric levels were discussed in Chapters 2–7. However, the mechanisms of energization by injected beams of lower atmospheric levels from the chromosphere to the photosphere are still being intensively investigated. This energization comes from a few mechanisms that define the temporal and spatial scales of optical emission observed in flares:

1. By hydrodynamic heating of a flaring atmosphere by particle beams depositing their energy in Coulomb collisions (discussed in Chapter 5) leading to strong thermal excitation and ionization of the ambient hydrogen (and other) atoms;
2. Direct nonthermal excitation and ionization of hydrogen atoms by impacts with beam electrons;
3. Radiative excitation and ionization of hydrogen atoms by high-energy emission from upper atmospheric levels (backwarming heating); and
4. Excitation and ionization by internal diffusive radiation (generated by radiative transfer) in all optically thick transitions.

Thermal collisional and external radiation rates are well defined and taken into account for most atoms emitting in an optical electromagnetic range. Nonthermal excitation and ionization rates by electrons or proton beams are defined less confidently and only for hydrogen and calcium atoms (Aboudarham and Hénoux, 1987; Zharkova and Kobylinskii, 1989; Fang *et al.*, 1993; Hénoux *et al.*, 1993; Zharkova and Kobylinskii, 1993). The definition of diffusive radiative rates is not that straightforward since it always involves simultaneous solutions of radiative transfer and steady state equations for given atoms and sets of atoms. These excitation and ionization processes are balanced by various deexcitation and recombination

Electron and Proton Kinetics and Dynamics in Flaring Atmospheres, First Edition. Valentina Zharkova.

processes, which also need to be considered in the same approach, in order to derive which ones will produce the optical emission observed from flares at various temporal scales from impulsive to gradual ones.

The investigation of radiative transfer effects has made significant progress in the past two decades with the development of the radiative codes MULTI (Carlsson, 1986) and others using it as a basis on which to consider non-LTE (local thermodynamic equilibrium) radiative transfer not only in line and continuous emission of hydrogen atoms, but in all other transitions of other atoms with a full coronal abundance (Uitenbroek, 2001). These radiative codes are very sophisticated in terms of atomic processes for each atom considered, which results in some simplification associated with the physical conditions in the flaring atmospheres considered, that is, temperature and density variations with depth are rather limited by a factor of less than 2, in order to keep the atomic processes intact and not to experience strong numeric instabilities caused by significant variations of atomic parameters across the atmospheric depth.

However, if one could explore hydrodynamic atmospheres produced by injection of either electron or proton beams, discussed in Chapter 5, it would be obvious that variations of temperature and density approached a few orders of magnitude across the whole flaring atmosphere and a factor 10 or higher for the chromosphere and photosphere. On top of this, to variations of physical conditions one needs to add the strong depth variations of electron or proton beam densities discussed in Chapter 4, which makes the task of including nonthermal excitation and ionization by these beams in the interpretation of optical emissions produced in such events a rather challenging task.

In this chapter we consider nonthermal excitation and ionization of hydrogen atoms by beam electrons alongside all other processes leading to the activation or deactivation of a hydrogen atom discussed above. First, we will derive precise nonthermal excitation and ionization rates based on the most commonly used cross-section for hydrogen emission for a 10-level hydrogen atom. Second, we will solve the relevant system of integrodifferential equations describing hydrogen atom abundances, or departure coefficients, for different transitions and calculate the resulting hydrogen line and continuous emission from flaring atmospheres emitted at various time scales and compare these with some observations.

10.2
Nonthermal Excitation and Ionization Rates

The rates of excitation and ionization of a single hydrogen atom from state n to state n' is described by the coefficient $C_{nn'}$, where $n' = c$ for a continuum. Then, if inelastic scattering of electrons has a cross-section of $\sigma_{nn'}$, the volumetric excitation rates can be written as follows (Brown, 1972; Elwert and Haug, 1970; Leach and Petrosian, 1983):

$$C_{nn'} = n_{\mathrm{e}} \cdot B_{nn'} \,, \quad [\mathrm{s}] \tag{10.1}$$

where n_e is the electron density, $B_{nn'}$ is the activation rate per electron and hydrogen atom ($cm^3\,s^{-1}$), defined as

$$B_{nn'} = \sqrt{\frac{2}{m}} \int_{E_0}^{\infty} \sqrt{E} f(E) \sigma_{nn'}(E) dE \,, \tag{10.2}$$

where $f(E)$ is the electron distribution function, E_0 is the energy of transition $n - n'$, that is, $E_0 = E_{nn'}$ for line transitions, or $E_0 = I_0$ for ionization; here I_0 is the energy of ionization from level n.

Hence, for the calculation of nonthermal activation processes (excitation and ionization) one needs to know the electron distribution, the density of beam electrons and ambient particles at a given depth, and cross-sections of the radiative processes being considered.

10.2.1 Beam Electron Density

Let us assume that there is an elementary flaring burst (EFB) that is a magnetic flux tube standing perpendicular to the solar surface. The EFB is assumed to be filled with a hydrogen plasma (atoms, ions, and free, or thermal, electrons) with a total density distribution of $n_0(\xi)$, and an electron temperature distribution of $T(\xi)$, where ξ is the column density described in Chapters 4 and 5. We also assume that the ambient density and temperature are derived from simulations of hydrodynamic response discussed in Chapter 5.

As we showed in Chapters 3 and 4, the density of beam electrons at a given depth is strongly dependent on what energy loss mechanisms are considered for electron precipitation into deeper atmospheres. However, for lower atmospheric levels in the chromosphere, where hydrogen emission is formed, electron energy losses are dominated by collisions (Chapter 4). Thus, for simplicity, let us use a formula derived for the density of beam electrons from the continuity equation written for pure Coulomb collisions (Syrovatskii and Shmeleva, 1972a):

$$\begin{aligned} N(E,\xi) = {} & K E^{1/2} (E^2 + 2a\xi)^{-(\gamma+1)/2} \theta\left(\sqrt{E^2 + 2a\xi} - E_1\right) \\ & - \theta\left(E_2 - \sqrt{E^2 + 2a\xi}\right) , \end{aligned} \tag{10.3}$$

where γ is the index of beam energy flux, E_1 and E_2 are the lower and upper cutoff energies of beam electrons, and θ is Heavyside's function, that is, $\theta(x) = 1$ if $x \geq 0$ and $\theta(x) = 0$ if $x < 0$. K is a constant of integration defined by the electron beam energy flux at injection at the top boundary (Chapter 3). The lower energy cutoff is well accepted to be about or higher than 10 keV, as discussed in Chapters 3 and 4, if one assumes that a single power-law electron beam is injected.

However, the simulations of particle acceleration (Zharkova and Gordovskyy, 2005b) during a magnetic reconnection in a 3-D current sheet found from MHD simulations (see, e.g., Somov and Oreshina, 2000) reveal that, in fact, energy spectra of beam electrons are much more complicated. Electrons have been shown to gain energy distributions with positive power laws with spectral index κ for lower energies and negative power laws with spectral index γ with the maximum occurring at energies between 0.1 and 5 keV (for electrons). Thus, for analytical calculations of beam electron densities required for the calculation of nonthermal excitation rates this gives us the flexibility f to accept the lower cutoff energy as being equal to zero, which does not introduce a large error for electron energy distributions with maxima at a real cutoff energy of 10 keV. The upper cutoff energy is reasonably high, approaching a few hundred megaelectronvolts for some flares, as discussed in Chapter 4, and hence it can be accepted as equal to infinity.

Then by integrating expression (10.3) in all energies one can an obtain analytical expression for electron beam density variations with column depth ξ as follows (Zharkova and Kobylinskij, 1992) (Section 3.3.2 in Chapter 3):

$$N(\xi) = K(2a\xi)^{\frac{1-2\gamma}{4}} \times \begin{cases} \frac{2}{2\gamma-1}\frac{\Gamma\left(\frac{2\gamma+3}{4}\right)}{\Gamma(0.25)} \times \left\{\sum_{n=0}^{\infty}(-1)^n \frac{\Gamma(0.25+n)}{\Gamma\left(\frac{2\gamma+3}{4}+n\right)} \right. \\ \left. \times \left[t_1^{-(n+0.25)}(1+t_1)^{(1-\gamma)/2} - t_2^{-(n+0.25)}(1+t_2)^{(1-\gamma)/2}\right]\right\}, \\ 0 \leq \frac{1}{t_1} \leq 1 \\ \frac{\Gamma\left(\frac{2\gamma-1}{4}\right)\Gamma(0.75)}{2\Gamma\left(\frac{\gamma+1}{2}\right)} - \frac{2}{2\gamma-1}(1+t_2)^{(1-\gamma)/2}\frac{\Gamma\left(\frac{2\gamma+3}{4}\right)}{\Gamma(0.25)} \\ \times \sum_{n=0}^{\infty}(-1)^n \frac{\Gamma(0.25+n)}{\Gamma\left(\frac{2\gamma+3}{4}+n\right)} t_2^{-(n+0.25)}, \\ 1 \leq \frac{1}{t_1} \leq \frac{1}{t_2}, \end{cases} \tag{10.4}$$

where F_0 is the initial energy flux at the top boundary, m is the electron mass, K is the same normalization coefficient as in Section 3.3.2, $\Gamma(\gamma)$ is the gamma function, and the parameters t_1 and t_2 are described in Section 3.3.2. The plots of electron density variations with depth for the magnitude of the energy loss constant $a = 2.87 \times 10^{-12}\,\text{eV}^2\,\text{cm}^2$ are presented in Figure 3.4.

10.2.2
Nonthermal Hydrogen Excitation Rates

Let us now calculate the rates of excitation of a single hydrogen atom by a single electron beam having energy power law distributions obtained from the continuity equation (3.47). We simplify this formula to the view regarding the energy ratio of the lower cutoff energy to the energy of beam electrons by taking E out of the square root providing the approximate function

$$f(E) = C \times \frac{(E_0)^\gamma}{E^{\gamma+1}}, \tag{10.5}$$

where C is some constant, E_0 is the lower cutoff energy of beam electrons, γ is a spectral index of electrons, and we excluded $E^{1/2}$, required for the calculation of the electron beam density. Interestingly enough, a similar formula was deduced by Brown (1971) for a prolonged injection of beam electrons with $C = \gamma$, which was used for the calculations.

For the purpose of calculation of excitation rates we need to consider beam electron energies compatible with the energies of the considered radiative transition because of the cross-section of hydrogen transition used for thermal collisions (Johnson, 1972):

$$\sigma_{nn'} = \pi a_0 \frac{8\pi}{\sqrt{3}} \left(\frac{I_{\rm H}}{E_{nn'}}\right)^2 f_{nn'} g \frac{E_{nn'}}{E} . \tag{10.6}$$

Here a_0 is the Bohr radius, $I_{\rm H}$ is the full ionization energy of a hydrogen atom, $f_{nn'}$ is the oscillator strength for a given transition, and g is an appropriate Gaunt factor for this transition. Now we can calculate analytically the excitation rate of a hydrogen atom transition $n - n'$ by a single beam of electrons by substituting Eqs. (10.5) and (10.6) into Eq. (10.2) and integrating in the energy. Since electron beam energy is shown to change from zero to a lower cutoff energy as a positive power law and then from the lower cutoff to any energy as a negative power law, and this distribution smoothly shifts to lower energies with every precipitation depth (Chapter 3), it is reasonable to assume that at lower atmospheric depth beam electrons can approachthe energy of the considered radiative transition, thus setting $E_0 = E_{nn'}$.

The integration allows us to obtain the following expression for the hydrogen excitation rate by a beam electron:

$$B_{nn'} = \frac{8\pi^2 a_0^2}{\sqrt{3}} \sqrt{\frac{2}{m}} I_{\rm H}^2 E_{nn'}^{\gamma-1} f_{nn'} P(\gamma) , \tag{10.7}$$

where

$$P(\gamma) = \int E^{-\gamma-3/2} g dE . \tag{10.8}$$

For transitions between neighboring levels ($n' = n + 1$ we use semiempirical formulas for g derived in Golden and Sampson (1971), while for all other transitions we use the expressions for g from Sampson and Golden (1971). For $f_{nn'}$ we use the semiclassical Kramers formula.

Thus, for neighboring transitions ($n' = n + 1$) the hydrogen excitation rates by beam electrons are as follows:

$$B_{nn'} = \frac{4\sqrt{2}}{\sqrt{m}} \pi a_0^2 \gamma \frac{I_{\rm H}^2}{E_{nn'}^{1.5}} f_{nn'} \left\{ \frac{2(\gamma+1)}{(\gamma+0.5)^2(\gamma+1.5)^2} \right.$$
$$\left. + H_n \frac{\left[S_{nn'}^{r_n} + A_n(S_{nn'} - 1)\right]}{(\gamma+0.5)(\gamma+1.5)} + 0.19 H_n \frac{S_{nn'}^{2.5}}{(\gamma+3.5)} \right\} , \tag{10.9}$$

and for all other transitions:

$$B_{nn'} = 4\sqrt{\left(\frac{2}{m}\right)}\pi a_0^2 \gamma \frac{I_{\mathrm{H}}^2}{E_{nn'}^{1.5}} f_{nn'} \left\{ \frac{1}{(\gamma + 0.5)^2} + 2H_n \frac{\left[S_{nn'}^{r_n} + A_n(S_{nn'} - 1)\right]}{(\gamma + 0.5)(\gamma + 1.5)(\gamma + 2.5)} + 0.19 H_n \frac{S_{nn'}^{2.5}}{(\gamma + 3.5)} \right\} , \tag{10.10}$$

where $S_{nn'} = E_{nn'}/I_{n'}$, and the values of A_n, H_n, and r_n are taken from Table 2 in Sampson and Golden (1971).

10.2.3
Nonthermal Hydrogen Ionization Rates

A cross-section for continuous transitions from a level n is taken in the form proposed by Sampson and Golden (1971):

$$\sigma_{nc} = \pi a_0^2 \frac{128}{9} n^3 \frac{I_n^4}{E} \int_{I_{nc}}^{E} g_{bn} \left(\frac{E_{\mathrm{e}}}{h}\right) \bar{g} \frac{d E_{\mathrm{e}}}{E_{\mathrm{e}}^4} , \tag{10.11}$$

where I_{nc} is the ionization energy from level n, E_{e} is the energy within the continuum, g_{bn} is the classic Gaunt factor for the nth continuum, and $\bar{g}$ is the correction of the Gaunt factor by semiempirical expression (7) from Sampson and Golden (1971) for a Bethe approximation. By substituting expression (10.11) into expression (10.2) and performing integration one obtains the following formula for ionization rate B_{nc} per hydrogen atom and per beam electron:

$$B_{nc} = \frac{64}{9} a_0^2 \sqrt{\frac{6 I_{\mathrm{H}}}{m}} g_{bn} \gamma n^2 \Gamma^*(\gamma) , \tag{10.12}$$

where

$$\Gamma^*(\gamma) = \frac{2.5 - \gamma}{9(\gamma + 0.5)^2} + \frac{\gamma - 0.5}{4(\gamma + 1.5)^2} - \frac{5}{36(\gamma + 3.5)} + H_n \left[\left(\frac{1}{2 - r_n'} - \frac{1}{3 - r_n'} \right) \frac{1}{\gamma - r_n' + 3.5} - \frac{1}{(1 - r_n')(\gamma + 1.5)} + \frac{1}{(3 - r_n')(\gamma + 0.5)} \right] + \frac{0.19 H_n}{(\gamma + 1)(\gamma + 3.5)} . \tag{10.13}$$

The parameters H_n, r_n', and g_{bn} are the same ones used for ionization rates taken from Table 2 in Sampson and Golden (1971).

10.2.4
Comparison of Thermal and Nonthermal Excitation and Ionization Rates

The calculated hydrogen impact excitation and ionization rates are presented in Table 10.1 for five levels of a hydrogen atom for impacts with beam electrons (first line

Table 10.1 Comparison of hydrogen nonthermal excitation B_{nn} and ionisation B_{nc} rates caused by beam electrons with $\gamma = 3$ for the cross-sections of Johnson (1972) with Gaunt coefficients taken as Sampson and Golden (1971) (first rows for each levels) with the similar rates caused by thermal electrons with given temperature distribution (10 000 K – 2nd rows and 6000 K – 3rd rows). The numbers to be read as follows: $0.162{-}7 = 0.162 \times 10^{-7}$.

n	$n' = 2$	3	4	5	6	c
1	0.162-07	0.389-9	0.138-08	0.655-09	0.364-09	0.615-08
	0.206-12	0.611-14	0.108-14	0.371-15	0.172-15	0.709-15
	0.941-16	0.634-18	0.682-19	0.185-19	0.765-20	0.150-19
2		0.326-06	0.714-07	0.262-07	0.128-07	0.484-07
		0.635-07	0.646-08	0.170-08	0.702-09	0.166-08
		0.131-07	0.927-09	0.200-09	0.740-10	0.764-10
3			0.189-05	0.413-06	0.149-06	0.164-06
			0.117-05	0.157-06	0.724-07	0.609-07
			0.594-06	0.641-07	0.290-07	0.125-07
4				0.636-05	0.146-05	0.393-06
				0.541-05	0.819-06	0.335-06
				0.341-05	0.471-06	0.120-06
5					0.59-04	0.774-06
						0.950-06
						0.449-06

in each column) and thermal electrons for temperatures of 10 000 K (second line) and 6000 K (third line). The thermal impact excitation and ionization rates are calculated using the same cross-sections but for Maxwellian distributions of electrons. It can be seen that for lower atmospheric levels ($n \leq 5$) the impact excitation and ionization rates by beam electrons are much larger than the similar one by thermal electrons even for higher electron temperatures of 10 000 K. This emphasizes the importance of nonthermal impact excitation and ionization at the initial stages of solar flares before the flaring plasma becomes heated by a hydrodynamic response to very high temperatures well above 10 000 K. Now let is investigate how these rates compare over atmospheric depths of a flaring atmosphere heated by beam electrons, described in Chapter 5. For a comparison let us use the hydrodynamic model of flaring atmosphere heated by an electron beam with a spectral index of 3 calculated for 4 s after beam onset, which corresponds to the maximum temperature in a lower-density condensation where hydrogen emission is supposed to originate (between column depths of 2×10^{19} and 10^{23} cm^{-2}).

10.3 Hydrogen Emission Produced by Impacts with Beam Electrons

Now let us explore hydrogen emission from a flaring atmosphere being affected by injection of an electron beam with given initial parameters (energy flux F and spectral index γ plotted in Figure 10.1). The flaring atmosphere is assumed to

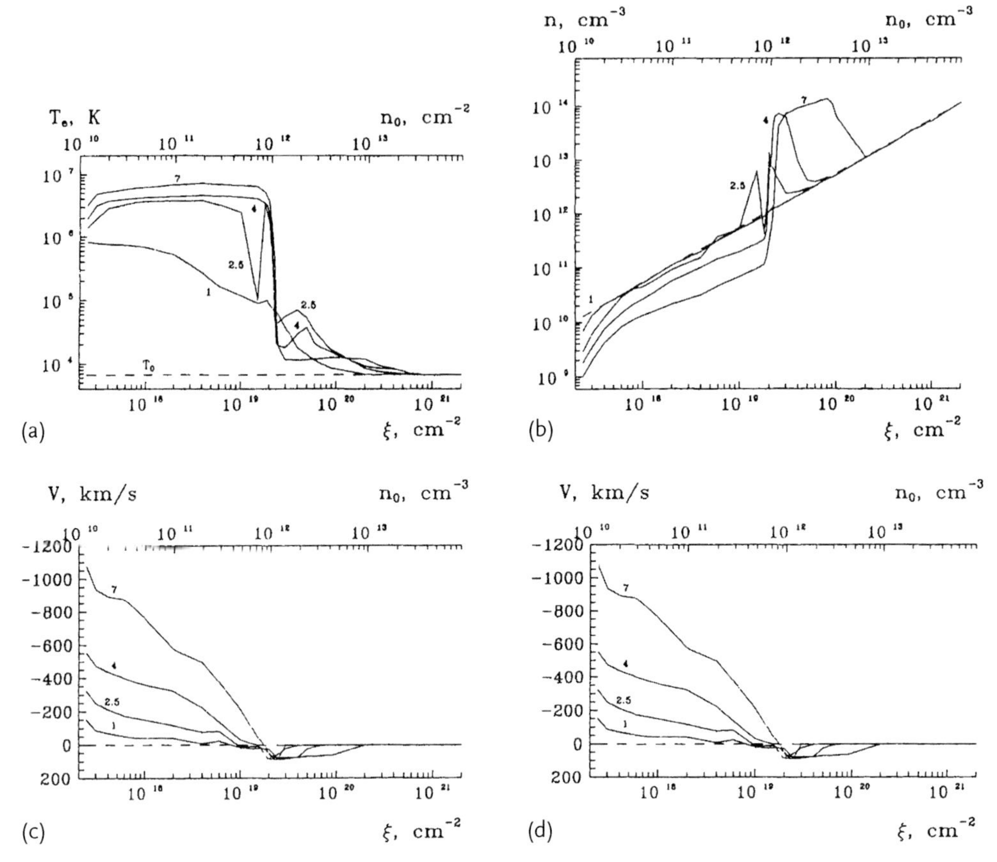

Figure 10.1 Column depth variations of ambient temperature (a), ambient density (b), and ambient macrovelocity (c) calculated from a hydrodynamic response to an injection of beam electrons with an initial energy flux of 10^{11} erg cm^{-2} s^{-1} and a spectral index of 3 and beam electron density for various beam parameters (d). In hydrodynamic model the numbers above the curves indicate seconds after beam injection onset. The beam density calculated from the continuity equation for the beam parameters (energy fluxes of F_0 and spectral indices) is indicated in the legend.

be that found from a hydrodynamic response to beam injection as described in Chapter 5, for example, it has variations with column depths of total density $n(\xi)$, electron temperature $T_e(\xi)$, ion temperature $T_i(\xi)$, and macrovelocities $V(\xi)$ (see Figure 10.1).

Let us now consider radiative processes in which a neutral hydrogen atom is involved in such a dynamic atmosphere; for example, let us solve joint equations of statistical equilibrium and radiative transfer in all the relevant transitions for the five levels plus a continuum hydrogen atom. Based on the ambient plasma parameters we calculate the degree of ionization and excitation of each level of hydrogen atoms by considering the statistical equilibrium of hydrogen atoms in all excited states, radiative transfer in all optically thick transitions, and the particle conservation equation as described below.

10.3.1
Equations of Statistical Equilibrium

Let us write the condition for statistical equilibrium at a given time t_0 of the upper level in the transition $i \to k$ as follows:

$$\frac{dn_i}{dt}(\xi) = n_i(\xi) \sum_{k,k\neq i}^{c} R_{ik}(\xi) + \sum_{k=1,k\neq i}^{c} n_k(\xi) R_{ki}(\xi) , \tag{10.14}$$

where $i = 1, N$ and N is the number of levels in a model atom ($N = 5$ in our consideration), $R_{ik}(\xi)$ is the total probability at a given column depth ξ of deactivation transitions from level i to other levels k, and $R_{ki}(\xi)$ is the total probability of the processes activating level i from other levels k. These probabilities can be written in the following form:

$$R_{ik}(\xi) = \begin{cases} n_e(\xi) C_{ik}(\xi) + B_{ik}\left[J_{ik}^{\text{dif}}(\xi) + J_{ik}^{\odot}(\xi)\right] + n_e^b(\xi) C_{ik}^b(\xi) , \\ n_e(\xi) C_{ik}(\xi) + A_{ik} , \\ A_{ci} + C_{ci}(\xi) n_e n^+(\xi) , \end{cases} \tag{10.15}$$

$$R_{ik}(\xi) = \begin{cases} n_e(\xi) C_{ki}(\xi) + A_{ki} , \\ n_e(\xi) C_{ki}(\xi) + B_{ki}\left[J_{ik}^{\text{dif}}(\xi) + J_{ik}^{\odot}(\xi)\right] + n_e^b(\xi) C_{ki}^b(\xi) , \\ n_e(\xi) C_{ic} + n_e^b C_{ic}^b(\xi) + F_{ic}^{\odot}(\xi) , \end{cases} \tag{10.16}$$

where A_{ik}, B_{ki}, and B_{ik} are the Einstein coefficients for spontaneous emission, induced emission, and absorption, C_{ik} and C_{ki} are impact deexcitation and impact excitation rates by thermal electrons, C_{ik}^b and C_{ki}^b are impact deexcitation and impact excitation rates by beam electrons, J_{ik} is the average emission intensity for the transition $i \to k$, A_{ci} and C_{ci} are the rates of spontaneous and three-body recombination, and $F_{ic}^{\odot}$ is the rate of photoionization from level i.

The equation for ionization equilibrium at depth ξ can be written in the following form:

$$\frac{dn_+}{dt}(\xi) = \sum_{i=1}\left[C_{ic} n_e(\xi) + B_{ic}(\xi)\left(J_{ic}^{\text{dif}} + J_{ic}^{\odot}\right) n_i(\xi)\right.$$
$$\left. -(C_{ci} n_e(\xi) + A_{ci} + B_{ci} J_{ic} n^+ n_e(\xi)\right] . \tag{10.17}$$

The rates of photoexcitation and photoionization are calculated using the following formulas:

$$F_{ik}^{\odot}(\tau_{ik}) = B_{ik} J_{ik}^{\odot}(\xi) = \frac{g_k}{g_i} \frac{c^2 A_{ki}}{2h(\nu_{ik}^0)^3} \frac{W_{ik}}{2\sqrt{\pi}}$$
$$\times \int_{-\infty}^{\infty} I_{ik}^{\odot}(x)\alpha(x) E_2[\alpha(x)\tau_{ik}(\xi)]dx , \tag{10.18}$$

where x is a dimensionless wavelength expressed in the Doppler half widths at a given depth.

$$F_{ic}^{\odot}(\tau_{ic}) = B_{ic} J_{ic}^{\odot}(\xi) = 4\pi k_{ic}^{0} \frac{W_{ic}}{2h} \times \int_{\nu_{ic}}^{\infty} I_{ic}^{\odot}(x) f_i(\nu) E_2[f_i(\nu)\tau_{ic}(\xi)] \frac{d\nu}{\nu} , \quad (10.19)$$

where ν_{ik}^{0} is the central frequency in the transition $i \to k$, g_i and g_k are the statistical weights of levels i and k, $E_2(y)$ is the exponential integral of the second kind, and $I_{ik}^{\odot}(x)$ and $I_{ic}^{\odot}(\nu)$ are the external intensity distributions over the relevant line profile within the frequency ν_{ik} and in the frequency of the ith continuum. W_{ik} and W_{ic} are the dilution factors of the relevant line and continuum, $\alpha(x)$ and f_i are the profiles of absorption coefficient in the line of the transition $i \to k$ and in the ith continuum, ν_{ic}^{0} is the frequency of the head of the ith continuum, $\tau_{ik}(\xi)$ and $\tau_{ic}(\xi)$ are optical depths in the central frequency/wavelength of the line and in the head of the ith continuum at a given column depth ξ, which are given by the following expressions: for the Lyman continuum,

$$\tau_{1c}(\xi) = k_{1c}^{0} \int_{\xi_0}^{\xi_{max}} \frac{n_H}{n_0}(\xi')[1 - X(\xi')]d\xi' ; \quad (10.20)$$

for lines of the Lyman series:

$$\tau_{1k}(\xi) = \frac{k_{1k}^{0}}{k_{ic}^{0}} \tau_{ic}(\xi) ; \quad (10.21)$$

for lines of the Balmer series:

$$\tau_{2k}(\xi) = \frac{k_{2k}^{0}}{k_{21}^{0}} \int_{\tau_{12}^{0}}^{0} \frac{n_2}{n_1}(t) dt . \quad (10.22)$$

Here k_{ik}^{0} and k_{ic}^{0} are the absorption coefficients in the line central frequency ν_{ik}^{0} and in the head of the ith continuum, ξ_0 is the initial column depth from which we consider radiative processes for hydrogen (in our case we accepted $\xi_0 = 10^{19}\,\text{cm}^{-2}$), $X(\xi) = n^{+}/n_H(\xi)$ is a distribution of the degree of ionization with depth, and $n_2/n_1(\xi)$ is the relative abundance of the second level.

10.3.2
Radiative Transfer Equations

In order to determine the average intensity of diffusive radiation in each transition, one must consider the partial opacity of the hydrogen plasma in line frequencies of the Lyman series, Lyman continuum, and the H_α line. Then one needs to write the radiative transfer equations in the following form (Ivanov and Serbin, 1984):

$$\mu \frac{d I_{ik}(\tau, \gamma, \mu)}{d\tau} = -\alpha(\gamma)[I_{ik}(\tau, \gamma, \mu) - S_{ik}(\tau, \mu)] \tag{10.23}$$

for $i = \overline{1, N-1}$ and $k = \overline{2, c}$, $\gamma = x$ for the lines, and $\gamma = \nu$ for (Lyman) the LC; α is the profile of the absorption coefficient in the lines or in the LC, $I_{ik}(\tau, \gamma, \mu)$ is the intensity of radiation in a given line or LC, τ is the optical depth in the relevant line, and S_{ik} is the source function for the lines ($k = 2, \ldots, N$) and the LC.

Under the assumption of complete redistribution over frequencies (CRF), the average intensity can be written using the following expressions:

for radiation in lines:

$$B_{ik} J_{ik}^{\text{dif}}(\tau) = \frac{A_{ki}}{2} \int_0^{\tau_{ik}^0} \frac{n_k}{n_i}(t) K(|\tau - t|) dt \,, \tag{10.24}$$

for radiation in LC:

$$B_{ic} J_{ic}^{\text{dif}}(\tau) = \frac{A_{c1}}{2} \int_0^{\tau_{ic}^0} \frac{n_e n_+}{n_1}(t) K(|\tau - t|) dt \,, \tag{10.25}$$

where the core function $K_1(|\tau|)$ is given by the formula

$$K_1(|\tau|) = F(T) \int_{\nu_{ic}}^{\infty} f_1 \nu^2 \exp\left[-\frac{h(\nu - \nu_{ic}^0)}{k T_i}\right] E_1(f_1(|\tau|)) d\nu \,. \tag{10.26}$$

Here $F(T)$ is a normalization factor, which is a function of the kinetic temperature of hydrogen atoms, h is Planck's constant, k is the Boltzmann constant, and ν_{ic}^0 is the frequency of the head of the Lyman continuum.

10.3.3
Conservation Equation for a Particle Number

The number or particles remains constant during the whole process of radiation, which results in the following particle number equation:

$$n_{\text{H}}(\xi) = n^+(\xi) + n^b(\xi) + \sum_{i=1}^{N} n_i(\xi) \,, \tag{10.27}$$

where n^+ is the abundance of hydrogen ions (protons), assumed to be $n^+ = n_e$ with n_e being the thermal electron abundance, and n_i is the number of electrons in the ith excited level. By assuming that the relative emission measure $n_e n^+/n_1$ is known, the abundance of free electrons/protons can be calculated from Eq. (10.27) as follows:

$$n_e(\xi) = \frac{\frac{n_e n^+}{n_1}(\xi)}{2\left[1+\sum_{i=2}^{N}\frac{n_i}{n_1}(\xi)\right]}\left\{\left[1.0 + \frac{4n_H(\xi)\left(1+\sum_{i=2}^{N}\frac{n_i}{n_1}(\xi)\right)}{\frac{n_e n^+}{n_1}(\xi)}\right]^{1/2} - 1.0\right\} . \quad (10.28)$$

The equations above describe the status of excitation and ionization of hydrogen atoms in a flaring atmosphere affected by the injection of beam electrons.

10.3.4 Method of Solution

Equations (10.14), (10.17), (10.23), and (10.26) define the system of integrodifferential equations to find non-LTE abundances of excitation level i and n_i/n_k and a relevant ionization measure $n_e n^+/n_1$.

First we calculate the states of a hydrogen atom at a fixed time t_0 for given distributions of total density $n_H(\xi)$ and $T(\xi)$ found from hydrodynamic solutions (Chapter 5), that is, let us set $(\partial n_i/\partial t)_{|t=t_0} = 0$. Then the equations above can be transformed into integral equations that can be solved using the Λ method (Ivanov and Serbin, 1984) by calculating in each transition the source functions $S(\tau)$ (related to the ratios n_i/n_k or $n_e n^+/n_1$) as follows:

$$S(\tau) = \frac{g(\tau)}{[1-\lambda+\lambda/2K(|\tau|)][1-\lambda+\lambda/2K(|\tau_0-\tau|)]} , \quad (10.29)$$

where $g(\tau)$ is the sum of all sources of atomic excitation at depth τ with the exception of diffusive radiation, λ is the probability of photon survival for a given line that is determined by the model atom (Zharkova, 1983), and $K(|\tau|)$ is the core defined by Eq. (10.26).

The following procedure is used for calculations.

1. We use the physical model of density and temperature obtained from hydrodynamic simulations (Somov *et al.*, 1981; Zharkova and Zharkov, 2007) at a time of 5 s corresponding to the maximum in electron energy flux. We assume that the initial hydrogen plasma had zero excitation and ionization before the electron beam was injected, for example $n_i/n_k = 0$, $n_e = 0$, and $X = 0$.
2. We calculate from Eqs. (10.20)–(10.22) the optical depths $\tau_{ik}(\xi_j)$ at given column depths ξ_j and the total optical thickness of the whole atmosphere ($\tau_{ik}^0 = \tau_{ik}^M$).
3. For transitions with total optical depths greater than 1 we calculate the intensities of diffusive radiation at every depth using Eqs. (10.24)–(10.25) and find for each transition the source functions, or relative abundances of the transitions in lines n_i/n_k and continua, $n_e n^+/n_1$, and depth distributions of a free

electron density n_e governed by the ionization balance equation. This will allow us to calculate the distribution of the hydrogen ionization degree X from Eq. (10.27).

4. Then we use the derived distributions n_i/n_k and n_e to compare them with those found in the previous step (in the first step they are compared with the initial values). If the difference is larger than required relative accuracy (accepted to be equal to 1%), to repeat the procedure from step 2 until solution convergence is achieved.

10.3.5 Accepted Parameters

We consider electron beams with the following parameters: initial energy fluxes $F_0 = 10^9$, 10^{10}, 10^{11}, and 10^{12} erg cm^{-2} s^{-1} and spectral indices $\gamma = 3$ and 5. We calculate the hydrodynamic responses of a flaring atmosphere to the injection of such beams up to 10 s after injection, or while the beam is being injected. Then from the hydrodynamic models we select the one with maximal heating by beam electrons, when the effect of thermal excitation and ionization is also maximal.

For calculation of the photoionization and photoexcitation rates by external radiation the intensities of chromospheric Ly_α line are taken from Brucker *et al.* (1976) and Vidal-Madjar *et al.* (1976), the distributions of the intensity in the chromospheric Ly continuum from Smith and Gottlieb (1974). The chromospheric intensities in Balmer lines and continua are calculated from Planck's formula for the excitation temperature of the quiet chromosphere. The absorption coefficients are accepted in Voigt form for lines and in the form $f = (\nu_{ic}/\nu)^3$ for the Ly continuum. The rates of collisional excitation and ionization of hydrogen atoms by thermal electrons is calculated from Golden and Sampson (1971) with the oscillation strengths taken from Johnson (1972) following the comparison by Zharkova (1990). The rates of nonthermal excitation and ionization are calculated as described in Sections 10.2.2 and 10.2.3. The rates of spontaneous emission are calculated following Burgess and Seaton (1964).

10.4 Hydrogen Excitation and Ionization

10.4.1 Comparison of Nonthermal and Thermal Excitation and Ionization Rates

The precipitation of electron beams deeper into the chromosphere and even the photosphere as revealed in Figure 4.10 in Chapter 4 raised the question of how these beam electrons interact with neutral atoms in the lower chromosphere and photosphere. Zharkova and Kobylinskii (1989, 1993) have compared the volume hydrogen nonthermal excitation and ionization rates with thermal ones and those by external radiation and concluded that nonthermal processes were by a factor 10^3–10^4 more effective in the lower chromosphere and photosphere, exciting hydrogen atoms to higher energetic levels or even fully ionizing them (see Figure 10.2).

In Figure 10.3 a comparison is presented for hydrogen volume ionization rates by thermal and nonthermal mechanisms for different electron-beam parameters and the physical conditions derived from hydrodynamic solutions. It can be seen that at chromospheric depths starting at a column depth of $\xi = 4 \times 10^{19}\,\text{cm}^{-2}$,

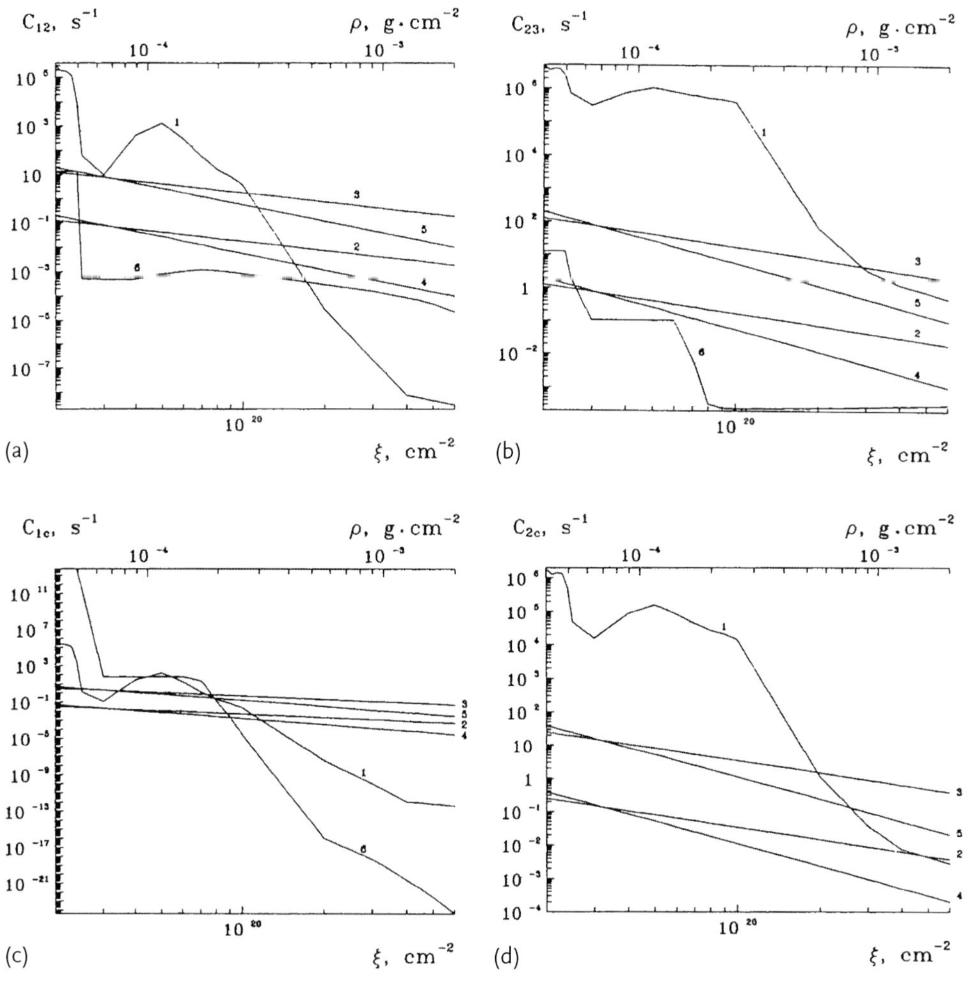

Figure 10.2 Column depth variations of hydrogen excitation rates for the following transitions: 1–2, Ly_α (a), 2–3, H_α (b), 1–c, Lyman continuum, and (c) 2–c, Balmer continuum (d). The numbers above the curves represent activation by 1: thermal electrons, 2: electron beam with $\gamma = 3$ and $F_0 = 10^9$, 3: electron beam with $\gamma = 3$ and $F_0 = 10^{11}$, 4: electron beam with $\gamma = 5$ and $F_0 = 10^9$, 5: electron beam with $\gamma = 5$ and $F_0 = 10^{11}$, 6: external UV radiation in Ly_α line.

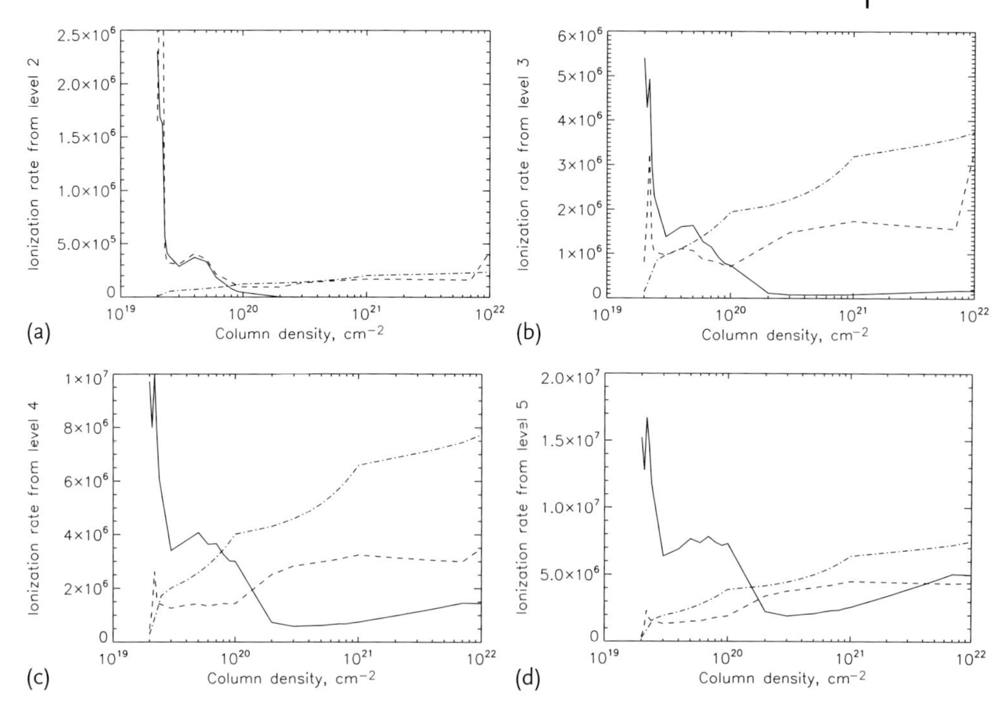

Figure 10.3 Variations of ionization rates with depth for atomic levels 2 and 4 (a,d) and for levels 3 and 5 (b,d) for thermal collisions (solid lines), external radiation (dashed line), and diffusive radiation (dotted–dashed lines).

or a mass depth of $2 \times 10^{-4}\,\mathrm{g\,cm^{-2}}$, nonthermal ionization by beam electrons is dominant by a few orders of magnitude; the same is valid for nonthermal and thermal excitation rates reported by Zharkova and Kobylinskii (1993). We emphasize that the ionization rates increase with the level numbers in a hydrogen atom, contrary to the assumption of Allred *et al.* (2005), who used a very simplified formula for the ionization rate based on the electron-beam heating function.

This additional nonthermal ionization and excitation is much higher while the beam is injected, in the first 10 s, than that by external radiation discussed in the section above. This nonthermal impact ionization leads to a fast increase of all hydrogen emissions including the H_α emission often observed simultaneously with HXRs (Zharkova *et al.*, 2007).

But most importantly, the nonthermal ionization caused by beam electrons leads to a significant increase (up to four orders of magnitude) of the ionization degree of the ambient plasma shown in Figure 10.4 (Allred *et al.*, 2005; Zharkova and Kobylinskii, 1991, 1993) and, as result, of hydrogen continuous emission in the Balmer and, especially, Pashen continua (Allred *et al.*, 2005; Zharkova and Kobylinskii, 1993). And because of the impact, ionization rates are much higher than hydrogen recombination rates (Zharkova and Kobylinskii, 1991), and, as a result of an increased optical thickness in the hydrogen continua (also confirmed by Zharkova

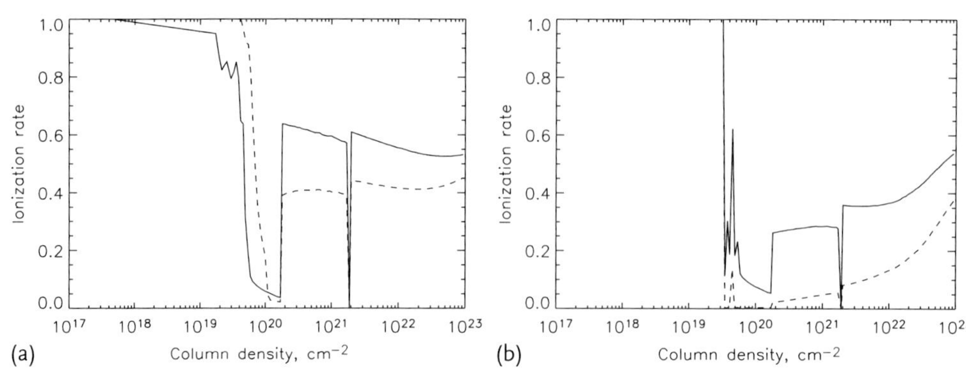

Figure 10.4 Variations in ionization degree with depth caused by electron beams. (a) Calculated for the hydrodynamic model 2 s after beam injection with a spectral index of 4 and an initial beam flux of $F_0 = 10^{10}$ erg cm^{-2} s^{-1} (dashed line) and $F_0 = 10^{11}$ erg cm^{-2} s^{-1} (solid line); (b) for beams with an initial flux of $F_0 = 10^{11}$ erg cm^{-2} s^{-1} and a spectral index of 6 after 2 s (dashed line) and 10 s (solid line) after beam injection.

and Kobylinskii, 1991), this can result in a much longer duration of this continu ous emission due to radiative transfer processes. Hence the radiation of the Pashen continuum of hydrogen atoms and of negative hydrogen ions appears as a white-light flare (Aboudarham and Hénoux, 1987; Hudson, 1972), which in the case of reduced heating by beam electrons leads to a gentle radiative cooling phase (Allred *et al.*, 2005).

By solving the full non-LTE problem for a hydrogen five-level-plus-continuum atom (Zharkova and Kobylinskii, 1991) one can find that the hard beam fluxes produce a much higher ionization than softer intense beams (Figure 10.4), and it increases toward the photosphere owing to the radiative transfer effects discussed in Section 10.4.4, while the precipitation of more intense hard electron beams to the lower chromosphere is strongly restricted by their increased ohmic losses, which prevents many beam electrons from reaching the photosphere (Zharkova and Gordovskyy, 2006).

However, since similar nonthermal ionization and excitation effects can be caused by proton beams (Hénoux *et al.*, 1993), these results must be tested with a refined proton-beam kinetics considering not only their particle–particle interactions, but also the wave–particle ones similar to those discussed by Gordovskyy *et al.* (2005). In X-class flares other kinds of particles should contribute to the momentum delivery to the lower atmosphere, like proton beams and quasithermal flows, or jets, as was reported for the Halloween 2003 flare (Zharkova and Zharkov, 2007).

10.4.2
Nonthermal Effects on Hydrogen Emission

In our radiative simulations we considered both radiative and collisional (thermal and nonthermal) processes. The resulting Ly_α and Ly_β line profiles calculated for different hydrodynamic models are plotted in Figure 10.5 and H_α line profiles in Figure 10.6. The distributions in the heads of Ba and Pa continua are plotted in Figure 10.7. We also simulated the time-dependent H_α line profiles for different times after beam onset (1, 2.5, 4, and 7 s) and the different beam parameters presented

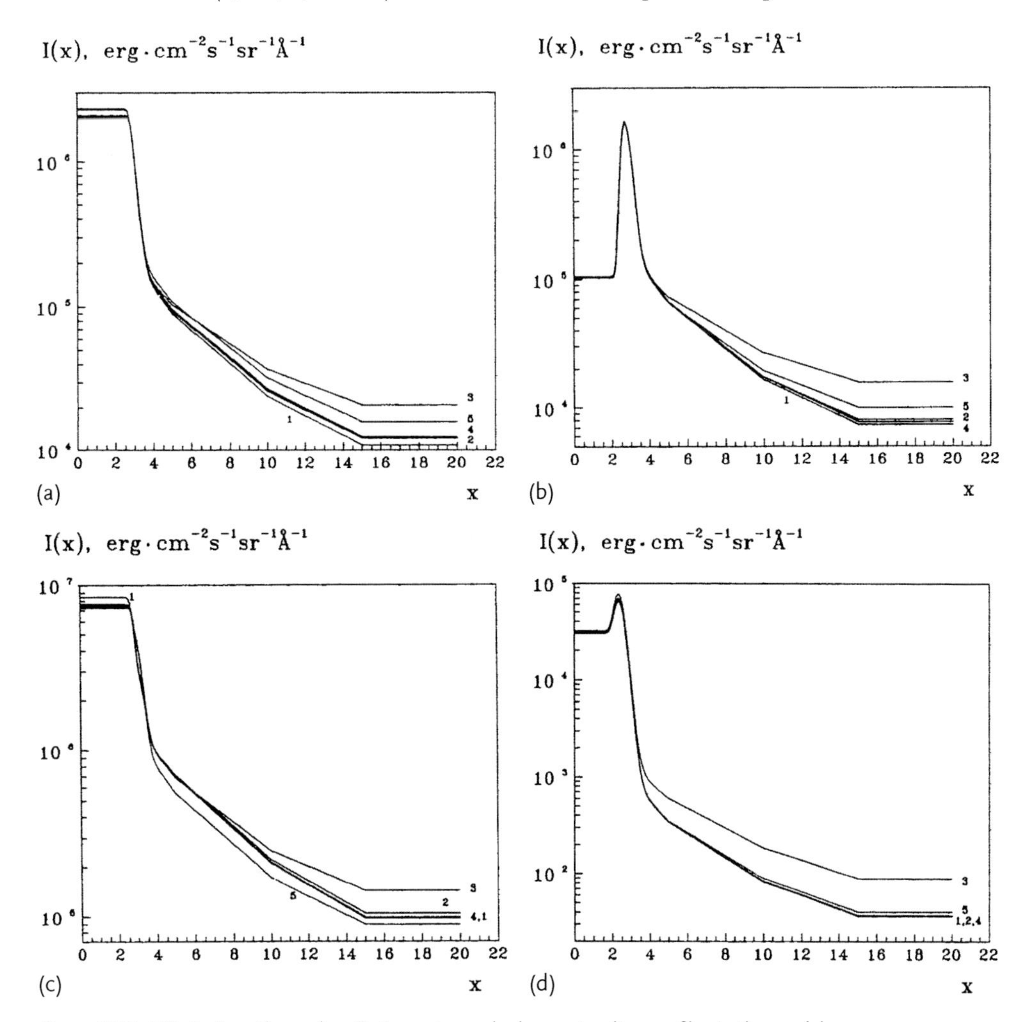

Figure 10.5 Effect of nonthermal excitation rates on hydrogen Ly_α-line profiles in the model of 1 s (a) and 4 s (b) and on Ly_β line for 1 s (c) and 4 s (d). The numbers near the figures correspond to the same beam parameters as in Figure 10.2.

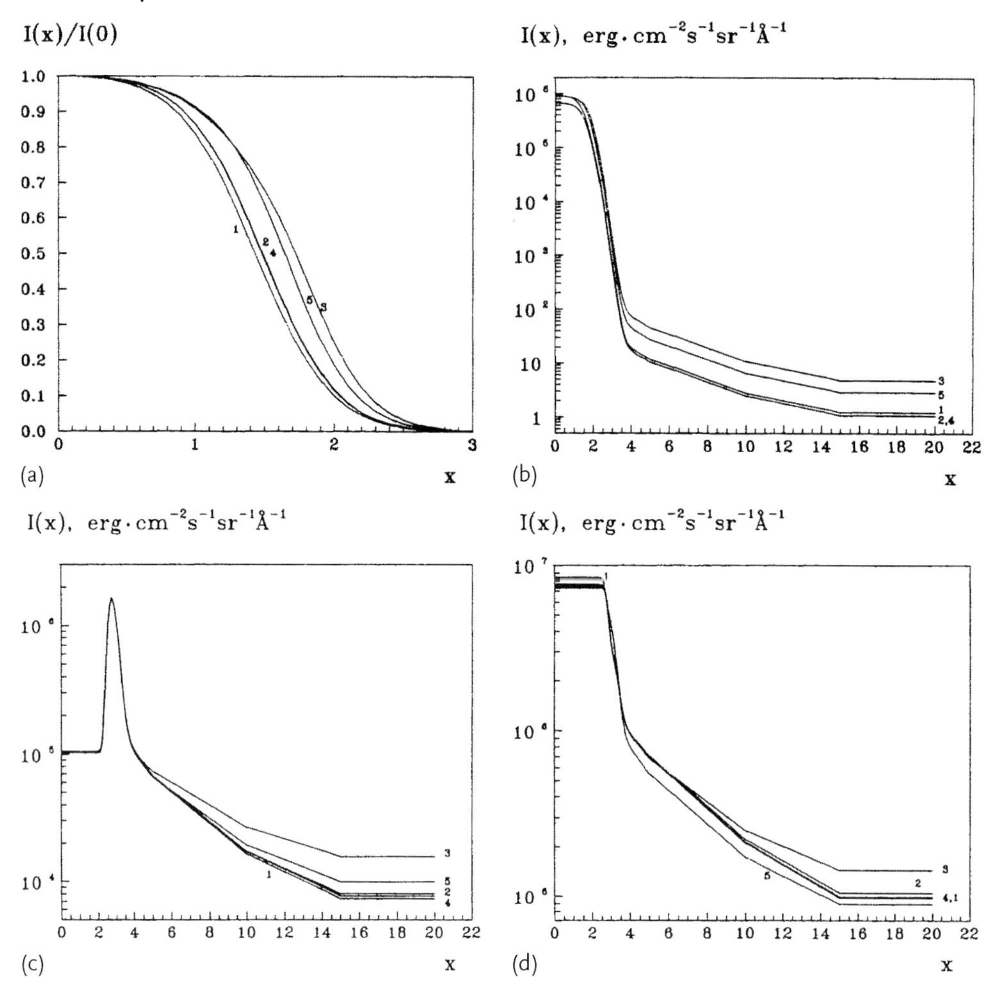

Figure 10.6 Effect of nonthermal excitation rates on H_α line profiles calculated for 1 s after beam onset in the core (a) and line wings (b) and at 2.5 s (c) and 4 s (d) after beam onset. The numbers near the figures correspond to the same beam parameters as in Figure 10.2.

in Figure 10.8 and compared them with the high time resolution observations in the kernel B1 by Wulser and Marti (1989) at 15:25:05 UT for the flare (insert date).

From the simulated line profiles one can see that impacts with beam electrons affect mostly line wings, which are likely to become more intense because of the increasing role of Stark's effect. For Ly-line wings impacts with beam electrons have been found to increase the wing intensity by a factor of 2–10 (for the most intense hard beams), while the line cores are mostly governed by thermal collisions. This is because Ly line cores are formed at higher levels of the atmosphere where temperatures are high, while the wing formation region is extended deeply into the

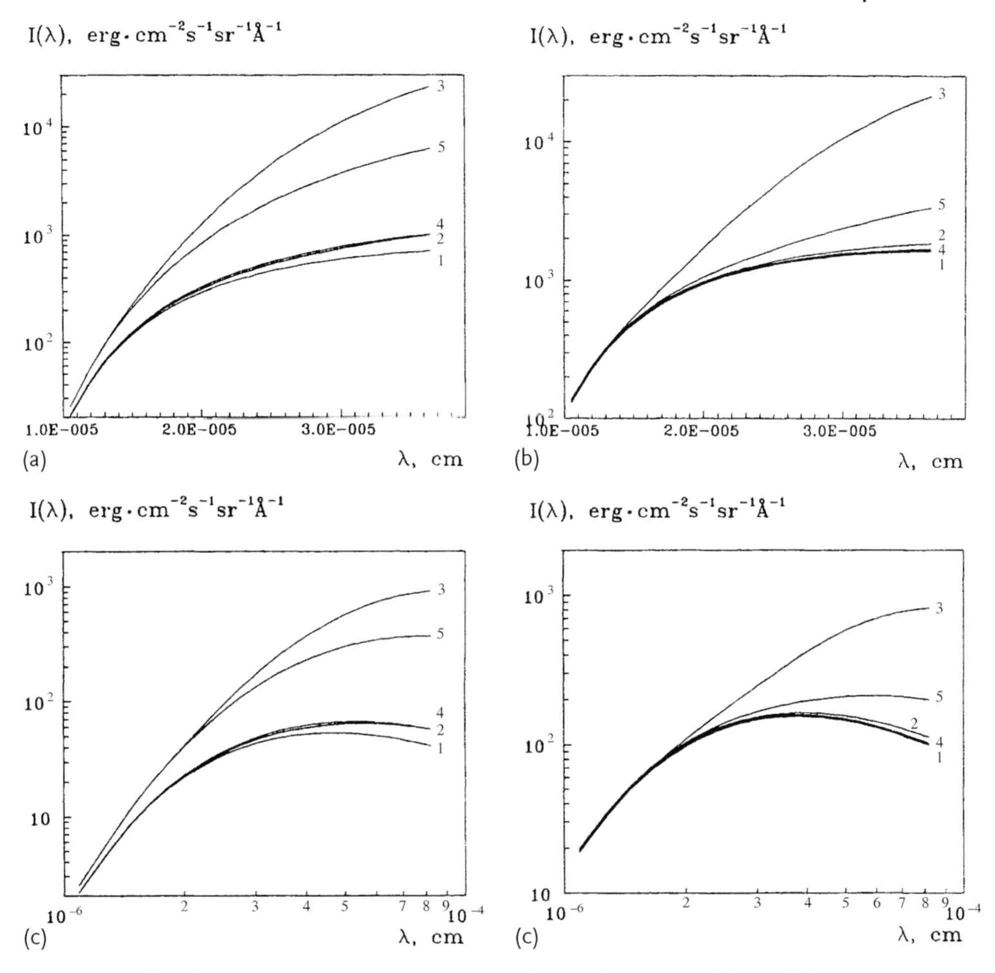

Figure 10.7 Effect of nonthermal ionization rates on emission: (a,b) in head of Balmer continuum calculated at 1 s (a) and 2.5 s (b) after beam onset; (c,d) head of Pashen continuum calculated at 1 s (c) and 2.5 s (d) after beam onset. The numbers near the figures correspond to the same beam parameters as in Figure 10.2.

lower chromosphere where impacts with electrons become very important in the very first seconds after beam injection.

The situation is slightly different with the H_α line. In general, H_α line profiles calculated for the first seconds after beam onset, that is, before the ambient plasma is heated by a hydrodynamic response, are found to reveal strong self-absorption in the central line intensities, wider line cores, and very extended wings. The wings of the H_α line are more affected by impacts with beam electrons showing increases in intensities up to one order of magnitude (strong Stark wings). The stronger and harder the beam, the higher the recorded increase in wing intensities. These effects

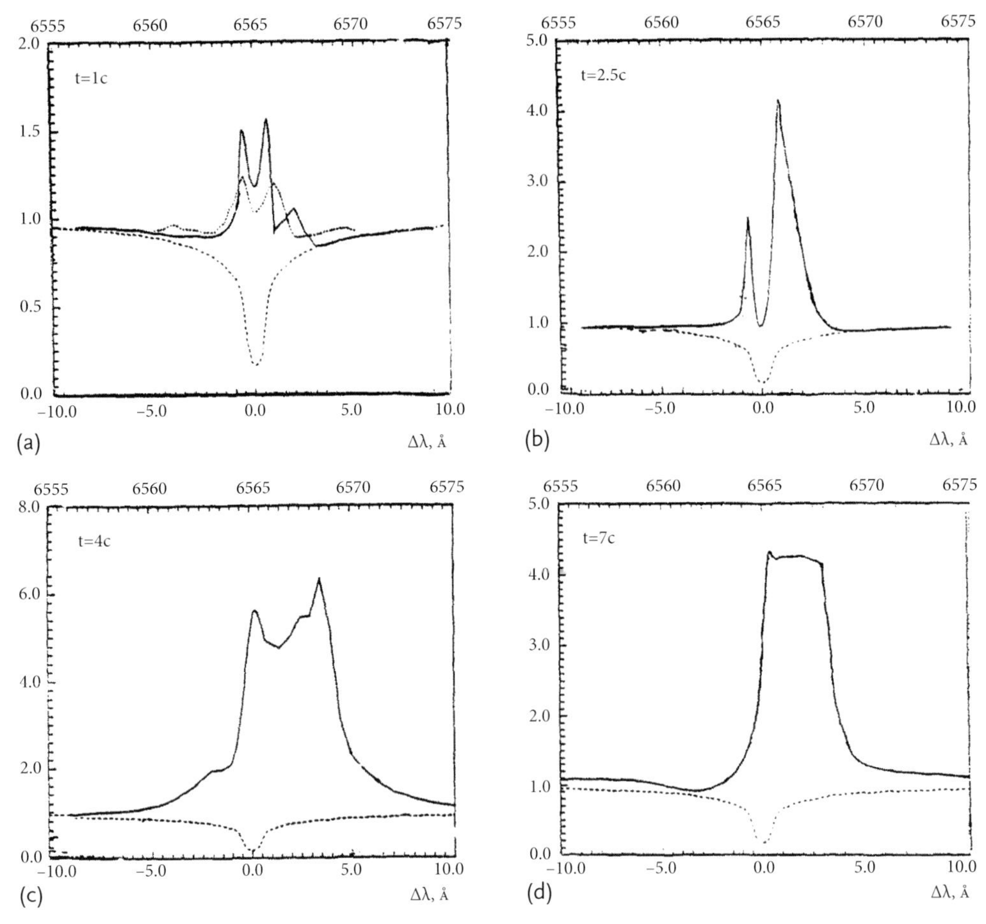

Figure 10.8 Resulting H_α profiles calculated for hydrogen atoms affected by electron beam with parameters $\gamma = 6$ and $F_0 = 10^{11}\,\text{erg}\,\text{cm}^{-2}\,\text{s}^{-1}$ for hydrodynamic models at $t = 1$ s (a), 2.5 s (b), 4 s (c), and 7 s (d). Solid lines: simulated profiles, dashed lines: residual absorption profiles, dotted line: profile observed in flare by Wulser and Marti (1989).

on the line profiles in the presence of a beam can also be detected from polarimetric observations as discussed in Chapter 9.

When H_α emission profiles are calculated for dynamic flaring models including the large Doppler motions of lower-density condensations and superimposed on the chromospheric absorption profiles as in Figure 10.8, we can reproduce strong red shifts in H_α profiles with an asymmetry of the blue and red peaks in the line cores. The H_α emission profile simulated at 1 s after beam onset (Figure 10.8a) reveals a well-defined emission profile with the central self-absorption and increased central intensity by a factor of 1.5, compared to an undisturbed atmosphere. The wing intensities are also increased at a distance of up to 4 Å in both wings for beams with $\gamma = 5$ and to 3 Å for beams with $\gamma = 3$. Also the central H_α intensity caused by harder beams is lower by a factor of 2 and has smaller self-absorption (Kobylinskii and Zharkova, 1996).

At 2.5 s after beam onset the ambient density and temperature increase due to hydrodynamic heating, which reduces the role of nonthermal effects on hydrogen emission. This produces the H_α line profiles with core intensity enhancements by a factor of 4 above the continuum with a very strong self-absorption in the central intensity, although the line wings are still well extended and have increased intensities compared to those for 1 s. This resembles the results by Aboudarham and Hénoux (1986) and Canfield and Gayley (1987), with the difference that their wings are extended only to 2 vs. 4 Å in our simulations. At 4 s of the hydrodynamic response there is a well-formed low-temperature condensation moving toward the photosphere. As result, the H_α line profiles reveal an increase in the central intensity, have a smaller self-absorption (the opacity of the ambient plasma in this line becomes lower), and a strong red shift. At 7 s, when the beam is turned off, the intensities in the core decrease while the wing intensities remain unchanged.

The effects of nonthermal impacts on hydrogen emission are somehow different from those simulated by Aboudarham and Hénoux (1986, 1987) and Fang *et al.* (1995), who reported a main increase in the intensities of the line cores without affecting at all the intensities in the wings. This discrepancy occurs because they used the volume excitation and ionization rates calculated for the beam densities at given depths found from the flux conservation approach (Brown, 1971; Emslie, 1978) discussed in Chapter 3. As a result, their electron-beam densities at higher atmospheric levels are significantly overestimated, as shown in Chapter 3 (see also Mauas and Gomez, 1997), leading to a shift of the main contribution of the impacts with beam electrons to the upper chromosphere, where the line core is formed.

In the present radiative simulations we used the model of electron precipitation derived from a continuity equation (Syrovatskii and Shmeleva, 1972b). This approach provides a very smooth distribution of beam electrons with precipitation depth shifting the main effect of the impacts of hydrogen atoms with beam electrons to lower chromospheric levels where the line wings are formed and where the ambient temperature is much lower than in the upper chromosphere, so that the additional impacts with beam electrons become very apparent. In fact, the extended wings are often observed as moustaches in H_α lines in compact flares as presented in Figure 10.8, revealing also a significant impact polarization in this emission as discussed in Chapter 11.

This increase of the intensity in wings is likely to be related to a substantial increase of the hydrogen ionization caused by the impacts with beam electrons, as shown in Figure 10.4. The ionization degree caused by beam electrons approaches magnitudes of 0.5–0.6, which is up to five orders of magnitude higher than it should be at a given kinetic temperature of the ambient plasma. Thus, there is a clear case of violation of the local thermodynamic equilibrium because the ionization temperature of a flaring atmosphere is higher by five orders of magnitude than the kinetic temperatures of ambient hydrogen atoms or thermal electrons.

This leads to an increase of the optical depth in continuum hydrogen emissions and to the occurrence of very extended continuous emission in all continua as shown above, and specifically in Pashen continua, seen as white-light flares. This continuous emission can last as long as the opacity of the visible Pashen continua remains higher than unity, until it approaches this limit.

10.4.3
Hydrogen Radiative Losses in Flares

The combined radiative losses of the main hydrogen lines and continua calculated for the hydrodynamic models at 1 and 7 s after beam onset are plotted in Figure 10.9. It can be observed that the hydrogen radiative losses are mainly defined by Lyman series in the transition region, by Balmer line and continuum in the chromosphere, and by Pashen and Brackett emission in the deeper atmosphere, in the photosphere.

The variations of the total hydrogen losses at the moments of beam injection are found to be strongly affected by beam parameters. For example, at the transition region the Lyman series radiative losses caused by harder beams are higher by an order of magnitude than those by softer beams. At deeper atmospheric levels the radiative losses in Pashen series become governed mainly by beam impacts and define the total radiative losses at the photospheric levels. This confirms the conclusions above that inelastic collisions with beam electrons are the main contributors to energy transport from beam electrons to ambient hydrogen atoms until either hydrodynamic heating increases the ambient temperatures or the beam injection stops.

10.4.4
Role of Backwarming Heating

A comparison of the radiative ionization rates for five-level hydrogen atoms with those caused by inelastic collisions with thermal electrons and by diffusive radiation formed by the action of radiative transfer, as shown in Figure 10.3 for levels 2–5, reveals a dominant effect of diffusive emission in the lower chromosphere (column depths $> 10^{20}$) because of a decrease of the continuous opacity for each level.

It can be seen that the radiative ionization rates increase for higher levels (4 and 5 in this model). But since the ambient plasma before a flare onset is almost neutral at this depth (the ionization degree is less than 10^{-6}), few hydrogen atoms are excited to higher energetic levels, and hence not many hydrogen atoms are ionized by the radiation coming from the chromosphere before any beam electrons are injected. However, after 1 s of beam injection most hydrogen atoms become strongly excited to higher energetic levels by beam electrons (Zharkova and Kobylinskii, 1993). This immediately increases the hydrogen ionization from these levels leading to the very high ionization degrees of 0.4–0.6 reported in Figure 10.4.

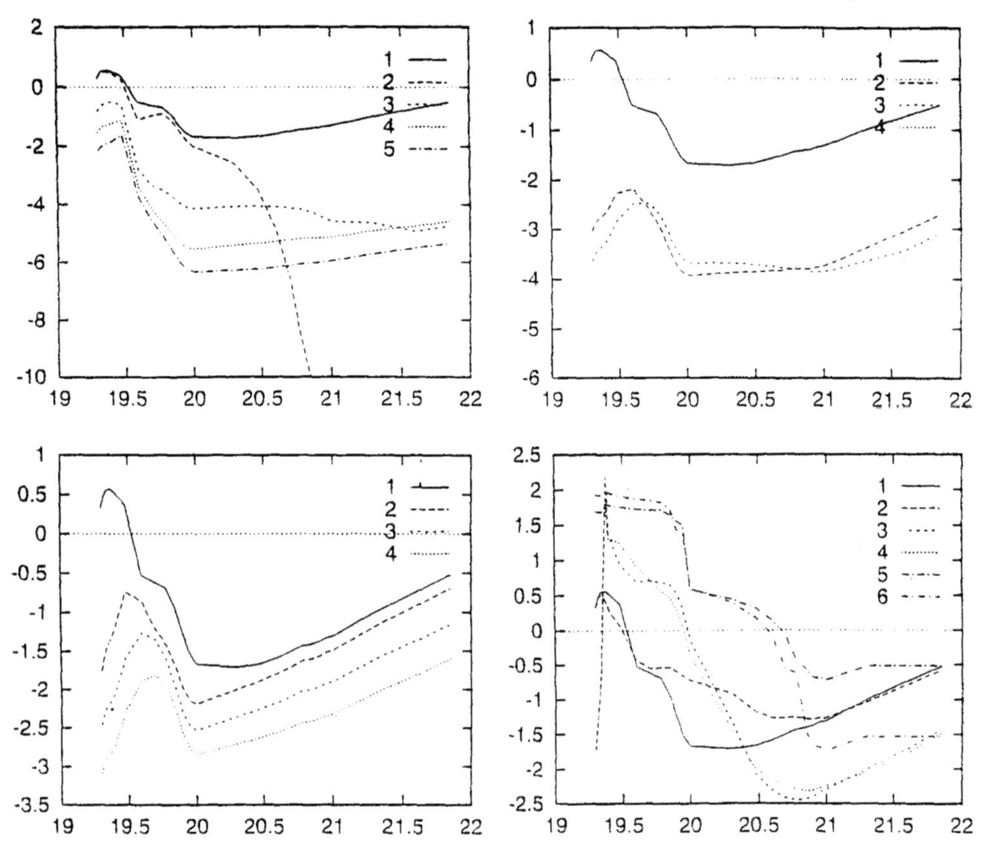

Figure 10.9 Hydrogen radiative losses calculated for the atmosphere model heated by a beam with parameters $\gamma = 5$ and $F_0 = 10^9\,\mathrm{erg\,cm^{-2}\,s^{-1}}$ for $t = 1$ s: (a) total radiative losses (1) and in Lyman series (2–5); (b) total (1) and for Balmer series (2–4); (c) the same as above for Pashen and Bracket series; (d) comparison of total hydrogen radiative losses for hydrodynamic models heated by beams with $1 - \gamma = 5$, $F_0 = 10^9\,\mathrm{erg\,cm^{-2}\,s^{-1}}$, $t = 1$ s; $2 - \gamma = 5$, $F_0 = 10^{11}\,\mathrm{erg\,cm^{-2}\,s^{-1}}$, $t = 1$ s; $3 - \gamma = 3$, $F_0 = 10^9\,\mathrm{erg\,cm^{-2}\,s^{-1}}$, 7 s; $4 - \gamma = 3$, $F_0 = 10^{11}\,\mathrm{erg\,cm^{-2}\,s^{-1}}$, 7 s; 5 – beam heating functions for beam with $F_0 = 10^9\,\mathrm{erg\,cm^{-2}\,s^{-1}}$ and $5 - \gamma = 5$ and $6 - \gamma = 3$ at $t = 2.5$ s.

As has been discussed by Donea *et al.* (2006) a backwarming heating can play some role in additional heating of the lower atmospheric levels, leading to white-light flares (Hudson, 1972; Metcalf *et al.*, 2003; Somov *et al.*, 1981) that often accompany the seismic emission in flares. A substantial fraction of the continuum emission from white-light flares from the overlying chromosphere has been proposed to be driven to the photosphere and to result in a "radiative backwarming" by recombination radiation recently included in radiative hydrodynamics by Allred *et al.* (2005).

The photosphere is assumed to absorb the part of this radiation that is emitted downward, and thus it is heated additionally to those heating by electrons and by external X-ray and EUV/UV radiation. The immediate effect of this absorption in the visible spectrum is mostly dissociation of H^- ions, which represents the predominant source of photospheric opacity. The radiation energy estimated from white-light flares is close to those measured with seismic emission (Donea *et al.*, 2006). This allows one to assume an additional source different from protons delivering energy to the lower atmosphere (the photosphere), leading to the acoustic emission that is closely associated with continuum emission.

However, Allred *et al.* (2005) have pointed out that for their hydrodynamic models, heating by electron beams of very moderate power, the backwarming heating is not very effective during the impulsive phase when the hydrogen emission is formed mostly by nonthermal ionization and external high-energy radiation (X-rays, UV, and EUV). The increased ionization immediately produces the strong emission in hydrogen lines and continua with a large opacity. This means that the radiation remains locked inside the atmosphere for a period of the difference between its recombination and impact excitation rates, discussed in Section 10.4.1.

After the beam heating has stopped (say, in 10 s), thermoconductivity continues to heat the lower atmosphere, increasing its emission in all lines and continua, leading to further increase of the line and continuous opacity, as reported by Zharkova and Kobylinskii (1993) and confirmed by Allred *et al.* (2005). Moreover, Allred *et al.* (2005) have reported that this additional heating does not increase the backwarming heating magnitude. The backwarming heating remains mild since it is governed by a high opacity of radiation.

However, the duration of this process can be increased. As a result, during all this time the backwarming heating can steadily affect the emission of the Ni line from which the seismic response is derived, and this definitely affects the results derived from the holographic method, as pointed out by Donea *et al.* (2006). However, this backwarming heating cannot account for the momentum delivered to the photosphere by some agents as reported by Zharkova (1999), Zharkova and Kosovichev (2000) and Zharkova and Zharkov (2007) for the seismic responses detected with the time-distance technique.

10.5 Interpretation of H_α Emission in 25 July 2004 Flare

10.5.1 Fast Changes of H_α Emission in the Main Flare Event

In the main flare event, the fast changes in the H_α intensity observed by VTT before the flare maximum are also correlated rather closely with the HXR emission observed by RHESSI (Figure 10.10). In order to search for fast intensity increases, for the given period of time the difference H_α intensity images. As a result, the intensity changes are found in a few bright kernels that appeared at the following times: 1 – 13:38:38 UT; 2 – 13:38:44 UT; 3 – 13:38:49 UT; 4 – 13:38:55 UT; 5,6,7 – 13:40:06 UT; 8,9 – 13:40:45 UT; 10 – 13:41:40 UT. These fast changes in the chromospheric emission are good spatial indicators of the locations where the initial energy detected from HXR spikes is deposited in the lower atmosphere.

10.5.2
Temporal and Spatial Evolution of the Main Flare Event

10.5.2.1 Temporal Evolution

Since TRACE instruments have a higher spatial resolution, the TRACE 195 Å images are used for the precise identification of the flaring loop elongation and the locations where the loops are likely to be embedded in the photosphere. This provides further information about the loop connectivity in the corona. The temporal evolution of the flaring loops is shown in Figure 10.11. The loops were brighter at 13:40:48 UT (image TRACE 195 Å) at the maximum of HXR emission, which was followed by a rather complex loop structure appearing at the moment 13:41:38 UT when the HXR emission started to decrease.

A superposition of the HXR sources (dashed and solid contours), H_α emission kernels with TRACE images at 195 and 1600 Å, is also presented in Figure 10.11.

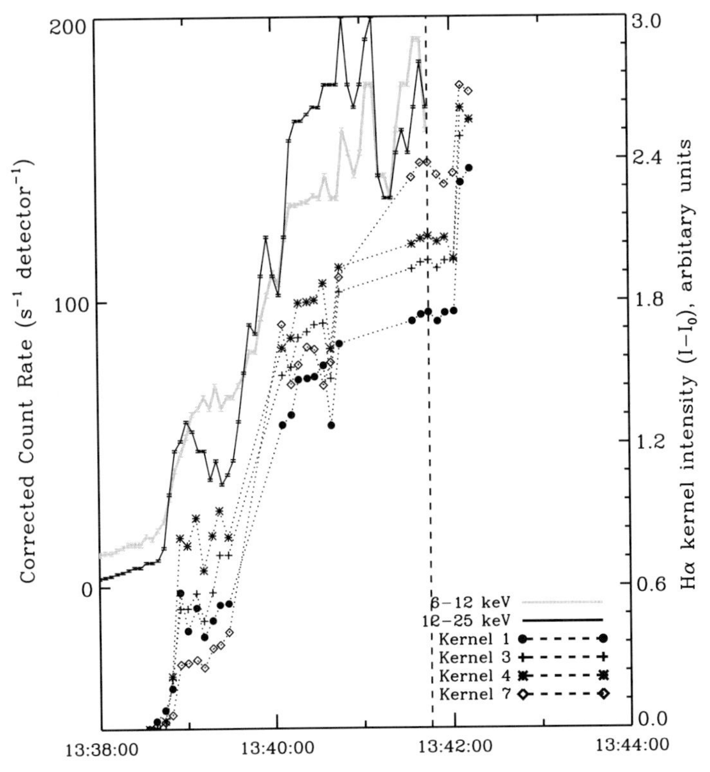

Figure 10.10 Temporal evolution of HXR flux (RHESSI) (left y-axis) and H_α intensity (right y-axis) for different kernels whose locations are shown in Figure 10.11. The dashed line shows the beginning of the RHESSI night time. I_0 is the initial intensity of the relevant kernels.

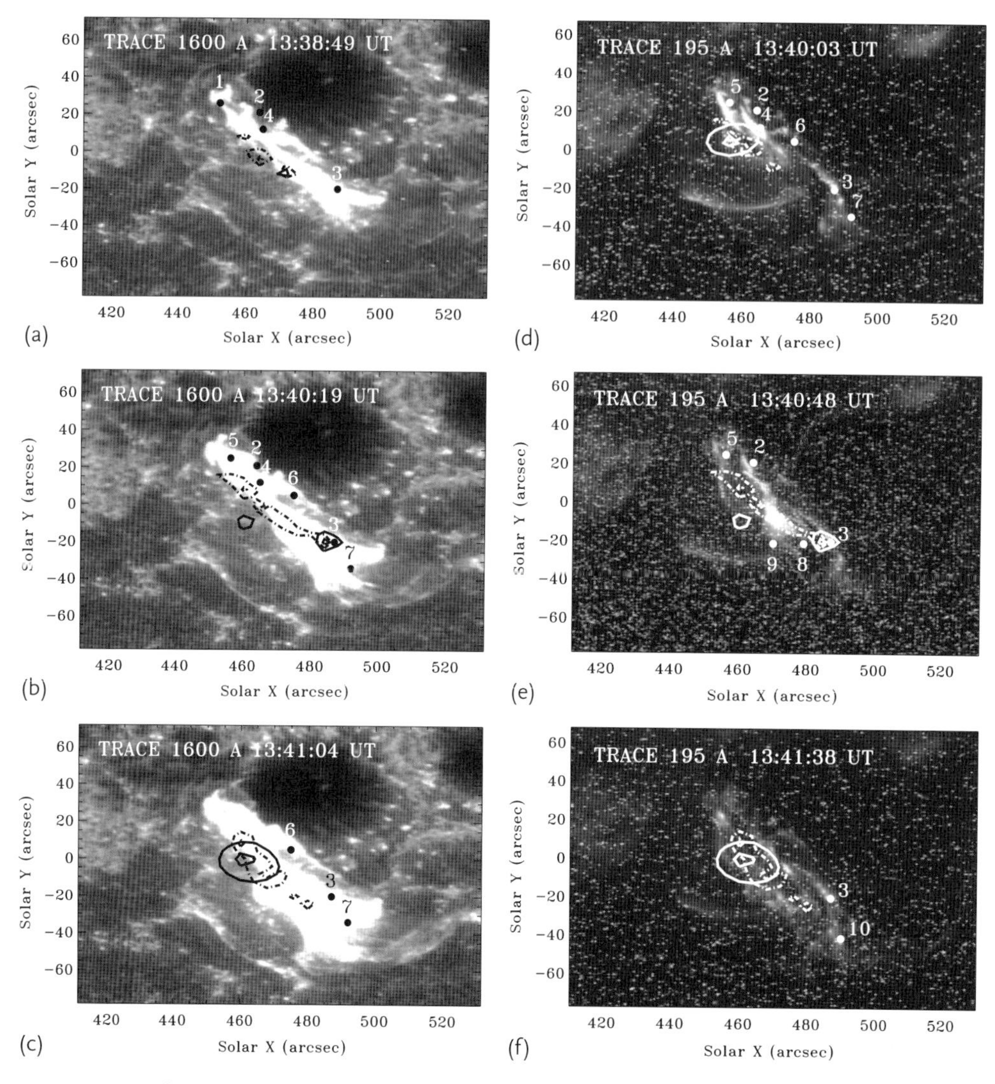

Figure 10.11 Main flare event: evolution of H_α emission kernels and HXR sources vs. TRACE image of the same region in 195 and 1600 Å. H_α emission kernels are marked as black points whose number corresponds to their order of onset. The HXR sources are plotted as contours at levels of 50 and 90% from their maximum. The dashed and solid contours represent 12–25 and 25–50 keV bands, respectively.

We used the HXR source images for 12–25 and 25–50 keV bands which were reconstructed with the PIXON algorithm (Metcalf *et al.*, 1996; Puetter and Pina, 1994) because of the sharper shape of the sources. Correct positions of the sources were confirmed by a comparison of the images reconstructed by both CLEAN and PIXON algorithms.

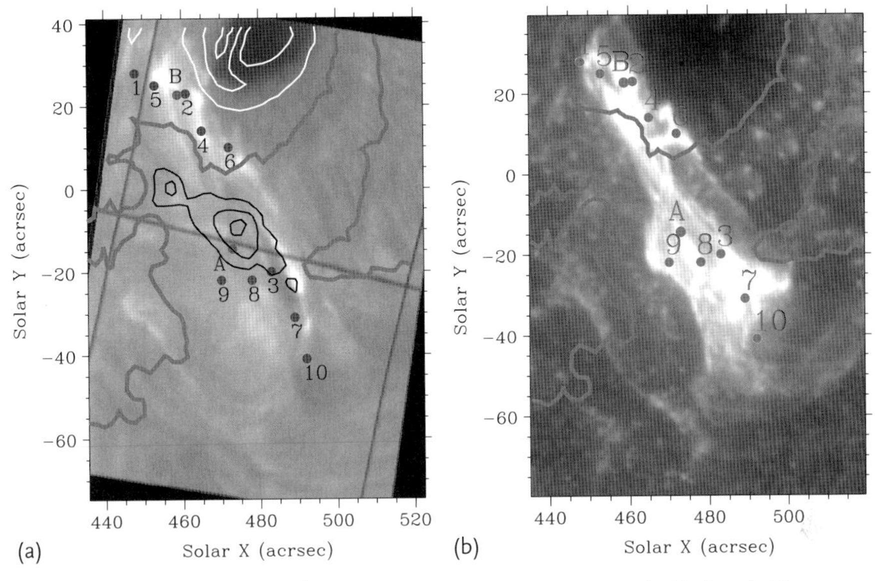

Figure 10.12 (a) shows the 1600 Å image at 13:40:02 UT with the locations of H_α emission (kernels) marked by white dots. "A" and "B" show locations of preflare kernels. (b) shows the same H_α kernels (black dots) overlaid on the H_α-line image. The black and white contours are negative and positive magnetic fields, correspondingly. A magnetic inversion line or neutral magnetic line (MNL) is marked by a gray line on both panels.

For the analysis of the magnetic structure of the flaring region SOHO/MDI magnetograms obtained with 1 min cadence are used. Figure 10.12a shows the longitudinal magnetogram at 13:40:02 UT overplotted with the locations of H_α emission kernels. The H_α sources appeared at the opposite sides of the neutral magnetic line (MNL), which, in turn, coincides with the filament location (Figure 10.12b).

10.5.2.2 Spatial Evolution

From the 1600 Å images by TRACE (Figure 10.11) and the magnetic field locations (Figure 10.12) one can see that the flare under investigation was, in fact, a two-ribbon flare. One of the ribbons is associated with the HXR sources, the other is not. The ribbons are located on opposite sides of the inversion line, or magnetic neutral line (MNL), which coincides with the filament (Figure 10.12b). This points to a set of loops embedded in the photosphere from opposite sides of the MNL with a strong shift of the embedded locations with opposite polarities strongly elongated with respect to the MNL.

In several footpoints, the HXR sources appear simultaneously (within an observational cadence of 5 s) with the H_α kernels, while in other footpoints the H_α emission is not associated spatially with the HXR sources and appear 10–20 s later than any HXR sources at the other locations.

10.5.2.3 Morphology of Flaring Kernels in the Main Event

The spatial locations of the HXR sources and H_α sources in relation to the MNL and their temporal variations often indicate the sites of energy deposition associated with a magnetic reconnection and the transport of this energy to deeper atmospheric levels by high-energy particles (Priest and Forbes, 2000). The magnetic reconnection events (MRs) are likely to occur in succession in each loop along the ribbon (Priest and Forbes, 2000).

Particles (electrons and protons) accelerated during this reconnection precipitate downwards into the loop legs, losing their energy in collisions with ambient particles (Brown, 1971) and in ohmic losses in the electric field induced by the particles themselves (Zharkova and Gordovskyy, 2005a, 2006). This process produces HXR emissions from the corona and chromosphere and H_α emissions in the chromosphere, while the thermo-conduction transfers (via a hydrodynamic response) heat from the beam to the whole atmosphere. Moreover, in the footpoints with HXR emission, there are signs of a transient increase, or reversible changes, of its magnetic field (Zharkova and Gordovskyy, 2005c), which was likely to have been caused by precipitating electrons (van den Oord, 1990).

The magnetic field changes recorded for this flare, along with the variations in the measured HXR photon flux, are presented in Figure 10.13, with a close-up plot of spectral index variations at lower energies (< 40 keV) and an electron flux corrected for ohmic energy losses. This shows that the line-of-sight (LOS) magnetic field changes were definitely associated with the flare phenomena, revealing a magnetic flux increase in close temporal correlation with the variations in the HXR photon flux. However, the cadence of magnetic field measurements was 1 min meaning these variations can only show us the magnetic field variations on a scale larger than 1 min. Thus, the LOS magnetic flux shows steady growth during the flare bursts, while only for the third burst, at about 13:40:48 UT, is there a noticeable magnetic field reduction (up to 26 G/pixel) resembling the reversible "transient" changes in magnetic field reported by Zharkova and Gordovskyy (2005c).

10.5.3 Resulting H_α Emission

The full H_α intensity and the intensities in all the other hydrogen lines and continua were calculated for the hydrodynamic models obtained for the very first moments of electron-beam injection at about 11:38:49 UT. We used a full non-LTE approach for the five-level-plus-continuum hydrogen atom (Zharkova and Kobylinskii, 1991, 1993). The full H_α line intensities simulated by taking into account the joint effect of radiation, thermal, and beam electrons with the parameters derived from HXR spectra and multiplied by the average kernel area are plotted in Figure 10.14 (solid curve). The observed evolution of the H_α emission in kernels 1, 3, 4, and 7 located on opposite sides of the magnetic neutral line (MNL) are plotted in Figure 10.14 (dashed and dotted–dashed curves).

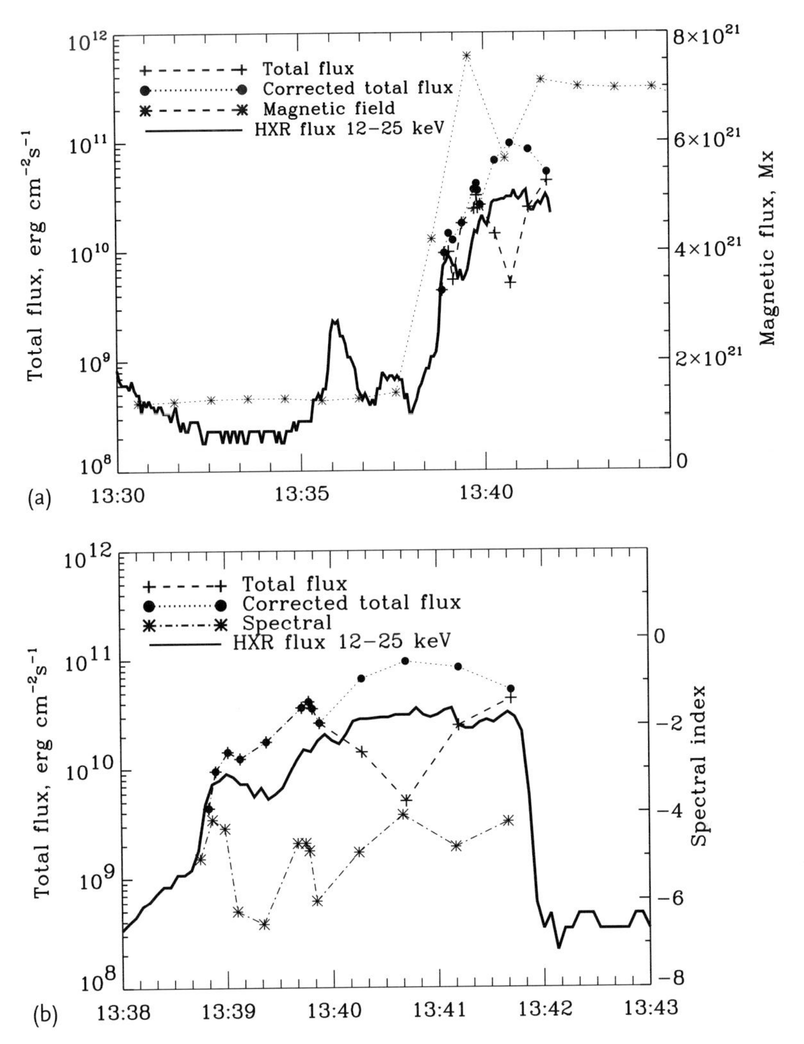

Figure 10.13 (a) Initial electron energy fluxes (dashed line with pluses), magnetic flux variations (dot-dashed line with asterisk), energy fluxes of beam electrons corrected for induced electric field losses (dotted line with black disks) in the main flare event plotted versus a light curve in an energy band of 12–25 keV (solid curve). (b) Close-in plots of the same initial electron energy fluxes and electron flux corrected for the induced electric field losses based on variations in spectral indices of HXR photon spectra at lower energies below 40 keV (dot-dashed line with asterisk) in the main flare event. Solid line: light curve in an energy band of 12–25 keV.

The simulations reveal a sharp increase of the full H_α emission immediately after the onset of an electron-beam injection likely caused by nonthermal hydrogen excitation by beam electrons (impulsive phase). A further increase in H_α intensity occurs owing to the temperature growth in the lower chromosphere caused by a hydrodynamic response owing to injection of other electron beams. The observed

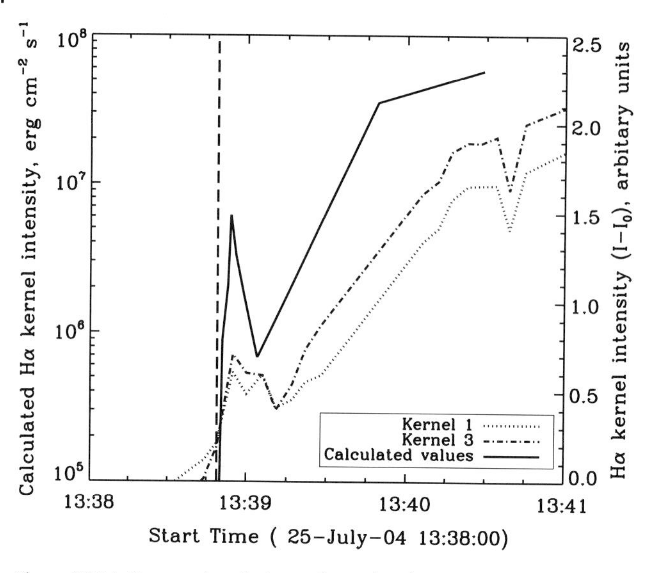

Figure 10.14 Temporal variations of simulated vs. calculated H_α line intensities in main flare event.

H_α emission in kernels 1 and 3 qualitatively fits rather well the theoretical emission increase in the impulsive phase starting at 13:38:49 UT. At the same time the H_α emission in kernels 4 and 7 reveal a delay of about 10–20 s between the simulated and observed H_α maxima.

This temporal delay, combined with the movement towards or out of the MNL of kernels 4 and 7 and the other two kernels 1 and 3 (Zharkova and Siversky, 2011), allows us to assume that the emission in them is likely to be caused by nonthermal excitation by different agents rather than by thermal excitation caused by a hydrodynamic response in the lower atmosphere. In sources 1 and 3 this emission was caused by beam electrons, while in kernels 4 and 7 this emission must have been caused by either the ambient electrons accelerated by plasma turbulence induced by the original precipitating beam (Hamilton and Petrosian, 1990; Zharkova and Siversky, 2011) or by protons accelerated at a magnetic reconnection of the loop as per the acceleration model in a reconnecting current sheet proposed by Zharkova and Gordovskyy (2004).

This links the processes of particle acceleration discussed in Chapter 2 with their precipitation into deep chromospheric levels, discussed in Chapters 3–5, with hydrodynamic heating of the ambient plasma by beam electrons and protons, as discussed in Chapter 6, and with nonthermal excitation and ionization by high-energy particles, discussed in this chapter.

11
H_α-Line Impact Polarization

11.1
Introduction

Observations of the linear polarization of spectral lines in solar flares provide unique information on the modes of energy transport from the corona to deeper layers during these dynamic events. The H_α line is the most observable line in solar flares, and significant properties of the energy transfer process can be derived from measurements of its polarization vector.

Linear polarization was shown to exist in about 30% of H_α-line observations (Firstova and Boulatov, 1996; Hénoux, 1991a) with a degree of polarization normally in the range of 3–5%, but in some cases not exceeding 10% (Chambe and Hénoux, 1979; Firstova and Boulatov, 1996; Hénoux, 1991a; Hénoux and Chambe, 1990). In many cases the highest degree of polarization does not correspond to the brightest areas of flares. In observations by Hénoux *et al.* (1990) the direction of plane of polarization coincides with the flare-to-disk center direction, whereas some observations by Firstova and Boulatov (1996) show the plane of polarization to be perpendicular to this direction.

The first interpretation of H_α-line polarization was made in an approximation of optically thin plasma, using Born cross-sections for line excitation by charged particles or external radiation (Hénoux and Chambe, 1990). The observed polarization was assigned to impact polarization or polarization by high-energy radiation (UV and EUV) as the Zeeman or Stark effects produce a polarization degree of about 0.5% (Chambe and Hénoux, 1979; Hénoux and Chambe, 1990). The authors have also shown that highly energetic particles (electron or protons) or highly energetic UV and EUV radiation produce negative polarization with the plane of polarization being mainly perpendicular to the solar center direction. On the other hand, a directed heat flux can produce positive polarization with the plane of polarization parallel to the solar center.

In order to explain the observed positive polarization in the H_α line, low-energy proton beams ($E \leq 200\,\text{keV}$) were used as the source of slow directed fluxes (Hénoux and Chambe, 1990; Hénoux *et al.*, 1993). Their simulations gave a reasonable degree of polarization; however, they did not take into account collective effects of proton beams on the ambient plasma, which can excite kinetic Alfvén waves simultaneously with H_α-line emission (Voitenko, 1996).

Recently simulations of impact polarization in H_α-line emission were done for proton beams precipitating into a flaring atmosphere and causing a redistribution in a population between the Zeeman excited states using the density matrix formalism (Vogt *et al.*, 1997). The collisional mechanisms of the 3rd level excitation by proton beams and by ambient plasma electrons were taken into account as well as the radiative mechanisms for incident and diffusive fields in H_α, Ly_α, and Ly_β lines. The calculated H_α-line polarization was found to be lower by up to an order of magnitude than those observed during a flare, fitting the observations only for the very weak emission occurring at the very beginning of a flare onset (Hénoux and Vogt, 1996; Vogt and Hénoux, 1996). Therefore, in order to fit the observations it is required to consider other agents of H_α-line polarization in flares.

Electron beams propagating in the fully ionized plasma of solar flares were proposed as such agents (Fletcher and Brown, 1995). Their simulations gave a polarization degree of about 5–7% but required electron beams with very high initial energy fluxes of 10^{13}–10^{15} erg cm^{-2} s^{-1}, which are by three orders of magnitude higher than typical fluxes, deduced from X-ray observations in solar flares.

In many flares the H_α-line emission is very bright and wide, so it is likely to be optically thick. Moreover, at chromospheric depths, where the magnetic field can reach 1000 G (Lozitskii and Baranovskii, 1993), the hydrogen atom levels are likely to be split. Therefore, for the interpretation of H_α-line polarization it is necessary to include these two effects. This can be done using the density matrix approach. It has been applied to a He D_3 line in solar prominences (Bommier, 1980; Landi Degl'Innocenti, 1982) and to flares (Vogt *et al.*, 1997). Later it was generalized for the transfer of polarized radiation of two-level atoms embedded in an optically thick magnetized medium with weak, intermediate, and strong magnetic fields (Bommier and Landi Degl'Innocenti, 1996; Bommier *et al.*, 1991; Landi Degl'Innocenti, 1996; Landi Degl'Innocenti *et al.*, 1990).

In this chapter the effect of electron-beam injection on H_α-line polarization during the impulsive phase of a flare is investigated in a loop with magnetized plasma (Zharkova and Syniavskii, 2000) using an approach similar to that of Vogt *et al.* (1997). First, for beam electrons with anisotropic scattering in the presence of a return current electric field and converging magnetic field, solutions to the time-dependent Boltzmann equation were used (Siversky and Zharkova, 2009b; Zharkova *et al.*, 1995). Second, a diffusive H_α radiation field for a five-level hydrogen model atom without fine structure was calculated in the full non-LTE approach as described by Zharkova and Kobylinskii (1989, 1991, 1993). And third, the density matrix technique was applied (Zharkova and Syniavskii, 2000) to the solution of a steady state equation in a flaring atmosphere with angular anisotropy caused by electron-beam impacts and external radiation following the method of Landi Degl'Innocenti (1985).

11.2 Basic Models

11.2.1 Physical Model

We consider an elementary magnetic flux tube filled by hydrogen plasma being affected by electron-beam injection. The beam causes a hydrodynamic response of the ambient plasma, which was calculated as described in Chapter 6 using the hydrodynamic response induced in the two-temperature ambient plasma by beam electrons (Somov *et al.*, 1981), developed further by Zharkova and Zharkov (2007) by adding radiative losses in hydrogen lines and continua following Kobylinskii and Zharkova (1996). The temperature, density, and macrovelocity variations in time and depth are as shown in Figure 10.1 in Chapter 10.

The main features of such a response can be briefly described as follows. The beam injection produces a strong depression of the transition region into deep atmospheric levels, which in turn causes a sharp increase in temperature and a decrease in total density at the coronal levels. This is followed by evaporation of the chromospheric plasma into the corona and by the formation of a dense cold condensation in the chromosphere that moves downward to the photosphere as a shock wave. We are interested in the chromospheric part of the hydrodynamic simulations, namely, in this cold and dense condensation in which the H_α-line emission is assumed to originate. We used the hydrodynamic models (temperature, density, macrovelocity) calculated for the first 10 s after electron-beam onset.

11.2.2 Kinetic Model

We considered electron beams injected from the corona into a flaring atmosphere in the direction of the magnetic field with normal distributions in pitch angles and time as described in Chapter 4. To find an electron population at each depth, we solved numerically the time-dependent Landau–Fokker–Planck equation for beam electrons, with energy power law and normal distributions in pitch angles and time, precipitating from the corona into deeper atmosphere in the presence of return current electric and converging magnetic fields (Siversky and Zharkova, 2009b; Zharkova and Gordovskyy, 2005a; Zharkova *et al.*, 1995). We considered beam electrons to lose their energy in collisions with charged particles and neutral hydrogen atoms, in ohmic heating and anisotropic scattering. This led to a dependence of the energy and pitch-angular variations on time and depth in the phase space of time, depth, energy and pitch angle where kinetic solutions exist. This resulted in the additional boundary conditions to be imposed on energy and pitch-angular variations with depth and time (for details see Chapter 4).

There are significant variations in the role of energy loss mechanisms throughout this precipitation. At upper levels in the corona, where the ambient plasma is fully ionized, Coulomb collisions dominate the other energy losses; in the transition region and upper chromosphere those are comparable to ohmic losses, with the latter causing a disruption of the initial beam, an appearance of the "return current beam", returning to the source in the corona. This in turn produces a split in the energy of the initial beam with an additional maximum at 30–35 keV.

In the lower chromosphere, where the ambient plasma is weakly ionized with an ionization degree of 10^{-4} (Zharkova and Kobylinskii, 1991), inelastic collisions of beam electrons with neutral atoms prevail over all other energy losses. Since the cross-sections of such collisions are much lower than those of Coulomb collisions, beam electrons lose less of their energy in these collisions and can precipitate downward to the photosphere, despite an increase in the total density. Electron beams, which are well directed along the magnetic field, still precipitate as well collimated beams into the lower atmospheric levels, but they are transformed into softer beams with wider angular distributions (Siversky and Zharkova, 2009b; Zharkova *et al.*, 1995).

The part of beam electrons that reaches the lower chromosphere can produce a dual effect on the atmosphere: first, the formation from the hydrodynamic response of a cold dense condensation, as described in Section 2.1; and second, additional excitation, ionization, and polarization of hydrogen atoms by inelastic collisions, as considered below.

11.2.3
Radiative Model

For polarimetric simulations we considered a three-level hydrogen model atom (total of nine sublevels: $1s_{1/2}$, $2s_{1/2}$, $2p_{1/2}$, $2p_{3/2}$, $3s_{1/2}$, $3p_{1/2}$, $3p_{3/2}$, $3d_{3/2}$, and $3d_{5/2}$) with a fine structure at the third and second levels caused by Zeeman splitting without level crossings. Only a Zeeman coherence, such as $\langle nLJM|\rho|nLJM'\rangle$, will be taken into account, whereas all other coherences with $n' = n$, $L' = L$, and $J' = J$ can be omitted following the estimation $\Gamma \ll \Delta E_{FS}/\hbar$, where $\Gamma \propto xB$ is the atomic energy of the Bohr frequency set, B is the magnetic induction in G, x is about unity, and ΔE_{FS} is the energy splitting from the spin–orbital interaction. To calculate the diffusive mean intensity and degree of ionization in these dynamic events, we used a five-level-plus-continuum model atom without a fine structure as described in Sections 10.3.1 and 10.3.2. The full non-LTE approach was applied for all the transitions at different depths and times, as described in the papers by Zharkova and Kobylinskii (1991, 1993). Collisions with thermal and beam electrons were considered along with radiative excitation/ionization and deactivation by external radiation in all considered transitions.

The ionization balance in a flaring atmosphere was found to be governed by hydrogen atom ionization. In the lower chromosphere the hydrogen ionization degree falls sharply down to a magnitude of 10^{-4} if only thermal electrons are considered. Beam electrons reaching this depth lose their energy, mostly in inelastic collisions with neutral hydrogen atoms (90–95%) (see also Aboudarham and Hénoux, 1986; Zharkova and Kobylinskii, 1989). These collisions increase the hydrogen ionization degree to about 10^{-1} (see Figure 2 in Zharkova and Kobylinskii, 1989). This in turn causes an enhancement in Lyman, Balmer, and Pashen line wings and heads of hydrogen continua.

11.2.3.1 Evaluation of Elementary Processes

Hydrogen atoms, embedded in a flaring atmosphere of a magnetic flux tube, are affected by depolarizing collisions with thermal electrons of the ambient plasma and by polarizing collisions with anisotropic beam electrons and external radiation. As the atomic levels in the magnetic field are split into sublevels of fine structure, it is important to evaluate the effect of each mechanism.

1. *Magnetic field effect*: Following Bommier and Landi Degl'Innocenti (1996), a weak magnetic field can partially destroy the coherence between magnetic sublevels, whereas Zeeman splitting should still not be taken into account (Hanle effect). In this case the following condition is valid:

$$\nu_{\mathrm{L}} = A\,, \tag{11.1}$$

where ν_{L} is the Larmor frequency and A is the Einstein spontaneous emission coefficient.
In the case of an "intermediate" magnetic field, the coherence between magnetic sublevels is absent, and in comparison with the Doppler width, Zeeman splitting is noticeable. This condition can be written as

$$A \ll \nu_{\mathrm{L}} < \Delta\nu_{\mathrm{D}}\,, \tag{11.2}$$

where $\Delta\nu_{\mathrm{D}}$ is the Doppler width. In the model flaring atmospheres from Chapter 8 the temperature lies in a range of 10^4–6×10^3 K, which results in $\Delta\nu_{\mathrm{D}} \approx 10^{12}$–$10^{10}\,\mathrm{s}^{-1}$. Magnetic fields in the chromosphere can vary from 500 to 1500 G, so $\nu_{\mathrm{L}} = (3\text{–}20) \times 10^8\,\mathrm{s}^{-1}$. Therefore, in the chromosphere, where H_α emission originates, these two regimes may occur and both should be considered. It can also be noticed that in the line formation region the Hanle effect will prevail in the line core, whereas in wings the "intermediate" magnetic field approximation is valid.
2. *Effects of collisions and radiation anisotropy*: Let us evaluate the role of depolarizing collisions with thermal electrons and of the anisotropy of incident radiation at these levels. The anisotropy of the incident radiation is described by the parameter R_{a}, which is determined by the nondiagonal elements of the density matrix. The isotropic part of radiation is described by the parameter R_{i}, which is determined by the diagonal elements. Therefore, a spectral line polarization, caused by anisotropy of radiation, is proportional to the ratio $R_{\mathrm{a}}/R_{\mathrm{i}}$. The effect of depolarizing collisions can be described by the parameter D, which is obviously proportional to the plasma density N:

$$D = N\sigma V\,, \tag{11.3}$$

where σ is the collisional cross-section of the H_α-line transitions and V is the relative velocity between atom and colliding particle. If σ is in units of $10^{-16}\,\mathrm{cm}^{-2}$ and V is in $\mathrm{km\,s}^{-1}$, then

$$D = N\sigma V \times 10^{-11}\,, \quad \mathrm{s}^{-1}\,. \tag{11.4}$$

The radiation anisotropy R_a can be described by the equation

$$R_a = W B_{lu} J , \tag{11.5}$$

where W is a dilution factor, B_{lu} is Einstein's absorption coefficient, and J is the mean radiation intensity. For strong optical lines with an effective temperature of 6000 K $R_a = W \times 10^6\,s^{-1}$. The condition $D = R_a$ leads to the following evaluation of the critical density where the collisional and radiative probabilities are comparable:

$$N_{crit} = W(\sigma V)^{-1} \times 10^{17} , \quad cm^{-3} . \tag{11.6}$$

Considering $W(\sigma V)^{-1}$ as being of 10^{-2}–10^{-3}, for a temperature range relevant to a flaring atmosphere, the critical density is $N_{crit} = 10^{14}$–$10^{15}\,cm^{-3}$. This corresponds to the temperature minimum region in the quiet atmosphere, while in flares this density can be reached at the lower chromospheric levels where H_α-line wing emission occurs.

11.3 Density Matrix Approach

For a description of the hydrogen atomic system interacting with an external radiation field and particles, following Berestetskij *et al.* (1989), Landi Degl'Innocenti (1983), and Bommier and Landi Degl'Innocenti (1996), let us consider the density matrix ${}^{nLJ}\rho^{A}_{MM'}$, where n is the main quantum number, L the orbital atomic momentum, J the total atomic momentum, M the total angle momentum projected on the magnetic field direction, and A stands for "atomic". The diagonal elements of the density matrix describe the populations of the atomic levels and the nondiagonal elements describe the coherence between two different levels.

11.3.1 Steady State Equation

A steady state equation for the density matrix can be written as follows:

$$\begin{aligned} \frac{d}{dt}\,{}^{n_1L_1J_1}\rho^{A}_{M_1M_1'} &= \frac{i}{\hbar}[E(n_1L_1J_1M_1) - E(n_1L_1J_1M_1')]\,{}^{n_1L_1J_1}\rho^{A}_{M_1M_1'} \\ &+ \sum_{\substack{M_2M_2'\\ L_2J_2\\ n_2}} P^{tot}_{n_1L_1J_1M_1M_1' \leftarrow n_2L_2J_2M_2M_2'}\,{}^{n_2L_2J_2}\rho^{A}_{M_2M_2'} \\ &- \sum_{\substack{M_2M_2'\\ L_2J_2\\ n_2}} P^{tot}_{n_2L_2J_2M_2M_2' \leftarrow n_1L_1J_1M_1M_1'}\,{}^{n_1L_1J_1}\rho^{A}_{M_1M_1'} , \end{aligned} \tag{11.7}$$

where $P^{\text{tot}}_{n_1 L_1 J_1 M_1 M_1' \leftarrow n_2 L_2 J_2 M_2 M_2'}$ and $P^{\text{tot}}_{n_1 L_1 J_1 M_1 M_1' \rightarrow n_2 L_2 J_2 M_2 M_2'}$ are respectively the total probabilities of the transitions $\langle n_1 L_1 J_1 M_1 M_1' | \rho^{\text{A}} | n_2 L_2 J_2 M_2 M_2' \rangle$ from and to the upper level. These in turn are determined by the sum of the probabilities of spontaneous P^{sp}, induced radiation P^{ind}, absorption P^{abs}, and collisional transitions P^{coll}, taking into account impacts with both thermal and beam electrons. The first member of the equation above describes the probability of transitions between sublevels, split in the magnetic field, the second member corresponds to the probability of the level excitation, and the third to the probability of its deexcitation.

11.3.2
Radiative Tensor

To describe the anisotropy of radiation field, let us introduce for H_α-line frequencies the radiation field tensor in a solid angle $\Omega = (\chi; \theta)$ on a plane perpendicular to the direction of the emitted quantum as follows:

$$\overline{\Phi}^{\text{rad}}_{ij} = \int \Phi^{\text{rad}}_{ij}(\omega) \frac{1}{\sqrt{\pi}} \exp\left[-\left(\frac{\Delta\omega}{\Delta\omega^{\text{D}}{}_{\text{H}_\alpha}} \right)^2 \right] d\omega \,, \tag{11.8}$$

where

$$\Phi^{\text{rad}}_{ij}(\omega) = \int \frac{d\Omega}{4\pi} I_\omega(\Omega) f_{ij}(\boldsymbol{\Omega}) \,; \quad i, j = x, y, z \,, \tag{11.9}$$

and f_{ij} is a transition matrix from the electron to the emitted photon coordinate system depending on the azimuthal χ and longitudinal θ angles. For calculations in the H_α-line frequencies we use a normalized radiative tensor:

$$\varphi^{\text{rad}}_{ij} = \frac{\overline{\Phi}^{\text{rad}}_{ij}}{\sum_{l=1}^{3} \Phi^{\text{rad}}_{ii}} \,. \tag{11.10}$$

In Ly_α and Ly_β lines, owing to their large opacity and assumed detailed balance, the radiative tensor has the following form:

$$\varphi^{\text{rad}}_{ij} = \frac{1}{3} \delta_{ij} \,, \tag{11.11}$$

where δ_{ij} is the Kronecker function.

The local radiation density can be described as

$$u^{\text{rad}} = \int_{(2\pi)} I_\omega(\Omega) d\boldsymbol{\Omega} \,, \tag{11.12}$$

with $I_\omega(\boldsymbol{\Omega})$ a local intensity at a solid angle Ω found from the non-LTE radiative transfer equation as described in Section 10.3.1.

11.3.3 Collisional Tensor

The collisional tensor describes the radiation anisotropy caused by collisions with both thermal and beam electrons. Thermal electrons have a Maxwellian distribution in energy, and their collisional tensor can be written as follows:

$$\varphi_{ij}^{\text{coll}} = \frac{1}{3}\delta_{ij} \,, \tag{11.13}$$

whereas for beam electrons one can use the following expression:

$$\overline{\Phi}_{ij}^{\text{coll}} = \int_{(4\pi)} \frac{d\boldsymbol{\Omega}}{4\pi} f_{ij}(\Omega) \int_0^\infty \rho_{\text{beam}}(t, \xi, v, \theta) v^2 dv \,, \quad i, j = x, y, z \,, \tag{11.14}$$

where ρ_{beam} is a normalized dimensionless electron beam distribution function, t is an injection time, ξ is the depth, and θ is pitch angle. For the calculations we use a normalized collisional tensor $\varphi_{ij}^{\text{coll}}$, with the normalization being similar to the radiative tensor above. Also we can introduce the local collisional density as follows:

$$u^{\text{coll}} = \int_{(4\pi)} \rho(\boldsymbol{\Omega}) d\boldsymbol{\Omega} \,, \tag{11.15}$$

where ρ is the sum of Maxwellian and power-law energy distributions for thermal and beam electrons, respectively, calculated as described in Section 2.2.

11.3.4 Probabilities of Radiative Transitions

The probability of spontaneous transition can be described as (Bommier and Sahal-Brechot, 1982)

$$P^{\text{spon}}_{n_1 L_1 J_1 M_1 M_1' \leftarrow n_2 L_2 J_2 M_2 M_2'} = D_{n_1 L_1 J_1 M_1 M_1' \leftarrow n_2 L_2 J_2 M_2 M_2'}(-m, -m') \times (2L_2 + 1) A(n_2 L_2 S \to n_1 L_1 S) \,, \tag{11.16}$$

where D is a matrix term presented in the form

$$D_{n_1 L_1 J_1 M_1 M_1' \leftarrow n_2 L_2 J_2 M_2 M_2'}(-m, -m') = \begin{pmatrix} J_1 & 1 & J_2 \\ -M_1 & -m & M_2 \end{pmatrix} \begin{pmatrix} J_1 & 1 & J_2 \\ -M_1' & -m' & M_2' \end{pmatrix} \times (2J_1 + 1)(2J_2 + 1) \times \begin{bmatrix} J_1 & 1 & J_2 \\ L_2 & S & L_1 \end{bmatrix}^2 \,, \tag{11.17}$$

with $m = M_2 - M_1$; $m' = M_2{}' - M_1$. For induced radiation the probability will be

$$P^{\text{ind}}_{n_1 L_1 J_1 M_1 M_1' \leftarrow n_2 L_2 J_2 M_2 M_2'} = D_{n_1 L_1 J_1 M_1 M_1' \leftarrow n_2 L_2 J_2 M_2 M_2'}(-m, -m') \times (2L_2 + 1) 3 u^{\text{rad}} \varphi^{\text{rad}}_{mm'} B(n_2 L_2 S \to n_1 L_1 S) \,, \tag{11.18}$$

and for absorbed radiation the probability is equal to

$$
\begin{aligned}
P^{\mathrm{abs}}_{n_1 L_1 J_1 M_1 M_1' \rightarrow n_2 L_2 J_2 M_2 M_2'} &= D_{n_1 L_1 J_1 M_1 M_1' \leftarrow n_2 L_2 J_2 M_2 M_2'}(-m, -m') \\
&\times (2L_2 + 1) 3 u^{\mathrm{rad}} \phi^{\mathrm{rad}}_{m m'} B(n_2 L_2 S \leftarrow n_1 L_1 S) \,, \qquad (11.19)
\end{aligned}
$$

with $A(n_2 L_2 S \rightarrow n_1 L_1 S)$, $B(n_2 L_2 S \rightarrow n_1 L_1 S)$, and $B(n_2 L_2 S \leftarrow n_1 L_1 S)$ being the Einstein coefficients of spontaneous emission, induced emission, and absorption.

11.3.5 Probabilities of Collisional Transitions

The first kind probability of a collisional transition (from lower to higher level) can be written as (Berestetskij *et al.*, 1989)

$$
\begin{aligned}
P^{\mathrm{coll}}_{n_1 L_1 J_1 M_1 M_1' \rightarrow n_2 L_2 J_2 M_2 M_2'} &= D_{n_1 L_1 J_1 M_1 M_1' \rightarrow n_2 L_2 J_2 M_2 M_2'}(-m, -m') \\
&\times \sigma_{\mathrm{eff}}(n_1 L_1 S \leftarrow n_2 L_2 S) u^{\mathrm{coll}} \phi^{\mathrm{coll}}_{m' m} \,, \qquad (11.20)
\end{aligned}
$$

where D are the matrix terms determined from Eq. (11.17).

For the second kind of transition (from higher to lower level) the probability is

$$
\begin{aligned}
P^{\mathrm{coll}}_{n_1 L_1 J_1 M_1 M_1' \leftarrow n_2 L_2 J_2 M_2 M_2'} &= D_{n_1 L_1 J_1 M_1 M_1' \leftarrow n_2 L_2 J_2 M_2 M_2'}(m, m') \\
&\times \sigma_{\mathrm{eff}}(n_1 L_1 S \rightarrow n_2 L_2 S) u^{\mathrm{coll}} \phi^{\mathrm{coll}}_{m m'} \,, \qquad (11.21)
\end{aligned}
$$

where σ_{eff} is the effective cross-section, which can be determined by the Born formulas. The matrix terms for the transitions from upper to lower level $D(m, m')$ are described by the equation above with m and m' having opposite signs.

11.3.6 Stokes Parameters

The Stokes parameters are linked to the density matrix as follows:

$$
\begin{aligned}
S_1 &\equiv U = i[\langle \boldsymbol{n}, -1|\rho|\boldsymbol{n}, +1\rangle - \langle \boldsymbol{n}, +1|\rho|\boldsymbol{n}, -1\rangle] \,, \\
S_2 &\equiv V = [\langle \boldsymbol{n}, +1|\rho|\boldsymbol{n}, +1\rangle - \langle \boldsymbol{n}, -1|\rho|\boldsymbol{n}, -1\rangle] \,, \\
S_3 &\equiv Q = -[\langle \boldsymbol{n}, -1|\rho|\boldsymbol{n}, +1\rangle + \langle \boldsymbol{n}, +1|\rho|\boldsymbol{n}, -1\rangle] \,. \qquad (11.22)
\end{aligned}
$$

Here S_1 describes a linear polarization along the ξ- and θ-axes, which are on the plane perpendicular to the photon direction $\boldsymbol{n}$. The probabilities of a linear polarization of a photon along these axes are equal, respectively, to $(1 + S_3)/2$ and $(1 - S_3)/2$. The values of $S_3 = \pm 1$ correspond to a full polarization in these directions. The parameter S_1 describes a linear polarization in the directions with angles of $\phi = \pi/4$ or $\phi = -\pi/4$, and the probabilities of a polarization in these directions are equal to $(1 + S_1)/2$ and $(1 - S_1)/2$.

Finally, the parameter S_2 describes a circular polarization with the probabilities of the photon having a left-handed and right-handed polarization equal to $(1 + S_2)/2$

and $(1 - S_2)/2$. These parameters are the characteristics of a photon's helicity. In astrophysical applications these parameters have been defined as $S_1 \equiv U$, $S_3 \equiv Q$, and $S_2 \equiv V$.

11.4
Results and Discussion

The results of simulations of the H_α-line polarization are presented in Figures 11.1–11.4. In Figure 11.1 the H_α-line polarization profiles are plotted vs. wavelengths at $t = 4.39$ s after electron-beam onset ($F_0 = 2 \times 10^9$ erg cm^{-2} s^{-1}; $\delta = 4, 7$) for viewing angle $\theta = \pi/3$ and different depths. In Figure 11.2 the linear polarization profiles are plotted for different viewing angles and two different times of beam injection for softer beams ($\delta = 7$). In Figures 11.3 and 11.4 the full (integrated in wavelength) linear polarization and intensity, respectively, caused by

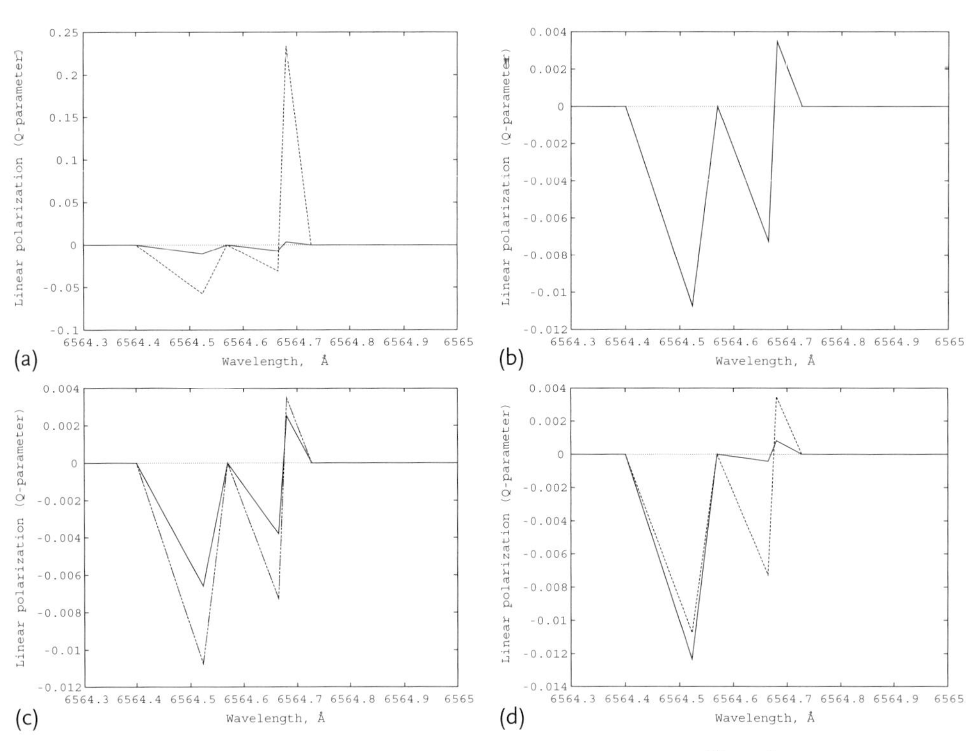

Figure 11.1 H_α-line polarization profiles (linear polarization – parameter Q) plotted for depths (a) 1.37–2.78×10^{20} cm^{-2}, (b) 4.96–7.36×10^{20} cm^{-2}, (c) 8.50–9.76×10^{20} cm^{-2}, and (d) 1.21–2.52×10^{21} cm^{-2} for a viewing angle of $\theta = \pi/2$ at the moment $t = 4.39$ s for beam parameters: solid line – spectral index $\delta = 4$, dashed line – $\delta = 7$; $F_0 = 2 \times 10^9$ erg cm^{-2} s^{-1}.

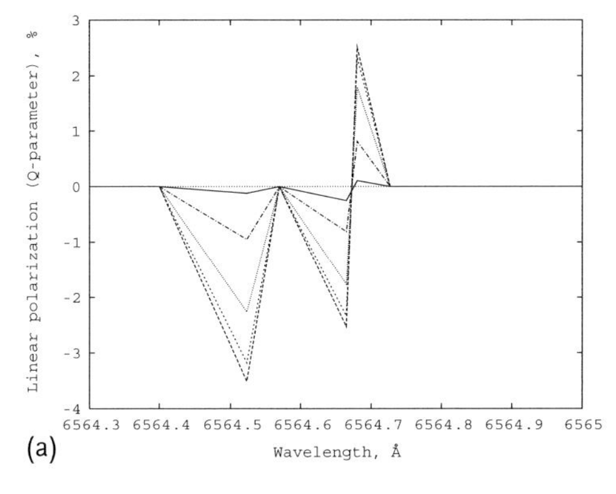

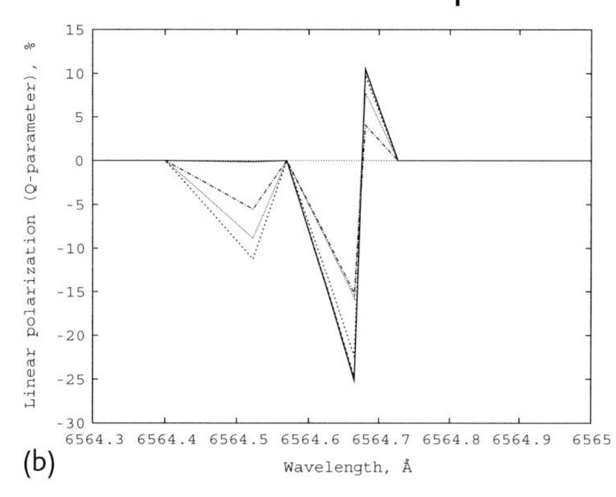

Figure 11.2 H_α line polarization profiles (linear polarization – parameter Q) plotted for depth $\xi = 7.36 \times 10^{20}\,\mathrm{cm}^{-2}$ at moments $t = 2.92\,\mathrm{s}$ (a) and $t = 5.94\,\mathrm{s}$ (b) for five viewing angles $\theta = (0.1, 0.2, 0.3, 0.4, 0.5)\pi$ (from top to bottom in each subfigure). The peaks correspond to the allowed transitions as in the text; beam parameters correspond to the softer beam from Figure 11.1.

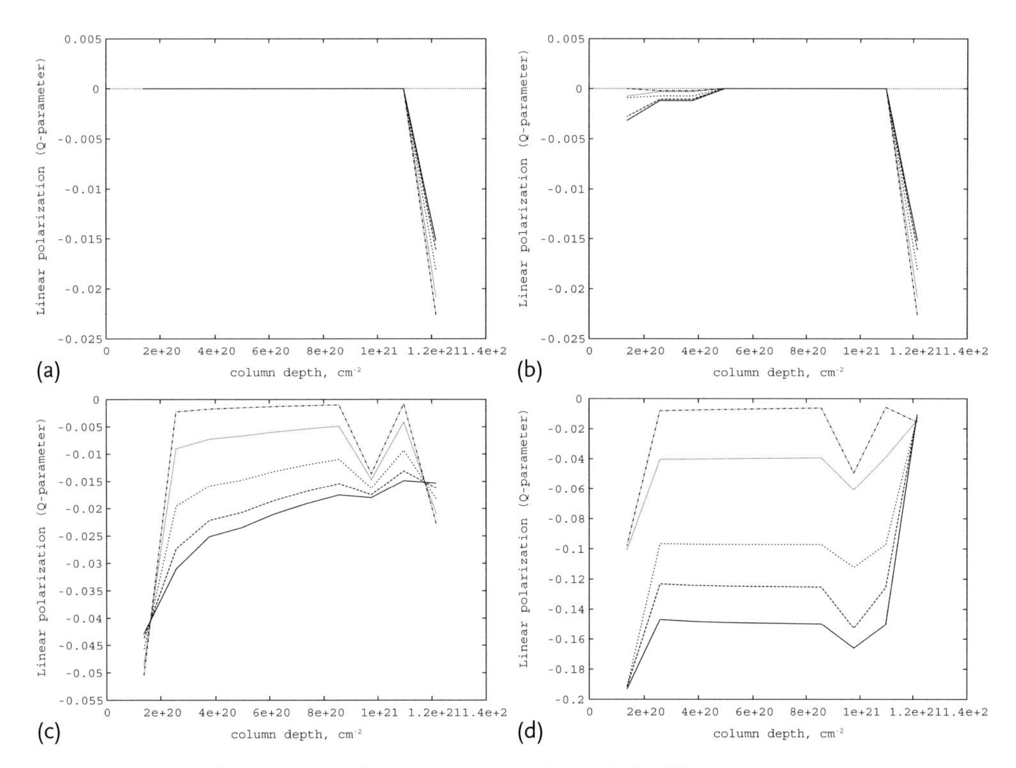

Figure 11.3 H_α-line full polarization (linear polarization – parameter Q) plotted vs. depth in a flaring atmosphere at moments $t = 0.01\,\mathrm{s}$ (a), $t = 1.46\,\mathrm{s}$ (b), $t = 2.92\,\mathrm{s}$ (c), and $t = 5.94\,\mathrm{s}$ (d) for different viewing angles $\theta = (0.1, 0.2, 0.3, 0.4, 0.5)\pi$ (from top to bottom in each subfigure). Beam parameters: $F_0 = 2 \times 10^{11}\,\mathrm{erg\,cm^{-2}\,s^{-1}}$, $\delta = 5$.

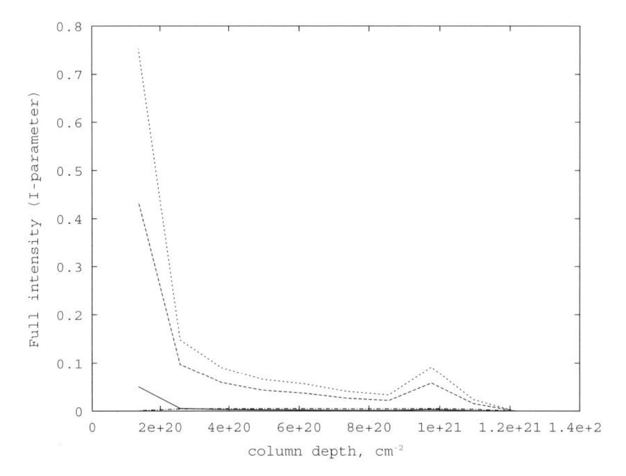

Figure 11.4 H$_\alpha$-line full intensity (parameter I) plotted vs. depth in a flaring atmosphere for different transitions of fine structure at moment $t = 4.39$ s (upper curve corresponds to $3d_{5/2} \rightarrow 2p_{3/2}$), then in descending order: to $3p_{3/2} \rightarrow 2s_{1/2}$ and $3d_{3/2} \rightarrow 2p_{3/2}$. Beam parameters are the same as in Figure 11.2.

the soft weak beam above ($\delta = 7$) are plotted vs. depth in a flaring atmosphere for viewing angle $\theta = \pi/3$ at different times.

The calculations show that impacts with beam electrons in a flaring atmosphere lead mostly to a linear polarization of H$_\alpha$-line emission, whereas a circular polarization is negligible. This is in agreement with most observations (Firstova and Boulatov, 1996; Hénoux and Chambe, 1990). It could be understood in terms that radiation with circular polarization is produced by small angular deviations of higher-energy electrons in collisions with ambient-plasma-charged particles (Berestetskij *et al.*, 1989). However, as the kinetics shows, in the lower chromosphere the number of such electrons is very small and they can not produce a noticeable effect. The absolute values of linear polarization in line profile vary in a wide range from a few tens of percent ($t = 1.46$ s, Figure 11.1b) to a maximum of 12–15% ($t = 4.39$ s, Figure 11.1d). The maximum (20–25%) of full H$_\alpha$-line polarization was also found for softer weaker electron beams ($F_0 = 2 \times 10^9 \,\text{erg}\,\text{cm}^{-2}\,\text{s}^{-1}$, $\delta = 7$), whereas for harder beams the degree of linear polarization is lower (5–8%).

This behavior is similar to that of HXR bremsstrahlung polarization (Zharkova *et al.*, 1995, 2010), although X-ray polarization is slightly decreased with an enhancement of the initial flux during injection whereas H$_\alpha$-line polarization steadily increases over time. This occurs owing to a difference in radiative cross-sections for radiation in these ranges: X-ray emission is a direct reproduction of electron-beam distributions; however, the H$_\alpha$ quanta are slightly delayed by the line opacity caused by a radiative transfer and line-broadening effects (Doppler and Stark ones).

Therefore, in the higher chromosphere and transition region where the return current effect causes a decrease in the number of beam electrons precipitating downward, it also decreases X-ray emission and polarization. In the lower chromosphere where the H$_\alpha$ emission originates, only a residual part of electrons that moved from higher to lower energy distributions can contribute to this emission. The number of these electrons increases with time and depth and their excitation and ionization rates prevail over the thermal ones. Thus, beam electrons can excite or ionize more and more hydrogen atoms throughout their precipitation. This accumulative effect results in a steady increase over time of the H_α-line polarization during a beam injection.

11.4.1 H_α-Line Polarization Profiles

The H_α-line profiles are rather asymmetric ones with the maxima corresponding to the main transitions for which $\Delta L = \Delta J$ ($3d_{5/2} - 2p_{3/2}$, $3d_{3/2} - 2p_{1/2}$, and $3p_{3/2} - 2s_{1/2}$). In order to demonstrate a pure polarization, all the broadening factors such as Doppler and Stark effects are excluded from these profiles in Figures 11.1 and 11.2; however, the calculations are done including these effects. As expected, the impact polarization appears only in the line cores whereas the wings are fully depolarized by collisions with thermal electrons.

At the initial phase of beam injection only the blue part of the line core (-0.1 Å from the central wavelength) of a H_α-line polarization profile is affected by impacts with beam electrons, as can be noticed from Figure 11.2a. This effect varies from -1% for a viewing angle of 0.1π to -3.5% for a viewing angle of 0.5π. With further beam penetration into the flaring atmosphere the polarization increases up to -12% in the blue part of the core and to -25% in the red part (at $+0.2$ Å from the central wavelength). More powerful soft beams produce higher negative polarization of about 20–25% at greater depths of $10^{20}\,\text{cm}^{-2}$ decreasing to a small percentage at a lower depth of $10^{21}\,\text{cm}^{-2}$ (Figure 11.2b).

At the greatest depth, where $\xi = 1 \times 10^{20}\,\text{cm}^{-2}$, radiation from the transitions $3d_{5/2} - 2p_{3/2}$ and $3p_{3/2} - 2s_{1/2}$ have a negative polarization, whereas in the transition $3d_{3/2} - 2p_{3/2}$ polarization is positive, reaching 25% (Figures 11.1a and 11.2b) when the beam reaches its maximum energy flux. Then there is a small decrease of positive polarization at depths of up to 3–5% for a viewing angle of 0.1π and up to 10% for a viewing angle of 0.5π; however, its contribution to the full line polarization is rather small.

These variations of H_α-line polarization profiles are believed to reflect a pitch-angular anisotropy of beam electrons at these depths. For radiation in the transitions ($3d_{5/2} - 2p_{3/2}$ and $3p_{3/2} - 2s_{1/2}$) more energetic beam electrons are required than for the lower-energy transition ($3d_{3/2} - 2p_{3/2}$). As the particle kinetics shows, a part of the initial electron beam is transformed into the secondary beam with lower-energy electrons returning to the corona. These lower-energy electrons are scattered to pitch angles between 90 and 180°, following the kinetic solutions by Zharkova *et al.* (1995). It produces an additional polarization with a plane of polarization perpendicular to the direction of a secondary beam but parallel the direction of the initial beam. For an observer it results in positive polarization in this transition at these levels. At lower depths an anisotropic scattering effect decreases and coincides with a decrease of the number of electrons scattered to larger pitch angles, as was shown from the kinetics. This leads to a decrease in this positive polarization to a small percentage. Therefore, full line polarization will be a sum of polarization degrees in each of these transitions and is dependent on the number of beam electrons in the initial and secondary beams.

11.4.2
Depth and Time Variations of H_α-Line Polarization

Beam electron impacts become noticeable in H_α-line polarization from a column depth of $\xi = 0.2 \times 10^{20}\,\mathrm{cm}^{-2}$, which coincides with those defined from the non-LTE simulations of hydrogen emission without atomic fine structure (Zharkova and Kobylinskii, 1991, 1993). At lower chromospheric levels, from $\xi = 1.2 \times 10^{21}\,\mathrm{cm}^{-2}$, anisotropic external radiation is a dominant mechanism in H_α-line polarization, producing polarization of about 2% (Figure 11.3a). The mean intensity of this external H_α radiation is higher by an order of magnitude than those of the diffusive H_α-line emission, produced in absence of beam electrons.

The effect of beam impacts becomes noticeable at upper atmospheric levels (Figure 11.3b) with an increase in time and energy flux. At time $t = 2.92\,\mathrm{s}$ (Figure 11.3c), being close to the beam's maximum energy flux, the impacts produce polarization comparable to and higher than that from external radiation. For instance, at a column depth of $2 \times 10^{20}\,\mathrm{cm}^{-2}$ it reaches 5% and then sharply decreases with depth to 0.5–1%.

At a column depth of $1 \times 10^{21}\,\mathrm{cm}^{-2}$ there is a secondary increase of polarization, caused by beam electrons, which has the same magnitude of about 2% as for external radiation.

This increase can be explained by the beam kinetics owing to the return current effect that causes a split of the initial beam at the upper levels into two beams moving downward with separate maxima in energy distributions (see, e.g., Siversky and Zharkova, 2009b; Zharkova *et al.*, 1995). At a depth of $1 \times 10^{21}\,\mathrm{cm}^{-2}$ the return current effect becomes negligible owing to the different reasons discussed in Siversky and Zharkova (2009b). These two beams become a single directed beam again with a higher abundance of lower-energy electrons that they gained from ohmic losses in addition to collisions. At this depth these electrons are responsible for the increase in the H_α-line polarization (Figure 11.3) and intensity (Figure 11.4).

From the time when a beam reaches its maximum flux ($F_0 = 4 \times 10^{14}\,\mathrm{cm}^{-2}\,\mathrm{s}^{-1}$), impacts with beam electrons at all levels prevail over other mechanisms including external radiation (Figure 11.3d). For a viewing angle of $\pi/2$, the polarization reaches 25% at the injection site, varying for other depths in a range of 15–20% with a secondary increase at the same level $1 \times 10^{21}\,\mathrm{cm}^{-2}$, as discussed above. Below this level H_α-line polarization falls to 2%, which reflects the fact that beam impacts become negligible and polarization is governed by external chromospheric radiation.

The H_α-line impact polarization has been shown to be strongly affected by a viewing angle with a large diversity of polarization from 5% for $\theta = 0.1\pi$ to 20%

for $\theta = 0.5\pi$ (Figure 11.3c,d). The bigger the viewing angle, the higher (in absolute value) the negative polarization, reaching its maximum for an angle of 90°.

11.4.3
Interpretation of Observational Features

Let us summarize the main features of the observations of H_α-line polarization in solar flares made by Hénoux and Chambe (1990), Hénoux *et al.* (1990), and Firstova and Boulatov (1996).

1. Linear polarization is found in about 30% of observed H_α spectra (Firstova and Boulatov, 1996).
2. The observed H_α-line polarization is positive in about 70% of observations with the plane of polarization being parallel to the solar center direction (Hénoux and Chambe, 1990). The plane of polarization in flares is close to the flare-to-solar center direction for parts of a flare with higher emission, whereas for weaker parts the plane of polarization is perpendicular to this direction (Firstova and Boulatov, 1996).
3. The absolute value of polarization does not exceed 10% (Hénoux and Chambe, 1990), although in some observations the average degree of polarization is about 7%, reaching 20% in regions with weaker emission and in adjacent areas of the weakly perturbed chromosphere (Firstova and Boulatov, 1996).

To explain these features, especially the positive H_α-line polarization, proton beams were considered as a source of low-energy and well-directed particle precipitation (Firstova and Boulatov, 1996; Hénoux and Chambe, 1990; Hénoux *et al.*, 1990). In this chapter we showed that by using our kinetic solutions for electron-beam precipitation in flaring atmospheres, these effects can be explained by impacts with beam electrons.

The observed magnitude of H_α-line polarization varies from a few to 25%, which is within the range found in our simulations for impact polarization caused by beam electrons. The question is whether the calculated plane of polarization can explain the observational features. As the dependence on pitch angle is rather essential (Figure 11.3), we discuss its possible consequences on polarization observed from different viewing angles.

If an observer measures the spectrum emitted from a flaring magnetic loop embedded in the solar atmosphere, the geometry of a loop standing on the solar surface is a crucial point in the observations. At the top of the loop beam electrons propagate horizontally to the solar surface and perpendicular to an observer looking from the top. This beam produces polarization with the plane of polarization perpendicular to the beam propagation but parallel to the observer; therefore, a positive polarization can be measured. On the other hand, at the feet of the loop, which stands vertically, the direction of electron precipitation will be parallel to the observer, and hence the produced polarization will be negative.

In real observations the loops can have inclinations to one side or another, and very often their top can be close to the solar surface, which may result in positive polarization of the H_α line. The areas projected from the top of the loop on the horizontal plane are normally bigger by a factor of two-thirds to three-fourths than those at the bottom. So the fact that 70% of flares have positive polarization in the observations by Firstova and Boulatov (1996) might indicate this projection effect as well as a dominance of inclined loops in the observations.

Another feature that is crucial for the interpretation of polarization is why for weaker parts of a flare it is higher and for brighter parts it is lower. We assume that the H_α-line emission, originating in the higher parts of the loop, is likely to be less affected by beam electrons as they are just starting their precipitation and the magnetic field at this depth is not high enough to give a strong splitting into a fine structure. However, this part of the loop could be the brightest one in H_α emission as the number of neutral hydrogen atoms, affected by beam electrons, is high enough and the optical thickness is less than or about unity, so all H_α-line emission is emitted from the volume. With further beam precipitation into the loop's feet the magnetic field increases while the H_α optical depth becomes much bigger (Kobylinskii and Zharkova, 1996; Zharkova and Kobylinskii, 1993). This results in weaker emitted H_α radiation caused by radiative transfer effects. However, it produces Zeeman splitting of a hydrogen atom with the resulting impact polarization up to 20–25% as was shown above.

The fact that only about one third of flares show noticeable polarization can also be explained by electron-beam kinetics. Only beams with moderate intensity and spectral index can precipitate to the lower chromosphere as a directed beam overcoming the effect of the return current at the transition region and upper chromosphere. Very intense and hard beams are completely disrupted by this effect at the transition region, whereas weak soft beams lose their energy in Coulomb collisions at upper chromospheric levels. It is likely that this number of 30% reflects the number of moderate-intensity electron beams generated at the site of beam injection.

Thus we suggest the following explanation for the observations by Firstova and Boulatov (1996). The authors could observe two magnetic loops with characteristic sizes of 6″ or 4200 km and 5″ or 3500 km, which were not perpendicular to the solar surface. In the upper parts of the loops, polarization is weaker but positive as the beam propagates horizontally, whereas in the feet the polarization is much higher and negative as the beam propagates toward the solar center.

11.5
Interpretation of Polarimetric H_α Observations

Observations of linear polarization in spectral lines from solar flares provide unique information on the directions of energy transport from the corona to deeper layers during these highly dynamic events. The H_α line is the most observable line in solar flares for ground-based instruments, and significant properties of energy transfer process can be derived from the measurements of its polarization vector.

Linear H_α polarization is not very often observed in large solar flares (Firstova and Boulatov, 1996; Firstova and Kashapova, 2002; Firstova *et al.*, 2003; Hénoux, 1991a) but more regularly in flaring events of much smaller scales called moustaches or Ellerman bombs (EB) with extended wings in the H_α-line profile. These events have sizes ranging from 5″ down to the diffraction limit of modern (1 m class) solar telescopes. Their resemblance in many spectral aspects to type II white-light flares (Hiei, 1986) suggests that H_α-line moustaches are small-scale appearances of impulsive heating similar to larger solar flares caused by electron-beam precipitation and nonthermal excitation of hydrogen atoms (Aboudarham and Hénoux, 1987; Bray and Loughhead, 1974; Severnyi, 1965; Zharkova and Kobylinskii, 1993).

Polarization observations in moustaches have not been carried out as actively as the investigation of their structure and motion (Bruzek, 1972; Georgoulis *et al.*, 2002; Koval, 1972; Kurokawa *et al.*, 1982; Nindos and Zirin, 1998; Severnyi, 1965). The observational results on polarization in moustaches and flares by different authors often conflict with each other in terms of magnitude of polarization and position of the polarization plane. The degree of polarization measured in the H_α-line center in flares ranges over 3–5%, in some rare cases exceeding 10% (Chambe and Hénoux, 1979; Firstova and Boulatov, 1996; Hénoux, 1991a; Hénoux and Chambe, 1990). In moustaches, considerable polarization also established the H_α-line center, seen to be about 7% (Firstova and Boulatov, 1996) or in the range of 3–10% (Babin and Koval, 1985, 1988; Firstova *et al.*, 1999; Kashapova, 2002). Other observations of 32 moustaches by a different instrument reported that polarization did not exceed 8% (Rust and Keil, 1992), with the bulk of them (about 87%) not exceeding 2%. A further spectropolarimetric investigation of 164 Ellerman bombs revealed their linear polarization in the H_α line to be less than 3% in 22 events (13.4% of the total number), less than 2% in 68% of the events, and up to 12% in the remainder ($\sim$ 20%) of these events (Kashapova, 2004). This ratio of the numbers of moustaches with high and low polarization significantly exceeds those obtained by Rust and Keil (1992).

Confusion also occurs over the direction of the plane of polarization measured by many observers. In observations by Hénoux *et al.* (1990) the direction of this plane was considered to coincide with the flare-to-disk center direction leading to a positive polarization, whereas some observations by Firstova and Boulatov (1996) showed the plane to be perpendicular to this direction, resulting in negative polarization. In different moustaches the polarization planes can very often have opposite orientations or even not have a well-defined direction on the solar disk center (Firstova, 1986; Firstova *et al.*, 1999).

The first interpretation of H_α-line polarization was made in an approximation of optically thin plasma, using Born cross-sections for line excitation by charged particles or external radiation (Hénoux, 1991a). The observed H_α-line polarization was assigned to impact polarization by either electron beam or thermal electrons or to polarization by high-energy radiation (UV and EUV) depending on the position of the polarization plane (Chambe and Hénoux, 1979; Hénoux, 1991a). The plane of polarization caused by suprathermal particles (electron or protons) or highly energetic UV and EUV radiation is expected to be mainly perpendicular to the solar center direction, that is, the polarization is expected to be negative. A directed heat flux produces positive polarization, with the plane of polarization being parallel to the solar center due to the lower particle energies.

To explain the observed positive polarization in the H_α line, low-energy proton beams ($E \leq 200\,\mathrm{keV}$) were referred to as the source of slow directed fluxes (Hénoux *et al.*, 1993; Vogt *et al.*, 1997). Impact polarization in H_α-line emission was computed for proton beams precipitating into a flaring atmosphere and causing a redistribution in population between the Zeeman split states using the density matrix formalism (Vogt *et al.*, 1997). Collisional excitation of hydrogen atoms by proton beams was taken into account as were radiative atoms for incident and diffusive fields in H_α, Ly_α, and Ly_β lines. The calculated H_α-line polarization was found to be lower by up to an order of magnitude than that observed during a flare, fitting the observations only for very weak emission at the very beginning of flare onset (Vogt *et al.*, 1997). However, the authors did not take into account the collective effects of proton beams on the ambient plasma, which can excite kinetic Alfvén waves simultaneously with H_α-line emission (Voitenko, 1996), which makes proton beam stability at such deep atmospheric levels questionable. Therefore, in order to fit the observations, we must consider other agents of H_α-line polarization in flares.

These agents are likely to be electron beams propagating in the plasma of solar flares at H_α-line depths as suggested by Fletcher and Brown (1995). Their simulations used stochastic kinetic solutions within a loss cone for electron-beam precipitation that gave a reasonable polarization degree of about 5–7%. However, this required electron beams with initial energy fluxes greater than $10^{13}\,\mathrm{erg\,cm^{-2}\,s^{-1}}$, which is three orders of magnitude higher than the typical energy fluxes deduced from X-ray observations in solar flares (Hoyng, 1975; Krucker and Lin, 2005).

More recent research has considered the effect of electron beams on H_α-line polarization for a 1-D atmosphere horizontally stratified with a moderate magnetic field (Zharkova and Syniavskii, 2000). The beam was assumed to be stationarily injected in the direction of the magnetic field with energy and pitch-angle diffusion causing nonthermal excitation and ionization of hydrogen atoms. H_α photons were considered to be emitted with photon momentum $\boldsymbol{K}$ toward the top of the atmosphere, which has a vertical axis in the direction of magnetic field $\boldsymbol{B}$. Then it was assumed that polarimetric observations were made from the atmosphere top, so the viewing angle coincided with the pitch angle.

For a solution of the steady state equation for a hydrogen atom with atomic superfine structure caused by magnetic field splitting, the density matrix technique was applied that took into account collisions with thermal and beam electrons and excitation/deexcitation by external and diffusive radiation (Zharkova and Syniavskii, 2000). The diffusive H_α radiation field for a five-level hydrogen model atom without this superfine structure was calculated in the full non-LTE approach as described by Zharkova and Kobylinskii (1991, 1993). The resulting H_α polarization was found to appear near the line core only, positive near the red wing (−0.25 Å) and line center but negative in the near blue wing (+0.25 Å), while the emission in the far wings was fully depolarized by collisions with thermal electrons. The full (integrated over a wavelength) H_α-line linear polarization, caused by weak or moderate electron beams and observed in the direction of the magnetic field, was shown to be negative, increasing from −2 to −20% with an increase of the viewing angle in a flaring loop, which in this geometry coincides with pitch angle.

The above results confirmed the conclusions by Hénoux and Chambe (1990) that electrons injected in the direction of a magnetic field produce H_α-line emission polarized in the plane perpendicular to magnetic field, that is, to the solar center direction, if the flaring loop is vertical. However, these results do not explain the variety of H_α-line polarization degrees and directions measured by different observers. Therefore, the motivation for the current section is to refine our previous simulations of H_α-line impact polarization in a vertical magnetic loop to explore the effect of an arbitrary viewing angle, that is, with a projection of the polarization plane from a spherical electromagnetic wave emitted from the loop located arbitrarily on the solar surface and observed in the plane of a flat 2-D solar image.

11.5.1 Revised Theoretical Model

11.5.1.1 Geometry of Impact Polarization

For the calculations of H_α impact polarization by electron beams we use an approach similar to that described by Zharkova and Syniavskii (2000) but with some changes in the polarization plane orientation. It is assumed that a 1-D flaring atmosphere stands vertically on the solar surface with a magnetic field directed along the vertical z-axis in the local Cartesian coordinate system XYZ. The observer is located at point O on the X'-axis in the Cartesian coordinate system $X'Y'Z'$ centered on the solar center with a viewing angle ψ (Figure 11.5) where the solar sphere is projected onto a flat solar disk (Figure 11.6). Similarly to Zharkova and Syniavskii (2000), full non-LTE simulation for hydrogen emission is performed using the density matrix approach for a three-level-plus-continuum model atom with superfine structure caused by Zeeman splitting in a moderate magnetic field. The ambient atmosphere is assumed to consist of the ambient hydrogen plasma with neutral atoms, ions, and thermal electrons.

Beam electrons are steadily injected from the atmosphere top along the vertical z-axis with the energy power-law spectra and Gaussian pitch-angle dispersion $\Delta\mu$ (Zharkova *et al.*, 1995, 2010), so that the electron energy momenta have a direction $\boldsymbol{P}$, meaning that the electrons in a beam which have different pitch angles θ and azimuthal angles φ for different electrons in the beam. A beam electron is scattered on an ambient plasma particle and emits a photon with momentum $\boldsymbol{K}$ at pitch angle ξ and azimuthal angle η, which is observed at viewing angle ψ in the flat 2-D solar image obtained by the observer at point O on the x-axis (Figure 11.5).

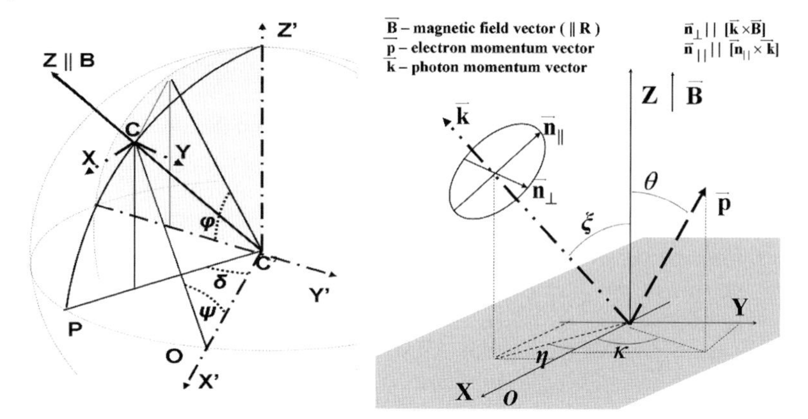

Figure 11.5 Position of polarization plane of a photon emitted in a vertical magnetic loop at a pitch angle of ξ and azimuthal angle of η. The viewing angle ψ is the angle between the local Cartesian system XYZ and the system $X'Y'Z'$ associated with the solar sphere center. The viewing angle, along with the loop position on the solar disk, defines the projection of a polarization plane in an XYZ system onto the observational plane $OY'Z'$ at the line-of-sight point O on the X'-axis for arbitrary electron momentum P and photon momentum K vectors (see Section 11.5.1.1 on orts, or unit vectors, of a polarization plane).

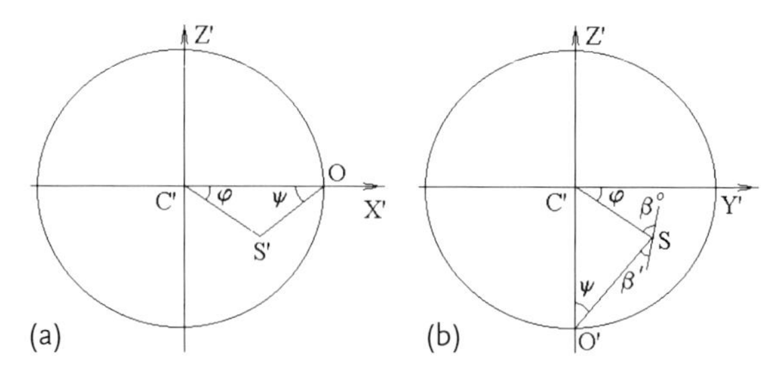

Figure 11.6 Layout of a slit position during observations on $Z'X'$ line-of-sight (LOS) plane, where the observer is located at point O (Figure 11.5), and on the $Z'Y'$ observation plane, where the flat solar disk image is obtained. Here β_0 is the angle between the slit and the object-to-disk center direction, β' is the angle between the slit and the object-to-LOS limb direction, and ψ is the viewing angle.

The resulting polarization plane is defined by the orthogonal unit vectors, or orts henceforth, $n_{\|}$ and $n_{\perp}$, with the former being parallel to the vector $\boldsymbol{B} \times \boldsymbol{K}$, that is, to the x-axis, and the latter is perpendicular to $\boldsymbol{B} \times \boldsymbol{K}$, where $\boldsymbol{B}$ is the magnetic field vector and $\boldsymbol{K}$ is the direction of the emitted photon. This polarization plane is projected onto a viewing plane at point O on the x-axis perpendicular to the direction OX, that is, a viewing angle ψ, which is a combination of flare location (heliolatitude and heliolongitude) on the solar sphere and the deviation of magnetic field from the local vertical.

11.5.1.2 Collisional Tensor Corrections

Electron-beam kinetics was considered taking into account energy and pitch-angular diffusion in Coulomb and inelastic collisions with ambient plasma ions, electrons, and neutral atoms in the presence of an electric field induced by a precipitating beam and a vertical magnetic field confining the ambient plasma (Siversky and Zharkova, 2009b; Zharkova *et al.*, 1995). The electron-beam-distribution functions were used for calculation of the collisional tensor Φ_{ij}^{coll} given above that includes the scattering indicatrix $f_{ij}(\Omega)$, where Ω is the scattering solid angle for beam electrons with pitch angle θ (index *i*) and azimuthal angle φ (index *j*). Electron precipitation is computed at each depth point along the local vertical from the kinetic approach with axial symmetry in φ and uniformly varying pitch angles on the assumption that in a single scattering process the angular changes are not very large, that is, $\Delta\mu/\mu \ll 1$. The scattering process is assumed to be very anisotropic according to the pitch-angular diffusion (see Chapter 4 and Siversky and Zharkova, 2009b) while confined within the pitch angles $\mu_{\min}$ and $\mu_{\max}$ at the ejection point with the limits increased to -1 and 1 during beam precipitation into the chromosphere and photosphere. The collisional tensor for thermal electrons is assumed isotropic, similarly to Zharkova and Syniavskii (2000).

11.5.1.3 Radiative Tensor Corrections

The non-LTE problem for a three-level-plus-continuum hydrogen model atom with levels split into a superfine structure by Zeeman effect was solved for the considered geometry using the same approach as in the paper by Zharkova and Syniavskii (2000). However, in the current approach we introduce a viewing angle effect on the H_α emission by considering the Maxwell tension tensor of a spherical electromagnetic wave (Berestetskij *et al.*, 1989) with an arbitrary viewing angle ψ instead of a plane wave in the *z*-direction with a viewing angle $\psi = 0$ considered by Zharkova and Syniavskii (2000). This means that the momentum $\boldsymbol{K}$ of a photon emitted at the pitch angle ξ has a plane of polarization inclined toward direction $\boldsymbol{K}$ at angle ψ with orthogonal orts, or unit vectors, $\boldsymbol{n}_\parallel$ and $\boldsymbol{n}_\perp$ defined by the vectors $n_\parallel = \boldsymbol{B} \times \boldsymbol{K}$ and $n_\perp = \boldsymbol{n}_\parallel \times \boldsymbol{K}$ as presented in Figure 11.5. The direction $n_\parallel$ is toward the observer located along the *x*-axis and the angle between this direction and *x*-axis is also the polarization viewing angle. Therefore, the observer measures the projection onto these orts of an electromagnetic wave electric vector, whose components are defined by the following equations:

$$
\begin{aligned}
I_x^{ij} &= I_0[\sin(\psi)\sin(\eta) - \cos(\psi)\cos(\eta)\cos(\xi)]\,, \\
I_y^{ij} &= -I_0[\sin(\psi)\cos(\eta) - \cos(\psi)\cos(\eta)\cos(\xi)]\,, \\
I_z^{ij} &= I_0\cos(\psi)\cos(\xi)
\end{aligned}
\tag{11.23}
$$

where I_0 is the intensity emitted along the direction $\psi = 0$, that is, along $\boldsymbol{K}$, and indices i and j refer to the directions ξ and η, respectively. This formalism is included in the diffusive radiative tensor Φ_{ij} described by Eq. (11.8) in Zharkova and Syniavskii (2000). The Stokes parameters are calculated for the orts $\boldsymbol{n}_\parallel$ and $\boldsymbol{n}_\perp$

taking into account a viewing angle ψ as follows:

$$I(\psi) = \iiint [I_{\|}(\lambda, \xi, \eta, \psi) + I_{\perp}(\lambda, \xi, \eta, \psi)] d\Omega\, d\lambda \,, \tag{11.24}$$

$$Q(\psi) = \iiint [I_{\|}(\lambda, \xi, \eta, \psi) - I_{\perp}(\lambda, \xi, \eta, \psi)] d\Omega\, d\lambda \,, \tag{11.25}$$

where the integration is over the solid angle Ω consisting of angles ξ and η presented in Figure 11.5 and the polarization is calculated as

$$P(\lambda, \psi) = \frac{Q(\psi)}{I(\psi)} \,. \tag{11.26}$$

11.5.1.4 **Viewing Angle Effect on the Observed Impact Polarization**

As shown in Zharkova and Syniavskii (2000) impact polarization appears only near line cores, whereas wings are fully depolarized by collisions with thermal electrons, which is in agreement with the observations by Firstova (1986). Calculations also reveal that simulated H_α-line profiles are rather asymmetric with the maxima corresponding to the main transitions for which $\Delta L = \Delta J$ ($3d_{5/2} - 2p_{3/2}$, $3d_{3/2} - 2p_{1/2}$, and $3p_{3/2} - 2s_{1/2}$). The full line polarization is the sum of polarization degrees in the transitions and is dependent on a number of beam electrons. In this section we concentrate on the observable values such as the full H_α-line polarization and its sign, or the polarization plane orientation.

As can be seen from Figure 11.5, the polarization plane in the system of coordinates XYZ of a flaring loop is always located on a plane perpendicular to those formed by the emitted photon direction $\boldsymbol{K}$ and magnetic field direction $\boldsymbol{B}$. The parallel ort $n_{\|}$ belongs to the plane $\boldsymbol{B} \times \boldsymbol{K}$, and $n_{\perp}$ is orthogonal to this. If the observer measures a photon from the top of a flaring atmosphere at angle ξ, the full line polarization is negative, varying from -3 to -25% as reported by Zharkova and Syniavskii (2000). Let us consider the more likely situation with the flaring atmosphere being located in the Cartesian coordinate system XYZ somewhere on the solar disk away from the disk center and the observer being located at point O of the Cartesian coordinate system $X'Y'Z'$ as described in Section 11.5.1.1. As mentioned before, in real observations the viewing angle is a combination of a loop's location on the solar sphere and a magnetic field deviation from the local vertical, which can be separated only by supporting observations of the flare position on a flat solar disk image.

The integrated H_α-line linear impact polarization caused by intense electron beams with soft ($\gamma = 6$) and hard ($\gamma = 3$) energy spectra calculated taking into account its projection in the direction of the observer, or a viewing angle, is presented in Figure 11.7 (solid and dashed lines) for those times when the beam is on.

It is seen that the linear polarization observed from smaller to larger viewing angles gradually changes from negative to positive magnitudes for all initial beam fluxes and spectral indices. For smaller ($\leq 50°$) viewing angles the full H_α polarization is negative and for larger ($\geq 60°$) viewing angles it is positive, being higher in

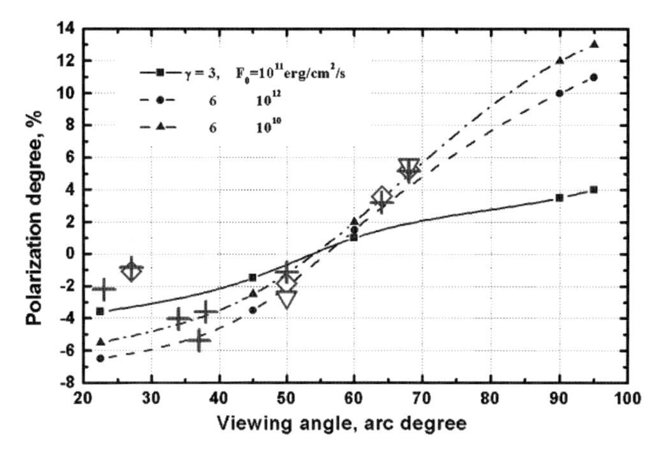

Figure 11.7 H_α-line polarization as a function of position angle ψ caused by electron beams with a spectral index of $\gamma = 7$, $F_0 = 10^{10}\,\text{erg}\,\text{cm}^{-2}\,\text{s}^{-1}$ (dashed line with triangles), $\gamma = 7$, $F_0 = 10^{12}$ (dashed line with circles), $\gamma = 4$, $F_0 = 10^{11}\,\text{erg}\,\text{cm}^{-2}\,\text{s}^{-1}$ (solid line with squares). The crosses, diamonds, and triangles are the observations for different moustaches, discussed in the text. For a color version of this figure, please see the color plates at the beginning of the book.

absolute value than the negative ones. In an angle range of 50–60° the polarization changes sign for all electron beam parameters. The slope of the polarization curve vs. the viewing angle is smaller for lower initial energy fluxes ($10^{10}\,\text{erg}\,\text{cm}^{-2}\,\text{s}^{-1}$) and softer beams ($\gamma = 6$) and increases with beam intensity (or its initial flux) growth and spectral index decrease from 6 to 3 (compare the upper and lower curves).

Now let us see how these predictions accord with observations. Apparently, comparing theoretical polarizations with measured ones can provide estimates for parameters of an electron beam causing this H_α impact polarization. Their comparison with those derived from HXR bremsstrahlung emission in upper atmospheric levels can provide important information about energy transport by electron beams from the uppermost to the deeper chromospheric levels of a flaring atmosphere.

11.5.2
Results of Observations

For comparison with the theoretical predictions above, the profiles of Stokes parameters Q/I and U/I of moustaches in the H_α hydrogen line obtained from June to August 1999 from the Large Solar Vacuum Telescope are used (Skomorovsky and Firstova, 1996). The basic parameters of the instrument and the method of Stokes parameter observation and processing are described in detail by Kashapova (2002). Since moustaches are known to be very sensitive to seeing conditions and the image resolution varies from 0.5 to 1.2 arcsec, relatively short exposure times (0.05 and 0.1 s) are selected in order to minimize image tremor. In addition, normalization of the H_α-line profiles to the continuum in each strip is carried out,

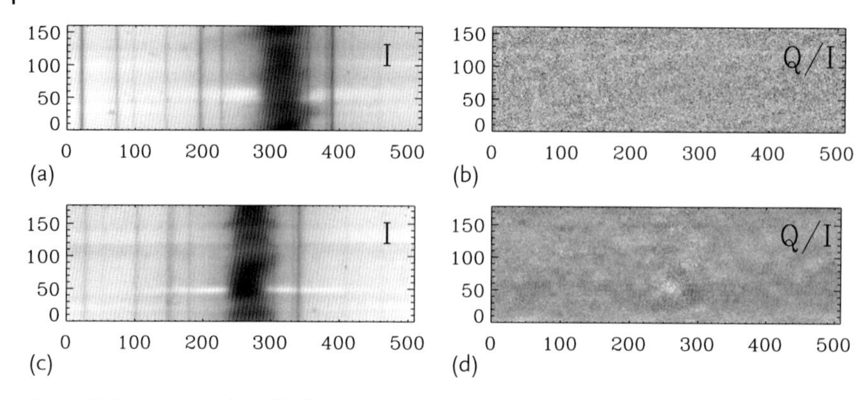

Figure 11.8 Two examples of Stokes parameters I (intensity) (a,c) and Q/I (linear polarization) (b,d) measured in moustaches with insignificant polarization (a,b) and noticeable polarization (c,d).

which helps to decrease the instrumental polarization (Firstova, 1986). The sample Stokes parameters I and Q/I of moustaches are presented in Figure 11.8, where Figure 11.8a shows a moustache without polarization and Figure 11.8c demonstrates those with noticeable polarization seen as a brighter source between 200 and 300 pixels (x-axis) around the 50th pixel (y-axis).

These data were previously analyzed by Kashapova (2004) without consideration of moustache locations on the solar disk. The Stokes parameters Q/I are converted from a slit coordinate system to a coordinate system related to the object-to-observer direction according to the layout shown in Figure 11.6.

It can be seen in Figure 11.6b that the angles β_0 and β' are related to the viewing angle ψ and latitude angle φ as

$$\psi = \beta_0 + \beta' - (90^\circ - \varphi) . \tag{11.27}$$

Here ψ' is the angle from the north–south vertical toward the slit on a flat solar image, under which the angle arc CS with length ψ is seen on the spherical solar surface. This is close to the viewing angle ψ presented in Figure 11.5 within the applicability of flat disk coordinates instead of spherical ones because of the triangles $C'OS'$ and $C'O'S'$ similarity on the ZX and ZY planes (compare Figure 11.6a,b). The angle φ is the polar angle on the solar disk that corresponds to a heliolatitude with its sign in the relevant hemispheres.

As stated above, in 68% of moustaches the H_α-line polarization is found to be less than 2%, that is, it is probably not caused by impacts but rather by chromospheric radiation (Zharkova and Syniavskii, 2000), in 13% of moustaches it ranges from 2 to 3%, and in the remainder it reaches 12% (Kashapova, 2004). This high-to-low polarization ratio obtained for different moustaches in various active regions and on different dates significantly exceeds that reported by Rust and Keil (1992) for a single active region. In the present paper we count the number of moustaches with a polarization exceeding and less than 2% in different zones of the solar disk

Table 11.1 Relative number of moustaches with polarization less than 2%.

	Distance from solar center, present paper				Rust and Keil (1992)
	10 ± 10°	30 ± 10°	50 ± 10°	70 ± 10°	
Rate, %	69	57	85	59	87

center. Each zone has a size of 20° centered on the viewing angle, that is, a zone with a viewing angle of 50° includes events located between 40 and 60°.

The ratios of the moustache numbers observed in a given zone with low polarization not exceeding 2% are presented in Table 11.1 and, naturally, the remaining moustaches have a linear polarization of 2–12%.

From Table 11.1 one can see that the number of moustaches with a noticeable polarization degree varies significantly with distance from the solar disk center, or with the moustache's position on the solar sphere. The number of moustaches with significant polarization (> 2%) does not increase considerably toward the solar limb, as suggested by Hénoux *et al.* (1993). However, it does not drop to zero for viewing angles greater than 70° as predicted by the evaporative model (Fletcher and Brown, 1995). Remarkably, the ratio of moustaches in a zone of 50° is very close to the results obtained by Rust and Keil (1992). This resemblance is rather surprising, keeping in mind that the observations by Rust and Keil (1992) were obtained in a single active region, while the Irkutsk data were obtained on different dates in various active regions located in the southern, northern, eastern, and western hemispheres of the Sun.

In Figure 11.7 the measured degrees of polarization are plotted vs. the central zone angles on top of the theoretical curves with crosses, diamonds, and triangles presenting the results for different moustaches by Kashapova (2004). It can be seen that for viewing angles of less than 60°, the polarization degrees are negative, approaching magnitudes of −8–9% at angles smaller than 20°, although for some events located near 30° it does not exceed 2%. For viewing angles greater than 60° the degrees of polarization are positive, approaching magnitudes of 12–15% for angles of 90° (limb). In the zone between 50 and 60° the polarization degree changes signs passing through zero magnitude.

These observations provide a very important property of the polarization magnitude and its sign, or direction, which is dependent on the position of moustaches on the solar disk. Both the polarization degree magnitude and its sign change in a very similar way for all the events considered. The higher degrees of polarization are observed toward the center or limb of the solar disk, being negative for locations around the solar center and changing to positive ones toward the solar limb. The agreement of the data with the simulations supports the proposed model of electron transport and suggests the beams as promising sources of impact polarization in the chromosphere.

11.5.3 Observational Recommendations

A comparison of the simulated polarization degrees and signs with the observed ones presented in Figure 11.7 by crosses, diamonds, and triangles (see Section 11.5.2 for details) reveals for most observations a surprisingly good fit of the theoretical curves for a spectral index of $\gamma > 4$ and any initial energy fluxes $F = 10^{10}$ or 10^{12} erg cm^{-2} s^{-1} presented in Section 11.5.1.4, keeping in mind the difference in the instruments and observing conditions. The absence of polarization measurements for beams with $F = 10^{11}$ erg cm^{-2} s^{-1} and $\gamma = 4$ is likely to be caused by additional ohmic energy losses in the induced electric field that are much higher for harder beams, as was concluded by Zharkova *et al.* (1995), Zharkova and Gordovskyy (2005a), and Siversky and Zharkova (2009b). As a result, more precipitating electrons return to the coronal source at the top and fewer precipitate downward to the chromosphere, which obviously does not produce an observable impact polarization.

The only three observations near the solar disk center have a degree of polarization of -1–2%, which is lower than the theoretical predictions for this viewing angle. However, these three observations may be of the inclined loops often reported by observers. In order to fit them to the theory one can move the observational points toward the theoretical curves and deduce the angle of magnetic field inclination from the local vertical, which for these observations can be about 25–30°.

Based on the comparison, one can introduce observing constraints on observations for detectable H_α impact polarization with future instruments with high spatial and spectral resolutions. The theoretical curves in Figure 11.7 reveal that the softer and more intense a beam is, the higher is the degree of H_α polarization produced by it. For viewing angles less than 50° the polarization produced by beam electrons is negative, and for angles higher than 60° it is positive. Hence, there is a zone at a viewing angle between 50 and 60° where the impact polarization is about zero, since it is obscured by the viewing conditions of this radiation despite the presence of beam electrons in the flaring atmosphere.

This condition can be quantified from Eq. (11.27) for zero impact polarization as the following inequality: $50° \leq \psi \leq 60°$. Thus, in order to observe impact polarization, the latitude angle φ, the angle between the slit and slit-to-disk center direction β_0, and the angle between the slit and slit-to-limb direction β' (Section 11.5.2) are to be restricted as follows:

$$\beta_0 + \beta' + \varphi > 150° \, , \quad \text{or} \quad \beta_0 + \beta' + \varphi \leq 140° \, . \tag{11.28}$$

If the observer can comply with the limits above for slit positioning and latitude angles, this can ensure observation of measurable impact polarization, if it is present in the event under investigation.

11.6 Conclusions

Using the density matrix approach, we calculated the H_α-line impact polarization produced by beam electrons with hydrogen atoms embedded in a flaring atmosphere with a magnetic field. The following was established:

1. H_α-line profiles show a linear polarization in a range of -2–25% only in line cores where a "weak magnetic field" approximation is valid; the wings are fully depolarized by collisions. The circular polarization is negligible.
2. Linear polarization in a H_α line varies with depth, being negative and higher for transitions with larger J and positive, about 25%, for smaller J. From a depth of 7×10^{20} cm^{-2} the polarization in all transition changes to a negative one of about 2–10%.
3. In the absence of electron beams at lower depths from 1.2×10^{21} cm^{-2} the H_α-line polarization is determined by the external chromospheric radiation and reaches about 2–3%.
4. The full H_α-line polarization caused by impacts with beam electrons is negative, varying from -5–10% for viewing angle $\theta = 0.1\pi$ to -18–25% for $\theta = \pi/2$.
5. In the upper chromosphere from the first few seconds electron-beam injection causes a growth of the full H_α-line polarization to 2–3% and increases to 10–25% for the maximum energy flux of the beam.
6. The secondary maximum of linear polarization appears in the lower chromosphere at a column depth of 1.0×10^{21} cm^{-2}, which reflects the effect of a return current on the precipitation of beam electrons.

The effect of a viewing angle on H_α-line impact polarization by beam electrons was also investigated. The H_α polarization profiles were found to be affected by electron beams only near line cores, whereas wings are fully depolarized by collisions with thermal electrons, which causes the appearance of extended wings in the H_α line like those observed in "moustaches".

The full (integrated in wavelength) H_α-line linear polarization, caused by moderate electron beams, varies in a range of 2–15% and can be either negative or positive depending on the position of the flaring loop on the solar disk, that is, its heliolongitude. For viewing angles less than 50° the H_α-line impact polarization is negative, increasing up to -10% toward angles of 20° or smaller.

For viewing angles greater than 60° the measured impact polarization becomes positive, sharply increasing up to 15% toward the limb and beyond. In the 50–60° zone the observed polarization degree crosses a zero point despite the actual presence of beam electrons in a flaring atmosphere.

This imposes constraints on the slit and moustache locations that allow for observations of measurable impact polarization signatures if they occur in these events, that is, the slit positioning angles β_0 and β' are to be restricted to $\beta_0+\beta'+\varphi > 150°$ or $\beta_0 + \beta' + \varphi \leq 140°$, where β_0 and β' are the slit angles toward the solar center and limb, respectively, and φ is the heliolatitude.

The theoretical predictions fit remarkably well the available observations of H_α-line linear polarization in moustaches, or Ellerman bombs, located in different positions on a solar disk. This fit allows observers to estimate the parameters of an electron beam causing this polarization and to compare them with those derived from HXR bremsstrahlung emission, which can provide important information on the energy transport mechanisms in flaring events on the Sun.

Therefore, we have shown that electron beams can be the agents producing observed features in H_α-line polarization during the early phases (the first tens of seconds) of solar flares. Polarimetric observations can provide valuable information on beam kinetics at different depths. To discriminate the effect of electron beams from those by proton beams, additional simulations of proton kinetics and complex multiwavelength observations are required.

12
Sunquakes Associated with Solar Flares

12.1
First Sunquake of 9 July 1996 Flare

The flare of 9 July 1996, the first one to produce a detectable sunquake, was the only significant X-ray flare observed in 1996 when the solar activity was at the minimum of its 11-year cycle (Kosovichev and Zharkova, 1998). This was a fairly moderate flare classified as X2.6/1B, with a corresponding X-ray flux of 0.26 erg cm^{-2} s^{-1}. More energetic flares are expected to quake the Sun stronger. However, previous attempts (Braun and Duvall, 1990; Haber *et al.*, 1988) to detect the seismic signal of solar flares from ground-based observations were inconclusive.

The X-ray impulse of this flare detected by BATSE (Burst and Transient Source Experiment) (Fishman *et al.*, 1993) on board the Compton Gamma Ray Observatory began to increase at 09:07:49 UT and reached a sharp maximum at 09:09:40 (Figure 12.1). The magnetic field measurements from MDI show that the flare energy release was associated with an emerging flux of opposite polarity in the active region NOAA 7978 a few hours prior to the flare. The flare occurred when MDI was operating in the full-disk mode taking 1024×1024 pixel 60 s average Doppler velocity observations every minute.

By analyzing the MDI dopplergrams, Kosovichev and Zharkova (1998) have detected strong localized upward and downward mass flows occurring during X-ray impulses. The downward flow is shown in Figure 12.1; it occupied two or three pixels of the solar image on the CCD, which corresponds to a linear size of about 3–5 Mm. The velocity impulse (Figure 12.1) was almost as sharp as that in the X-ray flux. However, the maximum velocity of $\sim$ 1.5 km s^{-1} was observed approximately 1 min after the X-ray maximum, at 09:11:00. It is likely that the actual velocity impulse is smoothed in the 60 s average dopplergrams and that in reality it was significantly stronger and sharper. We use the parameters of the velocity impulse for estimations of the momentum carried in this impact.

By simply displaying the MDI dopplergrams in sequence, Kosovichev and Zharkova (1998) have detected a circular wave packet propagating outward from the flare location as plotted in Figure 12.2. The wave was first detected at about 09:30 UT, approximately 20 min after the flare onset in HXR, at a distance of $\approx$ 18 Mm from the flare site. The wave was clearly observed in the sequence of

Electron and Proton Kinetics and Dynamics in Flaring Atmospheres, First Edition. Valentina Zharkova.

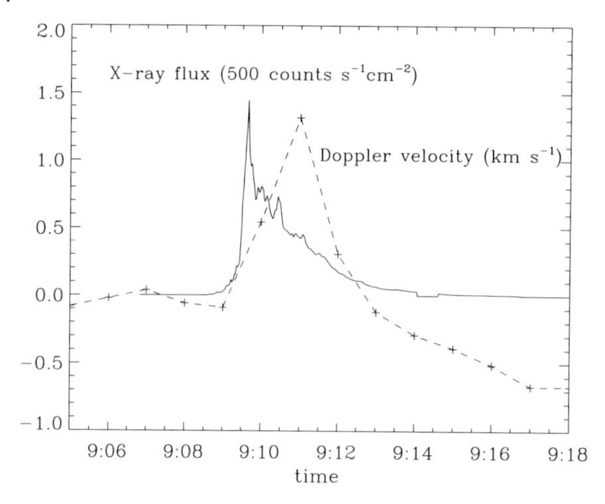

Figure 12.1 The 1.024 s averages of the X-ray flux in an energy range of 25–100 keV of the 9 July 1996 solar flare from BATSE flare monitor (solid curve), and the 1 min averages of the Doppler velocity of downflowing plasma in flare core from MDI (dashed curve).

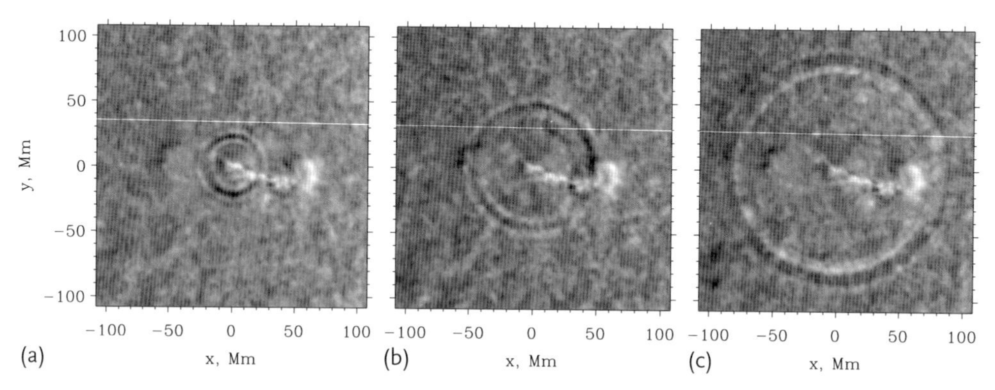

Figure 12.2 Remapped and filtered dopplergrams at 9:36 (a), 9:46 (b), and 9:56 (c), 25, 35, and 45 min following X-ray flare. The flare signal was extracted from the first three components ($m = 0$, 1 and 2) of the azimuthal Fourier transform by applying a 5 min wide Gaussian filter along the time-distance ridge of Figure 12.4a. Then the signal combined from these three components was multiplied by a factor of 4 to enhance it and superimposed on the remapped dopplergrams. Adopted from Kosovichev and Zharkova (1998).

dopplergrams for about 35 min, until 10:05 UT. After that, the wave amplitude dropped rapidly and the wave was lost in the ambient noise. The wave was not easily visible in individual dopplergrams and required some image improvement in order to show the ripples in Figure 12.2.

12.1.1 Methods of Sunquake Detection

Seismic waves are assumed to be created by deposition of a momentum to the solar surface creating acoustic waves, or magnetoacoustic waves propagating in the solar interior as shown in Figure 12.3. In order to detect these waves and to investigate what has happened during a flare in the solar interior one can explore dopplergrams obtained at the photospheric level with sufficiently small cadence (1 min or better) and high spatial resolution.

12.1.1.1 Time–Distance Diagram Technique

Let us use the 1 min cadence MDI dopplergrams for the hour starting from 11:00:00 UT. Since the typical oscillation frequencies associated with quakes (Donea and Lindsey, 2005; Kosovichev and Zharkova, 1998) are higher than background oscillations (3 mHz), we need to apply the frequency filtering centered at 5–6 mHz in order to increase the signal-to-noise ratio.

After this preprocessing the MDI dopplergrams are remapped to polar coordinates with centers around the locations of the holographic seismic sources X1–X4 (Donea and Lindsey, 2005), and the new velocity distributions $v(r, \theta, t)$ are obtained as a function of time t and their distance r from the center (defined with the accuracy of a single pixel). Then we extract the velocities $v_0^0(r, t)$ for every r averaged over various angles θ for $m = 0$ and measure for the different times the horizontal displacements of the propagating waves in a data cube of 120 Mm.

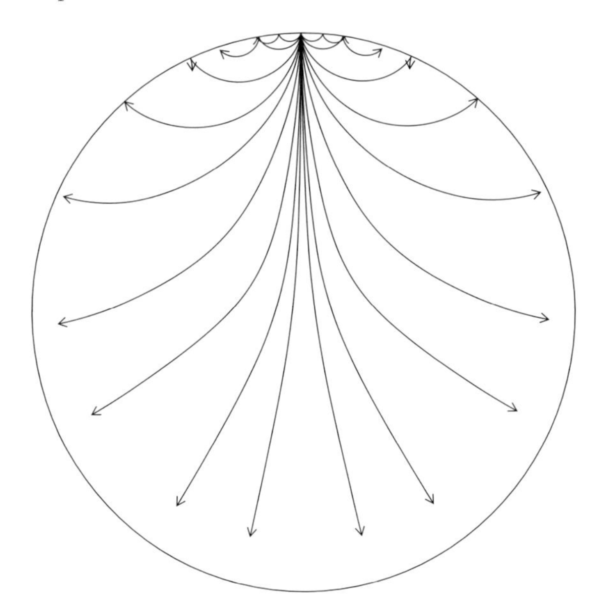

Figure 12.3 Samples of theoretical ray paths of acoustic waves excited by a flare and propagating through the solar interior 1.5 h after the flare. Adopted from Kosovichev and Zharkova (1998).

The obtained velocity distributions in an area of radius 120 Mm are fit by a circular wave using the angular Fourier transform for the angle θ (Christensen-Dalsgaard and Thompson, 2003) as follows:

$$v(r, \theta, t) = \sum_{m=1,3} v_m^0(r, t) e^{im\theta t} , \tag{12.1}$$

where $m = 0$ denotes a circular wave, $m = 1$ a dipole wave, $m = 2$ a quadrupole wave, and so on. First, one extracts the velocities $v_0^0(r, t)$ for every r averaged over various angles θ for $m = 0$, then tries to find the velocities $v_1^0(r, t)$ for $m = 1$, and so on.

After the fits are found for either m value, one can evaluate the horizontal displacements v_0 of the propagating waves for the different times, that is, $v_0^0(r_1, t_1)$, $v_0^0(r_2, t_2)$, or the spatial average of the temporal cross-covariance. The point-to-point covariance is defined as follows (Duvall *et al.*, 1993; Kosovichev and Duvall, 1997):

$$C(r_1, r_2, t) = \sum_{t'} \phi(r_1, t')\phi(r_2, t' + t) , \tag{12.2}$$

and the positive time lag part of the cross-covariance can be written as

$$C(r_1, r_2, t) \approx A \exp\left[-\frac{\delta\varpi^2}{4}(t - t_g)\right] \cos\left[\varpi_0(t - t_p)\right] , \tag{12.3}$$

where t_p and t_g are the phase and group travel times from r_1 to r_2, $\varpi_0 = c_s/k$ is the fundamental angular frequency, c_s is the sound speed. $k = \sqrt{l(l+1)}/R_\odot$ is the horizontal wave number, $R_\odot$ is the solar radius and $\delta\varpi = 2\pi/T$ is the frequency shift defined by the total time duration T of the event. Then by plotting along the y-axis the positive time lags of the velocities and along the x-axis the distances from the source center we can obtain a time-distance (TD) diagram of the horizontal disturbance propagation in time.

12.1.1.2 **Acoustic Holography Method**

Acoustic holography is also applied to calculate egression power maps from observation. The method (Braun and Lindsey, 1999, 2000; Donea *et al.*, 1999; Lindsey and Braun, 2000) works by essentially "backtracking" the observed surface signal, $\psi(\mathbf{r}, t)$, using Green's function, $G_+(|\mathbf{r} - \mathbf{r}'|, t - t')$, which prescribes the acoustic wave propagation from a point source.

This allows us to reconstruct egression images showing subsurface acoustic sources and sinks:

$$H_+(\mathbf{r}, z, t) = \int dt' \int_{a<|\mathbf{r}-\mathbf{r}'|<b} d^2\mathbf{r}' G_+(|\mathbf{r} - \mathbf{r}'|, t - t')\psi(\mathbf{r}', t') , \tag{12.4}$$

where a, b define the holographic pupil. The egression power is then defined as

$$P(z, \mathbf{r}) = \int dt |H_+(\mathbf{r}, z, t)|^2 dt . \tag{12.5}$$

Usually, Green's functions built by using the geometrical optics approach are applied for given pupil dimensions. Then for the egression power calculations the continuum intensity data obtained as $\psi(\boldsymbol{r}, t)$ are additionally required from either MDI or GONG.

12.1.2
Results from First Sunquake Detection

To construct seismograms of the solar flare of 9 July 1996, the dopplergrams of the 124×124-Mm region around the flare were tracked to remove the solar rotation and then remapped to polar coordinates centered at the point of the initial velocity impulse. A difference filter was applied to remove background velocities. After that the data were Fourier-transformed with respect to the azimuthal angle and the Fourier coefficients were plotted as functions of the angular distance from the initial point and time. Figure 12.4a shows the first coefficient of the Fourier transform corresponding to the azimuthally averaged signal (azimuthal order $m = 0$). Figure 12.4b shows the theoretical ridges obtained for three bounces of the wave packet with different frequencies as simulated by Kosovichev and Zharkova (1995).

This circular wave displays a set of ridges with a positive slope that begins about 18 Mm from the flare at 09:32 UT and reaches $\approx$ 50 Mm at 09:48. The velocity of the wave packet increased from $\approx$ 30 to $\approx$ 100 km s^{-1} as the wave moved from 20

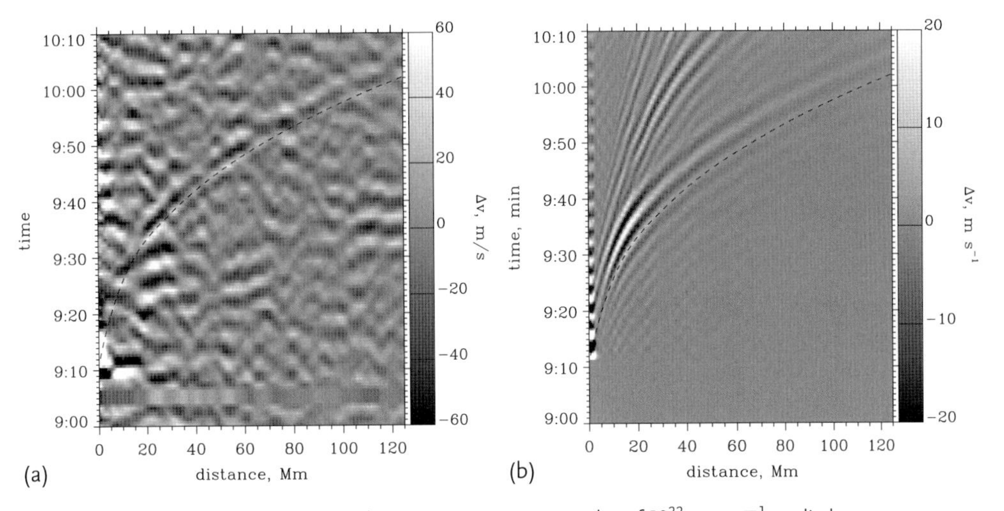

Figure 12.4 TD diagrams representing the first azimuthal components of the Doppler velocities observed in the 9 July 1996 flare (Kosovichev and Zharkova, 1998) (a) and theoretical (b) flare seismogram constructed from 1 min velocity differences. The theoretical flare seismogram computed for a localized momentum impulse of 10^{22} g cm s^{-1} applied to the solar surface at 09:11 UT. The dashed curves in both panels show the theoretical TD relation for acoustic rays initiated at the flare core at 09:11 UT. Adopted from Kosovichev and Zharkova (1998).

to 120 Mm from the epicenter. The maximum amplitude of this wave was approximately 50 m s^{-1}.

Kosovichev and Zharkova (1998) have also analyzed the dipole ($m = 1$) and quadrupole components ($m = 2$) of the flare wave. While the dipole component of the flare wave does not have a significant signal, the quadrupole component shows ridges at distances from 15 to 40 Mm from the flare. By applying a 5 min wide Gaussian filter along the TD ridge, the signal of the first three components of the wave was extracted. The combined signals of these components, multiplied by a factor of 4, are shown above in Figure 12.2 at three instances separated by 10 min. The wave amplitude is not uniform: initially it was higher in the north–south direction than in the east–west direction, and it was changing at later times. The deviation from the azimuthal symmetry could result from anisotropy of the momentum impulse, which would be the case if the impact was not normal to the surface, or from scattering on large-scale inhomogeneities in the active region.

Later, another method, a computational seismic holograph, was applied to the helioseismic MDI observations of this flare to image its seismic source (Donea *et al.*, 1999) centered on the composite umbra of the $\beta\gamma\delta$-configuration sunspot at the heart of the active region AR7978. The dimensions of the seismic source were similar to the composite dimensions of the two oppositely polarized umbrae of the sunspot. The source was clearly visible in the 2.5–4.5 mHz spectrum and even more pronounced in the 5–7 mHz spectrum. These results showed that sunquakes, or ripples, can also be accompanied by seismic emission in the megahertz range, the intensity of which varies significantly with depth in a flaring atmosphere indicating different physical processes taking place at different atmospheric levels.

12.1.3
Discrepancies between the Parameters Derived and the Basic Flare Theory

Kosovichev and Zharkova (1998) have compared the observational results of theoretical models of the flare seismic response, which predicted similar kinds of ripples (Kosovichev and Zharkova, 1995). The theoretical TD diagram for the 1 min velocity differences shown in Figure 12.4b reveals several sets of ridges. The lowest ridge corresponds to waves appearing at a given distance on the surface after the first bounce, the second ridge above the first ridge corresponds to two bounces, and so on. Recall that the theoretical ray paths of acoustic waves from a flare source are illustrated in Figure 12.3.

In the observed TD diagram (Figure 12.4a), one can easily see only the single first ridge corresponding to the first bounce. Also, the amplitude of the first ridge decays much faster in the model than it does in observations, indicating that the speed of the seismic wave was higher than that assumed in the very simple model used (Kosovichev and Zharkova, 1995).

The maximum amplitude of the theoretical seismic wave matches the observation if the flare momentum transported to the photosphere in a downward propagating shock wave was about 3×10^{22} g cm s^{-1}. However, the momentum in the flare core, $\rho S v^2 \tau$, estimated from the MDI data (Figures 12.1 and 12.2) for a density, $\rho \sim 10^{-8}\,\mathrm{g\,cm^{-3}}$, flare area, $S \sim 10^{17}\,\mathrm{cm^2}$, flow velocity, $v \sim 10^5\,\mathrm{cm\,s^{-1}}$, and impact duration, $\tau \sim 10^2\,\mathrm{s}$, is only $\sim 10^{21}\,\mathrm{g\,cm\,s^{-1}}$.

This estimation is consistent with the momentum determined from previous H_α observations (Zarro *et al.*, 1988). This suggests that additional sources of momentum and energy may have played a significant role in initiating the seismic wave during the impulsive phase of the flare. It is quite possible that the subsurfaces are heated strongly by the flare thermal wave propagating ahead of the shock (Kosovichev, 1986; Somov *et al.*, 1981; Zharkova and Zharkov, 2007) or directly by high-energy electrons and protons (Zharkova, 1999; Zharkova and Kosovichev, 2001). If this is the case, then the seismic flare source may be located in subsurface layers, causing a stronger impact on the Sun.

Kosovichev and Zharkova (1998) have also found that the observed travel time of a seismic wave is $\approx 2\,\mathrm{min}$ shorter than the time predicted by the quiet Sun model. This might be explained if the acoustic waves propagated faster in the flare region because of the higher plasma temperature and strong magnetic field, or if the seismic wave was initiated before 09:11 UT when the flare shock induced by beam electron reached the photospheric level.

The seismic ridges in Figure 12.4a associated with this flare are similar to the ridge pattern of the TD diagrams obtained by cross-correlating the oscillation signal between different regions on the solar surface (Duvall *et al.*, 1993). However, to detect the TD ridges, the cross-correlation function must be averaged over several hours of data. Thus, flares provide a unique opportunity to do TD seismology of active regions based on localized impulsive sources.

However, a comparison of the observed ripples with the theoretical model of seismic ripples (Kosovichev and Zharkova, 1995) revealed that the momentum required to produce the observed seismic response ($\sim 2 \times 10^{22}\,\mathrm{g\,cm\,s^{-1}}$) is about one order of the magnitude higher than that of $\sim 10^{21}\,\mathrm{g\,cm\,s^{-1}}$ observed from plasma downflow in the MDI dopplergrams shown in Figure 12.1.

Also the travel time required for the hydrodynamic shock produced by electron beam heating to travel to the photosphere is more than 2 min, while the time at which the helioseismic response seen in the TD diagrams coincides closely (within a minute) with the time of the HXR impulse. This raises a question about additional sources able to deliver the required momentum to the solar photosphere within a very short time scale coinciding with the start of HXR impulses.

12.2 Observations of Other Sunquakes

Donea and Lindsey (2005) anticipated that significant seismic activity could also be excited by relatively small flares (either C or M class) and provide sufficient energy to be measured as seismic sources in localized footpoints. This was recently confirmed by numerous observations of such seismic emission using the holographic method (Donea *et al.*, 2006; Moradi *et al.*, 2007). These sources of seismic emission associated with the medium flares were visible in the 2.5–3.5 mHz and even more pronounced in the 5–7 mHz spectrum (Donea *et al.*, 2006) and often were found to be cospatial with white-light flares (Donea *et al.*, 2006; Moradi *et al.*, 2007). In most cases the seismic sources were close to the locations of HXR emission and their onset times. Although for the flares of 6 and 10 April 2001 there was a delay of up to 3 min from the onset of HXR emission and the start of seismic emission (Beşliu-Ionescu *et al.*, 2005).

The TD diagram technique was further applied to some X-class flares allowing for the detection of observed seismic ripples associated with three other flares (16 August 2004, 15 January 2005, and 23 July 2002) (Kosovichev, 2006, 2007b). The author reported multiple sunquakes produced by each of these flares, which originated simultaneously in different spatial positions coinciding with the locations of strong downward motions detected by MDI dopplergrams and occurring within a time scale of up to 1 min close to the initiation of HXR emission. This supports the suggestion that these sunquakes are somehow related to the precipitation of high-energy particles and their energy deposition via a hydrodynamic response into a flaring atmosphere and the solar interior.

This wide variety of sunquakes for flares of different significance provides a unique opportunity to test existing theories of particle acceleration and transport into deeper atmospheric layers, hydrodynamic heating, and nonthermal excitation and ionization on MDI Doppler signals produced by the Ni 6768 Å line in order to understand the origin of sunquakes.

Flare of 9 September 2001 The detection of seismic waves emitted from the $\beta\gamma\delta$ active region NOAA 9608 on 9 September 2001 was associated with a quite impulsive solar flare of type M9.5 as reported from the helioseismic holography by Donea *et al.* (2006) by computing time series of the egression power maps in 2-mHz bands centered at 3 and 6 mHz. The 6-mHz images show an acoustic source associated with the flare about 30 Mm across in the east–west direction and 15 Mm in the north–south direction nestled in the southern penumbra of the main sunspot of AR9608.

This image coincides closely with three white-light flare kernels that appear in the sunspot penumbra. The close spatial correspondence between white-light and acoustic emission allowed the authors to assume that the acoustic emission was driven by heating of the lower photosphere (Donea *et al.*, 2006). The authors suggested that in the case where the direct heating of the low photosphere by protons or high-energy electrons was unrealistic, the strong association between the acoustic source and cospatial continuum emission could be regarded as evidence supporting the backwarming hypothesis (Donea *et al.*, 2006), in which the low photosphere is heated by high-energy radiation from the overlying chromosphere.

Flare of 10 March 2001 On 10 March 2001 active region NOAA 9368 produced an unusually impulsive solar flare of class M5.6 in close proximity with the solar limb also associated with a type II radio burst and coronal mass ejection. So far it is the least energetic flare in soft X-rays from all those associated with sunquakes. Holographic analysis of the flare shows a compact acoustic source strongly correlated with impulsive HXR, visible continuum, and radio emission (Beşliu-Ionescu *et al.*, 2006b).

TD diagrams of seismic waves emanating from the flare region also show faint signatures, mainly in the eastern sector of the active region. The strong spatial coincidence between the seismic source and the impulsive visible continuum emission reinforces the theory that a substantial component of the seismic emission seen is a result of sudden heating of the low photosphere associated with the observed visible continuum emission. Furthermore, the low-altitude magnetic loop structure inferred from potential field extrapolations in the flaring region suggests that there is a significant inverse correlation between the seismicity of a flare and the height of magnetic loops, in which particle beams precipitate from the corona.

Flare of 6 April 2004 An acoustic source was also detected in active region AR9415 during the X5.4-class flare. The holographic seismic signature of this flare at 6 mHz was very strong and revealed a double seismic source evolving for about 15 min and associated with the footpoints of coronal loops (Beşliu-Ionescu *et al.*, 2005). The seismic sources were located in the penumbrae of the leading sunspots of AR9415 and were spatially extended rather than dotlike sources. The northern seismic source at the egression power maps for 6 mHz was found lagging the southern one by approximately 3 min (Beşliu-Ionescu *et al.*, 2005).

Flare of 23 July 2002 Analysis of the helioseismic effects in the photosphere observed by MDI/SOHO and high-energy images from RHESSI taken during the solar flare of 23 July 2002 occurred close to the solar limb at a latitude of $\sim 73°$ show that they are closely associated with the sources of HXR emission but not in the locations of γ-ray emission (Beşliu-Ionescu *et al.*, 2006b; Kosovichev, 2007b). Kosovichev (2007b) suggested that this demonstrates that hydrodynamic and helioseismic responses (sunquakes) are more likely to be caused by a series of multiple impulses of high-energy electrons moving in the photosphere at supersonic speed rather than by high-energy protons. These moving sources are assumed to play a crucial role in the formation of the anisotropic wave front of sunquakes observed for this flare (Kosovichev, 2007b).

12.3
Sunquakes Associated with the Flare of 28 October 2003

The X17.2 flare occurred on 28 October 2003 in the very active region NOAA 10486 at the location 18E20S. It started as observed by GOES from 9:41 UT and lasted until 11:24 UT with the maximum at 11:10 UT in soft X-rays and until 18:00 UT in H_α emission. The active region NOAA 10486 had a complex delta sunspot and produced dramatic flare activities in the descending phase of solar cycle 23 with three X-class flares, that is, an X17 flare on 28 October 2003, an X10 flare on 29 October 2003, and an X8.3 flare on 2 November 2003, along with many weaker ones (Wang *et al.*, 2005).

There were three sunqukes detected as ridges in TD diagrams (Kosovichev, 2006; Zharkova and Zharkov, 2007; Zharkova *et al.*, 2005b) and four sunquakes detected with the holographic method (Donea and Lindsey, 2005) with the seismic emission at frequencies from 3 to 7 mHz as described below in Section 12.3.2. Let us investigate below in detail their timing and locations.

12.3.1
Hard X-Ray, γ-Ray Emission, and Accelerated Particles in the Earth's Orbit

The flare of 28 October 2003 started at before 11:00 UT and was observed only by KORONAS (Arkhangelskaja *et al.*, 2006) and INTEGRAL (Gros *et al.*, 2004; Kiener, 2005; Tatischeff *et al.*, 2005) while the RHESSI instruments did not start their observations until 11:06:00 UT. The passage through the polar orbit of the KORONAS satellite began at 11:00 UT, which made it possible to have about 20 min of uninterrupted observations of HXR and γ-ray emission of the flare and to register high-energy particles in the Earth's orbit. The flare produced a very strong CME and interplanetary shock wave whose onset time was 11:01:39 UT (Kuznetsov *et al.*, 2006).

The flare light curves in γ-rays measured by the SONG instrument aboard KORONAS are plotted in Figure 12.5 (courtesy of the KORONAS team) (Arkhangelskaja *et al.*, 2006). Similar light curves were observed in the γ-continuum (2.8–3.7 and 7.6–10 MeV) and γ-lines (2.22, 4.44, and 6.13 MeV) by the instruments aboard the INTEGRAL satellite (Figure 5 in Gros *et al.*, 2004). These light curves revealed three distinct phases in the flare evolution: a short impulsive one, phase A (under 1 min), with a sharp increase of the continuum emission in both channels with photons detected up to 15 MeV, the SPI energy limit; and two longer phases B and C, with a further sharp increase of the 2.8–3.7 MeV continuum emission (phase B) and a much smoother increase of the line emission in all three lines (phase C).

The flare HXR emission was observed by KORONAS (Arkhangelskaja *et al.*, 2006) and INTEGRAL (Gros *et al.*, 2004; Kiener, 2005; Tatischeff *et al.*, 2005) from the very start at 11:01:00 UT, while the RHESSI started observations only at 11:06:00 UT (Hurford *et al.*, 2006). The images of the sources of HXR (200–300 keV) and γ-ray emission (2.2 MeV) obtained by RHESSI (see Figure 12.6) reveal that at least after 11:06:00 UT, when RHESSI started to observe, there were two footpoints with HXR and γ-ray sources, which more or less have close spatial locations (Hurford *et al.*, 2006). The spectral indices of the proton energy spectra deduced from the ratios of ^{12}C and ^{16}O lines observed in phase B, or after 11:06:00 UT, from INTEGRAL vary from 3 to 3.8 (Kiener, 2005; Tatischeff *et al.*, 2005). This is close to those of 2.8 ∓ 0.4 reported also after 11:06:00 UT by the RHESSI measurements from the deexcitation line 2.22 MeV and positron annihilation line 511 keV with a total number of protons of about 10^{33} (Share *et al.*, 2004).

Hence there are ample indications from the high-energy emission observations in this flare about a few events of particles arriving at the loop footpoints at various

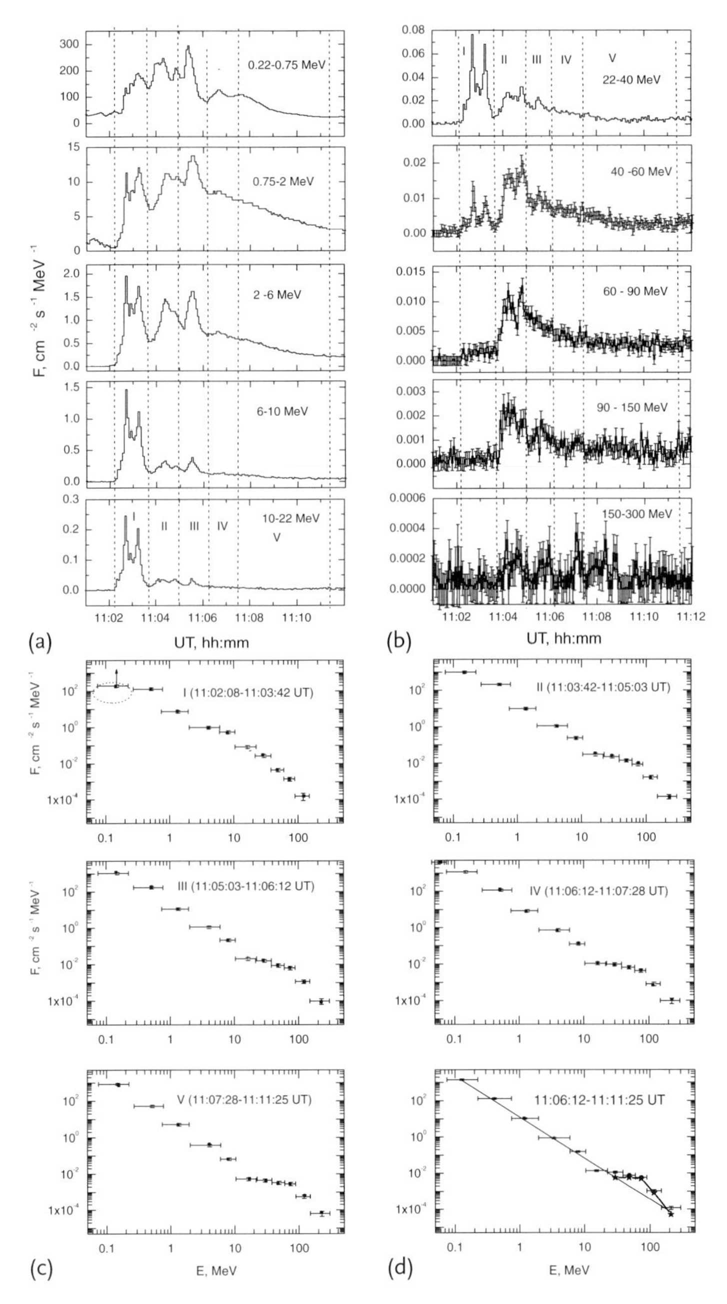

Figure 12.5 Light curves (a,c) and differential spectra (b,d) in different energy bands obtained by SONG (Solar Neutron and Gamma-ray) instrument aboard KORONAS during the whole flare duration. Courtesy of Dr. V. Kurt, Moscow University, and the KORONAS team (Arkhangelskaja *et al.*, 2006).

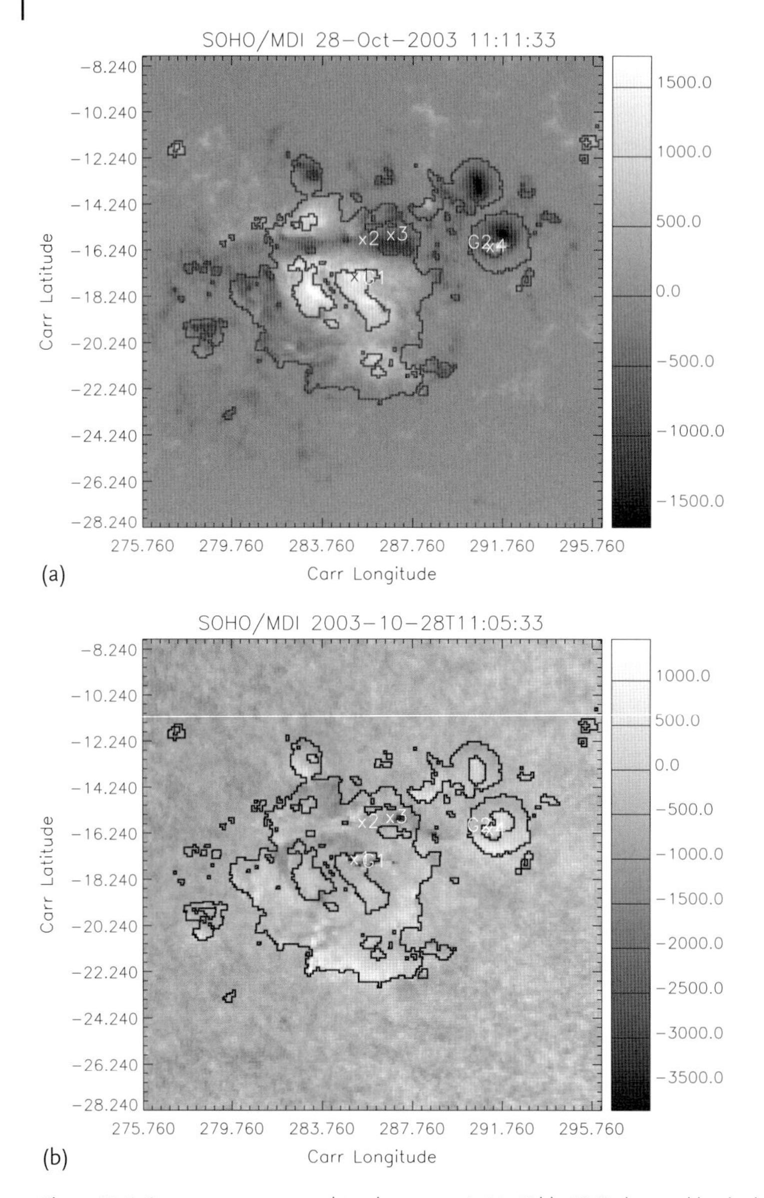

Figure 12.6 Sunspot group over plotted onto an MDI magnetogram 11:11:33 UT (a) and on a dopplergram (b) with the locations of the seismic sources marked by X1–X4 (columns 4–6 in Table 12.1) detected by the holographic method (Donea and Lindsey, 2005) with the γ-ray sources observed by RHESSI marked as G1 and G2 (Hurford *et al.*, 2006).

times starting from 11:02:00 UT until 11:07:00 UT. In order to establish a connection between high-energy particles and seismic source agents, we can now look in more detail at the velocities of vertical and horizontal displacements, or the ridges, associated with these seismic waves in the flare of 28 October 2003.

12.3.2
Observed Seismic Sources

For detection of the seismic sources we use the 1 min cadence MDI dopplergrams starting from 11:00:00 UT until 13:00:00 UT. Since the typical oscillation frequencies associated with quakes are higher than the background oscillations (3 mHz) (Donea and Lindsey, 2005; Kosovichev and Zharkova, 1998), the frequency filtering centered at 5–6 mHz is applied in order to increase the signal-to-noise ratio.

In order to detect seismic wave centers precisely, we selected areas of 20×20 pixels around the locations provided by Donea and Lindsey (2005) (Table 12.1, the last three columns). Then for every four sources (X1–X4) from Table 12.1 we selected a center of ripple location in the square of 20×20 pixels around the location provided by the holographic method and built TD ridges. This resulted in 400 TD diagrams, which were then visually investigated for the pixel locations, with the detectable ridges denoting the seismic wave propagation that is considered to belong to the source. The summation of these pixels with the detectable ridges were then used to define the total areas for each quake.

In general, Zharkova and Zharkov (2007) have detected 11 locations in MDI dopplergrams with downward motions larger than $1\ \mathrm{km\,s^{-1}}$ (Figure 12.7) using the automated technique (Zharkov *et al.*, 2005), while only three of them, marked as sources S1, S2, and S3, have revealed detectable ridges, or quakes (Figure 12.8a,c,e). The deduced locations for seismic sources S1, S2, and S3 with detectable ridges are summarized in Table 12.1 (columns 1–3) comparing them to sources X1–X4 by Donea and Lindsey (2005) (columns 4–6).

The downward velocities V_{down} directly measured for these three sources from the dopplergrams were $2.15\ \mathrm{km\,s^{-1}}$ for S1, $2.0\ \mathrm{km\,s^{-1}}$ for S2, and $1.75\ \mathrm{km\,s^{-1}}$ for S3, while for the other 8 of 11 sources the downward velocities were lower than $1.6\ \mathrm{km\,s^{-1}}$ (Table 12.1 and Figure 12.7). As seen from Figure 12.9, the durations of these downward motions do not exceed 1.5–2.0 min, starting at or after 11:05:00 UT, with the maximum a minute later and then decreasing back to the preflare magnitude for another 30–60 s.

The TD diagrams for the detected sources S1–S3 with theoretical ray paths (solid white lines) are presented in Figure 12.8. The start times of the seismic waves in the sources are slightly different, varying from about 11:05:00–11:06:00 UT for

Table 12.1 Heliographic locations L_0 and B_0 of the three seismic sources detected by the TD technique (columns 1–3) and four seismic sources (columns 4–6) in the 28 October 2003 flare reported using the holographic method (Donea and Lindsey, 2005).

Our source	L_0°	B_0°	DL05	L_0°	B_0°
S1	287.28	−15.96	X3	287.05	−15.78
S2	284.72	−17.62	X1	285.45	−17.61
S3	291.00	−16.64	X4	291.46	−16.43
			X2	285.86	−16.01

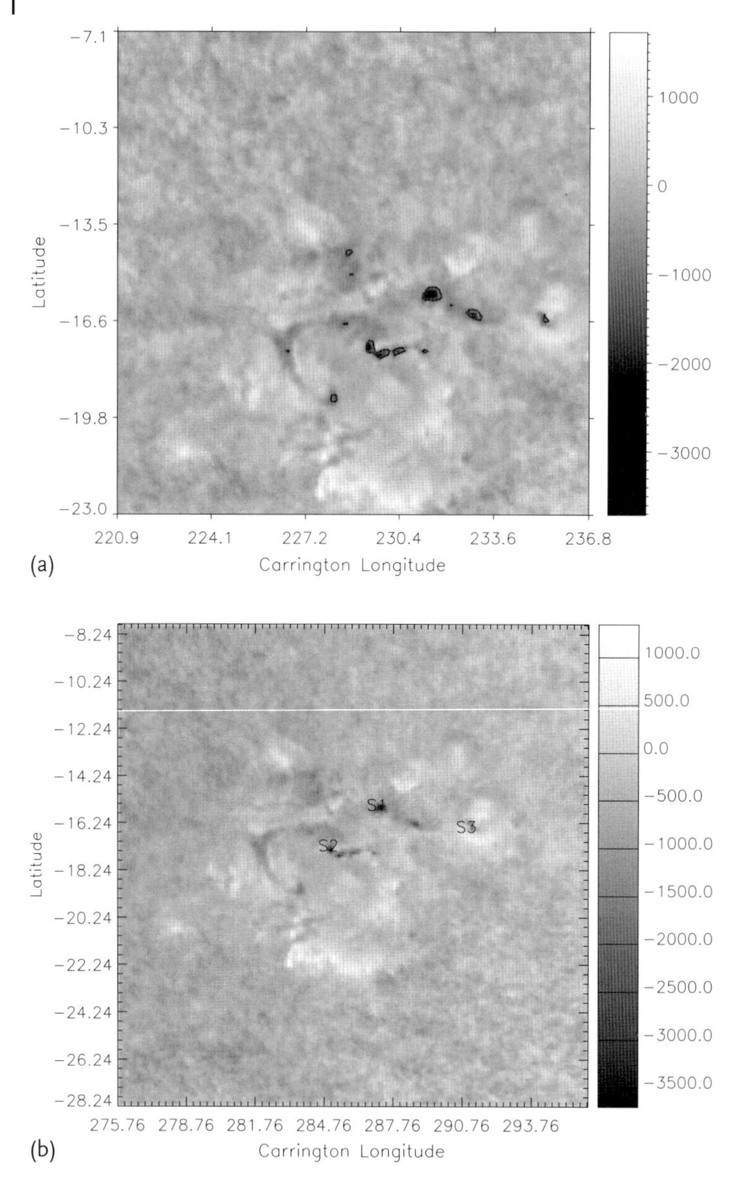

Figure 12.7 Locations of 11 Doppler sources (red contours) with downward motion higher than $1\,\mathrm{km\,s^{-1}}$ (a) but only three of them with detectable seismic responses in locations specified in columns 1–3 of Table 12.1 (b). For a color version of this figure, please see the color plates at the beginning of the book.

source S1 to 11:06:00 UT for sources S2 and S3. These times are close to the second maximum in both HXR and γ-rays, coinciding with high-energy protons (> 200 MeV) (S2, S3) (Arkhangelskaja *et al.*, 2006; Share *et al.*, 2004).

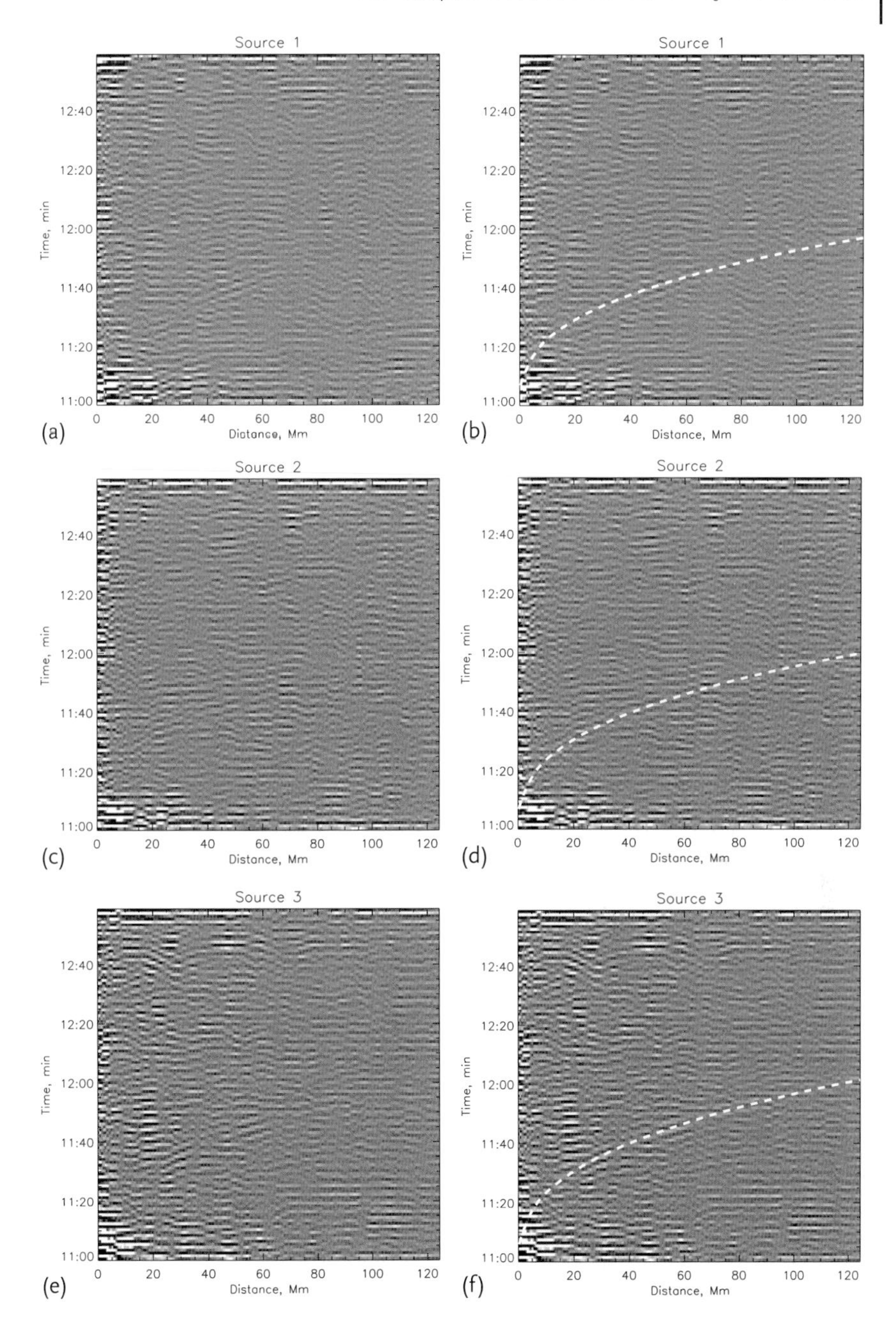

Figure 12.8 Frequency-filtered TD diagrams for seismic sources: S1 without ray path (a) and S1 with ray path (b), similar to sources S2 (c,d) and S3 (e,f).

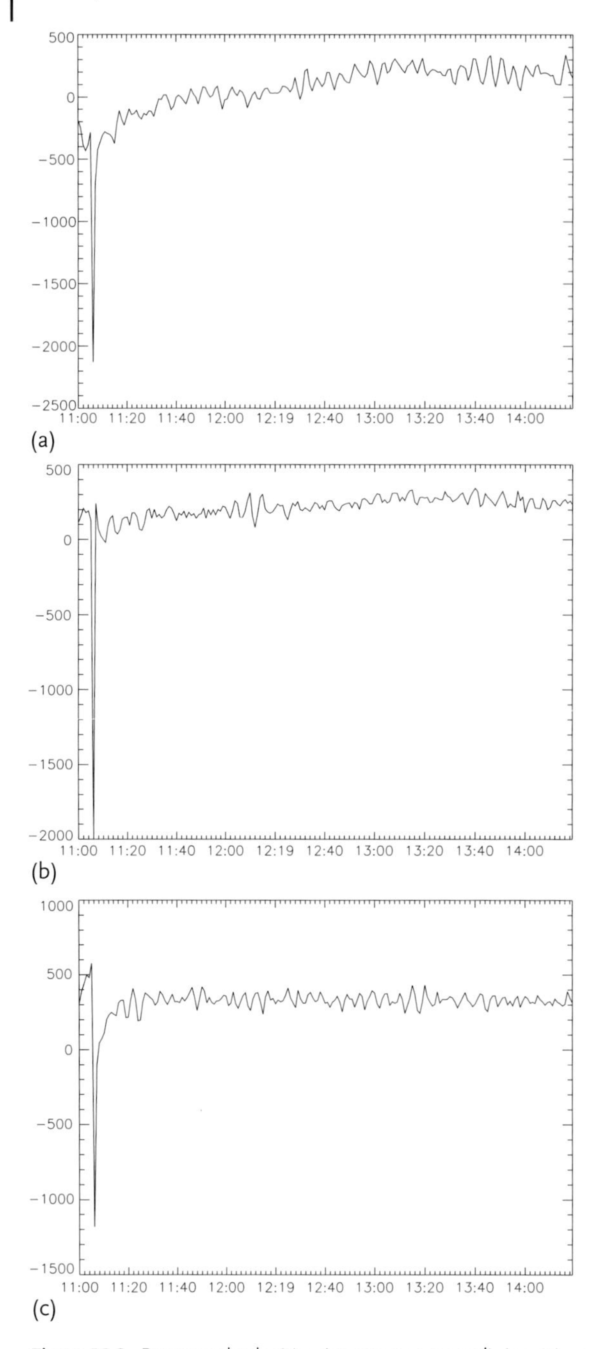

Figure 12.9 Downward velocities (meters per second) (y-axis) vs. time (minutes) (x-axis) in sources S1 (a), S2 (b), and S3 (c) in the center-of-gravity locations with the downward Doppler motions depicted in Figure 12.7b and Table 12.1.

Table 12.2 Areas A, downward (vertical) velocities V_{vert}, duration of observed downflow motions in source location $T_{impulse}$, average horizontal velocity V_{horiz}, and *Momenta* deduced for 28 October 2003 flare from TD diagram for three seismic sources S1–S3 (Figure 12.8).

Our sources	area, A, cm^2	V_{vert}, cm s^{-1},	$T_{impulse}$, s	V_{horiz}, cm s^{-1}	Momenta, g cm s^{-1}
S1	5.05×10^{17}	2.15×10^5	90	40.8×10^5	4.1×10^{22}
S2	3.34×10^{17}	2.00×10^5	70	38.0×10^5	3.8×10^{22}
S3	2.22×10^{17}	1.75×10^5	70	35.4×10^5	3.1×10^{22}

Source S1 (Figure 12.8a,b), corresponding to source X3 (Donea and Lindsey, 2005), reveals the highest horizontal velocity, about $42\,\mathrm{km\,s^{-1}}$, at the very start that increased in 50 min up to $140\,\mathrm{km\,s^{-1}}$ at 120 Mm. The ridge in source S2 (Figure 12.8c,d), or source X1 in Donea and Lindsey (2005), was slightly smoother compared to source 1, that is, the seismic wave was slower, its velocity varied from 38 to $128\,\mathrm{km\,s^{-1}}$ in 53 min. The ridge in source S3 (Figure 12.8e,f), source X4 in DL05 shows the lowest-velocity variations from 34 to $114\,\mathrm{km\,s^{-1}}$ in 60 min.

The areas of the seismic sources, defined by the presence of downward motions and by detectable ridges, were about $5.05 \times 10^{17}\,\mathrm{cm^2}$ for source S1, $3.34 \times 10^{17}\,\mathrm{cm^2}$ for source S2, and $3.22 \times 10^{17}\,\mathrm{cm^2}$ for source S3. Then, using the downward velocities above (Table 12.2), the momenta required to cause the observed seismic waves are as follows: $4.2 \times 10^{22}\,\mathrm{g\,cm\,s^{-1}}$ (S1), $4.0 \times 10^{22}\,\mathrm{g\,cm\,s^{-1}}$ (S2), and $3.7 \times 10^{22}\,\mathrm{g\,cm\,s^{-1}}$ (S3).

The locations of seismic sources S1–S3 found from TD diagrams are slightly different than those reported from the holographic method (Donea and Lindsey, 2005), which can be the result of different sensitivities of the techniques applied. The locations of sources S1–S3 coincide with the dark spots inside the sunspot umbras, while there are no indications for the fourth source in this area from the umbra locations. However, the TD technique does not produce distinct ridges for source X2, which was seen after 11:07:00 UT (Donea and Lindsey, 2005). Moreover, there is no downward motion at velocities higher than $1\,\mathrm{km\,s^{-1}}$ detected in the vicinity of source X2, although this source is close to an extended area of source S1.

This can mean that source X2 is either weak and, thus, not detected by the TD technique or it is an artifact of the holographic technique, like the interference of seismic waves produced by other sources (possibly S1 and S2). The question arises, then, as to why the seismic emission is not detected in the other 7 of 11 sources with downward velocities in Figure 12.7 if the holographic technique allows us to detect this weak source. The answer to this question is yet to be found.

12.3.3
Comparison of Momenta Delivered by Beams and Hydrodynamic Shocks

The theoretical calculations of momenta delivered by beams of different kinds and hydrodynamic shocks are presented in Chapter 6. Here we will estimate the momenta from observations and compare them with the theoretical momenta delivered by electron or proton beams, or by shocks caused by them in the ambient atmosphere.

12.3.3.1 Momenta Estimated from Dopplergrams

Let us apply this evaluation to electron or proton beams whose parameters are deduced from the HXR and γ-ray observations by CORONAS, INTEGRAL, and RHESSI (Arkhangelskaja *et al.*, 2006; Hurford *et al.*, 2006; Share *et al.*, 2004).

Source 1 The HXR photon spectra associated with source 1 had an upper energy of 50 MeV and a lower-energy cutoff of about 10 keV. The photon spectral index in this plot was about $\delta_{\text{high}} = 3.5{-}4.0$ in energy range of 70 keV to 60 MeV and $\delta_{\text{low}} = 1.5$ for an energy range of 10–70 keV (Figure 12.5). Since this was a very strong flare, we assumed that the precipitating beam had induced a very strong electric field and its HXR emission was dominated by ohmic energy losses (Zharkova and Gordovskyy, 2005a, 2006).

This allows us to deduce that at higher energies the spectral index γ of a precipitating electron beam must be nearly the same as the one δ_{high} in the mean spectrum above. However, if the beam was dominated by collisions, then the beam spectral index γ is higher by one than the mean flux index δ_{high}. So we can consider these two opposite cases. The difference between spectral indices δ_{high} and δ_{low} can provide us with the beam initial energy flux F_0 for a selected spectral index of an electron beam, that is, for beams with γ varying from 3.5–5, the index difference varies from 2.0–2.5.

Hence, from Figure 12 in Zharkova and Gordovskyy (2006) one can deduce the initial energy flux of beam electrons that can vary from $3 \times 10^{12}\,\text{erg}\,\text{cm}^{-2}\,\text{s}^{-1}$ for $\gamma = (3.5{-}4) \times 10^{11}\,\text{erg}\,\text{cm}^{-2}\,\text{s}^{-1}$ for $\gamma = 5$ (Zharkova and Gordovskyy, 2005a). Let us calculate the momenta delivered to the photosphere by electron beams with such parameters using a technique similar to that of Gordovskyy *et al.* (2005). For a flare area of about $10^{17}\,\text{cm}^2$ defined by the area of downward Doppler motion in Figures 12.7 and 12.9 the momentum delivered to this area by electron beam was about $2 \times 10^{22}\,\text{g}\,\text{cm}\,\text{s}^{-1}$.

The momentum delivered by the hydrodynamic (HD) shock caused by this beam can be calculated by solving the HD problem using the updated heating functions as described in Gordovskyy *et al.* (2005). This allows us to estimate the momentum delivered by such a shock to be about $8 \times 10^{21}\,\text{g}\,\text{cm}\,\text{s}^{-1}$ within a time scale of 2 min after the flare onset at 11:01 UT.

Sources 2 and 3 Sources 2 and 3 correspond to locations with both HXR and γ-ray emission observed by RHESSI (Share and Murphy, 2006). The total number of protons responsible for the observed emission is found to be about 10^{33} particles, and the lowest energy recorded is about 30 MeV, or $\approx 4.8 \times 10^{-5}$ erg. Then the momentum delivered by such a proton beam will be about 2.2×10^{22} g cm s^{-1}. For a lower cutoff energy of 2 MeV this momentum at the top of the loop can be increased to 5.6×10^{24} erg cm s^{-1}. Obviously, the momentum estimated in this proton source is comparable to those measured from the TD diagrams in Section 12.3.2.

12.3.3.2 Comparison with Momenta from Seismic Responses

Let us try to establish the agents delivering the momenta reported in Table 12.2 by investigating the parameters of high-energy particles associated with each source. Sources S2 and S3 correspond to the locations in HXR and γ-ray emission observed by RHESSI (Share *et al.*, 2004) that are slightly dislocated but still coincide with a few of the 11 downward sources reported in Figure 12.7 (Section 12.3.2) that can be explained by the full trajectory asymmetries at electron and proton acceleration in this flare (Zharkova and Gordovskyy, 2004, 2005c).

The spectral indices of the proton energy spectra observed after 11:06:00 UT in phase B by INTEGRAL (from ratios of ^{12}C and ^{16}O lines) vary from 3 to 3.8 (Kiener, 2005; Tatischeff *et al.*, 2005) or by the RHESSI measurements (from the deexcitation line 2.22 MeV and positron annihilation line 511 keV) vary 2.8 ∓ 0.4 with a total number of protons observed of about 10^{33} (Share *et al.*, 2004) and a lowest energy of about 30 MeV, or $\approx 4.8 \times 10^{-5}$ erg. Then the momentum delivered by such a proton beam is about 4.3×10^{22} g cm s^{-1}, which is close to those derived in sources S2 and S3 (Table 12.2). Hence protons can be the agents carrying a sufficient momentum in these two sources if it can be transferred to hydrodynamic shocks.

For source S1 there are no HXR or γ-ray RHESSI images that would indicate either earlier energy deposition in this location prior to the RHESSI payload's being deployed or to lower energies of the agents delivering the momentum. A higher upper cutoff energy in the photon spectra point to the presence of some energetic protons alongside electrons appearing upon acceleration by an electric field in reconnecting current sheets (RCSs) when the trajectory asymmetries are partial for particles with opposite charges (Zharkova and Gordovskyy, 2004, 2005b,c).

12.3.3.3 Momenta Carried by Hydrodynamic Shocks

Since the downward motion in source S1 occurred no later than 11:05:00 UT and was the strongest of all the sunquakes, Zharkova and Zharkov (2007) suggested that the required momentum in S1 was delivered by a hydrodynamic shock caused by either high-energy electrons observed by KORONAS in the first X-ray maximum or by lower-energy protons ($<$ 200 MeV) accompanying these electrons. The electrons should have the upper cutoff energy of 100 MeV and a lower cutoff energy of about 18 keV (Arkhangelskaja *et al.*, 2006), while the protons could be thermal protons of separatrix jets (Gordovskyy *et al.*, 2005) that do not produce noticeable

Table 12.3 Momenta gained by various types of beams during their acceleration in a current sheet.

	Fast beam		Slow beam	
	Electrons	Protons	Electrons	Protons
Energy flux, $erg\,cm^{-2}\,s^{-1}$	1.8×10^{11}	2.3×10^{13}	2×10^{8}	4×10^{11}
Momentum flux, $g\,cm^{-1}\,s^{-2}$	$< 10^{2}$	1.6×10^{4}	$< 10^{-2}$	10^{3}

HXRs and γ-rays but still carry a large momentum and can cause a very strong hydrodynamic shock (see Table 12.3).

If the agents are subrelativistic electrons with indices of 1.5 below 100 keV and 4.5–5.0 above, then their initial energy flux can vary from $8 \times 10^{11}\,erg\,cm^{-2}\,s^{-1}$ for $\gamma = (3.5–4) \times 10^{11}\,erg\,cm^{-2}\,s^{-1}$ for $\gamma = 5$ (Siversky and Zharkova, 2009b; Zharkova and Gordovskyy, 2005a). The hydrodynamic response caused by electron beams with $\gamma = 3, 5$ was calculated in Chapter 6 (Figures 6.2 and 6.3).

As can be seen from Figure 6.7, even a powerful electron beam with a spectral index of about 3.5 has a much lower momentum ($8 \times 10^{21}\,g\,cm\,s^{-1}$) and produces a hydrodynamic shock much higher in the upper chromosphere. Thus, its travel time to the photosphere will be about 250 s (compare the depth and densities of the lower-temperature condensations plotted in the lower graphs in Figure 6.7) before it reaches the photosphere and initiates the quake in S1.

A hard proton beam of moderate power ($10^{12}\,erg\,cm^{-2}\,s^{-1}$ in the graph in Figures 6.7 and 6.8) can deliver a higher momentum ($> 4.5 \times 10^{22}\,g\,cm\,s^{-1}$) and form a shock much deeper in the lower chromosphere than for an electron beam. Hence the shock travel time (< 60 s) to the photosphere can be close to the onset times of HXRs and γ-ray emission as measured in the TD diagrams (Figure 6.8).

Therefore, in the flare of 28 October 2003, 11 sources were detected with downward motions higher than $1\,km\,s^{-1}$; 3 of them (S1–S3) have seismic waves detected with the TD diagram technique. The 3 new seismic sources started around 11:05:00–11:06:00 UT and had slightly different heights of ridges (or different horizontal velocities), pointing to the different agents producing these seismic responses. The seismic sources are assumed to be caused by precipitating protons mixed with electrons, both occurring owing to the trajectory asymmetries at acceleration by a super-Dreicer electric field in RCSs (Zharkova and Gordovskyy, 2004, 2005b,c).

12.4 Seismic Sources Observed by GONG in 14 December 2006 Flare

In this study Matthews *et al.* (2011) used high-resolution multiwavelength signatures of an X-class solar flare accompanied by a sunquake, including the first detection of a quake in GONG data by TD methods.

12.4.1 Flare Morphology and Evolution

The flare on 14 December 2006 originated from AR 10930 and occurred at approximately 22:00 UT. The X-ray flux for the event peaked at the GOES X1.5 level around 22:12 UT. The event was observed by all instruments on the *Hinode* spacecraft (Kosugi *et al.*, 2007) by the Solar Optical Telescope (SOT, Tsuneta *et al.*, 2008). The SOT observed the flare throughout its duration with the Broad-band Filter Imager (BFI) in the G-band and Ca II H line, with a 2 min cadence and 0.1″ resolution. The Narrow-band Filter Imager (NFI) similarly provided Stokes I and V measurements with an approximately 2 min cadence and 0.16″ resolution in the Fe I 6302 Å line. Standard corrections were made to the data for CCD gain, readout defects, dark current, and pedestal. The data were aligned using cross-correlation and subpixel registration on a large field of view (FOV) similarly to the method described in Gosain *et al.* (2009). The alignment was verified by making running difference images, which showed random orientation of dipolar features in the FOV.

RHESSI observed the flare from its beginning until approximately 22:25 UT. THE CLEAN algorithm was used to produce images in 77 energy bins over five 1 min time intervals from 22:08:26–22:13:26 UT in order to perform imaging spectroscopy in 1 keV bands from 3 to 40 keV, 3 keV bands from 40 to 100 keV, 5 keV bands from 100–150 keV, and 10 keV bands from 150–250 keV. The authors then fitted the spectrum using a combination of a thermal-plus-thick-target model. For spatial comparison with the WL emission, photospheric magnetic field, and egression power, the authors produced images using the Pixon algorithm at 1 min intervals from 22:08 to 22:12 UT.

GONG observations during the flare of 14 December 2006 were made in the photospheric Ni I 6768 Å line with a 1 min cadence. The observations included full disk dopplergrams, line-of-sight (LOS) magnetograms, and intensity images. The pixel size of the intensity images is 2.5 arcsec. In this study we made use of the dopplergrams and intensity images to analyze the acoustic signatures of the flare through both the TD diagram technique (TD method; Kosovichev and Zharkova, 1998) and acoustic holography (e.g., Lindsey and Braun, 2000). To compensate for atmospheric distortion the dopplergram cleaning procedures outlined in Lindsey and Donea (2008) were applied.

12.4.2 Photospheric and Chromospheric Signatures for the 14 December 2006 Flare

The flare exhibits extended ribbon emission that propagates through the penumbrae of both the northern and southern spots. This emission is seen in multiple wavelengths displaying the temporal and spatial evolution of the flare in the G-band, the Fe I 6302 Å line, the Stokes V component of the Fe I 6302 Å line, the Ca II H line, and the TRACE WL channel (Matthews *et al.*, 2011). As can be seen from their Figure 1, the morphology and location of the emission seen in the G-band and Fe I Stokes I and V components is almost identical. The emission in the

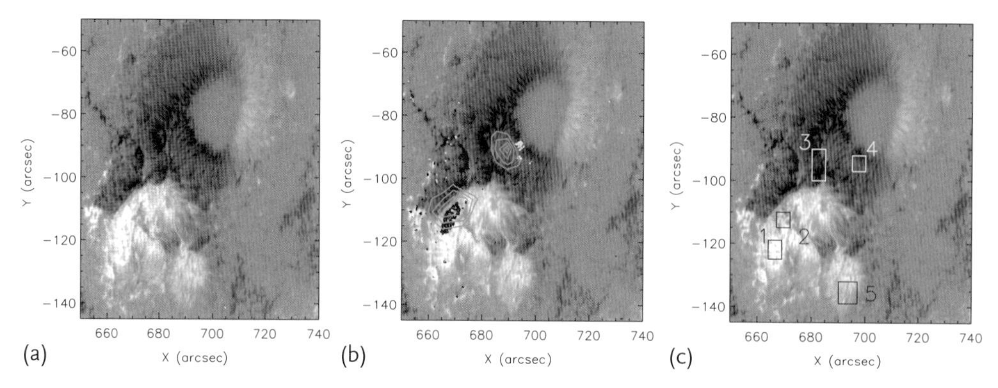

Figure 12.10 Longitdinal magnetic field image showing the location of 50–100 keV HXR emission seen by RHESSI (red contours), the magnetic reversals (black and white contours), and the regions where magnetic variations and photospheric velocities were measured. Magnetic variations were measured in regions 1–4 and shown in Figure 12.11. Horizontal photospheric velocities were measured in all five regions, with region 5 providing a reference location outside of the main flare disturbance. For a color version of this figure, please see the color plates at the beginning of the book.

G-band is somewhat stronger than in Fe I Stokes I and V, but clear reversals of the field are seen in the Stokes V component, which can be used as a proxy for the LOS magnetic field. The Ca II H line difference images show that the photospheric emission forms a more compact subset of the chromospheric emission, but that it is spatially coincident, forming two loops whose magnetic interaction is likely to have initiated the flare.

Figure 12.10 shows the temporal evolution of the flare in the G-band, Fe I Stokes I and V, and Ca II H in the regions within the boxes. The intensity is normalized by area, and while it is clear that the G-band intensity in the south is on the order of twice that in the north, the evolution in the north is more impulsive and occurs over a more compact region. Four locations were found with magnetic field transients (Figure 12.10c), three of which have a close spatial correspondence with the locations of HXR and white-light emission, while the fourth one (located in the northeast) did not show any spatial relations with either of these. Furthermore, the temporal evolution of the magnetic field in these sources (Figure 12.11) was substantially different showing clear magnetic transients in sources 1 and 4 that were cospatial with HXR emission, a smaller magnetic transient in source 2 combined with the step-type magnetic field variations, while in source 3 only step-type magnetic changes are recorded.

12.4.3 Photospheric Velocities

Matthews *et al.* (2011) take a sequence of Stokes V images captured in the Fe I 6302 Å line in the period from 22:00 to 23:00 UT. Using the whole FOV they registered this image series to the first image in the time series using cross-correlation

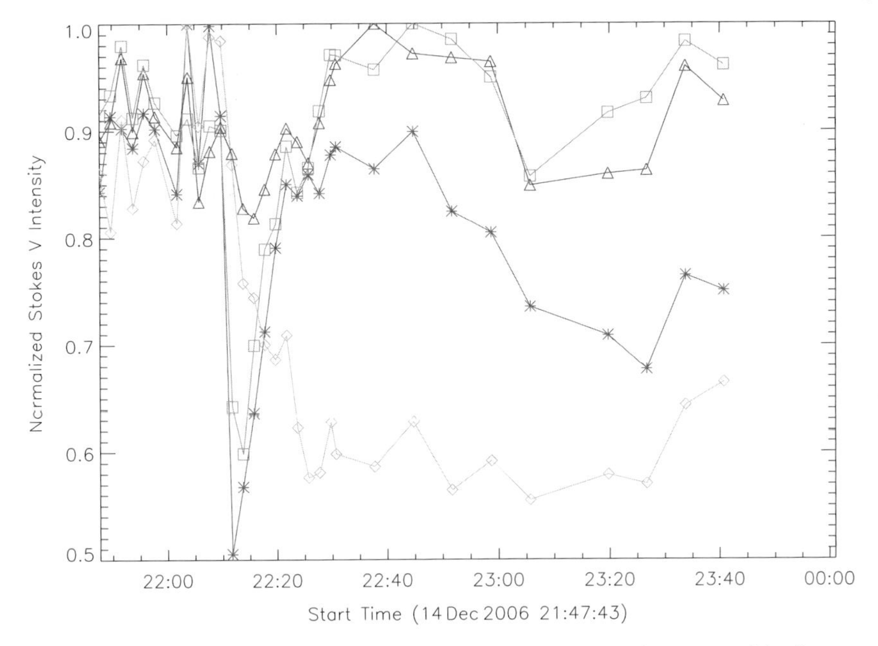

Figure 12.11 Normalized Stokes V intensity (longitudinal magnetic field) in regions 1 (red asterisks), 2 (green diamonds), 3 (blue triangles), and 4 (orange squares) as indicated in Figure 12.10. For a color version of this figure, please see the color plates at the beginning of the book.

and subpixel interpolation (e.g., November and Simon, 1988). Having performed the global alignment of the time sequence, Matthews *et al.* (2011) then looked at frame-to-frame changes in small regions centered on the locations of the largest changes associated with the magnetic changes, HXR and acoustic sources, as indicated in Figure 12.10. They also selected a region in the southwest of the active region away from the flare emission as a reference and used cross-correlation with subpixel interpolation between consecutive frames to determine the displacements Δx and Δy of the small-scale features within the boxes. These displacements are shown in Figure 12.12.

12.5
Observations of Solar Interior

Since there are no suitable SOHO/MDI observations available from the flare period, Matthews *et al.* (2011) have obtained intensity and dopplergram data from the GONG network (Harvey *et al.*, 1996). Both data sets used in their analysis consist of several hours of full-disk 1 min cadence intensity and LOS velocity observations starting at 22:00 UT on 14 December 2006. Each series was processed by tracking and derotating the region of interest centered on the active region using the Snod-

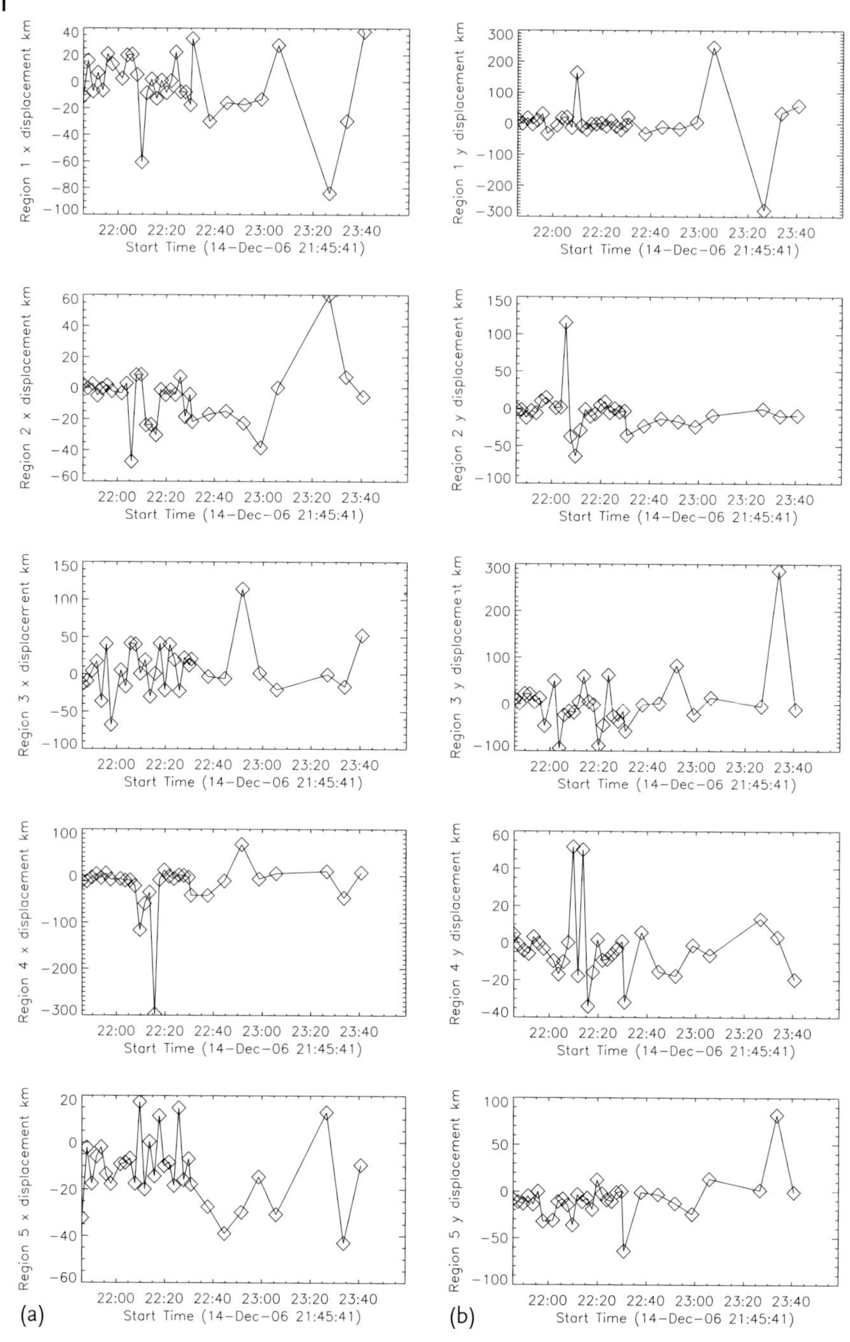

Figure 12.12 Horizontal displacements measured in regions shown in Figure 12.10: (a) x; (b) y.

grass rotation rate and then remapping the data onto a heliographic grid at 0.15° per pixel resolution. Also, in order to compensate for GONG's visibility condition, the authors used the cleaning procedures described in Lindsey and Donea (2008) for GONG intensity data and applied the parameters to correct the LOS velocity data.

To detect and analyze the solar quake associated with the flare, Matthews *et al.* (2011) used both TD analysis (Kosovichev and Zharkova, 1998) and acoustic holography (Donea *et al.*, 1999). In this work, Green's functions built using a geometrical optics approach were used; the pupil dimensions were defined as $a \approx 15$ Mm, $b \approx 55$ Mm, with the GONG continuum intensity data taken as $\psi(\mathbf{r}, t)$ for the egression power calculations.

12.5.1 Validation of Time–Distance Analysis with GONG

As we are not aware of the TD method's ever having been applied to GONG data, the authors tested the approach using the M9.5-type flare associated with active region NOAA 9608 that occurred on 9 September 2001 and two Halloween flares of 28 and 29 October 2003 (Zharkov *et al.*, 2011b). For these flares both GONG and SOHO/MDI (Scherrer *et al.*, 1995) LOS velocity data were available.

Both data sets were processed as described above, followed by the additional frequency filtering to select the signal in a 2 mHz band centered at 6 mHz. Using the fact that the ridge we were looking for corresponded to the first bounce acoustic wave packets normally seen in an interval of 15–150 Mm from the source, to improve the sensitivity of the technique for GONG data, we also applied a phase-speed-type filter designed to let through waves with phase speed, ω/k_h, between 20 and 90 km s^{-1}, unaffected by a Gaussian rolloff on each side.

The TD diagrams for the $m = 0$ component showed that the ridges representing the quakes are clearly seen in both the MDI and GONG data. Though less defined and obscured by significant noise contribution in GONG, the ridges were nevertheless definitely present for the flares of 28 October 2003 and 9 September 2001. Thus (Zharkov *et al.*, 2011b) conclude that GONG network data, though less sensitive to TD technique, can still respond to such analysis, that is, if ridges are seen in GONG data, they will also be observed using higher-quality satellite data.

12.5.2 Helioseismic Results

12.5.2.1 Egression Powers of Seismic Sources

Two GONG data sets are used in the analysis consisting of several hours of full-disk 1 min cadence intensity and LOS velocity observations starting at 22:00 UT on 14 December 2006 (Matthews *et al.*, 2011). Each series was processed by tracking and derotating the region of interest centered on the active region using the Snodgrass rotation rate and then remapping the data onto a heliographic grid at 0.15° per pixel resolution.

To compensate for the atmospheric contribution, Matthews *et al.* (2011) use the cleaning procedures described in Lindsey and Donea (2008) for GONG intensity data. Since both the intensity and velocity data come from the same instrument, the parameters extracted from the intensity series to correct the LOS velocity data were applied. In addition, the authors applied to the GONG velocity data an acoustic power map correction following the procedure described by Zharkov *et al.* (2011b) for egression power measurements.

The egression power snapshots computed from the corrected Dopplergram series are plotted in Figure 12.13 with the egression maps for two different times after

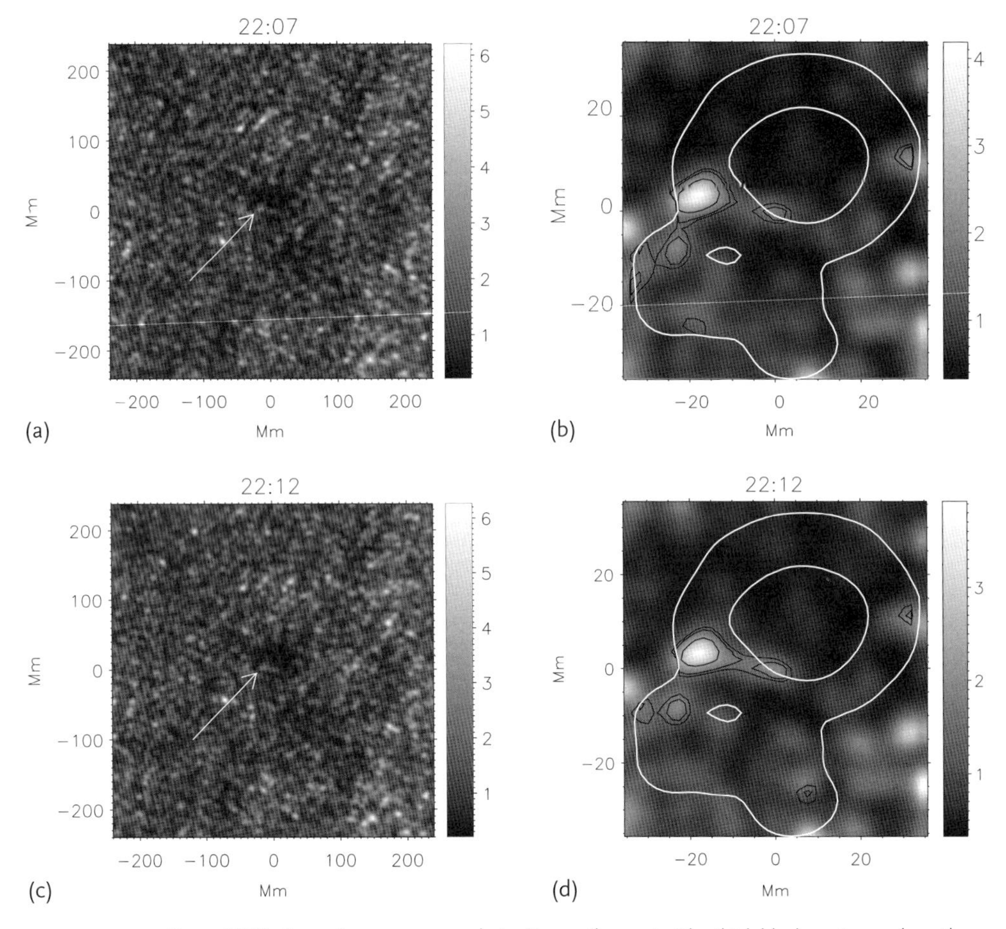

Figure 12.13 Egression power snapshots (4–6 mHz) (a,c) and close-ups (b,d) taken around 22:07 UT (a,b) and 22:12 UT (c,d) derived from GONG intensity data. The Carrington longitude is along the *x*-axis, latitude along the *y*-axis. The thick black contours show the egression power at 50, 70, and 90% of the quiet Sun power. The locations of four seismic sources coincide with those for four magnetic transients defined in Figure 12.10.

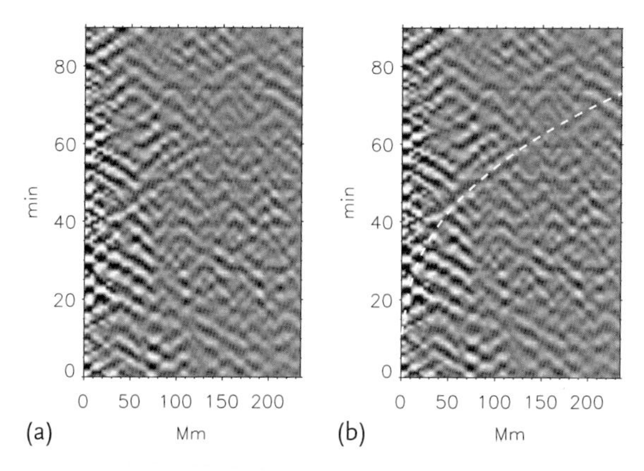

Figure 12.14 TD diagram obtained for the 14 December 2006 flare source 3 (a) with theoretical travel time ridge corresponding to $l = 1000$ overplotted using the white dashed line (b).

the flare onset (Figure 12.13a,c) with arrows indicating the location of the center where ripples measured from the TD diagram start (Figure 12.14) and the close-up of egression sources for the same times (Figure 12.13b,d). They show possible traces of acoustic activity at the four locations coinciding with the locations of fast magnetic field changes reported in Figures 12.10 and 12.11. It can be seen that the magnetic transients 1–4 and acoustic sources coincide rather well in space and in time with flare onset and duration.

Also Matthews *et al.* (2011) carried out a statistical validation of detected acoustic sources in comparison with background acoustic sources and proved the statistical significance of their egression sources showing that source 3 at 3 mHz has an egression power about 2.5 above the quiet Sun egression power, while at 6 mHz this cleared to 3.85. These maximum egression powers measured by GONG for this X1.1 flare are compatible or even higher than those detected in the MDI data for the upper M-class flares, for example, for the M9.5-class flare of 9 September 2001 (Donea *et al.*, 2006) or for the M6.5-class flare of 10 March 2001 (Martínez-Oliveros *et al.*, 2008a). This confirms that for the first time the authors detected from GONG data four seismic sources associated with the 14 December 2006 flare that are statistically significant.

12.5.2.2 **Time–Distance Diagram**

In order to detect ripples associated with acoustic waves occurring beneath regions 1–4 for the 14 December 2006 flare, first the authors searched for ridges in the TD plots in the locations of all regions showing magnetic changes. Time distance diagrams were computed from the GONG Dopplergrams at the locations 10×10 pixels around each region in the south and north. From 100 TD diagrams we searched for ridges corresponding to the propagation of a circular wave produced by a spherical seismic wave in the solar interior excited by the flare.

The only detectable TD ridge corresponding to a circular wave was found in the location of the largest egression power source 3 around 8.4° Carrington longitude and 5.25° latitude south (Figure 12.14a). The ridge is fitted reasonably well by the theoretical ridge (Figure 12.14b) corresponding to the spherical wave number $l = 1000$, which is the same as was observed in the first sunquake (Kosovichev and Zharkova, 1998), with the quake start time estimated at around 22:10 UT. There are some remnants of a ridge seen in TD diagrams for egression source 4, which we do not plot here. We are not able to reliably identify with the TD method any visible acoustic sources at the southern egression source locations 1 and 2, perhaps due to different physical conditions in these locations and the strong transient magnetic field variations and atmospheric contribution present in the GONG velocity observations (Lindsey and Donea, 2008; Zharkov *et al.*, 2011b).

Matthews *et al.* (2011) noted that the largest acoustic kernel seen in region 3 coincides with the steplike magnetic source 3 and with the location of the center of the TD source (Figure 12.14) and does not show any links to HXR of white-light sources, which is one of the very rare cases resembling the one reported recently for the flare of 15 February 2011 (Kosovichev, 2011), while the smaller northern egression source 4 coincides with the end of the white-light ribbon and HXR emission in a range of 40–100 keV.

The southern egression sources 1 and 2 are weak, but Matthews *et al.* (2011) also note that they are cospatial, with the locations of the strong magnetic changes seen in regions 1 and 2, which overlap spatially with the HXR emission locations for energies above 40 keV. Also the authors showed that egression sources 1–4 are closely correlated spatially with the ends (or footpoints) of the two flaring loops seen in Ca emission, which coincide temporally and spatially with the four locations of fast magnetic changes in this flare.

Hence, two of the magnetic sources (1 and 4) show transient-type magnetic field changes, also accompanied by HXR and white-light emission in all ranges, while the other two sources (2 and 3) show only the gradual steplike magnetic field changes and no white-light emission while source 2 shows signs of HXR emission below 40 keV.

12.5.2.3 **General Comments on Seismic Signatures in 14 December Flare**

Generally, there is a good spatial and temporal coincidence between HXR emission, photospheric emission as observed in the G-band, and the emission in the Fe I 6302 Å passband and transient changes in the longitudinal magnetic field. Matthews *et al.* (2011) found that photospheric emission is more compact than emission seen in the chromosphere, particularly in the TRACE WL and Ca II H observations, indicating that the energy deposition at this level occurs over smaller scales, which is consistent with findings by, for example, Fletcher *et al.* (2007) and Beşliu-Ionescu *et al.* (2005). In agreement with Watanabe *et al.* (2010), Matthews *et al.* (2011) found that the HXR emission and white light emission in G-band in this flare is asymmetric, with stronger emission appearing in the southern HXR source at the flare onset at 22:08 UT and then shifting to the North at later times.

The main sites of energy deposition in the lower atmosphere associated with this flare are sources 1 and 2 in the southern polarity of magnetic field and sources 3 and 4 in the northern one (Figure 12.10), possibly representing four footpoints of two interacting loops seen in the Ca II emission image. These sources depict fast transient variations of a magnetic field observed in locations 1 (south) and 4 (north), both associated with the two HXR sources at energies above 40–100 keV and fast step changes of the magnetic field in sources 2 (south) and 3 (north) from which source 2 is cospatial with HXR emission while source 3 does not show any links with either HXR or WL emission.

These four sites with magnetic variations are found to be correlated temporarily and spatially with the seismic sources detected with the holography method (Figure 12.13). The most extended egression power source appears in region 3, which is offset from the 40–100 keV HXR emission, while more compact egression sources appear in regions 1, 2, and 4, which are cospatial with HXR sources at 40–100 keV and white-light emission. The extended seismic source 3 in the north also revealed a noticeable ridge in the TD diagrams, indicating the presence of a circular ripple emanating from the central location of the egression map for this source. The cospatiality of seismic sources with the relevant magnetic sources leads the authors to believe they are real.

Observations from the SP by Hinode indicate that the underlying magnetic field strength is mixed within the regions where the four seismic sources are located, with the largest number of strong field concentrations ($>$ 3000 G) occurring predominantly in regions 2 and 4. The field inclination is also mixed within the vicinity of seismic sources, with the inclined field seen throughout the active region, but with more significant changes seen in regions 1–4 during the scan covering the time of the flare. This can be important for the generation of helioseismic signatures, as noted by Martínez-Oliveros *et al.* (2008b), since mode conversion will occur more effectively when the sound and Alfvén speeds coincide (Cally, 2000; Schunker *et al.*, 2008).

12.5.3 Summary of Observed Signatures in Sunquakes

The observations of sunquakes and seismic emission can be summarized as follows:

1. Many large X-class flares are acoustically inactive. For a few flares where seismic emission or ripples were detected, only a very small fraction, a few hundred thousandths, of the energy released during the flare is related to seismic emission that radiates into the subphotosphere or seismic waves propagating in the interior.
2. Much more seismic emission is detected in association with weaker M-class flares associated with rather hard photon spectra in HXRs including the most energetic seismic emission reported for the flare of 10 March 2001.
3. Seismic ripples and emission are always associated with strong, downward propagating shocks directly observed with MDI dopplergrams that start simultaneously with flare onset.

4. The shapes of seismic sources are often not circular but rather extended elliptic or even with slightly irregular shapes.
5. Seismic emission in M-class flares is often accompanied by the occurrence of white-light emission.
6. However, there are at least two flares (14 December 2006 and 15 February 2011) where well-detectable seismic sources are observed outside the locations of HXR and white-light emission.
7. Often there are multiple seismic sources in the same flare appearing in different surface locations either simultaneously or delayed by 3–5 min after the others.
8. The locations of egression power sources often coincide with the locations of fast magnetic field changes associated with a flare that is believed to indicate the footpoints of interacting loops.
9. The locations of these compact seismic sources often associated with flares can be associated with HXR emission and, occasionally, also with γ-ray emission (Halloween 2003 flares) in the footpoints of magnetic loops marking the precipitation of high-energy particles (beams or jets).
10. Sometimes there are also horizontal displacements (ripples) observed in the chromosphere (Moreton waves) and photosphere. This suggests that seismic flare signatures are driven by impulsive heating at the onset of a flare and linked to the hydrodynamic and magnetic responses of ambient plasma caused by these high-energy particles.

12.6 Theoretical Implications of Particle Kinetics and Dynamics Leading to Sunquakes

12.6.1 Topology of Particle Acceleration

Magnetic reconnection is well accepted as the sources of energy released in solar flares (Priest and Forbes, 2002). Current sheets are shown to accelerate the ambient neutral plasma just dragging it by a magnetic reconnection process. Recently, particle acceleration in RCSs occurring at the intersection of interacting loops and their ejection into loop legs was shown to be strongly related to the charge of the particle (electrons or protons) and the topology of the magnetic field configuration (Dalla and Browning, 2005; Wood and Neukirch, 2005; Zharkova and Agapitov, 2009; Zharkova and Gordovskyy, 2004, 2005b).

Therefore, both kinds of particles, protons and electrons, are simultaneously accelerated during this process and injected into loop legs with different velocities and energy distributions (Zharkova and Gordovskyy, 2005b,c), either into the same or opposite footpoints of interactive loops (Zharkova and Agapitov, 2009). There are four populations of particles produced by this process: power-law electrons, power-law protons, quasithermal (up to 10–20 keV) electrons, and quasithermal protons, which have energies much higher than thermal ones, up to 100 MeV (Gordovskyy *et al.*, 2005).

Hence, depending on the ratio of magnetic field components in the loop legs, there are either electron beams or proton beams that are fully separated and injected into the loop legs and accompanied by quasithermal protons or electrons. However, in most cases the condition for full separation is not satisfied (Zharkova and Gordovskyy, 2004), leading to a partial separation of electrons and protons into loop legs with one leg containing, for example, 70% of electrons and 30% of protons while the other one 70% of protons and 30% of electrons. The protons in the first leg have much lower velocities than the electrons, and their energy is not high enough for them to be detected in γ-ray emission, while in the second legs electrons have lower energies and cannot be observed or produce a weak soft X-ray emission without HXRs.

Therefore, the fact that only HXR emission is observed in a particular footpoint does not necessarily mean that there are only electrons precipitating into that footpoint, since they can be accompanied by higher-energy protons, nonthermal, or quasithermal ones with energies below a few megaelectronvolts. This can explain the two items from Section 12.5.3 – the fact why HXR emission is not always accompanied by γ-rays and if they appear in the same flare, they are separated spatially and delayed from HXRs in time as reported for the 23 July 2002 flare (Hurford *et al.*, 2003a,b).

12.6.2
Particle Precipitation

While electron precipitation is rather well investigated in the pure collisional approach discussed in Chapters 3 and 4 following pure collisional (Brown, 1971; Syrovatskii and Shmeleva, 1972a) or in joint collisional and ohmic-loss approaches (Siversky and Zharkova, 2009b; Zharkova and Gordovskyy, 2005a; Zharkova *et al.*, 1995), proton precipitation is only considered in a pure collisional approach (Emslie *et al.*, 1998; Lin *et al.*, 2003; Schrijver *et al.*, 2006) while their particle–wave interactions with the ambient plasma as a whole (Tamres *et al.*, 1989; Voitenko, 1996) were not taken into account until recently, as discussed in Chapter 6 following the approach by Gordovskyy *et al.* (2005).

Let us first summarize the effects of electron precipitation in order to establish their relevance for the items from the observation summary in Section 12.5.3. Precipitation of power-law electron beams is governed by the Fokker–Planck equation considering Coloumb collisions and ohmic losses in a self-induced electric field described in Chapter 4. Particle energy losses at given precipitation depths strongly depend on the initial beam parameters, and thus their density varies significantly with depth (see Figure 3.7 from Zharkova and Gordovskyy (2005a)).

At depths in the corona with or without an electric field (Figure 3.7) the beam densities are lower for a beam spectral index of 3 than for an index of 6. This difference becomes even more noticeable if ohmic losses in the self-induced electric field are taken into account (see Figures 4.28–4.30 in Chapter 4). This reflects the fact that intense and moderate soft electron beams lose most of their energy in the corona and only a small part of it in the chromosphere, while intense and moderate hard electron beams lose their energy mostly in the chromosphere and reach even the photosphere (compare Figures 4.28 and 4.30). This happens because an electric field in the corona and chromosphere induced by softer beams is higher than one induced by harder beams, so for softer beams there are more beam electrons returning to the source and fewer precipitating electrons moving downward to the chromosphere. This results in double power-law photon spectra being observed from flares, which was often reported by SMM and recently confirmed by RHESSI observations (Hurford *et al.*, 2003a; Sui *et al.*, 2002).

These depth variations of beam electron densities explain remarkably well the items in Section 12.5.3 stating that harder beams of moderate power produce much stronger seismic emission than softer ones. This happens because softer beams lose most of their energy in the upper atmospheric levels producing strong HXRs and few of them can reach the lower atmosphere. Also those that reach it can only deliver a small amount of their energy to the lower chromosphere and photosphere, as noted in Section 12.5.3 and shown in Figure 4.28 in Chapter 4, while hard beams of moderate intensity, which do not induce a strong electric field, manage to precipitate to much deeper layers and to deliver much more energy into the lower atmosphere via hydrodynamic shocks and induced electromagnetic fields leading to higher measured seismic emission and ripples.

Magnetic field convergence can contribute to higher HXR and radio emission at the upper atmospheric levels, reducing a part of the energy delivered to lower atmospheric levels in loops with magnetic convergences of up to 40. These variations become insignificant for HXR emission even with convergence parameters that change from 40 to 100. Therefore, magnetic mirroring cannot contribute significantly to the formation of sunquakes, but it can be a factor in an increase of HXR emission and radiative ionization of atoms at lower atmospheric levels (Zharkova, 2008a).

Proton-beam kinetics have not been investigated in the same detail as the kinetics of electron beams, apart from a few attempts in the past decade to consider proton precipitation in Coulomb and inelastic collisions (Emslie *et al.*, 1998; Fang *et al.*, 1995; Gordovskyy *et al.*, 2005; Hénoux, 1991b). Recently, attention has been devoted to proton-beam stability during their precipitation into a flaring atmosphere (Tamres *et al.*, 1989; Voitenko, 1996) and their role in additional plasma heating and particle acceleration at lower atmospheric levels (Gordovskyy *et al.*, 2005).

Findings about simultaneous proton and electron acceleration is encouraging to explore proton kinetics more carefully by including particle–wave interactions in addition to collisions, which will be discussed in a forthcoming paper. However, we have estimated already the possible heating by proton beams (Zharkova and Zharkov, 2007) and applied this heating to the calculation of hydrodynamic responses caused by these beams, as discussed in Chapter 6.

12.6.3
Plasma Responses to High-Energy Particles

12.6.3.1 Hydrodynamic Shocks and Deposited Momenta

In Chapter 5 we considered hydrodynamic responses of the 1-D solar atmosphere to the injection of electron or proton beams and jets by taking into account the continuity, momentum, and energy equations for ambient electrons and protons/ions as described by Zharkova and Zharkov (2007) (see also Allred *et al.*, 2005; Fisher *et al.*, 1985a,b,c; Somov *et al.*, 1981).

Observations usually show that for some flares the momenta deposited in upward and downward motions can be nearly equal (see, e.g., Zarro *et al.*, 1988, and references therein), while for many other flares only the upward motions (blue shifts) can be observed without the noticeable downward ones (red shifts) (Antonucci *et al.*, 1982). As the estimations show, the particle energies upon ejection from an RCS (Table 12.3), the momentum flux $M(x)$ from a reconnection region is carried mainly by protons (both beams, either "fast" or quasithermal ones called "slow" jets). The beam protons accelerated by a super-Dreicer electric field carry a momentum flux of $\sim 1.5 \times 10^4\,\mathrm{g\,cm^{-1}\,s^{-2}}$ into a loop with some area A, while separatrix jet protons carry about $1 \times 10^3\,\mathrm{g\,cm^{-1}\,s^{-2}}$. Hence, for a flare duration of 10 s and area of $A = 10^{18}\,\mathrm{cm^{-2}}$ beam protons deliver to the photosphere a momentum of about $10^{22}\,\mathrm{g\,cm\,s^{-1}}$, and "slow" protons precipitating to areas larger by a factor of 500 can deliver a momentum of about $5 \times 10^{23}\,\mathrm{g\,cm\,s^{-1}}$.

These estimations can be considered a lower limit of the momenta carried by protons since their ejection time can be longer than the 10 s accepted here. Hence the estimated momentum carried by "slow" protons is $\sim 5 \times 10^{23}\,\mathrm{g\,cm\,s^{-1}}$ and by "fast" protons it is $\sim 10^{22}\,\mathrm{g\,cm\,s^{-1}}$, which is in reasonable agreement with the magnitude reported from the helioseismic response (Kosovichev and Zharkova, 1998). These slow proton beams (jets) are good invisible candidates to be added to electron beams for the interpretation of energy and temporal constraints induced by the first sunquake.

On the other hand, estimations of the momentum delivered by each kind of particle to the lower chromosphere found from Table 12.3 for the accepted flaring area A show that beam electrons only carry a momentum of about $2 \times 10^{20}\,\mathrm{g\,cm\,s^{-1}}$. This momentum, delivered by an electron beam multiplied by the flare area and time when it was imposed for the flare of 28 October 2003, was found to be only marginally sufficient to deliver the required momentum of $(3–4) \times 10^{22}$ measured with TD diagrams (Zharkova and Zharkov, 2007).

Hence, electron beams might be marginally sufficient for smaller M- and C-class flares (and smaller sunquakes) as reported in other observations (Donea *et al.*, 2006; Kosovichev, 2007b), which revealed a temporal delay of 3–5 min in the appearance of some seismic sources with respect to the onset of HXR sources. This was assumed to be caused by a slower travel time of the shock produced by these electrons because they are formed higher in the upper chromosphere. Also, electron beams slightly mixed with jet protons were used to explain two out of three seismic sources in the X-class flare of 28 October 2003 (Zharkova and Zharkov, 2007).

The puzzle with the multiple seismic responses associated solely with the locations of HXR emission detected for the 23 July 2002 flare reported by Kosovichev (2007b) has a few plausible explanations. First, the flare and, especially, the source with γ-ray emission, where these multiple sources are supposedly recorded, are located very close to the limb. Thus, the Doppler observations produce more like a horizontal displacement rather than the vertical one required to identify the seismic source from TD diagrams.

Second, our previous research on magnetic fields in the locations of observed HXR sources for this flare revealed that only a few locations in the photosphere with high magnetic field strengths were associated with the loop legs embedded in the photosphere (Zharkova *et al.*, 2005a). Some of these magnetic sources emerged just before the flare, but they did not move significantly in the photosphere (Zharkova *et al.*, 2005a). On the contrary, the authors showed that it was the H_α emission that moved inside the loop from one HXR source to another that coincided with the movement of an HXR source along the loop reported by Krucker *et al.* (2003). The latter more likely reflects electron beam precipitation observed from the side but not from the top as in the solar disk center. This assumption is confirmed by the moving HXR intensity in a semicircular loop (Krucker *et al.*, 2003) that is well defined by the Coulomb and ohmic losses predicted by Zharkova and Gordovskyy (2006).

Moreover, the seismic source is only observed in the HXR source as reported by Kosovichev (2007b), which coincides with the middle magnetic source reported by Zharkova *et al.* (2005a), where the magnetic field shows the presence of a magnetic transient likely to be induced by a propagating electron beam. This confirms that there was a single impact to the photosphere from the shock caused by this beam. Again, similarly to the flare of 28 October 2003, from the strong H_α emission produced in this loop one can easily deduce that this was not a pure electron beam causing such an intensity increase. Possibly, this increase was caused by an electron beam mixed with lower-energy protons coming from separatrix jets.

12.6.3.2 Hydrodynamic Shocks Formed by Mixed Beams

We established in Chapter 4 that including a consideration of ohmic losses, in addition to Coulomb collisions, leads to strong soft electron beams' depositing a bulk of their energy in the corona and upper chromosphere, producing strong soft and HXRs and not producing observable hydrodynamic shocks. This explains why most large X-class flares are acoustically inactive or their detected seismic emissions or ripples are only a very small fraction, a few hundred thousandths, of the energy released during the flare. Therefore, we need to consider more energetic hard beams of electrons and protons as well as quasineutral jets that can definitely deliver shocks in the lower atmosphere.

The hydrodynamic responses of a flaring atmosphere to the injection of electron, proton, and mixed beams are simulated in Chapter 6 and briefly reproduced here in relation to the seismic response associated with them. In order to clarify the linear depth of this second shock formation, the macrovelocities of ambient plasma are plotted vs. a linear depth in Figure 12.15, with zero corresponding to the photospheric level. Note that there is a very significant difference in macrovelocities caused by different agents of chromospheric evaporation into the corona and low-temperature hydrodynamic shocks formed in the chromosphere and moving to the photosphere and beneath.

The macrovelocity variations caused by a pure electron beam from lower to high initial energy fluxes and higher spectral indices (Figure 12.15a) reveal rather high velocities (up to $10^8\,\mathrm{cm\,s^{-1}}$) of chromospheric evaporation occurring in the upper chromosphere that start from a column depth of $5\times10^{19}\,\mathrm{cm^{-2}}$) and lower-temperature shock formed immediately below this depth. The shock moves downward toward the photosphere at velocities of up to $10^7\,\mathrm{cm\,s^{-1}}$ and terminates at

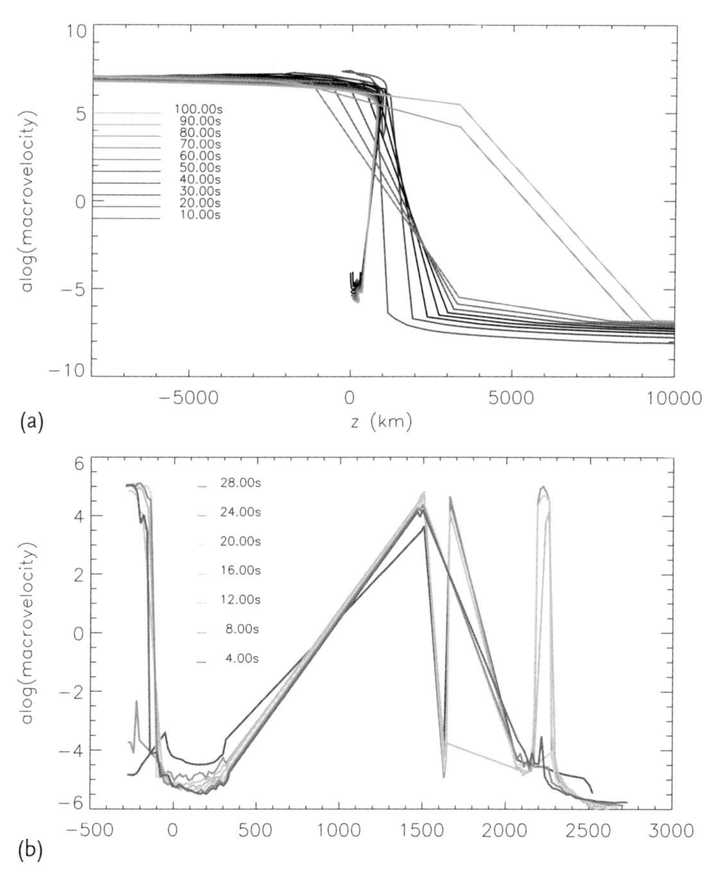

Figure 12.15 Macrovelocity profiles with shocks vs. linear depth above the photosphere formed at different times after the injection of the following agents: (a) a pure electron beam with parameters $2\times10^{13}\,\mathrm{erg\,cm^{-2}\,s^{-1}}$, $\gamma = 4.9$; (b) 70% of proton beam and 30% of electron beam with total energy flux of $3\times10^{13}\,\mathrm{erg\,cm^{-2}\,s^{-1}}$. For a color version of this figure, please see the color plates at the beginning of the book.

a depth just shy of 10^{21} cm^{-2}, with a velocity of up to 10^5 cm s^{-1} that on a linear scale is just a few hundred meters above the photosphere. As indicated earlier in Chapter 7 (Siversky and Zharkova, 2009b; Zharkova and Zharkov, 2007), this is obviously related to the fact that beam electrons deposit the maximum of their energy owing to collisions and ohmic losses in the corona and upper chromosphere.

When the initial energy flux of injected particles is increased on the assumption that 30% of the energy is brought by protons and 70% by electrons in beams precipitating into a flaring atmosphere, the energy deposition maximum shifts from the lower corona to the upper chromosphere. This reduces the macrovelocities of chromospheric evaporation into the corona from 10^8 cm s^{-1} in the first 10 s to 10^5 cm s^{-1} after 130 s. At the same time, the lower-temperature shock is now split into two shocks: one moving from the lower corona toward a chromospheric column depth of 10^{19} cm^{-2} and terminating at a depth of 10^{20} cm^{-2}, and the second one appearing at a column depth of 4×10^{20} cm^{-2} and terminating at a depth of 5×10^{21} cm^{-2}, or just above or at the photosphere. The maximum velocity approaches 10^6 cm s^{-1} in the first shock and 10^5 cm s^{-1} in the second one.

Further increase of the initial energy flux and momentum by considering the beams of protons (70% of the energy) and electrons (30% of the energy) (Figure 12.15b) while keeping the velocities of chromospheric evaporation at the same level (up to 10^8 cm s^{-1}) as for electrons, makes the first lower-temperature shock much wider, spreading from a column depth of 8×10^{19}–10^{22} cm^{-2}, with the second shock starting at a depth of about a few hundred meters below the photosphere and having a velocity of a few units of 10^5 cm s^{-1}. By increasing further the percentage of protons in precipitating beam one can obtain the lower-temperature shock formed well below (a few thousand meters) the photosphere.

Note that with a larger contribution of protons in hydrodynamic heating, the second shock is terminated below the photosphere in the solar interior at depths down to 500 km and deeper. These prolonged movements of shocks in the solar interior at supersonic velocities are reported to explain the observed horizontal displacements (Figure 12.12) seen in the chromosphere and photosphere (as discussed in Section 12.4.3 for the flare of 14 December 2005) and the seismic waves seen in many flares (Section 12.7.4).

Also this occurrence of either single or double hydrodynamic shocks (in the chromosphere and the photosphere, respectively), one of which keeps moving in the chromosphere toward the photosphere and the other one beneath the photosphere, can explain the observational features related to seismic responses, or sunquakes, occurring in the solar interior and their occasional links with Moreton waves occurring in the chromosphere or any other wave motions seen in the chromosphere and photosphere as reported in Section 12.4.3.

12.7 Nonthermal Ionization and Backwarming Heating

12.7.1 Hydrogen Nonthermal Excitation and Ionization

The precipitation of electron beams deeper into the chromosphere and even the photosphere as revealed in Figure 4.10 raises the question of how these beam electrons interact with neutral atoms in the lower chromosphere and photosphere. In Chapter 10 we compared the volume hydrogen nonthermal excitation and ionization rates with thermal ones and those by external radiation and concluded nonthermal processes to be by a factor of 10^3 10^4 more effective in the lower chromosphere and photosphere at exciting hydrogen atoms to higher energetic levels or even fully ionizing them.

In Figure 10.2 the hydrogen volume ionization rates by thermal and nonthermal mechanisms are compared for different electron-beam parameters and the physical conditions derived from hydrodynamic solutions. It can be seen that at chromospheric depths starting from a column depth of $\xi = 4 \times 10^{19}\,\mathrm{cm}^{-2}$ or mass depth of $2 \times 10^{-4}\,\mathrm{g\,cm}^{-2}$ nonthermal ionization by beam electrons is dominant by a few orders of magnitude; the same is true for nonthermal and thermal excitation rates reported by Zharkova and Kobylinskii (1993).

It must be emphasized that nonthermal ionization rates increase throughout all chromospheric depths with the level numbers of hydrogen atoms. This is contrary to Allred *et al.* (2005), who assume that ionization rates are constant for all levels and use a very simplified formula for the ionization rate based on the electron-beam heating function with the deposition maximum occurring higher in the upper chromosphere than it should occur for the continuity equation approach, as pointed out in Chapter 3 (Figure 3.7).

Additional nonthermal ionization and excitation are much higher in the first 10 or 20 s while the beam is being injected than those by thermal electrons or by external radiation of the overlying atmosphere, as discussed in the section above. This nonthermal impact ionization leads to a fast increase of all hydrogen emissions including the H_α emission often observed simultaneously with HXRs (Zharkova *et al.*, 2011b).

But most importantly, the nonthermal ionization caused by beam electrons leads to a significant increase (up to four or five orders of magnitude) of the ionization degree of the ambient plasma. And because impact ionization rates are much higher than hydrogen recombination rates (Zharkova and Kobylinskii, 1991) and because of an increased optical thickness in the hydrogen continua, this can result in a much longer duration of this continuous emission because of radiative transfer processes. Hence the radiation of Pashen continua of hydrogen atoms and of negative hydrogen ions appears as a white-light flare (Aboudarham and Hénoux, 1987; Hudson, 1972) that in the case of reduced heating by beam electrons leads to a gentle radiative cooling phase, as reported by Allred *et al.* (2005).

Solving the full non-LTE problem for a hydrogen five-level-plus-continuum atom (Zharkova and Kobylinskii, 1991) one can find that hard beam fluxes produce much higher ionization than softer intense beams (Figure 10.4), and that this ionization increases toward the photosphere owing to the radiative transfer effects discussed in Section 10.4.4. Therefore, it is no surprise that these hard electron beams of moderate power produce simultaneously both seismic emissions and white-light flares in the M-class flares reported by Beşliu-Ionescu *et al.* (2005) and Donea *et al.* (2006), while precipitation of more intense hard electron beams to the lower chromosphere is strongly restricted by their increased ohmic losses, which prevent many beam electrons from reaching the photosphere (Zharkova and Gordovskyy, 2006). This again agrees with the conclusions about weaker seismic emission in the flares with strong HXRs reported by Beşliu-Ionescu *et al.* (2005).

However, since similar nonthermal ionization and excitation effects can be caused by proton beams (Fang *et al.*, 1995), these results must be tested with a refined proton-beam kinetics, considering not only their particle–particle but also wave–particle interactions similar to those discussed by Gordovskyy *et al.* (2005). In X-class flares other kinds of particles should contribute to momentum delivery to the lower atmosphere, like proton beams and quasithermal flows, or jets, as was reported for the Halloween 2003 flare (Zharkova and Zharkov, 2007), the 14 December 2006 flare (Matthews *et al.*, 2011), and for 15 February 2011 flare (Kosovichev, 2011; Zharkov *et al.*, 2011a).

12.7.2 The Role of Backwarming Heating

A comparison of radiative ionization rates for five-level hydrogen atoms with those caused by inelastic collisions with thermal electrons and by diffusive radiation formed by the action of radiative transfer, as shown in Figure 10.3 for levels 2–5, reveals a dominant effect of diffusive emission in the lower chromosphere (column depths $> 10^{20}$) because of a decrease of the continuous opacity for each level.

It can be seen that radiative ionization rates increase for higher levels (4 and 5 in this model). But since the ambient plasma before a flare onset is almost neutral at this depth (the ionization degree is less than 10^{-6}), few hydrogen atoms are excited to higher energetic levels, and, hence, few hydrogen atoms are ionized by the radiation coming from the chromosphere before any beam electrons are injected. However, after 1 s of beam injection most hydrogen atoms become strongly excited to higher energetic levels by beam electrons (Chapter 10). This immediately increases hydrogen ionization from these levels, leading to the very high ionization degrees of 0.4–0.6 shown in Figure 10.4.

As has been discussed by Donea *et al.* (2006), backwarming heating can play some role in the additional heating of lower atmospheric levels, leading to white-light flares (Hudson, 1972; Hudson *et al.*, 2006) often accompanying the seismic emission in flares. A substantial fraction of the continuum emission from white-light flares from the overlying chromosphere is proposed to be driven to the photosphere and to result in a "radiative backwarming" by recombination radiation, recently included in radiative hydrodynamics by Allred *et al.* (2005) and causing the enhanced continuous emission seen in the visible continuous (Pashen) spectrum as a white-light flare.

The photosphere is assumed to absorb the part of this radiation that is emitted downward, and thus it becomes heated also by this external X-ray and EUV/UV radiation in addition to the heating caused by electron or proton beams. The immediate effect of this absorption in the visible spectrum is mostly dissociation of H^- ions, which represents the predominant source of the photospheric opacity. The radiation energy estimated from white-light flares is close to that measured by seismic emission (Donea *et al.*, 2006). This allows researchers to assume (Donea *et al.*, 2006) that this is an additional source of heating delivering the energy to the photosphere and leading to acoustic emission, which is different from either electrons or protons but closely associated with continuum emission.

However, Allred *et al.* (2005) have pointed out that for their hydrodynamic models the heating by electron beams of a very moderate backwarming power is not very effective during the impulsive phase when hydrogen emission is formed mostly by nonthermal ionization and external high-energy radiation (X-rays, UV, and EUV). The increased ionization immediately produces a strong emission in hydrogen lines and continua with a large opacity. This means that the radiation remains locked inside the atmosphere for a period of the difference between its recombination and impact excitation rates, as discussed in Section 12.7.1.

After the beam heating has stopped (say, in 10 s), thermoconductivity continues to heat further the lower atmosphere, increasing its emission in all lines and continua, leading to further increase of the line and continuous opacity, as reported by Zharkova and Kobylinskii (1993) and confirmed by Allred *et al.* (2005). Moreover, Allred *et al.* (2005) have reported that this additional heating does not increase the backwarming heating magnitude. The backwarming heating remains gentle since it is governed by a high opacity of the radiation.

However, the duration of this process can be increased. As a result, during all this time the backwarming heating can steadily affect the emission of the Ni line from which the seismic response is derived, and this definitely affects the results derived from the holographic method, as pointed out by Donea *et al.* (2006). However, this backwarming heating definitely cannot account for the magnitudes of the momentum delivered to the photosphere by any agents as reported above following the estimations by Zharkova (1999); Zharkova and Zharkov (2007) and Zharkova (2008a) for seismic responses detected using the TD technique (see Figure 12.16).

12.7.3 Ni-Line Emission

The formation region for the Ni line 6768 Å lies approximately within column depths of $(2.0–4.0) \times 10^{23}\,\text{cm}^{-2}$, or around 200 km for the quiet Sun (Figure 1 from Zharkova and Zharkov, 2007), which increases to $(4.0–6.0) \times 10^{23}\,\text{cm}^{-2}$, or around 180 km, for a flaring atmosphere heated by beam electrons, which agrees

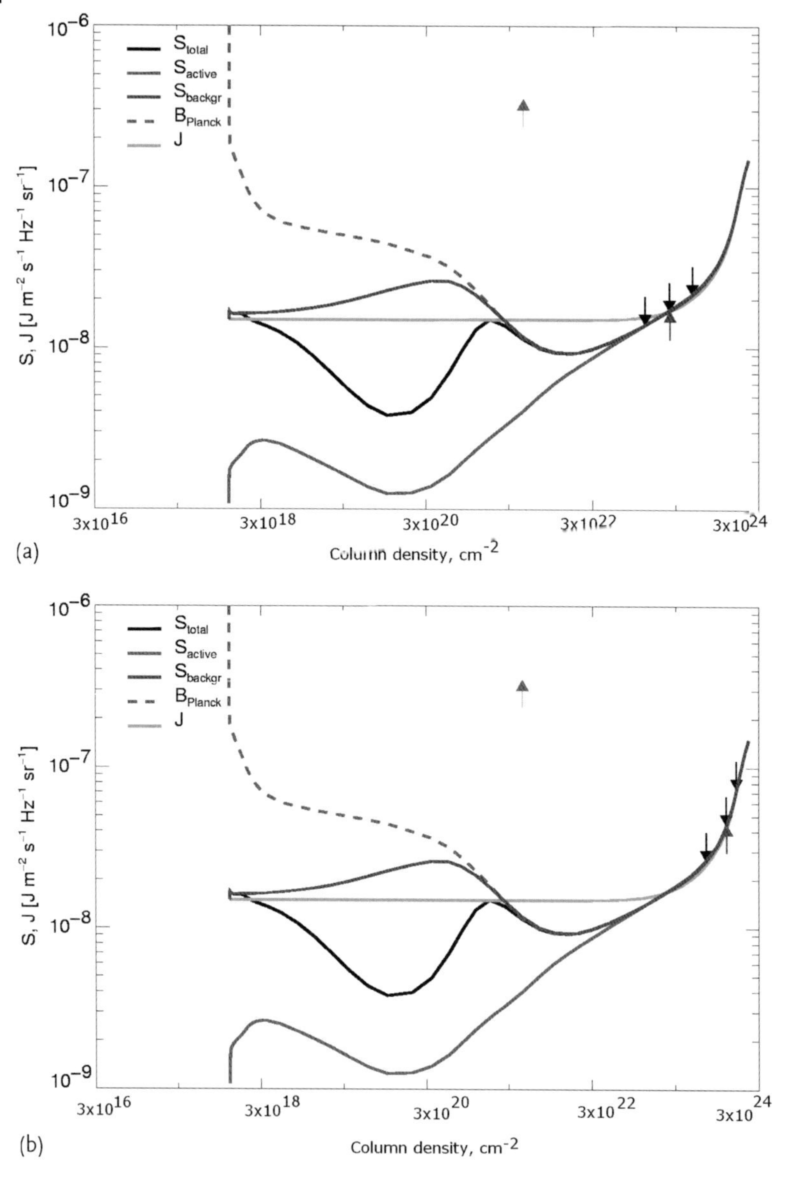

Figure 12.16 Source function distributions calculated for a quiet atmosphere (a) and flaring atmosphere (b) heated by a hard intense electron beam ($\gamma = 3$, $F_0 = 10^{12}$ erg cm^{-2} s^{-1} for the Ni-line transition 6768 Å (S_{active}, gray line), background elements (S_{backgr}, blue line), Planck function (S_{planck}, red dashed line), and total for all elements (S_{total}, black line), as well as the mean intensity J (yellow line) simulated using the full NLTE MULTI-based approach for the full coronal abundance of elements and some molecules (CO, C_2, CH, CN, 23 in total) (Uitenbroek, 2001; Zharkova and Kosovichev, 2002). For a color version of this figure, please see the color plates at the beginning of the book.

rather well with the estimations found from the previous non-LTE simulations (Baranovskij and Kurochka, 1989; Bruls, 1993; Zharkova and Kosovichev, 2002).

Most importantly, the Ni-line profiles become significantly distorted by the increased ionization degree in the ambient plasma, as shown in Figure 12.17 for the prebeam time (Figure 12.17a) and 10 s after beam injection with the parameters $F_0 = 10^{11}\,\mathrm{erg\,cm^{-2}\,s^{-1}}$ and $\gamma = 4$ (Figure 12.17b). The other lines display the

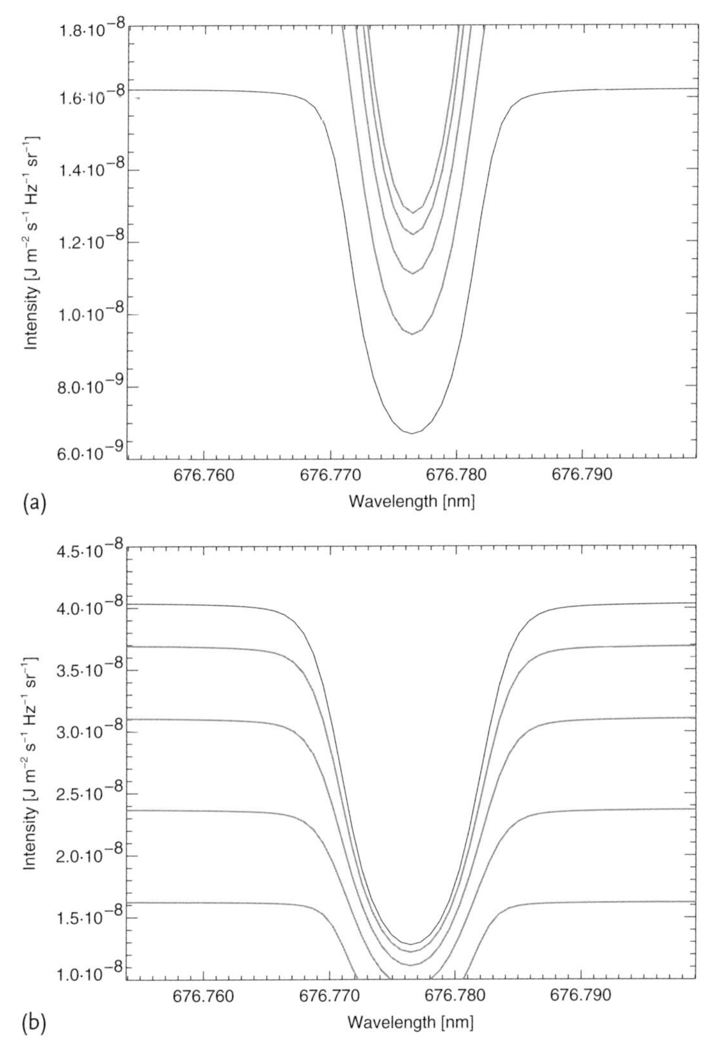

Figure 12.17 Ni-line profiles calculated for hydrodynamic models with initial energy fluxes of $10^{11}\,\mathrm{erg\,cm^{-2}\,s^{-1}}$ and a spectral index of 4 using full non-LTE code (Uitenbroek, 2001) and nonthermal hydrogen ionization degree caused by beam electrons at 0 s (a) and 10 s (b) after their injection. For comparison, the line profiles are plotted for 2, 4, 6, and 8 s in each graph from bottom to top, respectively. For a color version of this figure, please see the color plates at the beginning of the book.

Ni-line profile between these times with a 2 s cadence. This increase of the Ni-line intensity caused by nonthermal ionization is likely to be observed with the holographic method as seismic emission appearing from depths above the Ni-line region. Hence seismic emission is mostly associated with changes in the ionization degree described in Figure 10.4 caused by nonthermal ionization by beam electrons in the first minute and by thermal and radiative ionization in the following minutes.

What is measured with the TD diagrams from dopplergrams are likely to be the direct signatures of hydrodynamic shocks caused by these beams. The column depths for the Ni-line region are shown to be closer to a hydrodynamic shock formed by the mixed proton/jet beam rather than by pure electron beams, as shown by Zharkova and Zharkov (2007). Also, the temporal profiles of the shock velocity edge at the greatest depths, increasing from 0.1 $km\,s^{-1}$ (at 1 s) up to a few kilometers per second and then decreasing back to zero, resembles remarkably well the observed temporal profiles of Doppler velocity changes in observed solar flares (Kosovichev, 2007b; Zharkova and Zharkov, 2007).

Therefore, the difference in methodology of different seismic techniques (TD and holography) can result in different levels of seismic emission measured by them. For example, with the holographic method, in addition to Doppler variations of the Ni-line wings, one also obtains the Ni-line intensity variations at higher atmospheric levels, which increases owing to the increased hydrogen ionization, while these changes cannot by registered by the TD technique, which measures only the Doppler velocities. This discrepancy was first reported for the four seismic sources seen in the 28 October 2003 flare with the holographic technique (Donea and Lindsey, 2005), while only three sources were detected with the TD technique (Zharkova and Zharkov, 2007). This was also found in a number of other flares (Beşliu-Ionescu *et al.*, 2005; Donea *et al.*, 2006).

As discussed above, for weaker flares of class M or C this shock can be caused by moderately hard electron beams slightly mixed with lower-energy jet protons in order to provide the measured momentum at the photospheric level. These shocks are formed much higher in a flaring atmosphere, and their travel time to the region of Ni-line formation can be delayed by up to 5 min compared to the time of X-ray and γ-ray emission. Observations of such delayed shocks in the seismic emission are reported by Beşliu-Ionescu *et al.* (2006b) and Donea *et al.* (2006).

For stronger X-class flares with faster ripples detected from the TD diagrams these shocks must be formed in the loop legs mostly by precipitating protons (beams and jets), possibly mixed with high-energy electrons, which produces lower HXR emission than in the electron-dominated legs. These shocks need less than 60 s to reach the Ni-line-formation region that allows the detection of the seismic response nearly simultaneously with the HXRs, as reported for the 28 October 2003 flare (Donea and Lindsey, 2005; Zharkova and Zharkov, 2007). However, as can be seen from Figure 12.15, the shocks produced either by electrons only or by electrons mixed with protons can reach the photospheric level and even travel some distance in the solar interior. And the depth where this shock terminates can be crucial for the observation of ripples and other signatures of seismic activity in flares, as discussed in Section 12.7.4.

12.7.4
Generation of Seismic Response by a Pinpoint Source

Investigations with a simple polytrope model the seismic response of the solar interior to a momentum deposited by a high-frequency spherical pinpoint source at some depth in the solar interior. The simulations show that after the initial reflection of acoustic waves in the neighborhood of the source, the wavefront surfaces are formed at a considerable horizontal distance from the source. It was found that the closer the source is to the surface, the larger this distance will be.

Samples of these simulations are shown in Figure 12.18, which plots snapshots taken at different times of a wavefield generated by a high-frequency spherical point source at 6.5 mHz, located near the surface and calculated for a simple polytrope model. The depth is plotted along the y-axis, with 0.14 Mm approximately corresponding to the surface.

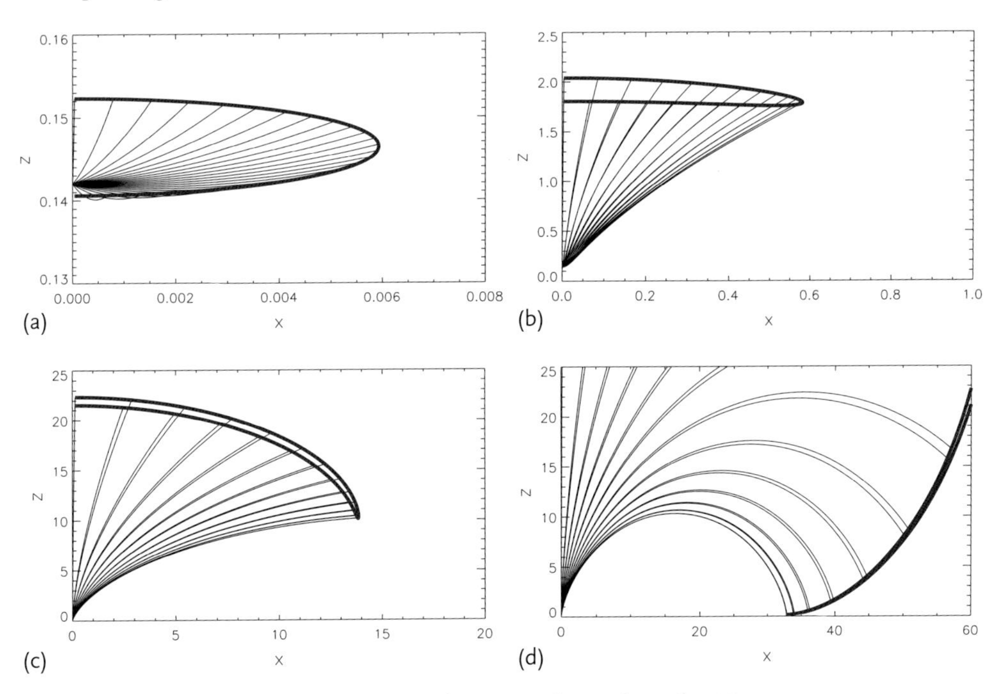

Figure 12.18 Wavefield generated by a high-frequency spherical point source, at 6.5 mHz, located near the surface as calculated for a simple polytrope model taken at different times: (a) 0.17 min; (b) 0.33 min; (c) 11.67 min; (d) 25.00 min. Depth is plotted along the y-axis, with 0.14 Mm approximately corresponding to the surface. The rays emanating from the source in the positive y-direction are plotted in black. The wavefront indicating the arrival of the perturbation is plotted in red. For a color version of this figure, please see the color plates at the beginning of the book.

This can explain why shocks deposited close to the surface (supposedly produced by weaker electron beams) may not produce wavefronts that are observable in the vicinity of a flare but well beyond it. The shocks generated by mixed or proton beams, however, can deposit their momentum much deeper in the solar interior and, thus, produce wavefronts in a vicinity of the flare, which can be observed within a standard datacube of $120 \times 120\ \text{Mm}^2$. However, more work is required to investigate the variety of different sources causing acoustic waves and their propagation in the solar interior including moving sources or sources with a magnetic field inside them.

12.7.5
Magnetic Field Change During Flares

Increasing attention is now being paid to the hypothesis that Lorentz-force transients resulting from a reconnection in coronal loops can be significant contributors to a flare's acoustic emission. Let us explore the existing status of the research in this domain. Zharkova and Kosovichev (2001) and Kosovichev and Zharkova (2001) first considered magnetic field changes during the flare of 14 July 2000 that was acoustically inactive as far as the helioseismic analysis to date has been able to determine.

This research was carried out in greater detail for the flare of 23 July 2002 by Zharkova *et al.* (2005a). The authors established two types of magnetic changes in HXR kernels: irreversible and reversible, or transient-type, ones. The former reveals a steplike magnetic energy change from one status to another, likely reflecting the magnetic energy deposited in the process of magnetic reconnection that initiates a flare. The latter shows a transient type of magnetic energy change that reflects the changes of magnetic field induced by the Lorentz force of the precipitating charged particles (Zharkova *et al.*, 2005a).

Sudol and Harvey (2005) also measured localized transients in the LOS magnetic field in a variety of flares, including the acoustically active flare of 29 October 2003. Donea *et al.* (2006) found a strong local transient in the LOS magnetic signature of the M9.5-class flare of 10 March 2001 coincident with the source region of transient acoustic emission. Hence, transient shifts in magnetic signatures have been detected in a number of flares, some of which were acoustically active and others of which were not.

Hudson *et al.* (2008) suggested that the transient shifts occurring in magnetic signatures during the impulsive phases of acoustically active flares are the result of a flare-related magnetic reconnection and the source of a flare acoustic emission. They estimated the mechanical work that would be done on the photosphere by a sudden shift in magnetic inclination consistent with observed magnetic signatures and found these energies to be roughly consistent with energy estimates based on sunquake observations (Donea *et al.*, 2006).

This outcome produced a very positive confirmation of the hypothesis by Kosovichev and Zharkova (2001) and Zharkova *et al.* (2005a) about transients' being the result of a Lorentz force induced by high-energy particle beams via a self-induced electric field, discussed in Section 12.6.2 (see also Zharkova and Gordovskyy, 2006, and RHESSI nugget 25). In most cases these two effects act jointly, that is, the same flux tube can be reconnecting and simultaneously producing high-energy particles precipitating into its deeper layers. Hence, in a wider sense, it can be accepted that the transients are indirect results of magnetic reconnection, which produces beams of accelerated particles (Zharkova and Agapitov, 2009; Zharkova and Gordovskyy, 2005a,b).

However, the basic scenario of these transients works as follows: electron (or proton) beams injected into a flux tube are rather dense and intense, inducing an electric field equal to their total charge (Zharkova and Gordovskyy, 2006). At the same time they gyrate about the flux tube magnetic field, and as a result of their gyration, they in turn induce their own magnetic field that can have either the same polarity as the flux tube field or one opposite to it (depending on a sign of the tube field and particle charge). The conclusions by Hudson *et al.* (2008) confirm a different way than those of Zharkova and Zharkov (2007) that during their precipitation into a flaring atmosphere high-energy particles carry sufficient energy to account for observed seismic signatures.

The fact is that transient changes in a magnetic field are not found in all footpoints associated with interacting magnetic loops, but approximately in half of them, as shown for the 2002 July 23 (Zharkova *et al.*, 2005a) and the 2006 December 14 flares (Matthews *et al.*, 2011), for example, in those footpoints where electrons precipitate. This can be understood in terms of electron-beam kinetics, described in Chapters 3 and 4, namely, precipitating beam electrons induce a strong electrostatic electric field (Zharkova and Gordovskyy, 2006) matched by the solenoidal electric and magnetic fields induced by these beam electrons (van den Oord, 1990). The electrostatic electric field turns more and more beam electrons back to the injection site, replenishing the number of electrons dragged into a current sheet and ejected into loop legs, while the solenoidal magnetic field is likely the one that will be measured as a transient magnetic field in flare kernels with HXR emission.

Given the fact that during a stationary injection of beam electrons into a flaring atmosphere these fields are established within 0.1 s and exist as long as the beam injection continues, as discussed in Chapter 4, these electric and magnetic fields appear nearly simultaneously with injection onset. In turn, the electron injection itself indicates the start of a magnetic reconnection process, and thus its termination can also be derived from the time when HXR emission disappears.

It can be noted from the distribution of electrons stationarily injected into a flaring atmosphere (Figure 4.20 of Chapter 4) that the electron density increases in the chromosphere where the ambient density increases and where low-temperature condensation forms (Chapter 6). It looks like this condensation is the place where the lower-energy beam electrons lose the bulk of their energy in collisions and ohmic losses and turn back to the corona. This means that this is the location where the solenoidal electric and magnetic fields induced by these beams reach their maximum and, thus, can produce the maximal effect on the ambient plasma combined with other effects produced by this condensation, that is, moving at a supersonic velocity as a shock toward the photosphere.

In kernels without HXR emission the magnetic field changes are more steplike, either increasing or decreasing from the initial state before a flare onset. These magnetic changes are likely to be induced by a magnetic diffusion process during reconnection in the loop legs where protons precipitate. Protons are not likely to induce solenoidal electric or magnetic fields because during their precipitation they drag in with them ambient electrons, which easily neutralize their charge. Hence, in these footpoints there is none of the electrostatic electric field induced by a precipitating proton beam, and thus no solenoidal magnetic field is produced either.

Therefore, it seems that a shock generated in different footpoints of interacting loops induced by electrons or by proton beams can have rather different electrodynamic signatures and, thus, will result in different seismic responses in the solar interior. However, the question remains as to how this energy is deposited in the photosphere in order to result in solar quakes and how it is converted into the acoustic signatures associated with sunquakes. This requires a further comparison in the near future of the effects by hydrodynamic shocks with those caused by the transient magnetic field variations.

12.8 Conclusion

In this chapter we considered a variety of seismic responses produced by different flares and tried to establish some links between these responses and high-energy particles supposedly injected into flaring atmospheres during these flares. The close correlation of seismic sources with HXR and sometimes γ-ray sources suggests that a flare's seismic signatures are driven by impulsive heating by high-energy particles at the onset of the flare and linked to hydrodynamic responses caused by any magnetic field changes induced by accelerated particles.

We established that soft electron beams deposit a bulk of their energy in the corona and upper chromosphere, producing strong soft and HXRs and not producing observable hydrodynamic shocks. This explains why most large X-class flares are acoustically inactive or their seismic emission or ripples detected are only a very small fraction, a few hundred thousandths, of the energy released during the flare. Much more seismic emission is detected in association with weaker M-class flares having rather hard photon spectra in HXRs including the most energetic seismic emission reported for the flare of 10 March 2001. This is linked again to electron beams or electrons with harder energy spectra plus proton beams and their ability to deposit their energy at the lower atmospheric levels, which can only be valid for moderately hard electron beams, which do not lose much of their energy in the corona and deposit a substantial part of it in the chromosphere.

The fact that the seismic emission in M-class flares is often accompanied by the occurrence of white-light emission can also be explained by nonthermal hydrogen excitation and ionization by beam electrons in the first minute after a beam injection and backwarming heating by radiation from the chromosphere occurring in the next few minutes after the beam is stopped.

The X-class flares that produce seismic responses are more likely to have mixed beams (electrons, protons, or jets) responsible for these responses, developing two strong shocks: one in the chromosphere and another in the solar interior a few hundred meters below the photosphere. The second shock, depositing its energy below the photosphere, causes the seismic responses seen as sunquakes, while the first shock produced in the chromosphere can cause EIT, Moreton, or other waves seen in the upper atmosphere.

The first theoretical investigation into the nature of solar quakes developed for a pin-source momentum deposition in a polytrope model revealed that the solar interior depths where shocks are deposited are important for defining where seismic ripples can be seen: either close to the point of deposition (e.g., flare location) if deposited deeply in the interior, or far away from this point if deposited at the surface and just below it.

More theoretical simulations are required for understanding particle kinetics, especially those of protons and mixed electron/proton beams and the hydrodynamic responses caused by them, possibly in 3-D models, in order to account for the elliptic ripples and nonregular seismic sources reported from observations. Also, it is very important to consider the effects of magnetic fields induced by precipitating beam electrons on the generation of seismic emission, which can help to explain further the variety of sunquakes observed in association with solar flares.

References

Aboudarham, J. and Hénoux, J.C. (1986) *Astron. Astrophys.*, **168**, 301.

Aboudarham, J. and Hénoux, J.C. (1987) *Astron. Astrophys.*, **174**, 270.

Allred, J.C., Hawley, S.L., Abbett, W.P., and Carlsson, M. (2005) *Astrophys. J.*, **630**, 573.

Altyntsev, A.T., Fleishman, G.D., Huang, G., and Melnikov, V.F. (2008) *Astrophys. J.*, **677**, 1367.

Amano, T. and Hoshino, M. (2010) *Phys. Rev. Lett.*, **104**(18), 181102.

Amato, E. and Blasi, P. (2005) *Mon. Not. R. Astron. Soc.*, **364**, L76.

Antonucci, E., Gabriel, A.H., Acton, L.W., Leibacher, J.W., Culhane, J.L., Rapley, C.G., Doyle, J.G., Machado, M.E., and Orwig, L.E. (1982) *Sol. Phys.*, **78**, 107.

Arber, T.D. and Haynes, M. (2006) *Phys. Plasmas*, **13**(11), 112105.

Arkhangelskaja, I.V., Arkhangelsky, A.I., Kotov, Y.D., Kuznetsov, S.N., and Glyanenko, A.S. (2006) *Sol. Syst. Res.*, **40**, 302.

Arzner, K. and Benz, A.O. (2005) *Sol. Phys.*, **231**, 117.

Arzner, K. and Vlahos, L. (2004) *Astrophys. J. Lett.*, **605**, L69.

Asai, A., Masuda, S., Yokoyama, T., Shimojo, M., Isobe, H., Kurokawa, H., and Shibata, K. (2002) *Astrophys. J. Lett.*, **578**, L91.

Asai, A., Nakajima, H., Shimojo, M., White, S.M., Hudson, H.S., and Lin, R.P. (2006) *Pub. Astron. Soc. Japan*, **58**, L1.

Aschwanden, M.J. (2005) *Physics of the Solar Corona. An Introduction with Problems and Solutions*, 2nd edn., Praxis Publishing Ltd., Chichester and Springer, New York, Berlin.

Aschwanden, M.J. (2007) *Astrophys. J.*, **661**, 1242.

Aschwanden, M.J. and Schwartz, R.A. (1996) *Astrophys. J.*, **464**, 974.

Aschwanden, M.J., Bynum, R.M., Kosugi, T., Hudson, H.S., and Schwartz, R.A. (1997) *Astrophys. J.*, **487**, 936.

Aschwanden, M.J., Metcalf, T.R., Krucker, S., Sato, J., Conway, A.J., Hurford, G.J., and Schmahl, E.J. (2004) *Sol. Phys.*, **219**, 149.

Aulanier, G., DeLuca, E.E., Antiochos, S.K., McMullen, R.A., and Golub, L. (2000) *Astrophys. J.*, **540**, 1126.

Aurass, H. and Mann, G. (2004) *Astrophys. J.*, **615**, 526.

Aurass, H., Vršnak, B., and Mann, G. (2002) *Astron. Astrophys.*, **384**, 273.

Babin, A.N. and Koval, A.N. (1985) *Bull. Crim. Astrophys. Obs.*, **72**, 122.

Babin, A.N. and Koval, A.N. (1988) *Izv. Ordena Trudovogo Krasn. Znam. Krymskoj Astrofiz. Obs.*, **80**, 110.

Bai, T. and Ramaty, R. (1978) *Astrophys. J.*, **219**, 705.

Baranovskij, E.A. and Kurochka, E.V. (1989) *Soln. Dann. Bull. Akad. Nauk SSSR*, **10**, 98.

Bastian, T.S. (1999) Proceedings of the Nobeyama Symposium, held in Kiyosato, Japan, 27–30 October 1998, (eds T.S. Bastian, N. Gopalswamy, and K. Shibasaki), NRO Report No. 479, Nobeyama Observatory, p. 211.

Bastian, T.S., Benz, A.O., and Gary, D.E. (1998) *Astron. Astrophys. Rev.*, **36**, 131.

Battaglia, M. and Benz, A.O. (2006) *Astron. Astrophys.*, **456**, 751.

Battaglia, M. and Benz, A.O. (2007) *Astron. Astrophys.*, **466**, 713.

Bell, T.E. (1978) *Astronomy*, **6**, 6.

Benz, A.O. (1977) *Astrophys. J.*, **211**, 270.

Electron and Proton Kinetics and Dynamics in Flaring Atmospheres, First Edition. Valentina Zharkova.

Benz, A.O. (ed.) (2002) *Plasma Astrophysics*, 2nd edn., Astrophysics and Space Science Library, vol. 279, Kluwer Academic Publishers, Dordrecht.

Benz, A.O., Grigis, P.C., Csillaghy, A., and Saint-Hilaire, P. (2005) *Sol. Phys.*, **226**, 121.

Benz, A.O., Perret, H., Saint-Hilaire, P., and Zlobec, P. (2006) *Adv. Space Res.*, **38**, 951.

Berestetskij, V.B., Liphshits, E.M., and Pitaevskij, L.P. (1989) *The Quantum Electrodynamics*, Nauka, Moscow.

Beşliu-Ionescu, D., Donea, A.-C., Cally, P., and Lindsey, C. (2005) *The Dynamic Sun: Challenges for Theory and Observations*, ESA Special Publication, vol. 600.

Beşliu-Ionescu, D., Donea, A.-C., Cally, P., and Lindsey, C. (2006a) *Rom. Astron. J.*, **16**, 203.

Beşliu-Ionescu, D., Donea, A.-C., Cally, P., and Lindsey, C. (2006b) *Solar Activity and its Magnetic Origin*, (eds V. Bothmer and A.A. Hady), IAU Symposium, vol. 233, Cambridge Univ. Press, Cambridge, p. 385.

Bianda, M., Benz, A.O., Stenflo, J.O., Küveler, G., and Ramelli, R. (2005) *Astron. Astrophys.*, **434**, 1183.

Birdsall, C.K. and Langdon, A.B. (1985) *Plasma Physics via Computer Simulation*, (eds C.K. Birdsall and A.B. Langdon), McGraw-Hill, New York.

Birn, J., Drake, J.F., Shay, M.A., Rogers, B.N., Denton, R.E., Hesse, M., Kuznetsova, M., Ma, Z.W., Bhattacharjee, A., Otto, A., and Pritchett, P.L. (2001) *J. Geophys. Res.*, **106**, 3715.

Biskamp, D. (1997) *Nonlinear Magnetohydrodynamics*, Cambridge Univ. Press, Cambridge.

Blandford, R.D. and Ostriker, J.P. (1978) *Astrophys. J. Lett.*, **221**, L29.

Bogachev, S.A. and Somov, B.V. (2005) *Astron. Lett.*, **31**, 537.

Bogachev, S.A. and Somov, B.V. (2007) *Astron. Lett.*, **33**, 54.

Boggs, S.E., Coburn, W., and Kalemci, E. (2006) *Astrophys. J.*, **638**, 1129.

Bohlin, J.D., Frost, K.J., Burr, P.T., Guha, A.K., and Withbroe, G.L. (1980) *Sol. Phys.*, **65**, 5.

Bohr, N (1915) *Les spectres et la structure de l'atome*, Ciel et Terre, Vol. 39, Bulletin of the Société Belge d'Astronomie, Brussels, 221.

Bommier, V. (1980) *Astron. Astrophys.*, **87**, 109.

Bommier, V. and Landi Degl'Innocenti, E. (1996) *Sol. Phys.*, **164**, 117.

Bommier, V. and Sahal-Brechot, S. (1982) *Sol. Phys.*, **78**, 157.

Bommier, V., Landi Degl'Innocenti, E., and Sahal-Brechot, S. (1991) *Astron. Astrophys.*, **244**, 383.

Braun, D.C. and Duvall, T.L., Jr. (1990) *Sol. Phys.*, **129**, 83.

Braun, D.C. and Lindsey, C. (1999) *Astrophys. J. Lett.*, **513**, L79.

Braun, D.C. and Lindsey, C. (2000) *Sol. Phys.*, **192**, 285.

Bray, R.J., and Loughhead, R.E. (1974) *The Solar Chromosphere*, Chapman and Hall, London.

Bret, A. (2009) *Astrophys. J.*, **699**, 990.

Brosius, J.W. and White, S.M. (2006) *Astrophys. J. Lett.*, **641**, L69.

Brown, J.C. (1971) *Sol. Phys.*, **18**, 489.

Brown, J.C. (1972) *Sol. Phys.*, **26**, 441.

Brown, J.C. (1974) Coronal Disturbances. IAU Symposium, vol. 57, (ed. G.A. Newkirk), Reidel, Dordrecht, Boston, p. 395.

Brown, J.C., and Bingham, R. (1984) *Astron. Astrophys.*, **131**, L11.

Brown, J.C. and Craig, I.J.D. (1984) *Astron. Astrophys.*, **130**, L5.

Brown, J.C. and Loran, J.M. (1985) *Mon. Not. R. Astron. Soc.*, **212**, 245.

Brown, J.C., Conway, A.J., and Aschwanden, M.J. (1998) *Astrophys. J.*, **509**, 911.

Brown, J.C., Emslie, A.G., and Kontar, E.P. (2003) *Astrophys. J.*, **595**, L115.

Brown, J.C., Emslie, A.G., Holman, G.D., Johns-Krull, C.M., Kontar, E.P., Lin, R.P., Massone, A.M., and Piana, M. (2006) *Astrophys. J.*, **643**, 523.

Browning, P. and Dalla, S. (2007) *Mem. Soc. Astron. Ital.*, **78**, 255.

Brucker, G.J., Ohanian, R.S., and Stassinopoulos, E.G. (1976) *IEEE Trans. Aerosp. Electron. Syst.*, **12**, 23.

Bruls, M.J.J.H. (1993) *Astron. Astrophys.*, **269**, 509.

Bruzek, A. (1972) *Sol. Phys.*, **26**, 94.

Büchner, J. (1996) *Adv. Space Res.*, **18**, 267.

Bulanov, S.V. and Syrovatskii, S.I. (1976) *Motion of Charged Particles in the Vicinity of a Magnetic-Field Zero Line*, Nauka, Moscow.

Burgess, A. and Seaton, M.J. (1964) *Mon. Not. R. Astron. Soc.*, **127**, 355.

Bykov, A.M. and Fleishman, G.D. (2009) *Astrophys. J. Lett.*, **692**, L45.

Bykov, A.M. and Toptygin, I.N. (1993) *International Cosmic Ray Conference 2*, International Cosmic Ray Conference, vol. 267, World Scientific, Singapore.

Cally, P.S. (2000) *Sol. Phys.*, **192**, 395.

Canfield, R.C. and Gayley, K.G. (1987) *Astrophys. J.*, **322**, 999.

Canfield, R.C., Gunkler, T.A., and Ricchiazzi, P.J. (1984) *Astrophys. J.*, **282**, 296.

Cargill, P.J. (1991) *Astrophys. J.*, **376**, 771.

Carlsson, M. (1986) *Uppsala Astronomical Observatory Reports, vol. 33.*

Chambe, G. and Hénoux, J.-C. (1979) *Astron. Astrophys.*, **80**, 123.

Chandra, R., Jain, R., Uddin, W., Yoshimura, K., Kosugi, T., Sakao, T., Joshi, A., and Deshpande, M.R. (2006) *Sol. Phys.*, **239**, 239.

Charikov, J.E., Guzman, A.B., and Kudryavtsev, I.V. (1996) *Astron. Astrophys.*, **308**, 924.

Charikov, Y.E., Dmitriyev, P.B., Koudriavtsev, I.V., Lazutkov, V.P., Matveev, G.A., Savchenko, M.I., and Skorodumov, D.V. (2004) *Multi-Wavelength Investigations of Solar Activity*, (eds A.V. Stepanov, E.E. Benevolenskaya, and A.G. Kosovichev), IAU Symposium, vol. 223, Cambridge Univ. Press, Cambridge, p. 429.

Chen, F.F. (1974) *Introduction to Plasma Physics*, Plenum Press, New York.

Chen, Q., Otto, A., and Lee, L.C. (1997) *J. Geophys. Res.*, **102**, 151.

Cheng, C.C., Orwig, L.E., and Tandberg-Hanssen, E. (1987) *Sol. Phys.*, **113**, 301.

Chernoff, H. and Lehmann, E.L. (1954) *Ann. Math. Stat.*, **25**, 256.

Chernov, G.P. (2006) *Space Sci. Rev.*, **127**, 195.

Christensen-Dalsgaard, J. and Thompson, M.J. (2003) *A Selective Overview*, Cambridge Univ. Press, Cambridge.

Chupp, E.L. and Benz, O.A. (1994) *Astrophys. J. Suppl.*, **90**, 511.

Cox, J.L., Jr. and Bennett, W.H. (1970) *Phys. Fluids*, **13**, 182.

Dahlburg, R.B., Antiochos, S.K., and Zang, T.A. (1992) *Phys. Fluids B*, **4**, 3902.

Dalla, S. and Browning, P.K. (2005) *Astron. Astrophys.*, **436**, 1103.

Dalla, S. and Browning, P.K. (2006) *Astrophys. J. Lett.*, **640**, L99.

Dalla, S. and Browning, P.K. (2008) *Astron. Astrophys.*, **491**, 289.

Datlowe, D.W. and Lin, R.P. (1973) *Sol. Phys.*, **32**, 459.

Daughton, W. (1999) *Phys. Plasmas*, **6**, 1329.

Daughton, W., Roytershteyn, V., Albright, B.J., Karimabadi, H., Yin, L., and Bowers, K.J. (2009) *Phys. Rev. Lett.*, **103**(6), 065004.

Dauphin, C. and Vilmer, N. (2007) *Astron. Astrophys.*, **468**, 289.

Davis, L. (1956) *Phys. Rev.*, **101**, 351.

Démoulin, P., Hénoux, J.C., Priest, E.R., and Mandrini, C.H. (1996a) *Astron. Astrophys.*, **308**, 643.

Démoulin, P., Priest, E.R., and Lonie, D.P. (1996b) *J. Geophys. Res.*, **101**, 7631.

Des Jardins, A., Canfield, R., Longcope, D., Fordyce, C., and Waitukaitis, S. (2009) *Astrophys. J.*, **693**, 1628.

Diakonov, S.V. and Somov, B.V. (1988) *Sol. Phys.*, **116**, 119.

Ding, M.D. (2003) *J. Korean Astron. Soc.*, **36**, 49.

Donea, A.-C. and Lindsey, C. (2005) *Astrophys. J.*, **630**, 1168.

Donea, A.-C., Braun, D.C., and Lindsey, C. (1999) *Astrophys. J. Lett.*, **513**, L143.

Donea, A.-C., Beşliu-Ionescu, D., Cally, P.S., Lindsey, C., and Zharkova, V.V. (2006) *Sol. Phys.*, **239**, 113.

Drake, J.F. and Lee, Y.C. (1977) *Phys. Rev. Lett.*, **39**, 453.

Drake, J.F., Biskamp, D., and Zeiler, A. (1997) *Geophys. Res. Lett.*, **24**, 2921.

Drake, J.F., Shay, M.A., Thongthai, W., and Swisdak, M. (2005) *Phys. Rev. Lett.*, **94**(9), 095001.

Drake, J.F., Swisdak, M., Che, H., and Shay, M.A. (2006a) *Nature*, **443**, 553.

Drake, J.F., Swisdak, M., Schoeffler, K.M., Rogers, B.N., and Kobayashi, S. (2006b) *Geophys. Res. Lett.*, **33**, 13105.

Drake, J.F., Opher, M., Swisdak, M., and Chamoun, J.N. (2010) *Astrophys. J.*, **709**, 963.

Dulk, G.A. (1985) *Annu. Rev. Astron. Astrophys.*, **23**, 169.

Dulk, G.A. and Marsh, K.A. (1982) *Astrophys. J.*, **259**, 350.

Duvall, T.L., Jr., Jefferies, S.M., Harvey, J.W., and Pomerantz, M.A. (1993) *Nature*, **362**, 430.

Eidman, V.Ya. (1958) *Sov. Phys. JETP*, **7**, 91.

Eidman, V.Ya. (1959) *Sov. Phys. JETP*, **9**, 947.

Ellison, D.C. *et al.* (2005) *International Cosmic Ray Conference 3*, Proceedings of the 29th International Cosmic Ray Conference, vol. 261, 3–10 August 2005, Pune India, (eds B. Sripathi, S. Gupta, P. Jagadeesan, A. Jain, S. Karthikeyan, S. Morris, and S. Tonwar), Tata Institute of Fundamental Research, Mumbai.

Elwert, G. and Haug, E. (1970) *Sol. Phys.*, **15**, 234.

Elwert, G. and Haug, E. (1972) *Space Sci. Rev.*, **13**, 761.

Emslie, A.G. (1978) *Astrophys. J.*, **224**, 241.

Emslie, A.G. (1980) *Astrophys. J.*, **235**, 1055.

Emslie, A.G. (1981) *Astrophys. J.*, **245**, 711.

Emslie, A.G. and Hénoux, J.-C. (1995) *Astrophys. J.*, **446**, 371.

Emslie, A.G. and Smith, D.F. (1984) *Astrophys. J.*, **279**, 882.

Emslie, A.G., Mariska, J.T., Montgomery, M.M., and Newton, E.K. (1998) *Astrophys. J.*, **498**, 441.

Emslie, A.G., Kontar, E.P., Krucker, S., and Lin, R.P. (2003) *Astrophys. J. Lett.*, **595**, L107.

Emslie, A.G., Bradsher, H.L., and McConnell, M.L. (2008) *Astrophys. J.*, **674**, 570.

Fang, C., Feautrier, N., and Hénoux, J. (1995) *Astron. Astrophys.*, **297**, 854.

Fang, C., Hénoux, J., and Gan, W.Q. (1993) *Astron. Astrophys.*, **274**, 917.

Fermi, E. (1949) *Phys. Rev.*, **75**, 1169.

Firstova, N.M. (1986) *Sol. Phys.*, **103**, 11.

Firstova, N.M. and Boulatov, A.V. (1996) *Sol. Phys.*, **164**, 361.

Firstova, N.M. and Kashapova, L.K. (2002) *Astron. Astrophys.*, **388**, L17.

Firstova, N.M., Boulatov, A.V., and Kashapova, L.K. (1999) *Polarization*, (eds K.N. Nagendra and J.O. Stenflo), Astrophysics and Space Science Library, vol. 243, Kluwer Academic Publishers, Boston, p. 451.

Firstova, N.M., Xu, Z., and Fang, C. (2003) *Astrophys. J. Lett.*, **595**, L131.

Fisher, G.H., Canfield, R.C., and McClymont, A.N. (1985a) *Astrophys. J.*, **289**, 414.

Fisher, G.H., Canfield, R.C., and McClymont, A.N. (1985b) *Astrophys. J.*, **289**, 425.

Fisher, G.H., Canfield, R.C., and McClymont, A.N. (1985c) *Astrophys. J.*, **289**, 434.

Fishman, G.J., Meegan, C.A., Wilson, R.B., Paciesas, W.S., Pendleton, G.N., Harmon, B.A., Horack, J.M., Kouveliotou, C., and Finger, M. (1993) *Astron. Astrophys. Suppl.*, **97**, 17.

Fleishman, G.D. and Kuznetsov, A.A. (2010) *Astrophys. J.*, **721**, 1127.

Fleishman, G.D. and Melnikov, V.F. (2003a) *Astrophys. J.*, **584**, 1071.

Fleishman, G.D. and Melnikov, V.F. (2003b) *Astrophys. J.*, **587**, 823.

Fletcher, L. and Brown, J.C. (1995) *Astron. Astrophys.*, **294**, 260.

Fletcher, L. and Hudson, H.S. (2002) *Sol. Phys.*, **210**, 307.

Fletcher, L., Metcalf, T.R., Alexander, D., Brown, D.S., and Ryder, L.A. (2001) *Astrophys. J.*, **554**, 451.

Fletcher, L., Hannah, I.G., Hudson, H.S., and Metcalf, T.R. (2007) *Astrophys. J.*, **656**, 1187.

Fletcher. L. *et al.* (2011) *Space Sci. Rev.*, **159**, 19.

Forbes, T.G. (1986) *Astrophys. J.*, **305**, 553.

Furth, H.P., Killeen, J., and Rosenbluth, M.N. (1963) *Phys. Fluids*, **6**, 459.

Georgoulis, M.K., Rust, D.M., Bernasconi, P.N., and Schmieder, B. (2002) SOLMAG 2002. *Proc. Magn. Coupling Sol. Atmos. Euroconf.*, (ed. H. Sawaya-Lacoste), ESA Special Publication, vol. 505, ESA Publications Division, Noordwijk, p. 125.

Giacalone, J. and Jokipii, J.R. (1996) *J. Geophys. Res.*, **101**, 11095.

Ginzburg, V.L. and Syrovatsky, S.I. (1961) *Prog. Theor. Phys. Suppl.*, **20**, 1.

Ginzburg, V.L. and Zhelezniakov, V.V. (1958) *Sov. Astron.*, **2**, 653.

Gluckstern, R.L. and Hull, M.H. (1953) *Phys. Rev.*, **90**, 1030.

Golden, L.B. and Sampson, D.H. (1971) *Astrophys. J.*, **163**, 405.

Gorbachev, V.S. and Somov, B.V. (1989) *Sov. Astron.*, **33**, 57.

Gordovskyy, M., Zharkova, V.V., Voitenko, Y.M., and Goossens, M. (2005) *Adv. Space Res.*, **35**, 1743.

Gosain, S., Venkatakrishnan, P., and Tiwari, S.K. (2009) *Astrophys. J. Lett.*, **706**, L240.

Grigis, P.C. and Benz, A.O. (2004) *Astron. Astrophys.*, **426**, 1093.

Grigis, P.C. and Benz, A.O. (2005) *Astron. Astrophys.*, **434**, 1173.

Grigis, P.C. and Benz, A.O. (2006) *Astron. Astrophys.*, **458**, 641.

Grigis, P.C. and Benz, A.O. (2008) *Astrophys. J.*, **683**, 1180.

Gros, M., Tatischeff, V., Kiener, J., Cordier, B., Chapuis, C., Weidenspointner, G., Vedrenne, G., von Kienlin, A., Diehl, R., Bykov, A., and Méndez, M. (2004) *5th INTEGRAL Workshop on the INTEGRAL Universe*, (eds V. Schoenfelder, G. Lichti, and C. Winkler), ESA Special Publication, vol. 552, p. 669.

Haber, D.A., Toomre, J., and Hill, F. (1988) *Advances in Helio- and Asteroseismology*, (eds J. Christensen-Dalsgaard and S. Frandsen), IAU Symposium, vol. 123, Reidel Publishing Co., Dordrecht, p. 59.

Hamilton, R.J. and Petrosian, V. (1987) *Astrophys. J.*, **321**, 721.

Hamilton, R.J. and Petrosian, V. (1990) *Astrophys. J.*, **365**, 778.

Hamilton, R.J. and Petrosian, V. (1992) *Astrophys. J.*, **398**, 350.

Hannah, I.G., Kontar, E.P., and Sirenko, O.K. (2009) *Astrophys. J. Lett.*, **707**, L45.

Harvey, J.W., Hill, F., Hubbard, R.P., Kennedy, J.R., Leibacher, J.W., Pintar, J.A., Gilman, P.A., Noyes, R.W., Title, A.M., Toomre, J., Ulrich, R.K., Bhatnagar, A., Kennewell, J.A., Marquette, W., Patron, J., Saa, O., and Yasukawa, E. (1996) *Science*, **272**, 1284.

Haug, E. (1997) *Astron. Astrophys.*, **326**, 417.

Hénoux, J.C. (1991a) *Solar Polarimetry*, (ed. L.J. November), National Solar Observatory, Sunspot, NM, p. 285.

Hénoux, J.C. (1991b) *Adv. Space Res.*, **11**, 7.

Hénoux, J.C. and Chambe, G. (1990) *J. Quant. Spectr. Radiat. Transf.*, **44**, 193.

Hénoux, J.-C. and Vogt, E. (1996) *JOSO Annu. Rep.*, **1995**, 60.

Hénoux, J.C., Chambe, G., Smith, D., Tamres, D., Feautrier, N., Rovira, M., and Sahal-Brechot, S. (1990) *Astrophys. J. Suppl.*, **73**, 303.

Hénoux, J.C., Fang, C., and Gan, W.Q. (1993) *Astron. Astrophys.*, **274**, 923.

Hiei, E. (1986) *Adv. Space Res.*, **6**, 227.

Hiei, E. (1987) *Sol. Phys.*, **113**, 249.

Hollweg, J.V. (1974) *J. Geophys. Res.*, **79**, 1357.

Holman, G.D. (1985) *Astrophys. J.*, **293**, 584.

Holman, G.D., Sui, L., Schwartz, R.A., and Emslie, A.G. (2003) *Astrophys. J. Lett.*, **595**, L97.

Holman, G.D., Aschwanden, M.J., Aurass, H., Bataglia, M., Grigiss, P.C., Kontar, E.P., Liu, P., Saint-Hillaire, P., and Zharkova, V.V. (2011) *Space Sci. Rev.*, **159**, 107.

Hoyng, P. (1975) Studies on hard X-ray emission from solar flares and on cyclotron radiation from a cold magnetoplasma. Ph.D. thesis, Utrecht, Rijksuniversiteit, Doctor in de Wiskunde en Natuurwetenschappen Dissertation, 157 p.

Hoyng, P., Knight, J.W., and Spicer, D.S. (1978) *Sol. Phys.*, **58**, 139.

Huba, J.D. and Rudakov, L.I. (2004) *Phys. Rev. Lett.*, **93**(17), 175003.

Hudson, H.S. (1972) *Sol. Phys.*, **24**, 414.

Hudson, H.S., Wolfson, C.J., and Metcalf, T.R. (2006) *Sol. Phys.*, **234**, 79.

Hudson, H.S., Fisher, G.H., and Welsch, B.T. (2008) *Subsurface and Atmospheric Influences on Solar Activity*, (eds R. Howe, R.W. Komm, K.S. Balasubramaniam, and G.J.D. Petrie), Astronomical Society of the Pacific Conference Series, vol. 383, Astronomical Society of the Pacific, San Francisco, p. 221.

Hurford, G.J., Schmahl, E.J., Schwartz, R.A., Conway, A.J., Aschwanden, M.J., Csillaghy, A., Dennis, B.R., Johns-Krull, C., Krucker, S., Lin, R.P., McTiernan, J., Metcalf, T.R., Sato, J., and Smith, D.M. (2002) *Sol. Phys.*, **210**, 61.

Hurford, G.J., Schwartz, R.A., Krucker, S., Lin, R.P., Smith, D.M., and Vilmer, N. (2003a) *Astrophys. J. Lett.*, **595**, L77.

Hurford, G.J., Schwartz, R.A., Krucker, S., Lin, R.P., Smith, D.M., and Vilmer, N. (2003b) in *Proceedings of the 28th International Cosmic Ray Conference*, 31 July–7 August 2003, Trukuba, Japan. Under the auspices of the International Union of Pure and Applied Phyics (UPAP), (eds T. Kajita, Y. Asaoka, A. Kawachi, Y. Matsubara, and M. Sasaki), p. 3203.

Hurford, G.J., Krucker, S., Lin, R.P., Schwartz, R.A., Share, G.H., and Smith, D.M. (2006) *Astrophys. J. Lett.*, **644**, L93.

Ichimoto, K. and Kurokawa, H. (1984) *Sol. Phys.*, **93**, 105.
Isobe, T., Feigelson, E.D., and Nelson, P.I. (1986) *Astrophys. J.*, **306**, 490.
Ivanov, V.D. and Kocharov, L.G. (1989) *Pis ma Astronom. Zh.*, **15**, 70.
Ivanov, V.V. and Serbin, V.M. (1984) *Sov. Astronom.*, **28**, 524.
Jiang, Y.W., Liu, S., Liu, W., and Petrosian, V. (2006) *Astrophys. J.*, **638**, 1140.
Johnson, L.C. (1972) *Astrophys. J.*, **174**, 227.
Jones, T.W. (1994) *Astrophys. J. Suppl.*, **90**, 969.
Kane, S.R., Chupp, E.L., Forrest, D.J., Share, G.H., and Rieger, E. (1986) *Astrophys. J. Lett.*, **300**, L95.
Karimabadi, H., Daughton, W., and Scudder, J. (2007) *Geophys. Res. Lett.*, **34**, 13104.
Karlický, M. and Bárta, M. (2008) *Sol. Phys.*, **247**, 335.
Karlický, M. and Kosugi, T. (2004) *Astron. Astrophys.*, **419**, 1159.
Karlický, M., Nickeler, D.H., and Bárta, M. (2008) *Astron. Astrophys.*, **486**, 325.
Kashapova, L.K. (2002) *Astron. Rep.*, **46**, 918.
Kashapova, L.K. (2004) *Multi-Wavelength Investigations of Solar Activity*, (eds A.V. Stepanov, E.E. Benevolenskaya, and A.G. Kosovichev), IAU Symposium, vol. 223, Cambridge Univ. Press, Cambridge, p. 459.
Kashapova, L.K., Zharkova, V.V., Chornogo, S.N., and Kotrc, P. (2006) Cosmic Particle Acceleration, 26th Meet. IAU, Joint Discussion 1, 16–17 August, 2006, Prague, Czech Republic, JD01, #47.
Kashapova, L.K., Zharkova, V.V., and Grechnev, V.V. (2007) *The Physics of Chromospheric Plasmas*, (eds P. Heinzel, I. Dorotovič, and R.J. Rutten), vol. 368, Astronomical Society of the Pacific, San Francisco, p. 437.
Kašparová, J., Karlický, M., Kontar, E.P., Schwartz, R.A., and Dennis, B.R. (2005a) *Sol. Phys.*, **232**, 63.
Kašparová, J., Karlický, M., Schwartz, R.A., and Dennis, B.R. (2005b) *Solar Magnetic Phenomena*, (eds A. Hanslmeier, A. Veronig, and M. Messerotti), Astrophysics and Space Science Library, vol. 320, Springer, Dordrecht, p. 187.
Kašparová, J., Kontar, E.P., and Brown, J.C. (2007) *Astron. Astrophys.*, **466**, 705.
Kendal, M.G. and Gobbons, J.D. (1990) *Rank Correlation Methods*, Arnold, London.
Kiener, J. (2005) Internal INTEGRAL Science Workshop, held at ESTEC from 18–21 January 2005, p. 87.
Kiplinger, A.L., Dennis, B.R., Frost, K.J., Orwig, L.E., and Emslie, A.G. (1983) *Astrophys. J. Lett.*, **265**, L99.
Klein, K. (1987) *Astron. Astrophys.*, **183**, 341.
Knight, J.W. and Sturrock, P.A. (1977) *Astrophys. J.*, **218**, 306.
Kobylinskii, V.A. and Zharkova, V.V. (1996) *Adv. Space Res.*, **17**, 129.
Koch, H.W. and Motz, J.W. (1959) *Rev. Mod. Phys.*, **31**, 920.
Kontar, E.P. and Brown, J.C. (2006) *Astrophys. J. Lett.*, **653**, L149.
Kontar, E.P. and MacKinnon, A.L. (2005) *Sol. Phys.*, **227**, 299.
Kontar, E.P., Brown, J.C., and McArthur, G.K. (2002) *Sol. Phys.*, **210**, 419.
Kontar, E.P., Emslie, A.G., Piana, M., Massone, A.M., and Brown, J.C. (2005) *Sol. Phys.*, **226**, 317.
Kontar, E.P., MacKinnon, A.L., Schwartz, R.A., and Brown, J.C. (2006) *Astron. Astrophys.*, **446**, 1157.
Kontar, E.P., Emslie, A.G., Massone, A.M., Piana, M., Brown, J.C., and Prato, M. (2007) *Astrophys. J.*, **670**, 857.
Kontar, E.P., Hannah, I.G., and MacKinnon, A.L. (2008) *Astron. Astrophys.*, **489**, L57.
Kontar, E. *et al.* (2011) *Space Sci. Rev.*, **159**, 301.
Kosovichev, A.G. (1986) *Izv. Ordena Trudovogo Krasn. Znam. Krymskoj Astrof. Obs.*, **75**, 8.
Kosovichev, A.G. (2006) *Sol. Phys.*, **238**, 1.
Kosovichev, A.G. (2007a) *New Solar Physics with Solar-B Mission*, (eds K. Shibata, S. Nagata, and T. Sakurai), Astronomical Society of the Pacific Conference Series, vol. 369, Astronomical Society of the Pacific, San Francisco, p. 325.
Kosovichev, A.G. (2007b) *Astrophys. J. Lett.*, **670**, L65.
Kosovichev, A.G. (2007c) Observations of helioseismic response to flare energy-release events, *New Solar Physics with Solar-B Mission*, (eds K. Shibata, S. Nagata, and T. Sakurai), Astronomical Society of the Pacific Conference Series, vol. 368, p. 423.
Kosovichev, A.G. (2011) *Astrophys. J. Lett.*, **734**, L15.

Kosovichev, A.G. and Duvall, T.L., Jr. (1997) *SCORe'96: Solar Convection and Oscillations and their Relationship*, (eds F.P. Pijpers, J. Christensen-Dalsgaard, and C.S. Rosenthal), Astrophysics and Space Science Library 225, Kluwer Academic Publishers, p. 241–260.

Kosovichev, A.G. and Zharkova, V.V. (1995) *Helioseismology*, ESA Special Publication, vol. 376, ESA, 341.

Kosovichev, A.G. and Zharkova, V.V. (1998) *Nature*, **393**, 317.

Kosovichev, A.G. and Zharkova, V.V. (2001) *Astrophys. J. Lett.*, **550**, L105.

Kosugi, T., Matsuzaki, K., Sakao, T. *et al.* (2007) *Sol. Phys.*, **243**, 3.

Koval, A.N. (1972) *Izv. Ordena Trudovogo Krasn. Znam. Krymskoj Astrof. Obs.*, **44**, 94.

Krall, N.A. and Trivelpiece, A.W. (1973) *Principles of Plasma Physics*, McGraw-Hill Kogakusha, Tokyo.

Krauss-Varban, D. and Burgess, D. (1991) *J. Geophys. Res.*, **96**, 143.

Krauss-Varban, D. and Wu, C.S. (1989) *J. Geophys. Res.*, **94**, 15367.

Krauss-Varban, D., Burgess, D., and Wu, C.S. (1989) *J. Geophys. Res.*, **94**, 15089.

Krucker, S. and Lin, R.P. (2005) *13th Cambridge Workshop on Cool Stars, Stellar Systems and the Sun*, (eds F. Favata, G.A.J. Hussain, and B. Battrick), ESA Special Publication, vol. 560, ESA, p. 101.

Krucker, S., Hurford, G.J., and Lin, R.P. (2003) *Astrophys. J. Lett.*, **595**, L103.

Krucker, S., White, S.M., and Lin, R.P. (2007) *Astrophys. J. Lett.*, **669**, L49.

Krucker, S., Battaglia, M., Cargill, P.J., Fletcher, L., Hudson, H.S., MacKinnon, A.L., Masuda, S., Sui, L., Tomczak, M., Veronig, A.L., Vlahos, L., and White, S.M. (2008a) *Astron. Astrophys. Rev.*, **16**, 155.

Krucker, S., Saint-Hilaire, P., Christe, S., White, S.M., Chavier, A.D., Bale, S.D., and Lin, R.P. (2008b) *Astrophys. J.*, **681**, 644.

Kundu, M.R., Grechnev, V.V., Garaimov, V.I., and White, S.M., (2001a) *Astrophys. J.*, **563**, 389.

Kundu, M.R., Nindos, A., White, S.M., and Grechnev, V.V., (2001b) *Astrophys. J.*, **557**, 880.

Kundu, M.R., Nindos, A., and Grechnev, V.V. (2004) *Astron. Astrophys.*, **420**, 351.

Kundu, M.R., Grechnev, V.V., White, S.M., Schmahl, E.J., Meshalkina, N.S., and Kashapova, L.K. (2009) *Sol. Phys.*, **260**, 135.

Kurokawa, H., Kawaguchi, I., Funakoshi, Y., and Nakai, Y. (1982) *Sol. Phys.*, **79**, 77.

Kurokawa, H., Takakura, T., and Ohki, K. (1988) *Pub. Astron. Soc. Japan*, **40**, 357.

Kuznetsov, A.A. and Zharkova, V.V. (2010) *Astrophys. J.*, **722**, 1577.

Kuznetsov, S.N., Kurt, V.G., Myagkova, I.N., Yushkov, B.Y., and Kudela, K. (2006) *Sol. Syst. Res.*, **40**, 104.

Landau, L.D. (1937) *Zh. Eksp. Teor. Fiz.*, **7**, 203.

Landau, L.D. and Lifshitz, E.M. (1960) *Electrodynamics of Continuous Media*, Nauka, Moscow.

Landi Degl'Innocenti, E. (1982) *Sol. Phys.*, **79**, 291.

Landi Degl'Innocenti, E. (1983) *Sol. Phys.*, **85**, 33.

Landi Degl'Innocenti, E. (1985) *Sol. Phys.*, **102**, 1.

Landi Degl'Innocenti, E. (1996) *Sol. Phys.*, **164**, 21.

Landi Degl'Innocenti, E., Bommier, V., and Sahal-Brechot, S. (1990) *Astron. Astrophys.*, **235**, 459.

Lang, K.R., Willson, R.F., Kile, J.N., Lemen, J., Strong, K.T., Bogod, V.L., Gelfreikh, G.B., Ryabov, B.I., Hafizov, S.R., Abramov, V.E., and Svetkov, S.V. (1993) *Astrophys. J.*, **419**, 398.

Langer, S.H. and Petrosian, V. (1977) *Astrophys. J.*, **215**, 666.

Lapenta, G., Brackbill, J.U., and Daughton, W.S. (2003) *Phys. Plasmas*, **10**, 1577.

Larosa, T.N. and Emslie, A.G. (1989) *Sol. Phys.*, **120**, 343.

Lau, Y.-T. and Finn, J.M. (1990) *Astrophys. J.*, **350**, 672.

Leach, J. and Petrosian, V. (1981) *Astrophys. J.*, **251**, 281.

Leach, J. and Petrosian, V. (1983) *Astrophys. J.*, **269**, 715.

Leroy, M.M. and Mangeney, A. (1984) *Ann. Geophys.*, **2**, 449.

Lin, R.P. and Hudson, H.S. (1971) *Sol. Phys.*, **17**, 412.

Lin, R.P., Dennis, B.R., Hurford, G.J., Smith, D.M., Zehnder, A., Harvey, P.R., Curtis, D.W., Pankow, D., Turin, P., Bester, M., Csillaghy, A., Lewis, M.,

Madden, N., van Beek, H.F., Appleby, M., Raudorf, T., McTiernan, J., Ramaty, R., Schmahl, E., Schwartz, R., Krucker, S., Abiad, R., Quinn, T., Berg, P., Hashii, M., Sterling, R., Jackson, R., Pratt, R., Campbell, R.D., Malone, D., Landis, D., Barrington-Leigh, C.P., Slassi-Sennou, S., Cork, C., Clark, D., Amato, D., Orwig, L., Boyle, R., Banks, I.S., Shirey, K., Tolbert, A.K., Zarro, D., Snow, F., Thomsen, K., Henneck, R., McHedlishvili, A., Ming, P., Fivian, M., Jordan, J., Wanner, R., Crubb, J., Preble, J., Matranga, M., Benz, A., Hudson, H., Canfield, R.C., Holman, G.D., Crannell, C., Kosugi, T., Emslie, A.G., Vilmer, N., Brown, J.C., Johns-Krull, C., Aschwanden, M., Metcalf, T., and Conway, A. (2002) *Sol. Phys.*, **210**, 3.

Lin, R.P., Krucker, S., Hurford, G.J., Smith, D.M., Hudson, H.S., Holman, G.D., Schwartz, R.A., Dennis, B.R., Share, G.H., Murphy, R.J., Emslie, A.G., Johns-Krull, C., and Vilmer, N. (2003) *Astrophys. J. Lett.*, **595**, L69.

Lindsey, C. and Braun, D.C. (2000) *Sol. Phys.*, **192**, 261.

Lindsey, C. and Donea, A.-C. (2008) *Sol. Phys.*, **251**, 627.

Litvinenko, Y.E. (1996a) *Sol. Phys.*, **167**, 321.

Litvinenko, Y.E. (1996b) *Astrophys. J.*, **462**, 997.

Litvinenko, Y.E. and Somov, B.V. (1993) *Sol. Phys.*, **146**, 127.

Liu, Y., Ding, M.D., and Fang, C. (2001) *Astrophys. J. Lett.*, **563**, L169.

Lozitskii, V.G. and Baranovskii, E.A. (1993) *Bull. Crim. Astrophys. Obs.*, **88**, 58.

Ma, Z.W. and Bhattacharjee, A. (1996) *Geophys. Res. Lett.*, **23**, 1673.

Mandt, M.E., Denton, R.E., and Drake, J.F. (1994) *Geophys. Res. Lett.*, **21**, 73.

Mann, G., Aurass, H., and Warmuth, A. (2006) *Astron. Astrophys.*, **454**, 969.

Mann, G., Warmuth, A., and Aurass, H. (2009) *Astron. Astrophys.*, **494**, 669.

Mariska, J.T. and McTiernan, J.M. (1999) *Astrophys. J.*, **514**, 484.

P. Martens, C.H. and Young, A. (1990) *Astrophys. J. Suppl.*, **73**, 333.

Martínez-Oliveros, J.C., Donea, A.-C., Cally, P.S., and Moradi, H. (2008a) *Mon. Not. R. Astron. Soc.*, **389**, 1905.

Martínez-Oliveros, J.C., Moradi, H., and Donea, A.-C. (2008b) *Sol. Phys.*, **251**, 613.

Masuda, S., Kosugi, T., Hara, H., Tsuneta, S., and Ogawara, Y. (1994) *Nature*, **371**, 495.

Matthews, S., Zharkov, S.I., and Zharkova, V.V. (2011) *Astrophys. J.*, **739**, 71.

P. Mauas, J.D. and Gomez, D.O. (1997) *Astrophys. J.*, **483**, 496.

McClements, K.G. (1987) *Astron. Astrophys.*, **175**, 255.

McClements, K.G. (1989) *Astron. Astrophys.*, **208**, 279.

McClements, K.G. (1992a) *Astron. Astrophys.*, **253**, 261.

McClements, K.G. (1992b) *Astron. Astrophys.*, **258**, 542.

McConnell, M.L., Smith, D.M., Emslie, A.G., Hurford, G.J., Lin, R.P., and Ryan, J.M. (2003) *Bull. Am. Astron. Soc.*, **35**, 850.

Melnikov, V.F., Gary, D.E., and Nita, G.M. (2008) *Sol. Phys.*, **253**, 43.

Melrose, D.B. (1968) *Astrophys. Space Sci.*, **2**, 171.

Metcalf, T.R., Hudson, H.S., Kosugi, T., Puetter, R.C., and Pina, R.K. (1996) *Astrophys. J.*, **466**, 585.

Metcalf, T.R., Alexander, D., Hudson, H.S., and Longcope, D.W. (2003) *Astrophys. J.*, **595**, 483.

Miller, J.A. and Roberts, D.A. (1995) *Astrophys. J.*, **452**, 912.

Miller, J.A., Larosa, T.N., and Moore, R.L. (1996) *Astrophys. J.*, **461**, 445.

Miller, J.A., Cargill, P.J., Emslie, A.G., Holman, G.D., Dennis, B.R., LaRosa, T.N., Winglee, R.M., Benka, S.G., and Tsuneta, S. (1997) *J. Geophys. Res.*, **102**, 14631.

Moradi, H., Donea, A.-C., Lindsey, C., Beşliu-Ionescu, D., and Cally, P.S. (2007) *Mon. Not. R. Astron. Soc.*, **374**, 1155.

Nagai, F. and Emslie, A.G. (1984) *Astrophys. J.*, **279**, 896.

Nakajima, H., Sekiguchi, H., Sawa, M., Kai, K., and Kawashima, S. (1985) *Pub. Astron. Soc. Japan*, **37**, 163.

Nindos, A. and Zirin, H. (1998) *Sol. Phys.*, **182**, 381.

Nocera, L., Skrynnikov, I.I., and Somov, B.V. (1985) *Sol. Phys.*, **97**, 81.

November, L.J. and Simon, G.W. (1988) *Astrophys. J.*, **333**, 427.

van den Oord, G.H.J. (1990) *Astron. Astrophys.*, **234**, 496.

Oreshina, A.V. and Somov, B.V. (2000) *Astron. Lett.*, **26**, 750.

Paterson, A. and Zirin, H. (1981) *Bull. Am. Astron. Soc.*, **13**, 821.

Petrosian, V. and Liu, S. (2004) *Astrophys. J.*, **610**, 550.

Petrosian, V., Donaghy, T.Q., and McTiernan, J.M. (2002) *Astrophys. J.*, **569**, 459.

Petrosian, V., Yan, H., and Lazarian, A. (2006) *Astrophys. J.*, **644**, 603.

Petschek, H.E. (1964) *The Physics of Solar Flares*, NASA Special Publication, vol. 50, NASA, Science and Technical Division, Washington, DC, 425 p.

Press, W.H., Teukolsky, S.A., Vetterling, W.T., and Flannery, B.P. (1997) *Numerical Recipes in C: The Art of Scientific Computing*, 2nd edn., Cambridge Univ. Press, Cambridge.

Priest, E.R. (1982) *Solar Magneto-Hydrodynamics*, Reidel Publishing Co., Hingham.

Priest, E.R. and Démoulin, P. (1995) *J. Geophys. Res.*, **100**, 23443.

Priest, E. and Forbes, T. (2000) *Magnetic Reconnection*, Cambridge Univ. Press, Cambridge.

Priest, E.R. and Forbes, T.G. (2002) *Astron. Astrophys. Rev.*, **10**, 313.

Priest, E.R. and Titov, V.S. (1996) *R. Soc. Philos. Trans. Ser. A*, **354**, 2951.

Priest, E.R., Bungey, T.N., and Titov, V.S. (1997) *Geophys. Astrophys. Fluid Dyn.*, **84**, 127.

Priest, E., Forbes, T., and Murdin, P. (2001) Magnetohydrodynamics, in *Encyclopedia of Astronomy and Astrophysics*, (ed. P. Murdin), article 5397, Institute of Physics Publishing, Bristol.

Pritchett, P.L. and Coroniti, F.V. (2004) *J. Geophys. Res. (Space Phys.)*, **109**, 1220.

Pritchett, P.L., Coroniti, F.V., and Decyk, V.K. (1996) *J. Geophys. Res.*, **101**, 27413.

Pryadko, J.M. and Petrosian, V. (1997) *Astrophys. J.*, **482**, 774.

Pryadko, J.M. and Petrosian, V. (1998) *Astrophys. J.*, **495**, 377.

Puetter, R.C. and Pina, R.K. (1994) *Exp. Astron.*, **3**, 293.

Ramaty, R. (1969) *Astrophys. J.*, **158**, 753.

Razin, V.A. (1960a) *Sov.-Izv. Vyssh. Uchebn. Zaved. Radiofiz. (Higher Education Newsletters, Radiophysics)*, **3**, 584.

Razin, V.A. (1960b) *Sov.-Izv. Vyssh. Uchebn. Zaved. Radiofiz. (Higher Education Newsletters, Radiophysics)*, **3**, 921.

Razin, V.A. (1963) *Proc. Natl. Acad. Sci.*, **49**, 785.

Ricchiazzi, P.J. and Canfield, R.C. (1983) *Astrophys. J.*, **272**, 739.

Rust, D.M. and Keil, S.L. (1992) *Sol. Phys.*, **140**, 55.

Saint-Hilaire, P. and Benz, A.O. (2005) *Astron. Astrophys.*, **435**, 743.

Samarskii, A.A. (2001) *The Theory of Difference Schemes*, Monographs and Textbooks in Pure and Applied Mathematics, vol. 240, Marcel Dekker Inc., New York, translated from the Russian.

Sampson, D.H. and Golden, L.B. (1971) *Astrophys. J.*, **170**, 169.

Schekochihin, A.A. and Cowley, S.C. (2007) *Turbulence and Magnetic Fields in Astrophysical Plasmas*, Springer.

Scherrer, P.H., Bogart, R.S., Bush, R.I., Hoeksema, J.T., Kosovichev, A.G., Schou, J., Rosenberg, W., Springer, L., Tarbell, T.D., Title, A., Wolfson, C.J., Zayer, I., and MDI Engineering Team (1995) *Sol. Phys.*, **162**, 129.

Schlickeiser, R. and Miller, J.A. (1998) *Astrophys. J.*, **492**, 352.

Schrijver, C.J., Hudson, H.S., Murphy, R.J., Share, G.H., and Tarbell, T.D. (2006) *Astrophys. J.*, **650**, 1184.

Schunker, H., Braun, D.C., Lindsey, C., and Cally, P.S. (2008) *Sol. Phys.*, **251**, 341.

Severnyi, A.B. (1965) *Sov. Astron.*, **9**, 171.

Shapiro, V.D. and Shevchenko, V.I. (1988) *Sov. Sci. Rev. E Astrophys. Space Phys. Rev.*, **6**, 425.

Share, G.H. and Murphy, R.J. (2006) *Washington DC American Geophysical Union Geophysical Monograph Series*, vol. 165, Astrophysics and Space Science Library, Kluwer Academic Publishers, Boston, p. 177.

Share, G.H., Murphy, R.J., Smith, D.M., Lin, R.P., Dennis, B.R., and Schwartz, R.A. (2003) *Astrophys. J. Lett.*, **595**, L89.

Share, G.H., Murphy, R.J., Smith, D.M., Schwartz, R.A., and Lin, R.P. (2004) *Astrophys. J. Lett.*, **615**, L169.

Shay, M.A. and Drake, J.F. (1998) *Geophys. Res. Lett.*, **25**, 3759.

Shay, M.A., Drake, J.F., and Swisdak, M. (2007) *Phys. Rev. Lett.*, **99**(15), 155002.

Shibata, K., Masuda, S., Shimojo, M., Hara, H., Yokoyama, T., Tsuneta, S., Ko-

sugi, T., and Ogawara, Y. (1995) *Astrophys. J. Lett.*, **451**, L83.
Shih, A.Y., Lin, R.P., and Smith, D.M. (2009) *Astrophys. J. Lett.*, **698**, L152.
Simnett, G.M. (1995) *Space Sci. Rev.*, **73**, 387.
Siversky, T.V. and Zharkova, V.V. (2009a) *J. Plasma Phys.*, **75**, 619.
Siversky, T.V. and Zharkova, V.V. (2009b) *Astron. Astrophys.*, **504**, 1057.
Skomorovsky, V.I. and Firstova, N.M. (1996) *Sol. Phys.*, **163**, 209.
Smith, E.V.P. and Gottlieb, D.M. (1974) *Space Sci. Rev.*, **16**, 771.
Somov, B.V. (ed.) (1992) *Physical Processes in Solar Flares*, Astrophysics and Space Science Library, vol. 256, Science, No. 5056/APR24, p. 556.
Somov, B.V. (ed.) (2000) *Cosmic Plasma Physics*, Astrophysics and Space Science Library, vol. 251, Kluwer Academic Publishers, Boston.
Somov, B.V. and Kosugi, T. (1997) *Astrophys. J.*, **485**, 859.
Somov, B.V. and Oreshina, A.V. (2000) *Astron. Astrophys.*, **354**, 703.
Somov, B.V., Spektor, A.R., and Syrovatskii, S.I. (1981) *Sol. Phys.*, **73**, 145.
Sonnerup, B.U.O. (1979) *Magnetospheric Boundary Layers*, (eds B. Battrick, J. Mort, G. Haerendel, and J. Ortner), ESA Special Publication 148, ESA Publication, p. 395.
Spicer, D.S. (1977) *Sol. Phys.*, **53**, 305.
Su, Y., Holman, G.D., Dennis, B.R., Tolbert, A.K., and Schwartz, R.A. (2009) *Astrophys. J.*, **705**, 1584.
Suarez-Garcia, E., Hajdas, W., Wigger, C., Arzner, K., Güdel, M., Zehnder, A., and Grigis, P. (2006) *Sol. Phys.*, **239**, 149.
Sudol, J.J. and Harvey, J.W. (2005) *Astrophys. J.*, **635**, 647.
Sui, L. and Holman, G.D. (2003) *Astrophys. J. Lett.*, **596**, L251.
Sui, L., Holman, G.D., Dennis, B.R., Krucker, S., Schwartz, R.A., and Tolbert, K. (2002) *Sol. Phys.*, **210**, 245.
Sui, L., Holman, G.D., and Dennis, B.R. (2005) *Astrophys. J.*, **626**, 1102.
Sui, L., Holman, G.D., and Dennis, B.R. (2007) *Astrophys. J.*, **670**, 862.
Summers, H.P. and McWhirter, R.W.P. (1979) *J. Phys. B At. Mol. Phys.*, **12**, 2387.
Sweet, P.A. (1969) *Annu. Rev. Astron. Astrophys.*, **7**, 149.
Syniavskii, D.V. and Zharkova, V.V. (1994) *Astrophys. J. Suppl.*, **90**, 729.
Syrovatskii, S.I. (1981) The physical nature of solar flares, in *Problems of Solar-Terrestrial Relationships*, (A82-17192 05-46), Ashkhabad, Izdatel'stvo Ylym, 21–41 (in Russian).
Syrovatskii, S.I. and Shmeleva, O.P. (1972a) *Sov. Astron. J.*, **49**, 334.
Syrovatskii, S.I. and Shmeleva, O.P. (1972b) *Sov. Astron. J.*, **16**, 273.
Takakura, T., Kosugi, T., Sakao, T., Makishima, K., Inda-Koide, M., and Masuda, S. (1995) *Pub. Astron. Soc. Japan*, **47**, 355.
Tamres, D.H., Melrose, D.B., and Canfield, R.C. (1989) *Astrophys. J.*, **342**, 576.
Tandberg-Hanssen, E. and Emslie, A.G. (1988) *The Physics of Solar Flares*, Cambridge Univ Press, Cambridge, New York.
Tatischeff, V., Kiener, J., and Gros, M. (2005) Proceedings of the 5th Rencontres du Vietnam, "New Views on the Universe", Hanoi, 5–11 August 2004, ArXiv Astrophysics e-prints.
Tindo, I.P., Ivanov, V.D., Mandel'Stam, S.L., and Shuryghin, A.I. (1970) *Sol. Phys.*, **14**, 204.
Tindo, I.P., Ivanov, V.D., Mandel'Stam, S.L., and Shuryghin, A.I. (1972a) *Sol. Phys.*, **24**, 429.
Tindo, I.P., Ivanov, V.D., Valníček, B., and Livshits, M.A. (1972b) *Sol. Phys.*, **27**, 426.
Tindo, I.P., Shurygin, A.I., and Steffen, W. (1976) *Sol. Phys.*, **46**, 219.
Titov, V.S. and Hornig, G. (2002) *Adv. Space Res.*, **29**, 1087.
Tramiel, L.J., Novick, R., and Chanan, G.A. (1984) *Astrophys. J.*, **280**, 440.
Trottet, G., Rolli, E., Magun, A., Barat, C., Kuznetsov, A., Sunyaev, R., and Terekhov, O. (2000) *Astron. Astrophys.*, **356**, 1067.
Tsiklauri, D. and Haruki, T. (2007) *Phys. Plasmas*, **14**(11), 112905.
Tsuneta, S., Ichimoto, K., Katsukawa, Y., Nagata, S., Otsubo, M., Shimizu, T., Suematsu, Y., Nakagiri, M., Noguchi, M., Tarbell, T., Title, A., Shine, R., Rosenberg, W., Hoffmann, C., Jurcevich, B., Kushner, G., Levay, M., Lites, B., Elmore, D., Matsushita, T., Kawaguchi, N., Saito, H., Mikami, I., Hill, L.D., and Owens, J.K. (2008) *Sol. Phys.*, **249**, 167.

Tsytovich, V.N. (1970) *Nonlinear Effects in Plasma*, Plenum Press, New York.
Uddin, W., Jain, R., Yoshimura, K., Chandra, R., Sakao, T., Kosugi, T., Joshi, A., and Despande, M.R. (2004) *Sol. Phys.*, **225**, 325.
Uitenbroek, H. (2001) *Astrophys. J.*, **557**, 389.
Vekstein, G.E. and Browning, P.K. (1997) *Phys. Plasmas*, **4**, 2261.
Verboncoeur, A.B., Langdon, J.P., and Gladd, N.T. (1995) *Comp. Phys. Commun.*, **87**, 199.
Vidal-Madjar, A., Roble, R.G., Mankin, W.G., Artzner, G., Bonnet, R.M., Lemaire, P., and Vial, J.C. (1976) *Atmospheric Physics from Spacelab*, (eds J.J. Burger, A. Pedersen, and B. Battrick), Astrophysics and Space Science Library, vol. 61, Astrophysics and Space Science Library, San Francisco, p. 117.
Vilmer, N., Krucker, S., Lin, R.P., and The Rhessi Team (2002) *Sol. Phys.*, **210**, 261.
Vilmer, N., MacKinnon, A.L., and Hurford, G.J. (2011) *Space Sci. Rev.*, **159**, 167.
Vogt, E. and Hénoux, J.C. (1996) *Sol. Phys.*, **164**, 345.
Vogt, E., Sahal-Brechot, S., and Hénoux, J.C. (1997) *Astron. Astrophys.*, **324**, 1211.
Voitenko, Y.M. (1996) *Sol. Phys.*, **168**, 219.
Voitenko, Y.M. (1998) *Sol. Phys.*, **182**, 411.
Wang, Y., Ye, P., Zhou, G., Wang, S., Wang, S., Yan, Y., and Wang, J. (2005) *Sol. Phys.*, **226**, 337.
Warmuth, A., Mann, G., and Aurass, H. (2007) *Cent. Eur. Astrophys. Bull.*, **31**, 135.
Warmuth, A., Mann, G., and Aurass, H. (2009) *Astron. Astrophys.*, **494**, 677.
Watanabe, K., Krucker, S., Hudson, H., Shimizu, T., Masuda, S., and Ichimoto, K. (2010) *Astrophys. J.*, **715**, 651.
White, S.M., Benz, A.O., Christe, S., Fárník, F., Kundu, M.R., Mann, G., Ning, Z., Raulin, J.-P., Silva-Válio, A.V.R., Saint-Hilaire, P., and 2 coauthors (2011) *Space Sci. Rev.*, **159**, 225.
Wild, J.P. and Hill, E.R. (1971) *Austral. J. Phys.*, **24**, 43.
Willson, R.F. and Holman, G. (2003) *Bull. Am. Astron. Soc.*, **35**, 836.
Wood, P. and Neukirch, T. (2005) *Sol. Phys.*, **226**, 73.
Wu, C.C. (1983) *Geophys. Res. Lett.*, **10**, 545.
Wulser, R.P. and Marti, H (1989) *Astrophys. J.*, **341**, 1088.
Zarro, D.M., Canfield, R.C., Metcalf, T.R., and Strong, K.T. (1988) *Astrophys. J.*, **324**, 582.
Zeiler, A., Biskamp, D., Drake, J.F., Rogers, B.N., Shay, M.A., and Scholer, M. (2002) *J. Geophys. Res. (Space Phys.)*, **107**, 1230.
Zharkov, S., Zharkova, V.V., and Ipson, S.S. (2005) *Sol. Phys.*, **228**, 377.
Zharkov, S.I., Green, L., Matthews, S., and Zharkova, V.V. (2011a) *Astrophys. J. Lett.*, **741**, L35.
Zharkov, S.I., Zharkova, V.V., and Matthews, S. (2011b) *Astrophys. J.*, **739**, 70.
Zharkova, V.V. (1983) *Astrom. Astrofiz.*, **49**, 32.
Zharkova, V.V. (1990) *Veroeffentlichungen der Universitaets-Sternwarte Kiel*, vol. 32, University Press, Kiel, p. 41.
Zharkova, V.V. (1999) *8th SOHO Workshop: Plasma Dynamics and Diagnostics in the Solar Transition Region and Corona*, Proceedings of the Conference held 22–25 June 1999 in CAP 15, 1–13 Quai de Grenelle, 75015 Paris, France. Sponsored by ESA, NASA, C.N.R.S.-I.N.S.U., Euroconferences, Institut d'Astrophysique Spatiale, Matra Marconi Space, SCOSTEP, Université Paris XI. ESA Special Publications 446, (eds J.-C. Vial and B. Kaldeich-Schümann), ESA Publisher, p. 727.
Zharkova, V.V. (2008a) *Sol. Phys.*, **251**, 641.
Zharkova, V.V. (2008b) *Sol. Phys.*, **251**, 665.
Zharkova, V.V. and Agapitov, O.V. (2009) *J. Plasma Phys.*, **75**, 159.
Zharkova, V.V. and Gordovskyy, M. (2004) *Astrophys. J.*, **604**, 884.
Zharkova, V.V. and Gordovskyy, M. (2005a) *Astron. Astrophys.*, **432**, 1033.
Zharkova, V.V. and Gordovskyy, M. (2005b) *Mon. Not. R. Astron. Soc.*, **356**, 1107.
Zharkova, V.V. and Gordovskyy, M. (2005c) *Space Sci. Rev.*, **121**, 165.
Zharkova, V.V. and Gordovskyy, M. (2006) *Astrophys. J.*, **651**, 553.
Zharkova, V.V. and Kobylinskii, V.A. (1989) *Sov. Astron. Lett.*, **15**, 366.
Zharkova, V.V. and Kobylinskii, V.A. (1991) *Sov. Astron. Lett.*, **17**, 34.
Zharkova, V.V. and Kobylinskij, V.A. (1992) *Kinemat. Phys. Celest. Bodies*, **8**, 34.
Zharkova, V.V. and Kobylinskii, V.A. (1993) *Sol. Phys.*, **143**, 259.

Zharkova, V.V. and Kosovichev, A.G. (2000) Proc. High Energy Solar Physics Workshop – Anticipating HESSI, ASP Conference Series, Vol. 206, (eds R. Ramaty and N. Mandzhavidze), ESA Publications, p. 77.

Zharkova, V.V. and Kosovichev, A.G. (2001) *SOHO 10/GONG 2000 Workshop: Helio- and Asteroseismology at the Dawn of the Millennium*, (eds A. Wilson and P.L. Pallé), ESA Special Publication 464, p. 259.

Zharkova, V.V. and Kosovichev, A.G. (2002) *Solar Variability: From Core to Outer Frontiers*, (ed. J. Kuijpers), vol. 2, The 10th European Solar Physics Meeting, 9–14 September 2002, Prague, Czech Republic, ESA Special Publication 506, ESA Publication Division, 1031–1034, ISBN 92-9092-816-6.

Zharkova, V.V. and Siversky, T.V. (2011) *Astrophys. J.*, **733**, 33.

Zharkova, V.V. and Syniavskii, D.V (2000) *Astron. Astrophys.*, **354**, 714.

Zharkova, V.V. and Zharkov, S.I. (2007) *Astrophys. J.*, **664**, 573.

Zharkova, V.V., Brown, J.C., and Syniavskii, D.V. (1995) *Astron. Astrophys.*, **304**, 284.

Zharkova, V.V., Brown, J.C., and Syniavskii, D.V. (1996) *Adv. Space Res.*, **17**, 81.

Zharkova, V.V., Zharkov, S.I., Ipson, S.S., and Benkhalil, A.K. (2005a) *J. Geophys. Res. (Space Phys.)*, **110**, 8104.

Zharkova, V.V., Zharkov, S., Gordovskyy, M., Share, G., and Murphy, R. (2005b) *AGU Spring Meeting Abstracts*, American Geophysical Union, p. A1, #SH51A-01.

Zharkova, V.V., Kashapova, L.K., Chornogor, S.N., and Andriyenko, O.V. (2007) *Adv. Space Res.*, **39**, 1483.

Zharkova, V.V., Kuznetsov, A.A., and Siversky, T.V. (2010) *Astron. Astrophys.*, **512**, A8.

Zharkova, V.V., Azner, K., Benz, A., Browning, P., Dauphin, G., Emslie, A.G., Kontar, E.P., Mann, G., Onopri, M., Petrosian, V., Turkmani, R., Vilmer, N., and Vlahos, L. (2011a) *Space Sci. Rev.*, **159**(1–4), 357, http://adsabs.harvard.edu/abs/2011SSRv..159..357Z.

Zharkova, V.V., Kashapova, L.K., Chornogor, S.N., and Andrienko, O.V. (2011b) *Mon. Not. R. Astron. Soc.*, **411**, 1562.

Zheleznyakov, V.V. and Zaitsev, V.V. (1970) *Sov. Astron.*, **14**, 47.

Zhitnik, I., Urnov, A., Zharkova, V., and Ivanchuk, V. (1991) *Solnechnye Dann. Bull. Akad. Nauk SSSR*, **11**, 101.

Zhitnik, I.A., Logachev, Y.I., Bogomolov, A.V., Denisov, Y.I., Kavanosyan, S.S., Kuznetsov, S.N., Morozov, O.V., Myagkova, I.N., Svertilov, S.I., Ignat'ev, A.P., Oparin, S.N., Pertsov, A.A., and Tindo, I.P. (2006) *Sol. Syst. Res.*, **40**, 93.

Zhu, Z. and Winglee, R.M. (1996) *J. Geophys. Res.*, **101**, 4885.

Index

Electron and Proton Kinetics and Dynamics in Flaring Atmospheres, First Edition. Valentina Zharkova.

k

l

m

n

o

p

r